Bioremediation of Industrial Waste for Environmental Safety

Gaurav Saxena • Ram Naresh Bharagava
Editors

Bioremediation of Industrial Waste for Environmental Safety

Volume I: Industrial Waste and Its Management

Springer

Editors
Gaurav Saxena
Laboratory of Bioremediation and
Metagenomics Research (LBMR)
Department of Microbiology (DM)
Babasaheb Bhimrao Ambedkar University
(A Central University)
Lucknow, Uttar Pradesh, India

Ram Naresh Bharagava
Laboratory of Bioremediation and
Metagenomics Research (LBMR)
Department of Microbiology (DM)
Babasaheb Bhimrao Ambedkar University
(A Central University)
Lucknow, Uttar Pradesh, India

ISBN 978-981-13-1890-0 ISBN 978-981-13-1891-7 (eBook)
https://doi.org/10.1007/978-981-13-1891-7

Library of Congress Control Number: 2018966132

This Springer imprint is published by the registered company Springer Nature Singapore Pte Ltd.
The registered company address is: 152 Beach Road, #21-01/04 Gateway East, Singapore 189721, Singapore

This book is truly dedicated to my parents for their unfailing patience, contagious love, forgiveness, selflessness, endless support, and nurturing and educating me to the date. Without them, I wouldn't be the person I am today.

Gaurav Saxena

This book is truly dedicated to my parents for their unfailing patience, contagious love, forgiveness, selflessness, and endless support; my wife for trusting me; and my kids for always being a hope to move forward in life.

Ram Naresh Bharagava

Foreword

Environmental pollution is a major problem of the world due to increasing industrializations. Industries play important roles in the national economy of every country, but they can also be the main sources for environmental pollution. Industrial wastes carry a variety of potentially toxic pollutants that can cause severe impacts on the environment and human health. Bioremediation is a promising eco-friendly and cost-effective method to tackle environmental pollution. It has many advantages over the conventional physico-chemical approaches that are expensive and cause secondary pollution.

Bioremediation is a US Environmental Protection Agency approved waste management technique that treats hazardous wastes using biological agents such as microbes and plants and, ultimately, restores the contaminated sites, whilst providing adequate protection for human health and safety to the environment. It is an active field of research; many efforts have been made to commercialize bioremediation technologies for waste treatment to protect the environment and public health. Currently, a number of commercial plants or microbe-based products are available in the market to provide low-cost, self-driven and eco-sustainable solutions to clean up contaminated sites.

Bioremediation of Industrial Waste for Environmental Safety: Industrial Waste and Its Management (Volume I), edited by Dr. Ram Naresh Bharagava and Mr. Gaurav Saxena, introduces the readers to the subject of industrial waste/pollutants bioremediation. This timely book covers different sustainable bioremediation approaches for a low-cost treatment and management of a number of industrial wastes. It provides comprehensive information on both established and novel treatment technologies and their value-added potentials. The editors' keen interest in environmental awareness and their focus on environmental protection made this book highly relevant to academia as well as industry. It will be helpful to scientists and professionals engaging in sustainable bioremediation. All the chapters are written by leading experts making excellent and outstanding contributions to this book. I congratulate the book editors for bringing out this valuable compilation with up-to-date knowledge in the field of industrial waste bioremediation. I wish a great success for this book as it will be of great value to the stakeholders, including researchers, academicians, students, environmentalists and policymakers.

Dr. Diane Purchase
Ph.D., FHEA, FIEnvSci

Honorary Secretary, *Committee of Heads of Environmental Sciences,* UK

Editor, *Environmental Science and Pollution Research*, a Springer Nature Journal

Coordinating Editor, *Environmental Geochemistry and Health*, a Springer Nature Journal

Associate Professor of Environmental Health and Biology
Department of Natural Sciences
Faculty of Science and Technology
Middlesex University
The Burroughs, Hendon, London NW4 4BT, England, UK

Preface

Environmental issues have been always at the forefront of sustainable development and have become a serious matter of concern in the twenty-first century. Environmental sustainability with rapid industrialization is one of the major challenges of the current scenario worldwide. Industries are the key drivers in the world economy, but these are also the major polluters due to the discharge of partially treated/untreated potentially toxic and hazardous wastes containing organic and inorganic pollutants, which cause environmental (soil and water) pollution and severe toxicity in living beings. Among the different sources of environmental pollution, industrial waste is considered as the major source of environmental pollution because industries use cheap and poorly or non-biodegradable chemicals to obtain the good quality of products within a short time period and in an economic way; however, their toxicity is usually ignored. Ensuring the safety of chemicals used in many industrial processes is a major challenge for environmental safety. The governments around the globe are also strictly advocating for the mitigation of environmental pollution due to industrial wastes to promote the sustainable development of our society with low environmental impact. Being a low cost and eco-friendly clean technology, bioremediation can be an eco-sustainable alternative to conventional technologies for the treatment and management of industrial wastes to protect the public health and environment.

Bioremediation is a waste management approach that utilizes microorganisms, plants or their enzymes to degrade/detoxify the organic and inorganic pollutants such as phenols, chlorophenols, petroleum hydrocarbons, polychlorinated biphenyls, organic solvents, azo dyes, pesticides, recalcitrant compounds, and toxic metals from contaminated soils and wastewaters. There has been an increasing concern regarding the release of various hazardous chemicals along with industrial wastes, which are considered as highly toxic for the environment and living beings. Some of these chemicals are listed as "priority pollutants" by the United States Environmental Protection Agency (USEPA) and other environmental pollution control agencies. The biological removal of a wide range of pollutants from contaminated sites requires our increasing understanding of different degradation pathways and regulatory networks to carbon flux for their degradation and detoxification, which is utmost important for environmental safety. Therefore, this book provides a comprehensive knowledge of the fundamental, practical and purposeful utilization of bioremediation technologies for the treatment and management of industrial wastes. The book describes the microbiological, biochemical and molecular aspects of biodegradation and bioremediation, including the use of "omics" technologies for the development of efficient bioremediation technologies for industrial wastes/pollutants to combat the forthcoming challenges.

This book *Bioremediation of Industrial Waste for Environmental Safety: Industrial Waste and Its Management (Volume I)* describes the toxicity of various organic and inorganic pollutants in industrial wastes, their environmental impact and bioremediation approaches for their treatment and management. For this book, many relevant topics have been contributed by the experts from different universities, research laboratories and institutes from around the globe in the area of biodegradation and bioremediation. In this book, extensive focus has been relied on the recent advances in bioremediation and phytoremediation technologies, including the use of various group of microbes for environmental remediation; terrestrial/aquatic plants for phytoremediation of toxic metals from contaminated soils/industrial wastewaters; constructed wetlands for degradation and detoxification of industrial wastewaters; microbial enzymes for degradation/detoxification of environmental pollutants; biosurfactants for remediation of petroleum polyaromatic hydrocarbons and heavy metals; biodegradation and bioremediation of azo dyes, organic solvents, pesticides, persistent organic pollutants and toxic metals from industrial wastes; bioremediation of industrial acid mine drainage (AMD), distillery wastewater, tannery wastewater, textile wastewater, oil refinery waste, plastic waste; bioremediation and phytoremediation of potentially toxic metals such as chromium and arsenic from contaminated matrix; nano-bioremediation technology for the decolourization of dyes in effluents; phytotechnologies for wastewater treatment and management; application of green synthesized nanoparticles (NPs) in degradation and detoxification of wastewaters; etc. Researchers working in the field of bioremediation, phytoremediation, waste treatment and management and related fields will find this compilation most useful for further study to learn about the subject matter. Further, to get richer in the knowledge on the subject, readers may please visit the second volume of this book series, *Bioremediation of Industrial Waste for Environmental Safety: Biological Agents and Methods for Industrial Waste Management (Volume II).*

At the end, we hope that the book will be of great value to researchers, environmental chemists and scientists, microbiologists and biotechnologists, ecotoxicologists, waste treatment engineers and managers, environmental science managers, administrators and policymakers, industry persons and students at bachelor's, master's and doctoral level in the relevant field. Thus, in this book, readers will find the updated information as well as the future direction for research in the field of bioremediation.

Lucknow, Uttar Pradesh, India — Gaurav Saxena
Lucknow, Uttar Pradesh, India — Ram Naresh Bharagava
May 2018

Acknowledgments

The edited book *Bioremediation of Industrial Waste for Environmental Safety: Industrial Waste and Its Management (Volume I)* is the outcome of a long dedicated effort of many individuals who directly or indirectly supported us during the compilation and upbringing of this valuable edition, many of whom deserve special mention.

The editors are firstly thankful to all the national and international contributing authors for their valuable submissions and cooperation and providing most up-to-date information on the diverse aspects of the subject regardless of their busy schedules; Dr. Diane Purchase, Middlesex University, London, England (United Kingdom), for writing a foreword for the book; Dr. G. D. Saratale, Dongguk University, Seoul (Republic of Korea), and Dr. Sikandar I. Mulla, Chinese Academy of Sciences (CAS), Xiamen (People's Republic of China), for the meaningful research collaboration, cooperation, and support; Dr. Jay Shankar Singh, Department of Environmental Microbiology (DEM), Babasaheb Bhimrao Ambedkar (Central) University (BBAU), India, for the better advice and helpful discussion on the subject; and Mr. Surya Pratap Goutam and Mr. Rajkamal Shastri, Doctoral Fellow, Department of Applied Physics; Roop Kishor, Doctoral Fellow, DEM, BBAU; and Mr. Akash Mishra, Doctoral Fellow, Defence Research and Development Organisation (DRDO)–Defence Institute of Bio-Energy Research (DIBER), Haldwani (India), for helping us in various ways during the book project.

We are extremely thankful to our publishing editors, Ms. Aakanksha Tyagi and Dr. Mamta Kapila, Springer Nature (India), for the encouragement, support, and valuable advice and skillful organization and management of entire book project; Ms. Raman Shukla, for the skillful management of book production; and Mr. John Ram Kumar for moving the book through the production process in an efficient and professional manner.

We are also heartily thankful to the Almighty God for helping us through the entire journey and making the experience enjoyable. Further, we hope that the book volume will be of great value to researchers in the area of bioremediation of industrial wastes and will go some way to make our planet safe and greener. At the end, we seek to learn more on the subject through the valuable comments, reviews, and suggestions from the readers, which can be directly sent to our e-mails: gaurav10saxena@gmail.com (GS) and bharagavarnbbau11@gmail.com (RNB).

Contents

About the Editors and Contributors

Editors

Gaurav Saxena is a Senior Doctoral Student, actively engaged in research at the Laboratory for Bioremediation and Metagenomics Research (LBMR), Department of Environmental Microbiology (DEM), Babasaheb Bhimrao Ambedkar (Central) University, Lucknow (UP) 2260 025, India.

Ram Naresh Bharagava is presently working as Assistant Professor in Department of Environmental Microbiology, Babasaheb Bhimrao Ambedkar University, Raebareli, Lucknow, Uttar Pradesh, India.

Contributors

Paula Alvarenga LEAF, Department of Sciences and Engineering of Biosystems, School of Agriculture, University of Lisbon, Lisbon, Portugal

GeoBioTec Research Center, Department of Geosciences, University of Aveiro, Aveiro, Portugal

Shweta Ambust Rhizospheric Biology Laboratory, Department of Environmental Microbiology, School for Environmental Sciences, Babasaheb Bhimrao Ambedkar University (A Central University), Lucknow, Uttar Pradesh, India

Fuad Ameen Department of Botany and Microbiology, Faculty of Science, King Saud University, Riyadh, Kingdom of Saudi Arabia

Christopher Chibueze Azubuike Department of Microbiology, Faculty of Science, University of Port Harcourt, Port Harcourt, Rivers States, Nigeria

Cassiano A. R. Bernardino Department of Civil Engineering, COPPE, Federal University of Rio de Janeiro, Rio de Janeiro, RJ, Brazil

Ram Naresh Bharagava Laboratory of Bioremediation and Metagenomics Research (LBMR), Department of Microbiology (DM), Babasaheb Bhimrao Ambedkar University (A Central University), Lucknow, Uttar Pradesh, India

Faizal Bux Institute for Water and Wastewater Technology, Department of Biotechnology and Food Technology, Durban University of Technology, Durban, South Africa

Paula M. L. Castro Centre of Biotechnology and Fine Chemistry – Associated Laboratory, Faculty of Biotechnology, Catholic University of Portugal, Porto, Portugal

Chioma Blaise Chikere Department of Microbiology, Faculty of Science, University of Port Harcourt, Port Harcourt, Rivers States, Nigeria

Eduardo Ferreira da Silva GeoBioTec Research Center, Department of Geosciences, University of Aveiro, Aveiro, Portugal

Anamika Das Centre for Environmental Science and Technology, School of Environment and Earth Sciences, Central University of Punjab, Bathinda, Punjab, India

Amar Jyoti Das Rhizospheric Biology Laboratory, Department of Environmental Microbiology, School for Environmental Sciences, Babasaheb Bhimrao Ambedkar University (A Central University), Lucknow, Uttar Pradesh, India

Ningombam Linthoingambi Devi Centre for Environmental Science, School of Earth, Biological and Environmental Sciences, Central University of South Bihar, Patna, Bihar, India

M. S. Dhanya Department of Environmental Sciences and Technology, School of Environment and Earth Sciences, Central University of Punjab, Bathinda, Punjab, India

Syed Ali Musstjab Akber Shah Eqani CAS Key Laboratory of Urban Environment and Health, Institute of Urban Environment, Chinese Academy of Sciences, Xiamen, People's Republic of China

Surya Pratap Goutam Advanced Materials Research Laboratory, Department of Applied Physics (DAP), School for Physical Sciences (SPS), Babasaheb Bhimrao Ambedkar University (A Central University), Lucknow, Uttar Pradesh, India

Vijay Jaswal Centre for Environmental Science and Technology, School of Environment and Earth Sciences, Central University of Punjab, Bathinda, Punjab, India

Arun Kalia Department of Environmental Sciences and Technology, School of Environment and Earth Sciences, Central University of Punjab, Bathinda, Punjab, India

Garima Kaushik Department of Environmental Science, Central University of Rajasthan, Ajmer, Rajasthan, India

Roop Kishor Laboratory of Bioremediation and Metagenomics Research (LBMR), Department of Microbiology (DM), Babasaheb Bhimrao Ambedkar University (A Central University), Lucknow, Uttar Pradesh, India

Manoj Kumar Department of Environmental Science, School of Earth, Environment and Space Studies, Central University of Haryana, Mahendergarh, Haryana, India

School of Environmental Sciences, Jawaharlal Nehru University, New Delhi, India

Rajesh Kumar Rhizospheric Biology Laboratory, Department of Environmental Microbiology, School for Environmental Sciences, Babasaheb Bhimrao Ambedkar University (A Central University), Lucknow, Uttar Pradesh, India

Vidya Laxmi Department of Environmental Science, Central University of Rajasthan, Ajmer, Rajasthan, India

Claudio F. Mahler Department of Civil Engineering, COPPE, Federal University of Rio de Janeiro, Rio de Janeiro, RJ, Brazil

Pradeep Kumar Mishra Department of Chemical Engineering, Indian Institute of Technology (IIT), Banaras Hindu University, Varanasi, Uttar Pradesh, India

Mathews Simon Mthembu Department of Biochemistry and Microbiology, Faculty of Science and Agriculture, University of Zululand, Richards Bay, South Africa

Sikandar I. Mulla Department of Biochemistry, Karnatak University, Dharwad, Karnataka, India

CAS Key Laboratory of Urban Environment and Health, Institute of Urban Environment, Chinese Academy of Sciences, Xiamen, People's Republic of China

Harichandra Z. Ninnekar Department of Biochemistry, Karnatak University, Dharwad, Karnataka, India

Luís A. B. Novo Centre of Biotechnology and Fine Chemistry – Associated Laboratory, Faculty of Biotechnology, Catholic University of Portugal, Porto, Portugal

GeoBioTec Research Center, Department of Geosciences, University of Aveiro, Aveiro, Portugal

Christine Akinyi Odinga Institute for Water and Wastewater Technology, Department of Biotechnology and Food Technology, Durban University of Technology, Durban, South Africa

Gideon Chijioke Okpokwasili Department of Microbiology, Faculty of Science, University of Port Harcourt, Port Harcourt, Rivers States, Nigeria

Diane Purchase Department of Natural Sciences, Faculty of Science and Technology, Middlesex University, London, UK

J. Rajesh Banu Department of Civil Engineering, Regional Center of Anna University, Tirunelveli, Tamilnadu, India

A. L. Ramanathan School of Environmental Sciences, Jawaharlal Nehru University, New Delhi, India

Jatinder Singh Randhawa Centre for Environmental Science and Technology, School of Environment and Earth Sciences, Central University of Punjab, Bathinda, Punjab, India

Gini Rani Centre for Environmental Science and Technology, School of Environment and Earth Sciences, Central University of Punjab, Bathinda, Punjab, India

Arpna Ratnakar Department of Environmental Science, School for Environmental Sciences, Babasaheb Bhimrao Ambedkar University (A Central University), Lucknow, Uttar Pradesh, India

Shalu Rawat Department of Environmental Science, School for Environmental Sciences, Babasaheb Bhimrao Ambedkar University (A Central University), Lucknow, Uttar Pradesh, India

Diptarka Roy Advanced Materials Research Laboratory, Department of Applied Physics (DAP), School for Physical Sciences (SPS), Babasaheb Bhimrao Ambedkar University (A Central University), Lucknow, Uttar Pradesh, India

Rijuta Ganesh Saratale Research Institute of Biotechnology and Medical Converged Science, Dongguk University-Seoul, Goyang-si, Gyeonggi-do, Republic of Korea

Ganesh Dattatraya Saratale Research Institute of Biotechnology and Medical Converged Science, Dongguk University-Seoul, Goyang-si, Gyeonggi-do, Republic of Korea

Department of Food Science and Biotechnology, Dongguk University-Seoul, Goyang-si, Gyeonggi-do, Republic of Korea

Gaurav Saxena Laboratory of Bioremediation and Metagenomics Research (LBMR), Department of Microbiology (DM), Babasaheb Bhimrao Ambedkar University (A Central University), Lucknow, Uttar Pradesh, India

Shiv Shankar Department of Environmental Science, School for Environmental Sciences, Babasaheb Bhimrao Ambedkar University (A Central University), Lucknow, Uttar Pradesh, India

Shikha Department of Environmental Science, School for Environmental Sciences, Babasaheb Bhimrao Ambedkar University (A Central University), Lucknow, Uttar Pradesh, India

Han-Seung Shin Department of Food Science and Biotechnology, Dongguk University-Seoul, Goyang-si, Gyeonggi-do, Republic of Korea

Sushil Kumar Shukla Centre for Environmental Sciences, Central University of Jharkhand, Ranchi, Jharkhand, India

Shailja Singh Department of Environmental Science, School for Environmental Sciences, Babasaheb Bhimrao Ambedkar University (A Central University), Lucknow, Uttar Pradesh, India

Sukhman Singh Centre for Environmental Science and Technology, School of Environment and Earth Sciences, Central University of Punjab, Bathinda, Punjab, India

Feroz Mahomed Swalaha Institute for Water and Wastewater Technology, Department of Biotechnology and Food Technology, Durban University of Technology, Durban, South Africa

Preeti N. Tallur Government Arts and Science College, Karwar, Uttara Kannada, Karnataka, India

Manjunatha P. Talwar Department of Biochemistry, Karnatak University, Dharwad, Karnataka, India

Vinod Kumar Tripathi Department of Farm Engineering, Institute of Agricultural Sciences, Banaras Hindu University, Varanasi, Uttar Pradesh, India

Anoop Yadav Department of Environmental Science, School of Earth, Environment and Space Studies, Central University of Haryana, Mahendergarh, Haryana, India

Anil Kumar Yadav Advanced Materials Research Laboratory, Department of Applied Physics (DAP), School for Physical Sciences (SPS), Babasaheb Bhimrao Ambedkar University (A Central University), Lucknow, Uttar Pradesh, India

Kadapakkam Nandabalan Yogalakshmi Centre for Environmental Science and Technology, School of Environment and Earth Sciences, Central University of Punjab, Bathinda, Punjab, India

Chapter 1
Introduction to Industrial Wastes Containing Organic and Inorganic Pollutants and Bioremediation Approaches for Environmental Management

Ram Naresh Bharagava, Gaurav Saxena, and Sikandar I. Mulla

Contents

Abstract Industrial wastes are one of the sources of environmental pollution. Industrial waste contains a variety of highly toxic organic and inorganic pollutants and thus may cause serious toxicity in the living organisms. Therefore, the adequate treatment and management of such hazardous wastes to protect the environment and public health. Bioremediation can be a suitable alternative to the physicochemical approaches, which are environmentally destructive and costly and may cause secondary pollution. It has been approved by the US Environmental Protection Agency (USEPA) as an eco-friendly waste management technique that

R. N. Bharagava (✉) · G. Saxena
Laboratory of Bioremediation and Metagenomics Research (LBMR), Department of Microbiology (DM), Babasaheb Bhimrao Ambedkar University (A Central University), Lucknow, Uttar Pradesh, India

S. I. Mulla
Department of Biochemistry, Karnatak University, Dharwad, Karnataka, India

CAS Key Laboratory of Urban Pollutant Conversion, Institute of Urban Environment, Chinese Academy of Sciences, Xiamen, People's Republic of China

G. Saxena, R. N. Bharagava (eds.), *Bioremediation of Industrial Waste for Environmental Safety*, https://doi.org/10.1007/978-981-13-1891-7_1

revitalizes the contaminated environment and promotes sustainable development. Therefore, this chapter introduces the toxicity profile of different industrial wastes containing various organic and inorganic pollutants and bioremediation technologies such as microbial bioremediation, phytoremediation, enzymatic remediation, electro-bioremediation, nano-bioremediation, etc. with limitations and challenges.

Keywords Industrial waste · Organic pollutants · Inorganic pollutants · Pollution · Toxicity · Bioremediation

1.1 Introduction

Industries are the key players in the national economies of many developing countries; however, unfortunately they are also the major polluters of the environment. Among the different sources of environmental pollution, industrial wastewater discharged from different industries is considered as the major source of environmental pollution (soil and water) (Goutam et al. 2018; Gautam et al. 2017; Saxena and Bharagava 2017; Saxena et al. 2015). Industrial wastewaters contain a variety of organic and inorganic pollutants that may cause serious environmental pollution and health hazards (Arora et al. 2014, 2018; Bharagava et al. 2017a; Maszenan et al. 2011; Megharaj et al. 2011).

The organic pollutants include phenols, chlorinated phenols, endocrine-disrupting chemicals, azo dyes, polyaromatic hydrocarbons, polychlorinated biphenyls, pesticides, etc. However, inorganic pollutants include a variety of toxic heavy metals such as cadmium (Cd), chromium (Cr), arsenic (As), lead (Pb), and mercury (Hg). The high concentration and poor biodegradability of recalcitrant organic pollutants and nonbiodegradable nature of inorganic metal pollutants in industrial wastewaters pose a major challenge for environmental safety and human health protection; thus, it is required to adequately treat industrial wastewater before its final disposal in the environment.

Bioremediation (the use of biological agents in environmental remediation) is considered as the suitable alternative to physicochemical treatment methods, which are environmentally destructive and create secondary pollution while environmental cleanup. It has been recognized by the US Environmental Protection Agency as an eco-friendly waste management technique. Bioremediation uses an array of microorganisms having diverse metabolic pathways to degrade/detoxify the organic and inorganic pollutants in contaminated matrix and, hence, is regarded as environmentally friendly, cost-effective method for wastewater treatment and management with simple structural setup, wider application, operational ease, and less sludge production (Bharagava et al. 2017b; Saxena and Bharagava 2016; Singh et al. 2011; Mendez-Paz et al. 2005; Pandey et al. 2007). Therefore, this chapter provides an overview on the various bioremediation techniques, which can be used for the treatment and management of industrial wastewaters to protect the environment and human health. In this chapter, merits and demerits are also discussed with future research prospects.

1.2 Industrial Wastes: Types and Characteristics

Industrial waste is of two types, i.e., solid and liquid, and often produced due to industrial activity and includes any material that is rendered useless during a product manufacturing process in industries. Liquid waste (i.e., wastewaters released from different industries) is considered as highly hazardous to living organisms and the environment as it carries a variety of potentially toxic pollutants. However, the nature and characteristics of industrial wastewater depend on the type of industry, production processes applied, and product quality. Industries often discharge high-strength wastewaters, which are characterized by high biochemical oxygen demand (BOD), chemical oxygen demand (COD), total dissolved solids (TDSs) and TSSs, and a variety of organic and inorganic pollutants (Saxena and Bharagava 2017). The nature and characteristics of different industrial wastewaters are presented in Table 1.1.

1.3 Pollutants in Industrial Wastes and Their Toxicity in Environment

The wastewaters discharged from different industries are considered the major sources of environmental pollution and toxicity in the living beings. A variety of highly toxic and recalcitrant pollutants are being discharged along with industrial wastewaters in the environment due to different industrial activities. Environmental pollutants are of two types: organic and inorganic. Organic pollutants mainly include phenols, nonylphenols, chlorinated phenols, azo dyes, phthalic esters, petroleum hydrocarbons, pesticides, persistent organic pollutants (POPs), etc. However, inorganic pollutants comprise a variety of highly toxic nonbiodegradable heavy metals such as arsenic (As), nickel (Ni), chromium (Cr), lead (Pb), mercury (Hg), and cadmium (Cd). A variety of organic and inorganic pollutants have been reported to cause serious soil and water pollution and severe toxic effects in living organisms (Chandra et al. 2008, 2011, 2015; Maszenan et al. 2011; Megharaj et al. 2011; Saxena and Bharagava 2015; Saxena et al. 2016). Hence, due to highly toxic nature, many of them have been regarded as priority pollutants by various environmental protection agencies such as the US Environmental Protection Agency (USEPA), Agency for Toxic Substances and Disease Registry (ATSDR), and World Health Organization (WHO). Table 1.2 represents the environmental hazards and toxic effects caused by various organic and inorganic pollutants in living organisms.

Table 1.1 Nature and characteristics of various types of industrial wastewater

Type of industrial wastewater	Nature and characteristics
Oil refinery wastewater	Poorly biodegradable and contains oil, various aromatic hydrocarbons (highly toxic to natural environment) such as BTEX, persistent organic pollutants (POPs) such as dioxins (highly toxic in nature), metals/metalloids, phenolics, salts, and surfactants
Pulp and paper mill wastewater	Highly intense dark brown color, BOD, SS and contains recalcitrant dioxins, furans, lignins, AOX, phenolic and chlorophenolic compounds especially pentachlorophenol (highly toxic to living beings and hazardous to environment)
Textile wastewater	Alkaline in nature and highly colored and often contains harmful residual dyes such as acidic, basic, reactive, disperse, azo, diazo, anthraquinone-based, and metal complex dyes (some carcinogenic in nature)
Tannery wastewater	Contains high organic loadings (BOD, COD, and TSS), salts (sodium, chloride, and sulfide), phenolic compounds, endocrine-disrupting chemicals such as nonylphenols and phthalates and other toxic metals especially chromium (highly toxic proven carcinogen)
Distillery wastewater (spent wash)	High organic loading (BOD, COD, and TDS) and contains phenolics and dark black color recalcitrant melanoidins (stop penetration of sunlight into water bodies and hence reduce photosynthesis)
Winery wastewater	Acidic in nature, variable flows and loadings, contains high content of organic matter, COD and TSS and organic fraction consist of sugars, alcohols, acids, and high molecular weight recalcitrant compounds such as polyphenols, tannins, and lignins
Pharmaceutical wastewater	Acidic in nature, has high COD and TDS, and contains many organic solvents, formulations, disinfectants, and many generic drugs such as antibiotics, analgesic, etc.
Abattoir (slaughter house) wastewater	Contains high levels of organic (COD is mainly in colloidal form) and coarse suspended matter and heavy metals, nutrients, pathogenic and nonpathogenic microorganisms, and detergents and disinfectants and sometimes pharmaceutical agents used for veterinary purpose
Agricultural wastewater	Alkaline in nature and contains high content of nitrogen, phosphorous, pesticides, and various toxic metals such as cadmium, lead, arsenic, etc.
Landfill leachate	Composition varies from landfill to landfill, generally colored, anoxic and has high TDS, COD, BOD and contains ammonia, phenols, benzene, toluene, chloride, iron, manganese, and other toxic metals such as lead, cadmium, zinc, arsenic, or chromium but little or no phosphorus
Acid mine drainage	Acidic in nature, contains high concentrations of iron, sulfate, copper, nickel, and toxic metals such as cadmium, lead, etc. (cause environmental damage)

Adapted from Saxena and Bharagava (2017)

Table 1.2 Toxicity of various organic and inorganic pollutants of industrial wastewater

Type of pollutants	Environmental pollution and toxicity profile
Organic pollutants	
Phenols	Most common pollutants of industrial wastewaters and associated with distilleries, pulp and paper mills, coal mines, oil refineries, wood preservation plants, pharmaceuticals, coke-oven batteries, herbicides, and pesticides as well as their wastewaters. It is also used in preparation of several chemicals such as alkylphenols, cresols, xylenols, phenolic resins, aniline, pesticides, explosives, dyes, and other compounds. Its acute exposure causes dryness of the throat and mouth, nausea, vomiting, and diarrhea, while chronic exposure causes methemoglobinemia, hemolytic anemia, profuse sweating, hypotension, arrhythmia, pulmonary edema, tachycardia, and dark-colored urine excreted due to lipid peroxidation and central nervous system disorders leading to collapse and coma and sometimes muscular convulsions with reduction in body temperature (hypothermia). Inhalation and dermal contact cause cardiovascular diseases and skin blisters, respectively, while ingestion can cause serious gastrointestinal damage, and oral administration may result in muscle tremors and death
Endocrine-disrupting chemicals	These are widely used as plasticizers in industries, in the manufacturing of plastic resins such as polyvinyl resins and cellulosic and polyurethane polymers. These disturb the delicate hormonal balance (endocrine system) and compromise the reproductive fitness of living beings and ultimately may lead to carcinogenic and mutagenic effects. Examples include dibutyl phthalate (DBP), benzyl butyl phthalate (BBP), bis(2-ethylhexyl) phthalate (DEHP), and nonylphenol (NP), 4-aminobiphenyl, hexachlorobenzene, and benzidine. These also cause irritation of the skin, conjunctiva, mucous membranes of oral and nasal cavities, testicular lesions, hypospadia, cryptorchidism in males and in females cause obesity, prolongation of the estrous cycle, and anovulation due to a decrease in the serum estradiol level
Chlorinated phenols	Chlorophenols are considered as the major environmental pollutants discharged along with wastewaters from pulp and paper mills, tanneries, distilleries, dye and paint manufacturing, and pharmaceutical industries, e.g., pentachlorophenol (PCP). It is widely applied as herbicides and fungicides and in wood protection, tanneries, distilleries, paint manufacturing, and pulp and paper mills. It is highly carcinogenic, teratogenic, and mutagenic in nature and causes toxicity to living beings by inhibiting oxidative phosphorylation, inactivating respiratory enzymes, and damaging mitochondrial structure. However, its high concentration can also cause obstruction in the circulatory system of the lungs, heart failure, and damage to the central nervous system
Azo dyes	Textile, leather, paint, acrylic, cosmetics, plastics, pharmaceutical, etc., industries use different dyes to color products. Azo dyes cause severe health hazards, such as skin irritation, digestive tract irritation, nausea, vomiting, liver and kidney damage, etc., in humans and animals

(continued)

Table 1.2 (continued)

Type of pollutants	Environmental pollution and toxicity profile
Petroleum hydrocarbons	Refinery industry wastewaters are the major sources of petroleum hydrocarbons in the environment. The most common petroleum hydrocarbons include aliphatic, branched, and cycloaliphatic alkanes, as well as monocyclic and polycyclic aromatic hydrocarbons (PAHs), which include naphthalene, fluorene, phenanthrene, anthracene, fluoranthene, pyrene, benzo[a]anthracene, and benzo[a]pyrene. Inhalation of hydrocarbons in humans can lead to criminal or violent behavior, development of memory and other cognitive deficits, cerebellar dysfunction, encephalopathy, weakness, dementia, depression of the central nervous system, metabolic acidosis, arrhythmia, or even a fatal malignant arrhythmia termed "sudden sniffing death." In addition, the aspiration of hydrocarbons causes a potentially fatal pneumonitis characterized by cough, wheezing, respiratory distress, and hypoxia. Dermal exposure can cause dermatitis, chemical burns, and defatting injury, whereas oral exposure can cause local irritation as well as vomiting, diarrhea, and abdominal pain. Thus, acute hydrocarbon exposure can result in a wide array of pathologies, such as encephalopathy, pneumonitis, arrhythmia, acidosis, and dermatitis. Intentional inhalation and accidental ingestion exposures with aspiration also lead to the greatest morbidity and mortality
	Melanoidins are released as environmental pollutants by various agro-based industries, especially from cane molasses-based distilleries and fermentation industries. Discharge of melanoidins containing distillery wastewater (DWW) into the environment causes several problems, such as reduction of sunlight penetration, decreased photosynthetic activity, and dissolved oxygen concentration, thereby posing deleterious effects to aquatic life. On land, it causes a reduction in soil alkalinity and inhibition of seed germination. The melanoidins containing DWW have severe toxic effects on fishes and other aquatic organisms
Pesticides	Pesticides are used in agricultural applications to control pests and enhance crop productivity. Pesticides (such as DDT, pyrethroids, organophosphates, carbamates, etc.), fungicides (used to kill fungi, such as hexachlorobenzene, benzothiazole, pentachlorophenol, etc.), herbicides (used to kill weeds, such as 2,4-D, atrazine, picloram, chlorophenoxy compounds, etc.), rodenticides (used to kill rodents, such as zinc phosphide, α-naphthylthiourea [ANTU], 4-hydroxycoumarin, 1,3-indandiones, etc.), and fumigants (used to kill pests, such as phosphine, dibromochloropropane, etc.) cause immune suppression, hormone disruption, diminished intelligence, reproductive abnormalities, cancer upon exposure to humans and animals and also directly or indirectly affect the nontarget organisms
Persistent organic pollutants (POPs)	POPs exhibit low water solubility but high lipid solubility and therefore bioaccumulate in fatty tissues and cause a variety of toxic effects in living beings, such as disruption of the endocrine system, neurobehavioral disorders, cancer (possibly including breast cancer), allergies, hypersensitivity, damage to the central and peripheral nervous systems, birth defects, learning disabilities, reproductive disorders, disruption of the immune system by altering the hormonal system, and mutagenicity and carcinogenicity

(continued)

Table 1.2 (continued)

Type of pollutants	Environmental pollution and toxicity profile
Inorganic pollutants	
Cadmium	It is used in rechargeable batteries, special alloy production, coatings, pigments, platings, as a plastic stabilizer and also present in tobacco smoke. Acute exposure of Cd causes abdominal pain, burning sensation, nausea, vomiting, salivation, muscle cramps, itai-itai disease (a combination of osteomalacia and osteoporosis), vertigo, and shock. Loss of consciousness and convulsions usually appear within 15–30 min, and gastrointestinal tract erosion; pulmonary, hepatic, or renal injury; and coma may develop depending on the route of poisoning. However, chronic exposure of Cd causes a depressive effect on levels of norepinephrine, serotonin, and acetylcholine. The mechanism of Cd toxicity is not well known but is assumed to cause cell damage through the generation of reactive oxygen species (ROS) that cause DNA damage
Chromium	Industries such as metallurgical, chemical, refractory brick, leather, wood preservation, and pigments and dyes are the major consumers of chromium. It causes severe health problems such as skin irritation, nasal irritation, ulceration, eardrum perforation, and lung carcinoma
Arsenic	Arsenic (As) causes severe disturbances of the cardiovascular and central nervous systems, bone marrow depression, hemolysis, hepatomegaly, melanosis, polyneuropathy, and encephalopathy, and exposure eventually leads to death. Ingestion may also cause black foot disease that is only reported in Taiwan
Lead	The contamination of soil and water with Pb mostly occurs from anthropogenic activities, industrial wastes, mining and smelting, and the past and present use of Pb in paints, batteries, gasoline, pesticides, and explosives. Pb mainly causes injury to the central nervous system (CNS) and causes headache, poor attention span, irritability, encephalopathy (characterized by sleeplessness and restlessness), loss of memory, acute psychosis, confusion, reduced consciousness, and dullness and also adversely affects the kidneys, liver, hematopoietic system, endocrine system, and reproductive system
Mercury	The major sources of Hg exposure include its use in dental amalgams, thermometers, sphygmomanometers, barometers, fossil fuel emissions, incandescent lights, batteries, ritualistic practices using mercury, and the incineration of medical waste. In humans and animals, it causes mental retardation, dysarthria, blindness, neurological deficits, loss of hearing, developmental defects, and abnormal muscle tone

Adapted from Saxena and Bharagava (2017)

1.4 Bioremediation Approaches for Industrial Wastes/Pollutants

The description of different bioremediation technologies for industrial wastes/pollutants is presented in Table 1.3.

Table 1.3 Bioremediation approaches for organic and inorganic pollutants of industrial wastewater

Bioremediation approaches	Application
Microbial bioremediation	Bioremediation is an eco-friendly waste management method that utilizes the inherent potential of microbes such as algae, fungi, and bacteria to remove organic and inorganic pollutants from industrial wastes
Phytoremediation	Phytoremediation is an eco-sustainable green technology that makes use of green plants for the in situ remediation of environmental pollutants, whether organic or inorganic in nature
Microbe-assisted phytoremediation	Microbe-assisted phytoremediation is a type of bioremediation wherein plant-associated bacteria like rhizobacteria and endophytes are used to enhance the efficiency of remediating plants in the stressed environment via increasing the bioavailability of heavy metals in soil and plant growth promotion
Enzymatic remediation	The use of a variety of catabolic enzymes in the degradation and detoxification of various organic and inorganic pollutants from contaminated wastewater
Electro-bioremediation	It is a popular hybrid technology that uses the combination of bioremediation and electrokinetics for the treatment of environmental pollutants. It involves the electrokinetic phenomena for the acceleration and orientation of transport of environmental pollutants and microbes for pollutant bioremediation
Electrokinetic-phytoremediation	It is a popular hybrid technology that combines phytoremediation with electrokinetic remediation to enhance metal mobility in contaminated soil and facilitate their plant uptake and thus phytoremediation
Microbial fuel cells	A microbial fuel cell (MFC) is a bio-electrochemical device that harnesses the power of respiring microbes to convert organic substrates present in wastewater directly into electrical energy and thus wastewater treatment
Constructed wetlands	*Constructed wetlands* are treatment systems that use natural processes involving *wetland* vegetation, soils, and their associated microbial assemblages to improve water quality
Nano-bioremediation	Integration of nanoparticles and bioremediation for sustainable remediation of environmental pollutants from contaminated matrix

1.4.1 Bioremediation

It is an eco-friendly remediation technique that uses the inherent ability of microbes such as algae, fungi, and bacteria to degrade/detoxify organic and inorganic pollutants from industrial wastewaters (Bharagava et al. 2018b). In bioremediation process, waste is converted into inorganic compounds such as carbon dioxide, water, and methane and thus leads to mineralization and detoxification (Reshma et al. 2011). Bioremediation chiefly depends on the metabolic capability of microbes to degrade/detoxify or transform the pollutants, which is also affected by the accessibility of pollutants and their bioavailability (Antizar-Ladislao 2010). It can be applied as both in situ (remediation at the site) and ex situ (remediation elsewhere) remediation technology. Bioremediation involves (Maszenan et al. 2011) bioattenuation (natural process of degradation and can be monitored by a decrease in pollutant concentration with increasing time), biostimulation (intentional stimulation of pollutant degradation by addition of water, nutrients, and electron donors or acceptors), and bioaugmentation (addition of laboratory-grown microbes with potential for degradation). A great deal of literature can be found in the public domain on the biodegradation and bioremediation of industrial waste/pollutants (see Maszenan et al. 2011; Megharaj et al. 2011; Saxena and Bharagava 2015; Saxena et al. 2016).

1.4.2 Phytoremediation

Phytoremediation is a low-cost and eco-sustainable in situ remediation technology. It is advantageous over the conventional physicochemical cleanup methods that require high-capital investment and labor, alter soil properties, and disturb soil microflora. Phytoremediation is a type of bioremediation wherein green plants with associated microbes are used for the removal of toxic metals from the contaminated matrix to safeguard the environment and public health. It involves different strategies such as phytoextraction, phytostabilization, phytodegradation, phytostimulation, phytovolatilization, and rhizofiltration to remove metal pollutants from the contaminated sites (Lee 2013; Chandra et al. 2015; Chirakkara et al. 2016). It can be commercialized, and income can be generated, if metals removed from contaminated sites could be utilized as "bio-ore" to extract usable form of economically viable metals (i.e., phytomining) (Chandra et al. 2015; Mahar et al. 2016). Bioenergy can be generated through the burning of plant biomass, and land restoration can be achieved for sustainable agricultural development or general habitation (Lintern et al. 2013; Stephenson and Black 2014; Mahar et al. 2016). The rationale, mechanisms, and economic feasibility of phytoremediation have been discussed elsewhere (Ali et al. 2013; Wan et al. 2016; Sarwar et al. 2017). A great deal of literature can be found in the public domain on the phytoremediation of heavy metals from contaminated matrix (Ali et al. 2013; Chandra et al. 2015; Mahar et al. 2016; Sarwar et al. 2017). However, extensive research is currently underway to testify the phytoremediation potential of hyperaccumulating plants in the field for the effective treatment and management of HM-contaminated sites.

1.4.3 Microbe-Assisted Phytoremediation

Exploiting plant-associated microbes with desired traits to enhance the phytoremediation efficiency of hyperaccumulating plants via increasing the bioavailability of metals in soil and plant growth promotion in the stressed environment is termed as microbe-assisted phytoremediation. Inoculation of plants with plant growth-promoting bacteria (PGPR) may be helpful in phytoremediation as they can suppress phytopathogens, tolerate abiotic stress, lower the metal toxicity to remediating plants through biosorption/bioaccumulation as bacterial cells have extremely high ratio of surface area to volume, as well as promote plant growth by secreting various hormones, organic acids, and antibiotics (Rajkumar et al. 2012; Ullah et al. 2015). Endophytes (bacteria that reside in the inner tissues of host plants) are also able to tolerate high metal concentration and hence lower phytotoxicity to remediating plants and help in growth promotion by various means and thus enhance phytoremediation efficiency (Ma et al. 2011, 2015). In addition, arbuscular mycorrhizal fungi (AMF, colonize plant roots) have been also reported to protect their host plants against heavy metal toxicity through their mobilization from soil and thus help in the phytoremediation (Marques et al. 2009; Khan et al. 2014). A great deal of literature can be found in the public domain on the microbe-assisted phytoremediation of heavy metals (Khan et al. 2014; Ma et al. 2011, 2015; Rajkumar et al. 2012; Ullah et al. 2015). Further, to ameliorate metal toxicity, plant growth promotion, and metal sequestration, extensive research efforts are also required to explore novel microbial diversity, their distribution, as well as functions in the autochthonous and allochthonous soil habitats for microbe-assisted phytoremediation of HM-contaminated sites.

1.4.4 Enzymatic Remediation

Bioremediation is an eco-friendly remediation technology that uses biological agents for the degradation and detoxification of organic and inorganic pollutants from industrial wastewaters. However, the efficiency of bioremediation chiefly depends on the ability of enzymes produced by microbes to catalyze the degradation and detoxification of environmental pollutants. A large number of enzymes from bacteria and fungi have been reported to be involved in the biodegradation and biodetoxification of toxic organic and inorganic pollutants. Enzymes such as ligninolytic enzymes (such as lignin peroxidase, LiP; manganese peroxidase, MnP; laccase), chrome reductase, monooxygenase, dioxygenase, and azo-reductase have been reported in the degradation and detoxification of various pollutants from industrial wastewaters. For instance, Bharagava et al. (2018a) reported the biodegradation of crystal violet dye by a ligninolytic enzyme-producing bacterium, *Aeromonas hydrophila*, isolated from textile wastewater. Paisio et al. (2012) also isolated and characterized a dioxygenase-producing *Rhodococcus* strain with phenol-degrading ability and used the bacterium in the biotreatment of tannery effluent.

1.4.5 Emerging Bioremediation Technologies

1.4.5.1 Electro-bioremediation

It is becoming an increasingly popular hybrid technology that uses the combination of bioremediation and electrokinetics for the treatment of environmental pollutants (Maszenan et al. 2011). It involves the electrokinetic phenomena for the acceleration and orientation of transport of environmental pollutants and microbes for pollutant bioremediation (Li et al. 2010; Maszenan et al. 2011). Electrokinetics involves the use of several phenomena like diffusion, electrolysis, electroosmosis, electrophoresis, and electromigration and uses weak electric currents of about 0.2–2 Vcm^{-1} (Saichek and Reddy 2005; Maszenan et al. 2011). A number of studies are available on the use of electro-bioremediation technology for contaminated soils (Wick et al. 2007; Martinez-Prado et al. 2014; Yan and Reible 2015). In addition, the applications, potentials, and limitations of electro-bioremediation technology have been reviewed by many authors (Wick et al. 2007; Maszenan et al. 2011; Gill et al. 2014).

1.4.5.2 Electrokinetic-Phytoremediation

Combining phytoremediation with electrokinetic remediation could be an excellent strategy to enhance metal mobility in contaminated soil and facilitate their plant uptake and thus phytoremediation. For instance, Mao et al. (2016) evaluated the feasibility of electrokinetic remediation coupled with phytoremediation to remove Pb, As, and Cs from contaminated paddy soil. Results revealed that the solubility and bioavailability of Cs and As were significantly increased by the electrokinetic field (EKF) and thereby lower the pH of contaminated soil. Furthermore, they observed that EKF significantly enhanced the bioaccumulation of As and Cs in plant roots and shoots and thus enhanced phytoremediation efficiency. However, the optimization of electrical parameters such as electrical field intensity, current application mode, distance between the electrodes, stimulation period, and their effect on the mobility and bioavailability of HMs are the associated key challenges (Mao et al. 2016). Further, application of electrokinetic-phytoremediation for the mixed contaminants (organic and inorganic) is also not reported so far.

1.4.5.3 Microbial Fuel Cells

A microbial fuel cell (MFC) is a bio-electrochemical device that harnesses the power of respiring microbes to convert organic substrates directly into electrical energy. MFC can be a suitable alternative to the conventional activated sludge process-based treatment systems in terms of energy consumption and excess sludge generation. MFCs offer several advantages over conventional treatment systems related to energy (like direct electricity generation, energy savings by anaerobic treatment due to elimination of aeration, low sludge yield), environmental (water reclamation, low-carbon footprint, less sludge generation), economic (revenue through energy and value-added products-chemicals, low operational costs), and operational benefits (self-generation of microorganisms, good resistance to environmental stress, and amenable to real-time monitoring and control) (Li et al. 2014; Gude 2016).

MFCs are environmentally friendly technologies as they can produce clean electricity directly from organic matter in wastewater without any need for separation, purification, and conversion of the energy products and function at mild operating conditions especially at ambient temperatures (Gude 2016). MFCs can produce up to 1.43 kWh/m^3 from a primary sludge or 1.8 kWh/m^3 from a treated effluent (Ge et al. 2015). MFCs consume only 0.024 kW or 0.076 kWh/kg COD in average (mainly for feeding and mixing in the reactor), about one order of magnitude less than activated sludge-based aerobic processes (~0.3 kW or 0.6 kWh/kg COD) (Zhang et al. 2013a, b). It means MFCs consume only about 10% of the external energy for their operation when compared with conventional-activated sludge process showing great potential for energy savings as well as possible energy recovery from wastewater treatment (Gude 2016).

1.4.5.4 Nano-bioremediation

Nano-bioremediation is the new concept that integrate the use of nanoparticles and bioremediation for sustainable remediation of environmental pollutants in contaminated matrix (Cecchin et al. 2017). For instance, Le et al. (2015) conducted a study for the degradation of a solution containing Aroclor 1248 (PCB) using nZVI (1000 mg/L) and subsequently using biodegradation with bacterium, *Burkholderia xenovorans*. The researchers obtained 89% degradation of the congeners after application of nZVI. Subsequently, they observed a biodegradation of 90% in the biphenyls produced after the dechlorination of PCB by bacterial metabolism. In this study, no toxic effect toward microorganisms by the nZVI was observed.

Further, Bokare et al. (2012) conducted a study into the feasibility of an integration of bioremediation process and reductive process through nanoparticles in a contaminated solution with triclosan (5 g/L). The researchers promoted a sequential degradation of the contaminant by subjecting it to an anaerobic dechlorination through the nanoparticles of Pd/Fe. Subsequently, further remediation is achieved by oxidation of the by-products through the application of the enzyme produced by *Trametes versicolor* (laccase-producing fungi). The results showed complete dechlorination of triclosan in 20 min after application, and its by-products totally oxidized by microbial enzyme. Thus, nano-bioremediation could be an excellent strategy for the remediation of contaminants in environmental matrix.

1.4.5.5 Constructed Wetlands

Constructed wetlands (CWs) are the eco-technological option for the treatment and purification of HM-rich wastewaters. These are the man-engineered systems constructed to utilize the natural processes of aquatic macrophytes with their associated microbial assemblages for wastewater treatment within a more controlled environment (Stottmeister et al. 2003; Khan et al. 2009). CWs are mainly vegetated with different wetland plants with high biomass, fast growth rate, and metal accumulation capacity such as *Phragmites australis*, *Typha latifolia*, *Canna indica*, *Stenotaphrum secundatum*, *Scirpus americanus*, *Scirpus acutus*, *Iris pseudacorus*, etc. for metal-rich wastewater treatment (Bharagava et al. 2017c). CWs have been proved to be successful in the removal of a variety of organic and inorganic pollutants such as metals, nutrients, fecal indicator bacteria and pathogens, and a wide range of micro-pollutants, such as pharmaceutical and personal care products (Zhang et al. 2015). However, the pollutant removal efficiency of CWs mainly depends on wastewater treatment rate, organic loading rate, hydrologic regime, hydraulic retention time, operational mode, and vegetation type (Zhang et al. 2015). The application of CWs in pollutant removal from wastewaters has been recently reviewed by many workers (see Vymazal 2010; Zhang et al. 2015; Bharagava et al. 2017a, b, c).

CWs may provide many ecological and economic benefits such as require low capital investment for construction, low electricity for operation, and less maintenance and provide wildlife habitat, as well as human recreational opportunities and a reuse and recycling option for wastewater treatment facility. CWs are more favored in developing countries due to easily available and less costly land and tropical environment, which help to flourish the microbial communities responsible for the degradation/detoxification of organic and inorganic contaminants in wastewaters and therefore high treatment efficiency (Zhang et al. 2015). Thus, increasing use of CWs can successfully remediate heavy metal pollution and solve various water quality issues in the world. In addition, integrating CWs with a microbial fuel cell (MFC) for wastewater treatment and electricity generation could be an innovative approach for the improved degradation of pollutants. According to a recent study, a maximum power density of 15.73 mW m^{-2} and maximum current density of 69.75 mA m^{-2} could be achieved during the treatment of synthetic wastewater containing methylene blue dye (1000 mg l^{-1}765) with 75% COD removal in an integrated CW-MFC system planted with an ornamental plant, *Canna indica* (Yadav et al. 2012). Moreover, CWs may have great potential for bioenergy production and carbon sequestration, if planted with energy crops. According to a study, the incineration of harvested biomass (16,737 kg with C content, 6185 kg) of *Ludwigia* sp. and *Typha* sp. recovered from a subtropical CW could produce 11,846 kWh for 1 month (Wang et al. 2011). However, the future research should be focused on (a) understanding of microbiological dynamics and correlation of biological and non-biological processes in CWs, (b) knowledge of element cycle dynamics that will help to understand the fundamental processes of greenhouse gas emission in CWs, and (c) understanding of microbial community and plant-microbe interactions to know the underlying mechanism of pollutant removal in CWs (Carvalho et al. 2017). Furthermore, researches are underway to expand the scope and efficacy of CWs for the treatment of metal-contaminated wastewaters.

1.5 Challenges in Bioremediation

Bioremediation has emerged as a low-cost alternative to conventional remediation technologies, which are environmentally destructive and costly and create secondary pollution and thus negatively affect the ecosystem. However, it may get restricted by several factors such as low or non-bioavailability of pollutants to microbes, toxicity of pollutants to microbes and remediating plants, lack of enzymes responsible for the degradation and detoxification of specific environmental pollutants, recalcitrant nature of environmental pollutants, and low biomass, toxicity of nanoparticles to microbes as in the case of nano-bioremediation, and slow growth rate of remediating plants as in the case of phytoremediation. Further, molecular techniques may advance the meaning of bio- and phytoremediation by developing transgenic microbes and plants for environmental remediation, but environmental risks such as invasion of exotic plants and loss of biodiversity, associated with transgenic organisms, make them less feasible for environmental decontamination. Moreover, the strict USA and Western countries' regulations on the use of these organisms also restrict their filed applications. These limitations are sufficient to discredit the applicability of bioremediation technologies and together constitute a major challenge in the way of success at field scale. However, future research efforts may provide new ways to make the bioremediation technologies more efficient for environmental remediation.

1.6 Concluding Remarks

Industrial wastewater is a major source of pollution and toxicity in environment, and bioremediation is an eco-friendly option to treat and manage such hazardous waste. To expand the scope and efficacy of bioremediation, the future research should be focused on (a) search for potential microbial degraders for environmental pollutants, (b) search for catabolic enzymes or genes for the enhanced degradation/detoxification of environmental pollutants, (c) development of transgenic microbes and designer plants using genetic engineering for effective bio- and phytoremediation, (d) selection of suitable plants for phytoremediation, and (e) search for novel rhizobacteria and endophytes for microbe-assisted phytoremediation. However, continued efforts are required to realize the economic feasibility of bioremediation technologies including phytoremediation at the field.

Acknowledgment The financial support as "Major Research Projects" (Grant No.: EEQ/2017/000407) from the "Science and Engineering Research Board" (SERB), Department of Science and Technology (DOST), Government of India (GOI), New Delhi, India, and University Grant Commission (UGC) Fellowship received by Mr. Gaurav Saxena for doctoral studies is duly acknowledged.

References

Ali H, Khan E, Sajad MA (2013) Phytoremediation of heavy metals-concepts and applications. Chemosphere 91(7):869–881

Antizar-Ladislao B (2010) Bioremediation: working with bacteria. Elements 6:389–394

Arora PK, Srivastava A, Singh VP (2014) Bacterial degradation of nitrophenols and their derivatives. J Hazard Mater 266:42–59

Arora PK, Srivastava A, Garg SK, Singh VP (2018) Recent advances in degradation of chloronitrophenols. Bioresour Technol 250C:902–909

Bharagava RN, Saxena G, Mulla SI, Patel DK (2017a) Characterization and identification of recalcitrant organic pollutants (ROPs) in tannery wastewater and its phytotoxicity evaluation for environmental safety. Arch Environ Contam Toxicol. https://doi.org/10.1007/s00244-017-0490-x

Bharagava RN, Chowdhary P, Saxena G (2017b) Bioremediation: an ecosustainable green technology: its applications and limitations. In: Bharagava RN (ed) Environmental pollutants and their bioremediation approaches, 1st edn. CRC Press, Taylor & Francis Group, Boca Raton, pp 1–22. https://doi.org/10.1201/9781315173351-2

Bharagava RN, Saxena G, Chowdhary P (2017c) Constructed wetlands: an emerging phytotechnology for degradation and detoxification of industrial wastewaters. In: Bharagava RN (ed) Environmental pollutants and their bioremediation approaches, 1st edn. CRC Press, Taylor & Francis Group, Boca Raton, pp 397–426. https://doi.org/10.1201/9781315173351-15

Bharagava RN, Mani S, Mullab SI, Saratale GD (2018a) Degradation and decolourization potential of an ligninolytic enzyme producing Aeromonas hydrophila for crystal violet dye and its phytotoxicity evaluation. Ecotoxicol Environ Saf 156:166–175

Bharagava RN, Purchase D, Saxena G, Mulla SI (2018b)Applications of metagenomics in microbial bioremediation of pollutants: from genomics to environmental cleanup. In: Das S, Dash H (eds) Microbial diversity in the genomic era, 1st edn. Academic Press, Elsevier. https://doi.org/10.1016/B978-0-12-814849-5.00026-5

Bokare V, Murugesan K, Kim JH, Kim EJ, Chang YS (2012) Integrated hybrid treatment for the remediation of 2,3,7,8-tetrachlorodibenzo-p-dioxin. Sci Total Environ 435–436:563–566

Carvalho PN, Arias CA, Brix H (2017) Constructed wetlands for water treatment: new developments. Water 9:397

Cecchin I, Krishna R, Thom A, Tessaro EF, Schnaid F (2017) Nanobioremediation: integration of nanoparticles and bioremediation for sustainable remediation of chlorinated organic contaminants in soils. Int Biodeterior Biodegrad 119:419–428

Chandra R, Bharagava RN, Rai V (2008) Melanoidins as major colourant in sugarcane molasses based distillery effluent and its degradation. Bioresour Technol 99:4648–4660

Chandra R, Bharagava RN, Kapley A, Purohit HJ (2011) Bacterial diversity, organic pollutants and their metabolites in two aeration lagoons of common effluent treatment plant (CETP) during the degradation and detoxification of tannery wastewater. Bioresour Technol 102(3):2333–2341

Chandra R, Saxena G, Kumar V (2015) Phytoremediation of environmental pollutants: an ecosustainable green technology to environmental management. In: Chandra R (ed) Advances in biodegradation and bioremediation of industrial waste. CRC Press, Boca Raton, pp 1–30

Chirakkara RA, Cameselle C, Reddy KR (2016) Assessing the applicability of phytoremediation of soils with mixed organic and heavy metal contaminants. Rev Environ Sci Biotechnol. https://doi.org/10.1007/s11157-016-9391-0

Gautam S, Kaithwas G, Bharagava RN, Saxena G (2017) Pollutants in tannery wastewater, pharmacological effects and bioremediation approaches for human health protection and environmental safety. In: Bharagava RN (ed) Environmental pollutants and their bioremediation approaches, 1st edn. CRC Press, Taylor & Francis Group, Boca Raton, pp 369–396. https://doi.org/10.1201/9781315173351-14

Ge Z, Wu L, Zhang F, He Z (2015) Energy extraction from a large-scale microbial fuel cell system treating municipal wastewater. J Power Sources 297:260–264

Gill RT, Harbottle MJ, Smith JWN, Thornton SF (2014) Electrokinetic-enhanced bioremediation of organic contaminants: a review of processes and environmental applications. Chemosphere 107:31–42

Goutam SP, Saxena G, Singh V, Yadav AK, Bharagava RN (2018) Green synthesis of TiO_2 nanoparticles using leaf extract of *Jatropha curcas* L. for photocatalytic degradation of tannery wastewater. Chem Eng J 336:386–396

Gude VG (2016) Wastewater treatment in microbial fuel cells e an overview. J Clean Prod 122:287–307

Khan S, Ahmad I, Shah MT, Rehman S, Khaliq A (2009) Use of constructed wetland for the removal of heavy metals from industrial wastewater. J Environ Manag 90:3451–3457

Khan A, Sharif M, Ali A, Shah SNM, Mian IA, Wahid F, Jan B, Adnan M, Nawaz S, Ali N (2014) Potential of AM fungi in phytoremediation of heavy metals and effect on yield of wheat crop. Am J Plant Sci 5:1578–1586

Le TT, Nguyen KH, Jeon JR, Francis AJ, Chang YS (2015) Nano/bio treatment of polychlorinated biphenyls with evaluation of comparative toxicity. J Hazard Mater 287:335–341

Lee JH (2013) An overview of phytoremediation as a potentially promising technology for environmental pollution control. Biotechnol Bioprocess Eng 18:431–439

Li T, Guo S, Wu B, Li F, Niu Z (2010) Effect of electric intensity on the microbial degradation of petroleum pollutants in soil. J Environ Sci 22:1381–1386

Li Y, Wu Y, Puranik S, Lei Y, Vadas T, Li B (2014) Metals as electron acceptors in single-chamber microbial fuel cells. J Power Sources 269:430–439

Lintern M, Anand R, Ryan C (2013) Natural gold particles in Eucalyptus leaves and their relevance to exploration for buried gold deposits. Nature Commun. Available at: www.nature.com/ncomms/2013/131022/ncomms3614ncomms3614.ht

Ma Y, Prasad MNV, Rajkumar M, Freitas H (2011) Plant growth promoting rhizobacteria and endophytes accelerate phytoremediation of metalliferous soils. Biotechnol Adv 29:248–258

Ma Y, Oliviera RS, Nai F, Rajkumar M, Luo Y, Rocha I, Freitas H (2015) The hyperaccumulator *Sedum plumbizincicola* harbors metal-resistant endophytic bacteria that improve its phytoextraction capacity in 1206 multi-metal contaminated soil. J Environ Manag 156:62–69

Mahar A, Wang P, Ali A, Awasthi MK, Lahori AH, Wang Q, Li R, Zhang Z (2016) Challenges and opportunities in the phytoremediation of heavy metals contaminated soils: a review. Ecotoxicol Environ Saf 26:111–121

Mao X, Han FX, Shao X, Guo K, McComb J, Arslan Z, Zhang Z (2016) Electro-kinetic remediation coupled with phytoremediation to remove lead, arsenic and cesium from contaminated paddy soil. Ecotoxicol Environ Saf 125:16–24

Marques APGC, Rangel AOSS, Castro PML (2009) Remediation of heavy metal contaminated soils: phytoremediation as a potentially promising clean-up technology. Crit Rev Environ Sci Technol 39(8):622–654

Martınez-Prado MA, Unzueta-Medina J, Perez Lopez ME (2014) Electrobioremediation as a hybrid technology to treat soil contaminated with total petroleum hydrocarbons. Rev Mex Ing Quím 13(1):113–127

Maszenan AM, Liu Y, Ng WJ (2011) Bioremediation of wastewaters with recalcitrant organic compounds and metals by aerobic granules. Biotechnol Adv 29:111–123

Megharaj M, Ramakrishnan B, Venkateswarlu K, Sethunathan N, Naidu R (2011) Bioremediation approaches for organic pollutants: a critical perspective. Environ Int 37:1362–1375

Mendez-Paz D, Omil F, Lema JM (2005) Anaerobic treatment of azo dye Acid Orange 7 under fed-batch and continuous. Water Res 39:771–778

Paisio CE, Talano MA, González PS, Busto VD, Talou JR, Agostini E (2012) Isolation and characterization of a Rhodococcus strain with phenol-degrading ability and its potential use for tannery effluent biotreatment. Environ Sci Pollut Res Int 19(8):3430–3439

Pandey A, Singh P, Iyengar L (2007) Bacterial decolorization and degradation of azo dyes. Int Biodeterior Biodegrad 59:73–84

Rajkumar M, Sandhya S, Prasad MN, Freitas H (2012) Perspectives of plant-associated microbes in heavy metal phytoremediation. Biotechnol Adv 30(6):1562–1574

Reshma SV, Spandana S, Sowmya M (2011) Bioremediation technologies. World Congress of Biotechnology, India

Saichek RE, Reddy KR (2005) Electrokinetically enhanced remediation of hydrophobic organic compounds in soil: a review. Crit Rev Environ Sci Technol 35:115–192

Sarwar N, Imran M, Shaheen MR, Ishaq W, Kamran A, Matloob A, Rehim A, Hussain S (2017) Phytoremediation strategies for soils contaminated with heavy metals: modifications and future perspectives. Chemosphere 171:710–721

Saxena G, Bharagava RN (2015) Persistent organic pollutants and bacterial communities present during the treatment of tannery wastewater. In: Chandra R (ed) Environmental waste management, 1st edn. CRC Press, Taylor & Francis Group, Boca Raton, pp 217–247. https://doi.org/10.1201/b19243-10

Saxena G, Bharagava RN (2016) Ram Chandra: advances in biodegradation and bioremediation of industrial waste. Clean Technol Environ Policy 18:979–980. https://doi.org/10.1007/s10098-0151084-9

Saxena G, Bharagava RN (2017) Organic and inorganic pollutants in industrial wastes, their ecotoxicological effects, health hazards and bioremediation approaches. In: Bharagava RN (ed) Environmental pollutants and their bioremediation approaches, 1st edn. CRC Press, Taylor & Francis Group, Boca Raton, pp 23–56. https://doi.org/10.1201/9781315173351-3

Saxena G, Chandra R, Bharagava RN (2016) Environmental pollution, toxicity profile and treatment approaches for tannery wastewater and its chemical pollutants. Rev Environ Contam Toxicol 240:31–69. https://doi.org/10.1007/398_2015_5009

Saxena G, Bharagava RN, Kaithwas G, Raj A (2015) Microbial indicators, pathogens and methods for their monitoring in water environment. J Water Health 13:319–339. https://doi.org/10.2166/wh.2014.275

Singh JS, Abhilash PC, Singh HB, Singh RP, Singh DP (2011) Genetically engineered bacteria: an emerging tool for environmental remediation and future research perspectives. Gene 480:1–9

Stephenson C, Black CR (2014) One step forward, two steps back: the evolution of phytoremediation into commercial technologies. Biosci Horiz 7:1–15

Stottmeister UA, Wießner P, Kuschk U, Kappelmeyer M, Kastner O, Bederski RA, Muller H, Moormann (2003) Effects of plants and microorganisms in constructed wetlands for wastewater treatment. Biotechnol Adv 22:93–117

Ullah A, Heng S, Munis M, Hussain F, Shah F, Xiyan Y (2015) Phytoremediation of heavy metals assisted by plant growth promoting (PGP) bacteria: a review. Environ Exp Bot. https://doi.org/10.1016/j.envexpbot.2015.05.001

Vymazal J (2010) Constructed wetlands for wastewater treatment. Water 2(3):530–549

Wan X, Lei M, Chen T (2016) Cost-benefit calculation 1383 of phytoremediation technology for heavy-metal-contaminated soil. Sci Total Environ 563–564:796–802

Wang YC, Ko CH, Chang FC, Chen PY, Liu TF, Sheu YS, shih TL, Teng CJ (2011) Bioenergy production potential for aboveground biomass from a subtropical constructed wetland. Biomass Bioenergy 35:50–58

Wick LY, Shi L, Harms H (2007) Electro-bioremediation of hydrophobic organic soilcontaminants: a review of fundamental interactions. Electrochim Acta 52(10):3441–3448

Yadav AK, Dash P, Mohanty A, Abbassi R, Mishra BK (2012) Performance assessment of innovative constructed wetland-microbial fuel cell for electricity production and dye removal. Ecol Eng 47:126–131

Yan F, Reible D (2015) Electro-bioremediation of contaminated sediment by electrode enhanced capping. J Environ Manag 15(155):154–161

Zhang F, Ge Z, Grimaud J, Hurst J, He Z (2013a) Long-term performance of literscale microbial fuel cells treating primary effluent installed in a municipal wastewater treatment facility. Environ Sci Technol 47(9):4941–4948

Zhang F, Ge Z, Grimaud J, Hurst J, He Z (2013b) In situ investigation of tubular microbial fuel cells deployed in an aeration tank at a municipal wastewater treatment plant. Bioresour Technol 136:316–321

Zhang DQ, Jinadasa KBSN, Gersberg RM, Liu Y, Tan SK, Ng WJ (2015) Application of constructed wetlands for wastewater treatment in tropical and subtropical regions (2000–2013). J Environ Sci 30:30–46

Chapter 2
Bioremediation: An Eco-friendly Sustainable Technology for Environmental Management

Christopher Chibueze Azubuike, Chioma Blaise Chikere, and Gideon Chijioke Okpokwasili

Contents

Abstract Evironmental pollution is a major public health concern due to the detrimental effects of pollutants to humans and to other living organisms. Chemical and physical methods of remediation are expensive and do not result in complete removal of pollutants. Moreover, both methods may lead to more pollution and site disruption, thus impacting negatively to humans and other biota in the immediate vicinity of the polluted site. Therefore, chemical and physical methods of remediation are not considered eco-sustainable. Unlike these methods, bioremediation, which relies on biological processes (mediated by different groups of living organisms), results in the permanent removal of pollutants. This chapter covers: the eco-sustainable features of bioremediation, pollutants that are susceptible to bioremediation, groups of organisms that play significant roles in bioremediation, and advantages of bioremediation. Furthermore, it highlighted some limitations of

C. C. Azubuike (✉) · C. B. Chikere · G. C. Okpokwasili
Department of Microbiology, Faculty of Science, University of Port Harcourt, Port Harcourt, Rivers States, Nigeria
e-mail: christopher.azubuike@uniport.edu.ng

G. Saxena, R. N. Bharagava (eds.), *Bioremediation of Industrial Waste for Environmental Safety*, https://doi.org/10.1007/978-981-13-1891-7_2

bioremediation and ways of overcoming the limitations. Together, the advantages of bioremediation techniques notably its cost-effectiveness at different scales of operation, the simplicity of operation, process monitoring, and its less destructive features to polluted sites during operation are amongst the features that make bioremediation an eco-sustainable technology for environmental management.

Keywords Environmental pollutants · Toxicity · Bioremediation · Eco-sustainable technology · Environmental management

2.1 Introduction

Pollution of environments by human activities either by direct or indirect means is a major challenge that has been of concern to the past and present generations. The act of pollution seems inevitable given that the world still relies on some of the activities that lead to environmental pollution as a major source of energy. Many acts and protocols have been enforced to regulate the rate at which environments are polluted. Due to the harmful effects of pollutants to living organisms, humans in addition to reducing the rate of pollution have devised means to mitigate the negative effects of pollutants should they manage and enter any environment. The process of restoring the environment to its normal state after pollution is known as remediation. There are many methods of remediation; however, biological, chemical, and physical methods are the major categories based on the nature of the remediating agent (Ghosal et al. 2016). The term bioremediation is used when biological agents are used in an attempt to reduce the level of a pollutant in any polluted site or an environment (Bharagava et al. 2017a; Saxena and Bharagava 2016). Therefore, bioremediation can be defined as the application of biological agents, processes, or mechanisms to restore polluted environment (Singh et al. 2011). The process of bioremediation can be achieved by any or combination of the followings: degradation, mineralization, and transformation (Saxena and Bharagava 2017). Degradation and mineralization are usually associated with organic pollutants. The former involves the sequential breakdown of pollutants into smaller and less toxic forms, whilst in the latter process, pollutants are broken down to carbon dioxide and water as end products under aerobic condition, or methane under anaerobic condition (Haritash and Kaushik 2009). In contrast, transformation is mostly associated with inorganic pollutants owing to their nature and origin. For example, some inorganic pollutants lack carbon, which would otherwise serve as energy source for microbes. Transformation involves modifying original pollutant into another chemically related form. The modified form is not always less toxic compared to the initial pollutant (Ghosal et al. 2016). Unlike transformation, degradation and mineralization are made possible due to microbes utilizing the carbons present in organic pollutants as an energy source for metabolic activities.

The sources and/or means by which pollutants are released into environments are by far greater than the classes of pollutants. In most cases, human activities contribute more to pollution compared to natural processes. Pollutants that are predominant in environments include hydrocarbons, chlorophenols, nitrophenols, heavy metals, herbicides, pesticides, halogenated compounds, solvents, and persistent organic pollutants (Arora et al. 2014, 2018; Vergani et al. 2016; Saxena and Bharagava 2015). The presence of these pollutant in an environment requires urgent attention as a consequence of the imminent threats they pose to human health and to the ecosystem (Huang et al. 2017). This chapter focuses on bioremediation as an eco-sustainable approach for environmental management. Thus, it covers features that make bioremediation eco-sustainable, bioremediation of organic and inorganic pollutants in different environments, major factors limiting bioremediation, and advantages of bioremediation.

2.2 Bioremediation Technology as an Eco-sustainable Technology for Environmental Management

Bioremediation is a sustainable technology for environmental management for many reasons. Firstly, the technology is vast (Fig. 2.1), simple and flexible to operate. It can be operated in situ or ex situ (Azubuike et al. 2016). This makes it possible for the technology to be used on different classes of pollutants. Secondly, it is effective

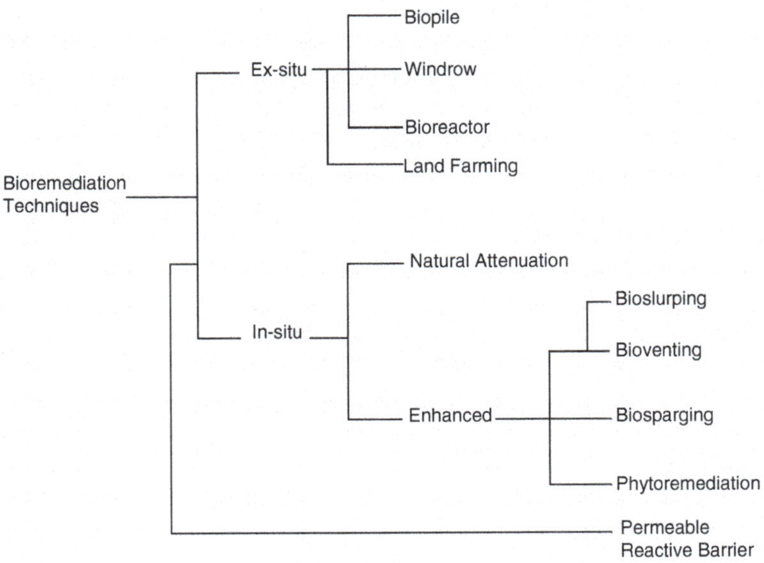

Fig. 2.1 Bioremediation techniques: the divergence of each technique is hypothetical; therefore, the left to right order of internal nodes are not the order of evolution (technique development). Permeable reactive barrier (PBR) is not the arbitrary tree root. It is a physical remediation technique with some elements of bioremediation hence the early hypothetical divergence. (From authors previous source: Azubuike et al. 2016)

with continual and reproducible successes at different scales of operation. (Philp and Atlas 2005). Thirdly, the effectiveness and efficiency of the technology can be monitored, measured, and confirmed using different approaches (Balba et al. 1998; Boopathy 2000; Maila and Cloete 2005; Khudur et al. 2015). Fourthly, the groups of living organisms including microbes that participate in pollutant removal occur naturally in the polluted environment and are economical to use to achieve remediation (Varjani 2017). Lastly, it is not harmful to the environment. Therefore, it is described as an eco-friendly or a green technology. Most importantly, the different approaches or techniques of bioremediation are aimed at achieving the same goal - reduction in pollutant concentration(s) in any polluted environment. In general, the flexibility, efficiency, effectiveness, economical, and eco-friendly nature of bioremediation are the features that contribute to its eco-sustainability.

2.3 Bioremediation Technology for Remediation of Environmental Pollutants

2.3.1 Crude Oil

Crude oil is a complex mixture of mostly hydrocarbons (aliphatics and aromatics) and other non-polar components such as resin and asphaltenes (Head et al. 2006). This class of organic pollutant contributes significantly to marine and terrestrial pollution because of oil exploration and transportation activities. Amongst the different hydrocarbons, aliphatic fractions are preferentially degraded over aromatic fractions (Das and Mukherjee 2007). This is probably due to aliphatic fractions having less complex structure and fractions, and are more bioavailable in a polluted environment (Demeter et al. 2017). Bioremediation has played a pivotal role in remediating environments polluted with crude oil (Wu et al. 2017). The 1989 *Exxon Valdez* oil spill (EVOS) is one of the notable remediation projects, which brought the application of bioremediation to limelight (Lim et al. 2016). Remediation of crude oil constituents by biological means has proved successful and sustainable over the years. Plants and microbial (fungal, algal, and bacterial) enzymes are the mediators of bioremediation of crude oil. Recently, Beškoski et al. (2017) reported >95% hydrocarbon removal from 1158 m^3 groundwater following in situ biostimulation and bioaugmentation with zymogenous microorganisms. Hydrocarbon removal during the study was evidenced by decreased concentration of total petroleum hydrocarbon and a substantial increase in total chemoorganoheterotroph and hydrocarbon-degrading microorganisms notably *Achromobacter*, *Bacillus*, *Micromonospora*, *Pseudomonas*, and *Rhodococcus*. Similarly, Gomez and Sartaj (2013) reported >82% total petroleum hydrocarbon during field-scale ex situ bioremediation of petroleum-contaminated soil in cold climate conditions. Interestingly, bioremediation approach is feasible for large-scale environmental management in the polar region such as the Antarctica, despite the challenges (low temperature and prohibition of the use of non-native microorganisms and bulking agents) associated with on-site remediation in such environment (McWatters et al. 2016).

Several plasmids and genes have been reported to play a key role in bioremediation of hydrocarbons. Examples of these plasmids are NAH7, OCT, pND140, pND160, Q15, and TOL, whilst that of degradative genes are *alkA*, *alkB*, *alkM*, *almA*, *assA1*, *assA2*, *LadA*, *nahA-M*, and *theA* (Throne-Holst et al. 2007; Liu et al. 2011; Abbasian et al. 2015; Wilkes et al. 2016; Varjani 2017). Although most genera of bacteria that degrade hydrocarbons are known to be specialist, very few have been reported to be a generalist, having an affinity for aliphatic and aromatic hydrocarbons. Some of the known generalist hydrocarbon degraders belong to the genera: *Acinetobacter*, *Marinobacter*, *Pseudomonas*, *Rhodococcus*, and *Roseobacter* (McGenity et al. 2012). *Pseudomonas* sp. (with 99.49% sequence similarity to *Pseudomonas brenneri*) isolated from crude oil-contaminated soil was reported to degrade polyaromatic hydrocarbons (PAH) and *n*-alkane (Guermouche M'rassi et al. 2015). Both *alk* and *nah* catabolic pathways were reported to be present in some psychrotrophic strains of *Pseudomonas* (Whyte et al. 1997).

Fungi, on the other hand, contribute greatly towards polyaromatic hydrocarbon removal. Lignin peroxidase, laccase, and manganese peroxidase of ligninolytic fungi are known to be essential for polyaromatic hydrocarbon degradation (Haritash and Kaushik 2009). Mycoremediation with *Pleurotus ostreatus* recorded high PAH degradation from creosote-contaminated soil (García-Delgado et al. 2015). Lignolytic fungi are known to secrete extracellular enzymes, which have low substrate specificity, thus making them a good candidate for biodegradation of wide range of compounds (Haritash and Kaushik 2009). More widely, genus of *Aspergillus*, *Fusarium*, *Trichocladium*, *Verticillium*, and *Acremonium* was reported to degrade 2–7-ringed PAH under micro-aerobic condition (Silva et al. 2009). Yeasts tend to predominate other fungi in polluted environments (Haritash and Kaushik 2009). *Rhodotorula glutinis* isolated from the contaminated stream was reported to have similar degradation capacity to *Pseudomonas aeruginosa* (Romero et al. 1998). Furthermore, algae possess degradative capabilities towards hydrocarbons especially PAH by using enzyme systems and metabolic pathways like that of bacteria and fungi (Cerniglia et al. 1980; Warshawasky et al. 1995). Examples of some genus of algae, which have been reported to degrade aromatic hydrocarbon, are *Agmenellum*, *Chlorella*, *Oscillatoria*, *Scenedesmus*, and *Selenastrum* (Cerniglia et al. 1979; Lei et al. 2007). Other algae belonging to the genus, *Nitzschia* and *Prototheca*, are known to accumulate PAH (Ueno et al. 2008; Hong et al. 2008), therefore reducing their concentrations in the polluted environment. Algal-bacterial microcosms exhibit synergistic relationship, which tends to favour aromatic hydrocarbon removal from contaminated samples (Borde et al. 2003).

In some anoxic environments (aquifers, marine sediments, submerged soil), aerobic biodegradation of hydrocarbons especially aromatics proceeds faster than anaerobic biodegradation (McGenity et al. 2012). This is attributable to hydrocarbon degradation been primarily dependent on aerobic respiration, where oxygen not only serves as electron acceptor but also as a co-substrate for initial attack (hydroxylation and oxygen-mediated ring cleavage) of aromatic hydrocarbons for degradation (McGenity et al. 2012; Ghosal et al. 2016). Nonetheless, other electron acceptors (nitrate, sulphate, ferric, and manganese ions) can initiate hydrocarbon degradation under anaerobic conditions (Foght 2008; Carmona et al. 2009), hence making bioremediation possible even in anoxic environment. In addition, some plants have been reported as bioremediators of hydrocarbons (Dadrasnia and Agamuthu 2013; Gregorio et al. 2014; Somtrakoon et al. 2014; Almansoory et al. 2015). Plants appear to be specialist bioremediator, with several mechanisms involved in plant-mediated remediation (phytoremediation).

2.3.2 Wastewater

Indiscriminate discharge of poorly treated wastewaters has adverse health consequences on aquatic and terrestrial organisms (Goutam et al. 2018; Bharagava et al. 2017b; Gautam et al. 2017; Saxena et al. 2015, 2016). Wastewaters may act as sinks for organic and inorganic contaminants that are recalcitrant to biodegradation (Heidrich et al. 2011). Treatment of wastewaters employs processes such as activated sludge (AS) or extended aeration activated sludge (EAAS), chemical precipitation, ion exchange, reverse osmosis, ultrafiltration, and other electrochemical technologies. Although these processes have their advantages, high cost of operation, incomplete purification, limited versatility, sludge generation, formation of secondary hazardous byproduct(s), and low nutrient removal limit the efficiency and effectiveness of wastewaters treatment by the conventional processes (Segneanu et al. 2013). More so, these processes are likely not to be effective in removing persistent contaminants, which might be present in wastewaters after treatment. Microbial process (biodegradation) has been reported to be efficient, cost-effective, and eco-friendly for removal of xenobiotics from wastewaters. It was observed that following effective application of EAAS process to wastewater treatment, gas chromatography-mass spectrometry (GC-MS) analysis revealed presence of recalcitrant compounds (PAH, pesticides, plasticizers, and other industrial and pharmaceutical compounds), which peaks (concentrations) were reduced after batch culture of the wastewater with a bacterium, *Serratia* sp. ISTVKR1 (Gupta and Thakur 2014). Similarly, inoculation of urban wastewater with microalgae consortia containing mostly of *Chlorella* sp. and *Scenedesmus* sp., resulted in increased biodegradation efficiency of ibuprofen by 40% (Matamoros et al. 2016). The use of microalgae (phycoremediation) in wastewater treatment is fascinating. This is attributed to their ability to use both organic and inorganic carbon sources under mixotrophic condition, which is normally the case with wastewater (Gupta et al. 2016). Therefore, microalgae play a crucial role in remediating wastewater due to their central role in autotrophic carbon dioxide fixation, thus providing additional carbon needed for wastewater treatment process (Das et al. 2016). It follows that advanced treatment of wastewaters after secondary clarification process using microbial pure culture or consortia results in enhanced removal of recalcitrant compounds from wastewaters prior to discharging into surrounding environment.

2.3.3 Organohalides

Halogenated organic compounds (organohalides) are organic compounds containing one or more halogen atoms. These groups of compounds are found useful as additives, biocides, flame retardants, and solvents for many industrial, environmental, and pharmaceutical applications. However, they are known as groundwater pollutant of concerns due to their toxicity and tendency to bioaccumulate and persist in an environment (Cooper et al. 2015; Koenig et al. 2015). Dioxins, polychlorinated biphenyls (PCBs), polybrominated diphenyl ethers (PBDEs), chlorinated and brominated phenols, and mono-, di-, tri-, tetra-, and polychlorinated saturated and unsaturated aliphatic hydrocarbons are examples of commonly known organohalides (Jugder et al. 2016). Dehalogenation process is crucial for a successful degradation of organohalides. This process leads to formation of less toxic and more degradable compound(s). Microbial dehalogenation process occurs preferentially under anoxic condition, which is typical of aquifers (Jugder et al. 2015). *Dehalobacter*, *Dehalococcoides*, *Dehalogenimonas*, and *Desulfitobacterium* have been recognized as versatile organohalide-respiring bacteria (ORB). They play a vital role during anaerobic microbial dehalogenation process referred to as organohalide respiration (Nijenhuis and Kuntze 2016). Reductive dehalogenases (RDases) are the class of enzymes that facilitate dehalogenation of organohalides. The formation of less toxic compound(s) following dehalogenation process is not always the case; therefore mineralization of organohalides and its byproducts is of interest during remediation. To achieve mineralization, at least three major mediators of the process are required. These mediators are electron donor, a dehalogenator, and a hydrocarbon oxidizer. The mediators can be supplied in tandem by using microbial consortia with known functionality and adaptability. The use of synthetic microbial community comprising *Clostridium* sp. Ma13 (a H_2 generator, which provides electron for reductive dehalogenation), *Dehalobacter* sp., (a dehalogenator), and *Desulfatiglans parachlorophenolica* strain DS (an aromatic oxidizer) is a promising approach for anaerobic mineralization of 2,4,6-tribromophenol, a priority pollutant of deep soil, aquifers, and groundwater (Li et al. 2015). Similarly, Garg et al. (2016) reported field-scale transformation of hexachlorocyclohexane to less toxic forms using microbial consortia consisting of *Sphingobium* spp. In addition, aerobic biodegradation of organohalides with enriched bacterial culture from municipal wastewater treatment plant resulted in mineralization of monobrominated and chlorinated benzoic acids both as a single compound and as a mixture (Gaza et al. 2015).

2.3.4 Plastics

Plastics are polymers of synthetic or microbial origins, which can be moulded into shapes. Examples of plastics widely produced by industries are polymers of amides, ethylene, propylene, urethane, styrene, and vinyl chloride (Krueger et al. 2015). These polymers are known to be recalcitrant to degradation due to their high molecular weight, complex three-dimensional structure, and hydrophobicity together contributing to their non-bioavailability to microbes (Kale 2015). The common practice of managing plastic pollution such as incineration has proved inadequate as it can lead to release of more toxic pollutants into the environment. Moreover, the application of incineration is limited to terrestrial ecosystem and will be difficult to apply in aquatic ecosystem, which acts as a sink for most pollutants including plastics. Furthermore, only a small percentage (<10%) of used plastics are being recycled (Sivan 2011). Therefore, microbial processes seem to be the likely option that will result in complete degradation of plastics in any environment. The major challenge however is the time it takes to completely degrade plastics by microorganisms.

Aerobic and anaerobic microorganisms from marine sediment colonized biodegradable plastics better compared to polyethylene, a synthetic counterpart (Nauendorf et al. 2016). Nevertheless, no sign of biodegradation was observed within the study period, 98 days. This further demonstrates that (1) biodegradable plastics are more likely to be degraded faster than synthetic plastics and (2) continual disposal of plastics into environments still poses a threat to both aquatic and terrestrial ecosystem, as complete degradation of plastics of any origin is yet to be confirmed.

Nevertheless, biodegradation of hydrolysable and non-hydrolysable plastics by different microbial groups has been described (Krueger et al. 2015). Few microbial groups namely, actinomycete (*Rhodococcus ruber* strain C208), bacterium (*Brevibacillus borstelensis* strain 707), and fungus (*Penicillium simplicissimum*), were reported to degrade polyethylene (Shah et al. 2008). A fungus *Lasiodiplodia theobromae* isolated from an endemic plant *Psychotria flavida* showed degradability potentials towards irradiated polypropylene film (Sheik et al. 2015). The degradability potential was evidenced by 0.3 mg decrease in the weight of the plastics. Exposure of plastics to high rays such as that of gamma, UV, and sunlight resulted in increased susceptibility to biodegradation (da Luz et al. 2015; Sheik et al. 2015). For example, UV light was reported to initiate dehalogenation of polyvinyl chloride (Gewert et al. 2015). Interestingly, mealworms (larvae of *Tenebrio molitor* Linnaeus) were reported to degrade and mineralize Styrofoam, a polystyrene product within 16-day study period (Yang et al. 2015a). Further, it was reported that gut bacteria (*Exiguobacterium* sp. strain YT2) of the mealworms played a crucial role in the degradation and mineralization of Styrofoam (Yang et al. 2015b).

2.3.5 Pesticides

Pesticides are typically used to control and/or prevent pests to improve the yield of agricultural products. However, nearby surface and groundwater bodies are prone to contamination with pesticides due to runoffs, which might originate from farmlands or industries. Although the use of pesticides with formulations that poses imminent danger to human and other life forms has been banned in most countries, some residues of pesticides or their metabolites (2,6-dichlorobenzamide, chlorpyrifos, lindane) are still detected as pollutant of groundwater (Guillén-Jiménez et al. 2012; Mattsson et al. 2015; Khalid et al. 2016). More so, they are known to be persistent in the contaminated environment and recalcitrant to remediation process (Nisha et al. 2015). Fortunately, microorganisms use some pesticides as carbon, nitrogen, and energy sources (Guillén-Jiménez et al. 2012; Ma et al. 2017), thus contributing to pesticides removal from the contaminated environment (Table 2.1). A microbial consortium consisting of *Streptomyces* was able to grow and remove organochloride mixture from liquid waste. The removal efficiencies were >90% for chlordane and methoxychlor and 40% for lindane (Fuentes et al. 2017). Likewise, Briceño et al. (2016) reported >70% removal of organophosphate-based pesticide (chlorpyrifos) and 50% removal of its intermediate 3,5,6-trichloro-2-pyridinol (TCP) from a liquid medium using both pure and mixed culture of *Streptomyces* spp. Specifically, genus of *Streptomyces* has been demonstrated to have a wide substrate range for pesticides and was reported to degrade and/or transform a variety of structurally different pesticides (Shelton et al. 1996). Biosurfactant addition to mixed bacterial culture

Table 2.1 Examples of some pesticides degraded by microorganisms

Nature/ source of sample	Pesticide	Degradation (%)	Microorganism	References
Wastewaters	Atrazine	72	*Aspergillus niger* AN 400	Marinho et al. (2017)
Effluent and sludge	N, N-dimethylformamide	>46	*Paracoccus* sp. MUK1 and MUK2	Nisha et al. (2015)
Wastewater	Chlorpyrifos		*Pseudomonas kilonensis* SRK1	Khalid et al. (2016)
Agricultural soil	Isoproturon	46	*Sphingomonas* sp. strain AK1	Li et al. (2016)
Soil	2,4-dichlorophenolacetic acid and 2-methyl-4-chlorophenoxy acetic acid	90	*Achromobacter* sp.	Xia et al. (2017)
Not specified	Endosulfan	60	*Pseudomonas aeruginosa*	Pradeep and Subbaiah (2016)
Soil	Atrazine	>90	*Ensifer* sp.	Ma et al. (2017)
Soil	Pentachlorophenol	98	*Citrobacter* sp.	WerheniAmmeri et al. (2016)

significantly improved the rate of chlorpyrifos degradation in soil and prevented accumulation of toxic intermediates (Singh et al. 2016). White-rot fungus *Phanerochaete sordida* has been demonstrated to degrade clothianidin, a neonicotinoid insecticide to the intermediate *N,N′*-methylurea with no neurological toxicity (Mori et al. 2017). The use of stable isotope probing showed that archaea also participate during degradation and mineralization of priority pollutant, pentachlorophenol (Tong et al. 2015). Furthermore, studies have shown that the use of microbial consortium comprising *Arthrobacter* sp. and other bacteria such as *Achromobacter*, *Advenella*, and *Variovorax* resulted in mineralization of diuron and further degradation of the more toxic intermediate (3,4-dichloroaniline) of the herbicide (Devers-Lamrani et al. 2014; Villaverde et al. 2017).

2.3.6 Heavy Metals

Heavy metal pollution is of imminent threat to humans due to the detrimental effects associated with them even at a very low concentration (Alvarez et al. 2017). The non-biodegradable nature of heavy metals contributes to their persistent in the environment and leads to bioaccumulation and biomagnification. Although biodegradation is not practical with heavy metals, other remediation processes mediated by living organisms (plants and microbes) are more eco-friendly compared to conventional physiochemical methods of metal remediation. Such methods are very expensive and might result in the generation of more toxic byproducts (Gupta and Diwan 2016). The presence of metals in an environment induces microbial resistance response, which results in transformation or immobilization of metals thus reducing their concentration in the environment (Gupta and Diwan 2016). Examples of some microorganisms involved in the bioremediation of heavy metals and their resistance mechanisms are outlined (Table 2.2). Selective application of bacteria (*Bacillus* sp., *Lysinibacillus* sp., and *Rhodococcus* sp.) resulted in the significant removal of metals (Al, Pb, and Cu) from multi-metal-contaminated soil (Emenike et al. 2017). Sulphate-reducing bacteria (SRB) has been demonstrated by Zhang et al. (2016) to play an important role during remediation of simulated wastewater contaminated with Antimony (Sb). It was observed that Sb was removed by bioprecipitation, which occurred via two main reactions. The first was a chemical reduction of pentavalent antimony (Sb(V)) to trivalent antimony (Sb(III)), which was mediated by hydrogen sulphide generated by the SRB. The second reaction was a reaction of Sb(III) with excess sulphide resulting in the formation of an insoluble antimony sulphide (Sb_2S_3) as a precipitate. Although it was reported that the SRB was not able to enzymatically reduce Sb(V) to Sb(III), its metabolic activities were responsible for >90% removal of aqueous Sb from the contaminated water. Consequently, biosorption of Sb by the SRB also contributed to the removal of Sb from the wastewater during the 11-day study period. Earthworms increase metal availability in contaminated soil (Dandan et al. 2007). Their activities aid in metal uptake by plants therefore contributing to reduction in metal concentration in contaminated site, and toxic effects to other living organisms (Yao et al. 2012; Lemtiri et al. 2016).

Table 2.2 Some heavy metals removed by bacteria and plants during their bioremediation and phytoremediation

Organisms	Heavy metal	Removal (%)	Resistance mechanism	References
Bacteria				
Bacillus flexus strain As-12	As	28–45	Oxidation and reduction	Jebeli et al. (2017)
Pseudomonas sp. B50D	Hg	85	Reduction, biosorption, biofilm, and siderophore production	Giovanella et al. (2017)
Bacillus sp.	Cd	56–60	Biosorption and transformation	Zhao et al. (2016)
Paenibacillus sp.	Cu	59		Govarthanan et al. (2016)
Plants				
Spartina maritime	As, Cu, Pb, Zn	19–65	Accumulation	Mesa et al. (2015)
Eichhornia crassipes	Fe, Zn, Cd, Cu, B, and Cr		Filtration	Elias et al. (2014)
Plectranthus amboinicus	Pb	50–100	Filtration	Ignatius et al. (2014)
Amaranthus paniculatus	Ni	25–60	Accumulation	Iori et al. (2013)

Several studies have reported that plants also have the capacity to reduce metal concentrations in the metal-contaminated environment (Table 2.2). Amongst the phytoremediation mechanisms, accumulation seems to be the predominant resistance mechanism of some plants to metals (Saxena et al. 2018; Kersten et al. 2017; Bharagava et al. 2017c; Belouchrani et al. 2016; Cojocaru et al. 2016; Chandra et al. 2015). *Alcea aucheri* hyperaccumulated Pb and Cd in the environment naturally polluted with metals (Ravanbakhsh et al. 2016). The plant not only accumulated metals but also translocated high concentration to other aerials parts and produced high biomass, thus making recovery of precious metals following remediation a viable option. Likewise, *Brassica* spp. have been reported as Cd hyperaccumulators (Cojocaru et al. 2016; Li et al. 2016). Plants involved in the removal of some heavy metals are summarized in Table 2.2.

2.4 Advantages and Disadvantages of Bioremediation

Bioremediation as a method of remediating polluted environment has many notable advantages compared to other methods of remediation, the foremost being its eco-friendly features, which include minimal site disruption, permanent waste removal, and elimination of long-term liability (Boopathy 2000). This makes it gain more public acceptance, as it does not impact negatively to the environment during remediation. The efficiency and effectiveness of bioremediation have been demonstrated at lab-, pilot-, and field-scale levels (Huang et al. 2015; Shannon et al. 2015; McWatters et al. 2016; Simpanen et al. 2016; Varjani and Upasani 2016). Being effective, bioremediation results in a reduction in pollutant concentration by one or combination of several mechanisms (degradation, transformation, mineralization, etc.). Efficiency, on the other hand, implies that expected outcome is achieved within a short period of time with minimal effort and resources. This is true for bioremediation; however, achieving the desired pollutant concentration during field studies requires control of some environmental factors that might limit the success of bioremediation. Some of these factors are nutrient availability, microbial population and activity, electron acceptors and donors, pH, temperature, and above all availability of pollutant to microbes. Nevertheless, these factors can be controlled by adopting viable approaches. For example, exogenous addition of nutrients enhanced pollutant removal, hence speeding up the rate of bioremediation (Nwogu et al. 2015; Chikere et al. 2015). In some instances, microbes may be present in sufficient number in the polluted environment but may not be well equipped to degrade the polluting substance. This challenge can be addressed by adding microbes with known degradability potentials. These exogenous additions of nutrient, and microbes to expedite bioremediation process have been, respectively, referred to as biostimulation and bioaugmentation. The latter has been reported to play a vital role in the remediation of recalcitrant, persistent, and xenobiotic pollutant (Cycoń et al. 2017). Depending on the limiting factor(s), there are many bioremediation techniques to choose from to achieve the desired goal. These techniques (Azubuike et al. 2016) have their unique features, with each tailored to overcome certain limitations. Pollutant non-bioavailability tends to reduce the rate of bioremediation. Addition of biosurfactant increases the solubility of sorbed and/or hydrophobic pollutant, and in turn increases mass transfer to microbes. It also increases microbial accessibility to pollutant by causing structural changes in the bacterial cell surface. These actions speed up bioremediation process and make it more efficient (Varjani and Upasani 2017).

Furthermore, bioremediation as a method is very simple to implement and cost-effective. Some notable examples are land farming and monitored natural attenuation. Both techniques of bioremediation require less equipment and manpower for operation and rely mostly on microbial aerobic and anaerobic processes to degrade pollutant. Importantly, techniques of bioremediation are flexible and can be made to operate in tandem. This simultaneous application of more than one technique of bioremediation tends to improve the efficiency of bioremediation process, by overcoming the weakness of a single technique (Cassidy et al. 2015; García-Delgado et al. 2015; Martínez-Pascual et al. 2015). In addition, bioremediation can be coupled to other chemical or physical methods of remediation (Boopathy 2000). Lastly, the process and success of pollutant removal during bioremediation can be monitored. Quantitatively, pollutant removal can be determined by using an analytical method such as gas chromatography, whilst the use of biological activities (bioindicators) provides qualitative information on pollutant removal. Enzymes, microorganisms, plants, and earthworms are some of the bioindicators for monitoring pollutant removal (Balba et al. 1998; Maila and Cloete 2005).

2.5 Conclusion

This chapter covered the effectiveness of bioremediation on different classes of pollutants. Many attributes make bioremediation an eco-sustainable technology for environmental management. Amongst these attributes are cost-effectiveness, efficiency, reliability, reproducibility, and above all the vast techniques used to achieve bioremediation. Different groups of organisms (bacteria, fungi, algae, plants, and other microorganisms) have the potentials to act as a bioremediator of pollutant. These organisms can act as either generalists or specialists towards pollutant removal. Despite the bioremediation potentials of these groups of organisms, factors such as the nature and concentration of pollutant, site of pollution, and other environmental factors (pH, temperature, nutrient and pollutant bioavailability, and availability of electron acceptors and donors) need to be considered for a successful bioremediation project. It is important to state that site characterization of a polluted site is crucial before executing any bioremediation project. This will help determine factors that might impede the success of bioremediation process and devise a strategy to overcome such factors, thus making the process more efficient.

Acknowledgement The authors are thankful to Dr. (Mrs.) Okeke-Ogbuafor NA, School of Natural and Applied Sciences, Marine Science and Technology, Newcastle University, United Kingdom, for proofreading of this chapter.

References

Abbasian F, Lockington R, Mallavarapu M, Naidu R (2015) A comprehensive review of aliphatic hydrocarbon biodegradation by bacteria. Appl Biochem Biotechnol 176:1–30. https://doi.org/10.1007/s12010-015-1603-5

Almansoory AF, Hasan HA, Idris M, Abdullah SRS, Anuar N (2015) Potential application of a biosurfactant in phytoremediation technology for treatment of gasoline-contaminated soil. Ecol Eng 84:113–120. https://doi.org/10.1016/j.ecoleng.2015.08.001

Alvarez A, Saez JM, Davila Costa JS, Colin VL, Fuentes MS, Cuozzo SA, Benimeli CS, Polti MA, Amoroso MJ (2017) Actinobacteria: current research and perspectives for bioremediation of pesticides and heavy metals. Chemosphere 166:41–62. https://doi.org/10.1016/j.chemosphere.2016.09.070

Arora PK, Srivastava A, Singh VP (2014) Bacterial degradation of nitrophenols and their derivatives. J Hazard Mater 266:42–59

Arora PK, Srivastava A, Garg SK, Singh VP (2018) Recent advances in degradation of chloronitrophenols. Bioresour Technol 250C:902–909

Azubuike CC, Chikere CB, Okpokwasili GC (2016) Bioremediation techniques-classification based on site of application: principles, advantages, limitations and prospects. World J Microbiol Biotechnol 32:180. https://doi.org/10.1007/s11274-016-2137-x

Balba MT, Al-awadhi N, Al-daher R (1998) Bioremediation of oil contaminated soil: microbiological methods for feasibility assessment and field evaluation. J Microbiol Methods 32:155–164
Belouchrani AS, Mameri N, Abdi N, Grib H, Lounici H, Drouiche N (2016) Phytoremediation of soil contaminated with Zn using Canola(*Brassica napus* L). Ecol Eng 95:43–49. https://doi.org/10.1016/j.ecoleng.2016.06.064
Beškoski VP, Miletić S, Ilić M, Gojgić-Cvijović G, Papić P, Marić N, Šolević-Knudsen T, Jovančićević BS, Nakano T, Vrvić MM (2017) Biodegradation of isoprenoids, steranes, terpanes, and phenanthrenes during in situ bioremediation of petroleum-contaminated groundwater. Clean (Weinh) 45:1–7. https://doi.org/10.1002/clen.201600023
Bharagava RN, Chowdhary P, Saxena G (2017a) Bioremediation: an ecosustainable green technology: its applications and limitations. In: Bharagava RN (ed) Environmental pollutants and their bioremediation approaches, 1st edn. CRC Press, Taylor & Francis Group, Boca Raton, pp 1–22. https://doi.org/10.1201/9781315173351-2
Bharagava RN, Saxena G, Chowdhary P (2017b) Constructed wetlands: an emerging phytotechnology for degradation and detoxification of industrial wastewaters. In: Bharagava RN (ed) Environmental pollutants and their bioremediation approaches, 1st edn. CRC Press, Taylor & Francis Group, Boca Raton, pp 397–426. https://doi.org/10.1201/9781315173351-15
Bharagava RN, Saxena G, Mulla SI, Patel DK (2017c) Characterization and identification of recalcitrant organic pollutants (ROPs) in tannery wastewater and its phytotoxicity evaluation for environmental safety. Arch Environ Contam Toxicol. https://doi.org/10.1007/s00244-017-0490-x
Boopathy (2000) Factors limiting bioremediation technology. Bioresour Technol 74:63–67
Borde X, Guieysse B, Delgado O, Munoz R, Hatti-Kaul R, Nugier-Chauvin C, Patin H, Mattiasson B (2003) Synergistic relationships in algal-bacterial microcosms for the treatment of aromatic pollutants. Bioresour Technol 86:293–300. https://doi.org/10.1016/S0960-8524(02)00074-3
Briceño G, Schalchli H, Mutis A, Benimeli CS, Palma G, Tortella GR, Diez MC (2016) Use of pure and mixed culture of diazinon-degrading *Streptomyces* to remove other organophosphorus pesticides. Int Biodeterior Biodegrad 114:193–201. https://doi.org/10.1016/j.ibiod.2016.06.018
Carmona M, Zamarro MT, Blazquez B, Durante-Rodriguez G, Juarez JF, Valderrama JA, Barragán MJ, García JL, Díaz E (2009) Anaerobic catabolism of aromatic compounds: a genetic and genomic view. Microbiol Mol Biol Rev 73:71–133. https://doi.org/10.1128/MMBR.00021-08
Cassidy DP, Srivastava VJ, Dombrowski FJ, Lingle JW (2015) Combining in situ chemical oxidation, stabilization, and anaerobic bioremediation in a single application to reduce contaminant mass and leachability in soil. J Hazard Mater 297:347–355. https://doi.org/10.1016/j.jhazmat.2015.05.030
Cerniglia CE, Gibson DT, Baalen CV (1979) Algal oxidation of aromatic hydrocarbons: formation of 1-Naphthol from naphthalene by *Agmenellum quadruplicatum* strain PR-6. Biochem Biophys Res Commun 88:50–58. https://doi.org/10.1016/0006-291X(79)91695-4
Cerniglia CE, Baalen CV, Gibson DT (1980) Metabolism of naphthalene by cyanobacterium *Oscillatoria* sp. strain JCM. J Gen Microbiol 116:485–494
Chandra R, Saxena G, Kumar V (2015) Phytoremediation of environmental pollutants: an ecosustainable green technology to environmental management. In: Chandra R (ed) Advances in biodegradation and bioremediation of industrial waste, 1st edn. CRC Press, Taylor & Francis Group, Boca Raton, pp 1–30. https://doi.org/10.1201/b18218-2
Chikere CB, Obieze CC, Okerentugba P (2015) Molecular assessment of microbial species involved in the biodegradation of crude oil in saline Niger Delta sediments using bioreactors. J Bioremed Biodegr 6:1–7. https://doi.org/10.4172/2155-6199.1000307
Cojocaru P, Gusiatin ZM, Cretescu I (2016) Phytoextraction of Cd and Zn as single or mixed pollutants from soil by rape (*Brassica napus*). Environ Sci Pollut Res 23:10693–10701. https://doi.org/10.1007/s11356-016-6176-5
Cooper M, Wagner A, Wondrousch D, Sonntag F, Sonnabend A, Brehm M, Schüürmann G, Adrian L (2015) Anaerobic microbial transformation of halogenated aromatics and fate prediction using electron density modeling. Environ Sci Technol 49:6018–6028. https://doi.org/10.1021/acs.est.5b00303

Cycoń M, Mrozik A, Piotrowska-Seget Z (2017) Bioaugmentation as a strategy for the remediation of pesticide-polluted soil: a review. Chemosphere 172:52–71. https://doi.org/10.1016/j.chemosphere.2016.12.129

da Luz JMR, Paes SA, Ribeiro KVG, Mendes IR, Kasuya MCM (2015) Degradation of green polyethylene by *Pleurotus ostreatus*. PLoS One. https://doi.org/10.1371/journal.pone.0126047

Dadrasnia A, Agamuthu P (2013) Diesel fuel degradation from contaminated soil by *Dracaena reflexa* using organic waste supplementation. J Jpn Petrol Inst 56:236–243. https://doi.org/10.1627/jpi.56.236

Dandan W, Huixin L, Feng H, Xia W (2007) Role of earthworm-straw interactions on phytoremediation of Cu contaminated soil by ryegrass. Acta Ecol Sin 27:1292–1298. https://doi.org/10.1016/S1872-2032(07)60030-4

Das K, Mukherjee AK (2007) Crude petroleum-oil biodegradation efficiency of *Bacillus subtilis* and *Pseudomonas aeruginosa* strains isolated from a petroleum-oil contaminated soil from North-East India. Bioresour Technol 98:1339–1345. https://doi.org/10.1016/j.biortech.2006.05.032

Das C, Naseera K, Ram A, Meena RM, Ramaiah N (2016) Bioremediation of tannery wastewater by a salt-tolerant strain of *Chlorella vulgaris*. J Appl Phycol. https://doi.org/10.1007/s10811-016-0910-8

Demeter MA, Lemire JA, Mercer SM, Turner RJ (2017) Screening selectively harnessed environmental microbial communities for biodegradation of polycyclic aromatic hydrocarbons in moving bed biofilm reactors. Bioresour Technol 228:116–124. https://doi.org/10.1016/j.biortech.2016.12.086

Devers-Lamrani M, Pesce S, Rouard N, Martin-Laurent F (2014) Evidence for cooperative mineralization of diuron by *Arthrobacter* sp. BS2 and *Achromobacter* sp. SP1 isolated from a mixed culture enriched from diuron exposed environments. Chemosphere 117:208–215. https://doi.org/10.1016/j.chemosphere.2014.06.080

Elias SH, Mohamed M, Nor-Anuar A, Muda K, Hassan MAHM, Othman MN, Chelliapan S (2014) Ceramic industry wastewater treatment by rhizofiltration system-application of water hyacinth bioremediation. Inst Integr Omics Appl Biotechnol J 5:6–14

Emenike CU, Liew W, Fahmi MG, Jalil KN, Pariathamby A, Hamid FS (2017) Optimal removal of heavy metals from leachate contaminated soil using bioaugmentation process. Clean (Weinh) 45:1500802. https://doi.org/10.1002/clen.201500802

Foght J (2008) Anaerobic biodegradation of aromatic hydrocarbons: pathways and prospects. J Mol Microbiol Biotechnol 15:93–120. https://doi.org/10.1159/000121324

Fuentes MS, Raimondo EE, Amoroso MJ, Benimeli CS (2017) Removal of a mixture of pesticides by a *Streptomyces* consortium: influence of different soil systems. Chemosphere 173:359–367. https://doi.org/10.1016/j.chemosphere.2017.01.044

García-Delgado C, Alfaro-Barta I, Eymar E (2015) Combination of biochar amendment and mycoremediation for polycyclic aromatic hydrocarbons immobilization and biodegradation in creosote-contaminated soil. J Hazard Mater 285:259–266. https://doi.org/10.1016/j.jhazmat.2014.12.002

Garg N, Lata P, Jit S, Sangwan N, Singh AK, Dwivedi V, Niharika N, Kaur J, Saxena A, Dua A, Nayyar N, Kohli P, Geueke B, Kunz P, Rentsch D, Holliger C, Kohler HPE, Lal R (2016) Laboratory and field scale bioremediation of hexachlorocyclohexane (HCH) contaminated soils by means of bioaugmentation and biostimulation. Biodegradation 27:1–15. https://doi.org/10.1007/s10532-016-9765-6

Gautam S, Kaithwas G, Bharagava RN, Saxena G (2017) Pollutants in tannery wastewater, pharmacological effects and bioremediation approaches for human health protection and environmental safety. In: Bharagava RN (ed) Environmental pollutants and their bioremediation approaches, 1st edn. CRC Press, Taylor & Francis Group, Boca Raton, pp 369–396. https://doi.org/10.1201/9781315173351-14

Gaza S, Felgner A, Otto J, Kushmaro A, Ben-Dov E, Tiehm A (2015) Biodegradation of chloro- and bromobenzoic acids: effect of milieu conditions and microbial community analysis. J Hazard Mater 287:24–31. https://doi.org/10.1016/j.jhazmat.2015.01.025

Gewert B, Plassmann MM, MacLeod M (2015) Pathways for degradation of plastic polymers floating in the marine environment. Environ Sci Process Impact 17:1513–1521. https://doi.org/10.1039/c5em00207a

Ghosal D, Ghosh S, Dutta TK, Ahn Y (2016) Current state of knowledge in microbial degradation of polycyclic aromatic hydrocarbons (PAHs): a review. Front Microbiol. https://doi.org/10.3389/fmicb.2016.01369

Giovanella P, Cabral L, Costa AP, de Oliveira Camargo FA, Gianello C, Bento FM (2017) Metal resistance mechanisms in gram-negative bacteria and their potential to remove Hg in the presence of other metals. Ecotoxicol Environ Saf 140:162–169. https://doi.org/10.1016/j.ecoenv.2017.02.010

Gomez F, Sartaj M (2013) Field scale ex-situ bioremediation of petroleum contaminated soil under cold climate conditions. Int Biodeterior Biodegrad 85:375–382. https://doi.org/10.1016/j.ibiod.2013.08.003

Goutam SP, Saxena G, Singh V, Yadav AK, Bharagava RN (2018) Green synthesis of TiO_2 nanoparticles using leaf extract of *Jatropha curcas* L. for photocatalytic degradation of tannery wastewater. Chem Eng J 336:386–396. https://doi.org/10.1016/j.cej.2017.12.029

Govarthanan M, Mythili R, Selvankumar T, Kamala-Kannan S, Rajasekar A, Chang YC (2016) Bioremediation of heavy metals using an endophytic bacterium *Paenibacillus* sp. RM isolated from the roots of *Tridax procumbens*. 3. Biotech 6:1–7. https://doi.org/10.1007/s13205-016-0560-1

Gregorio SD, Gentini A, Siracusa G, Becarelli S, Azaizeh H, Lorenzi R (2014) Phytomediated biostimulation of the autochthonous bacterial community for the acceleration of the depletion of polycyclic aromatic hydrocarbons in contaminated sediments. Biomed Res Int:891630. https://doi.org/10.1155/2014/891630

Guermouche M'rassi A, Bensalah F, Gury J, Duran R (2015) Isolation and characterization of different bacterial strains for bioremediation of n-alkanes and polycyclic aromatic hydrocarbons. Environ Sci Pollut Res 22:15332–15346. https://doi.org/10.1007/s11356-015-4343-8

Guillén-Jiménez F d M, Cristiani-Urbina E, Cancino-Díaz JC, Flores-Moreno JL, Barragán-Huerta BE (2012) Lindane biodegradation by the *Fusarium verticillioides* AT-100 strain, isolated from Agave tequilana leaves: Kinetic study and identification of metabolites. Int Biodeterior Biodegrad 74:36–47. https://doi.org/10.1016/j.ibiod.2012.04.020

Gupta P, Diwan B (2016) Bacterial exopolysaccharide mediated heavy metal removal: a review on biosynthesis, mechanism and remediation strategies. Biotechnol Rep 13:58–71. https://doi.org/10.1016/j.btre.2016.12.006

Gupta A, Thakur IS (2014) Biodegradation of wastewater organic contaminants using *Serratia* sp. ISTVKR1 isolated from sewage sludge. Biochem Eng J 102:115–124. https://doi.org/10.1016/j.bej.2015.02.007

Gupta S, Pandey RA, Pawar SB (2016) Microalgal bioremediation of food-processing industrial wastewater under mixotrophic conditions: Kinetics and scale-up approach. Front Chem Sci Eng 10:499–508. https://doi.org/10.1007/s11705-016-1602-2

Haritash AK, Kaushik CP (2009) Biodegradation aspects of polycyclic aromatic hydrocarbons (PAHs): a review. J Hazard Mater 169:1–15. https://doi.org/10.1016/j.jhazmat.2009.03.137

Head IM, Jones DM, Röling WFM (2006) Marine microorganisms make a meal of oil. Nat Rev Microbiol 4:173–182. https://doi.org/10.1038/nrmicro1348

Heidrich ES, Curtis TP, Dolfing J (2011) Determination of the internal chemical energy of wastewater. Environ Sci Technol 45:827–832. https://doi.org/10.1021/es103058w

Hong Y-W, Yuan D-X, Lin Q-M, Yang T-L (2008) Accumulation and biodegradation of phenanthrene and fluoranthene by the algae enriched from a mangrove aquatic ecosystem. Mar Pollut Bull 56:1400–1405. https://doi.org/10.1016/j.marpolbul.2008.05.003

Huang G, Liu F, Yang Y, Deng W, Li S, Huang Y, Kong X (2015) Removal of ammonium-nitrogen from groundwater using a fully passive permeable reactive barrier with oxygen-releasing compound and clinoptilolite. J Environ Manag 154:1–7. https://doi.org/10.1016/j.jenvman.2015.02.012

Huang D, Hu C, Zeng G, Cheng M, Xu P, Gong X, Wang R, Xue W (2017) Combination of Fenton processes and biotreatment for wastewater treatment and soil remediation. Sci Total Environ 574:1599–1610. https://doi.org/10.1016/j.scitotenv.2016.08.199

Ignatius A, Arunbabu V, Neethu J, Ramasamy EV (2014) Rhizofiltration of lead using an aromatic medicinal plant *Plectranthus amboinicus* cultured in a hydroponic nutrient film technique (NFT) system. Environ Sci Pollut Res 21:13007–13016. https://doi.org/10.1007/s11356-014-3204-1

Iori V, Pietrini F, Cheremisina A, Shevyakova NI, Radyukina N, Kuznetsov VV, Zacchini M (2013) Growth responses, metal accumulation and phytoremoval capability in Amaranthus plants exposed to nickel under hydroponics. Water Air Soil Pollut 224:1–10. https://doi.org/10.1007/s11270-013-1450-3

Jebeli MA, Maleki A, Amoozegar MA, Kalantar E, Izanloo H, Gharibi F (2017) *Bacillus flexus* strain As-12, a new arsenic transformer bacterium isolated from contaminated water resources. Chemosphere 169:636–641. https://doi.org/10.1016/j.chemosphere.2016.11.129

Jugder BE, Ertan H, Lee M, Manefield M, Marquis CP (2015) Reductive dehalogenases come of age in biological destruction of organohalides. Trends Biotechnol 33:595–610. https://doi.org/10.1016/j.tibtech.2015.07.004

Jugder BE, Ertan H, Bohl S, Lee M, Marquis CP, Manefield M (2016) Organohalide respiring bacteria and reductive dehalogenases: key tools in organohalide bioremediation. Front Microbiol 7:1–12. https://doi.org/10.3389/fmicb.2016.00249

Kale (2015) Microbial degradation of plastics: a review. J Biochem Technol 6:952–961. https://doi.org/10.1504/IJEP.2008.016895

Kersten G, Majestic B, Quigley M (2017) Phytoremediation of cadmium and lead-polluted watersheds. Ecotoxicol Environ Saf 137:225–232. https://doi.org/10.1016/j.ecoenv.2016.12.001

Khalid S, Hashmi I, Khan SJ (2016) Bacterial assisted degradation of chlorpyrifos: the key role of environmental conditions, trace metals and organic solvents. J Environ Manag 168:1–9. https://doi.org/10.1016/j.jenvman.2015.11.030

Khudur LS, Shahsavari E, Miranda AF, Morrison PD, Nugegoda D, Ball AS (2015) Evaluating the efficacy of bioremediating a diesel-contaminated soil using ecotoxicological and bacterial community indices. Environ Sci Pollut Res 22:14809–14819. https://doi.org/10.1007/s11356-015-4624-2

Koenig J, Lee J, Manefield M (2015) Aliphatic organochlorine degradation in subsurface environments. Rev Environ Sci Biotechnol 14:49–71. https://doi.org/10.1007/s11157-014-9345-3

Krueger MC, Harms H, Schlosser D (2015) Prospects for microbiological solutions to environmental pollution with plastics. Appl Microbiol Biotechnol 99:8857–8874. https://doi.org/10.1007/s00253-015-6879-4

Lei AP, Hu ZL, Wong YS, Tam NRY (2007) Removal of fluoranthene and pyrene by different microalgal species. Bioresour Technol 98:273–280. https://doi.org/10.1016/j.biortech.2006.01.012

Lemtiri A, Liénard A, Alabi T, Brostaux Y, Cluzeau D, Francis F, Colinet G (2016) Earthworms *Eisenia fetida* affect the uptake of heavy metals by plants *Vicia faba* and *Zea mays* in metal-contaminated soils. Appl Soil Ecol 104:67–78. https://doi.org/10.1016/j.apsoil.2015.11.021

Li Z, Yoshida N, Wang A, Nan J, Liang B, Zhang C, Zhang D, Suzuki D, Zhou X, Xiao Z, Katayama A (2015) Anaerobic mineralization of 2,4,6-tribromophenol to CO2 by a synthetic microbial community comprising *Clostridium*, *Dehalobacter*, and *Desulfatiglans*. Bioresour Technol 176:225–232. https://doi.org/10.1016/j.biortech.2014.10.097

Li R, Dörfler U, Munch JC, Schroll R (2016) Enhanced degradation of isoproturon in an agricultural soil by a *Sphingomonas* sp. strain and a microbial consortium. Chemosphere 168:1169–1176. https://doi.org/10.1016/j.chemosphere.2016.10.084

Li X, Zhang X, Yang Y, Li B, Wu Y, Sun H, Yang Y (2016) Cadmium accumulation characteristics in turnip landraces from China and assessment of their phytoremediation potential for contaminated soils. Front Plant Sci 7:1–10. https://doi.org/10.3389/fpls.2016.01862

Lim MW, Lau EV, Poh PE (2016) A comprehensive guide of remediation technologies for oil contaminated soil – present works and future directions. Mar Pollut Bull 109:14–45. https://doi.org/10.1016/j.marpolbul.2016.04.023

Liu CL, Wang WP, Wu YH, Zhou ZW, Lai QL, Shao ZZ (2011) Multiple alkane hydroxylase systems in a marine alkane degrader, *Alcanivorax dieselolei* B-5. Environ Microbiol 13:1168–1178. https://doi.org/10.1111/j.1462-2920.2010.02416.x

Ma L, Chen S, Yuan J, Yang P, Liu Y, Stewart K (2017) Rapid biodegradation of atrazine by *Ensifer* sp. strain and its degradation genes. Int Biodeterior Biodegrad 116:133–140. https://doi.org/10.1016/j.ibiod.2016.10.022

Maila MP, Cloete TE (2005) The use of biological activities to monitor the removal of fuel contaminants – perspective for monitoring hydrocarbon contamination: a review. Int Biodeterior Biodegrad 55:1–8. https://doi.org/10.1016/j.ibiod.2004.10.003

Marinho G, Barbosa BCA, Rodrigues K, Aquino M, Pereira L (2017) Potential of the filamentous fungus *Aspergillus niger* AN 400 to degrade Atrazine in wastewaters. Biocatal Agric Biotechnol 9:162–167. https://doi.org/10.1016/j.bcab.2016.12.013

Martínez-Pascual E, Grotenhuis T, Solanas AM, Viñas M (2015) Coupling chemical oxidation and biostimulation: effects on the natural attenuation capacity and resilience of the native microbial community in alkylbenzene-polluted soil. J Hazard Mater 300:135–143. https://doi.org/10.1016/j.jhazmat.2015.06.061

Matamoros V, Uggetti E, García J, Bayona JM (2016) Assessment of the mechanisms involved in the removal of emerging contaminants by microalgae from wastewater: a laboratory scale study. J Hazard Mater 301:197–205. https://doi.org/10.1016/j.jhazmat.2015.08.050

Mattsson MK, Liu X, Yu D, Kontro MH (2015) Depth, soil type, water table, and site effects on microbial community composition in sediments of pesticide-contaminated aquifer. Environ Sci Pollut Res 22:10263–10279. https://doi.org/10.1007/s11356-015-4224-1

McGenity TJ, Folwell BD, McKew BA, Sanni GO (2012) Marine crude-oil biodegradation: a central role for interspecies interactions. Aquat Biosyst 8:10. https://doi.org/10.1186/2046-9063-8-10

McWatters RS, Wilkins D, Spedding T et al (2016) On site remediation of a fuel spill and soil reuse in Antarctica. Sci Total Environ 571:963–973. https://doi.org/10.1016/j.scitotenv.2016.07.084

Mesa J, Rodríguez-Llorente JD, Pajuelo E, Piedras JMB, Caviedes MA, Redondo-Gómez S, Mateos-Naranjo E (2015) Moving closer towards restoration of contaminated estuaries: bioaugmentation with autochthonous rhizobacteria improves metal rhizoaccumulation in native *Spartina maritima*. J Hazard Mater 300:263–271. https://doi.org/10.1016/j.jhazmat.2015.07.006

Mori T, Wang J, Tanaka Y, Nagai K, Kawagishi H, Hirai H (2017) Bioremediation of the neonicotinoid insecticide clothianidin by the white-rot fungus *Phanerochaete sordida*. J Hazard Mater 321:586–590. https://doi.org/10.1016/j.jhazmat.2016.09.049

Nauendorf A, Krause S, Bigalke NK, Gorb EV, Gorb SN, Haeckel M, Wahl M, Treude T (2016) Microbial colonization and degradation of polyethylene and biodegradable plastic bags in temperate fine-grained organic-rich marine sediments. Mar Pollut Bull 103:168–178. https://doi.org/10.1016/j.marpolbul.2015.12.024

Nijenhuis I, Kuntze K (2016) Anaerobic microbial dehalogenation of organohalides-state of the art and remediation strategies. Curr Opin Biotechnol 38:33–38. https://doi.org/10.1016/j.copbio.2015.11.009

Nisha KN, Devi V, Varalakshmi P, Ashokkumar B (2015) Biodegradation and utilization of dimethylformamide by biofilm forming *Paracoccus* sp. strains MKU1 and MKU2. Bioresour Technol 188:9–13. https://doi.org/10.1016/j.biortech.2015.02.042

Nwogu TP, Azubuike CC, Ogubgue CJ (2015) Enhanced bioremediation of soil artificially-contaminated with petroleum hydrocarbons after amendment with *Capra aegagrus hircus* (goat) manure. Biotechnol Res Int:1–7. https://doi.org/10.1155/2015/657349

Philp JC, Atlas RM (2005) Bioremediation of contaminated soils and aquifers. In: Atlas RM, Philp JC (eds) Bioremediation: applied microbial solutions for real-world environmental cleanup. American Society for Microbiology. (ASM) Press, Washington, DC, pp 139–236

Pradeep V, Subbaiah UM (2016) Use of Ca-alginate immobilized *Pseudomonas aeruginosa* for repeated batch and continuous degradation of Endosulfan. 3 Biotech 6:124. https://doi.org/10.1007/s13205-016-0438-2

Ravanbakhsh M, Ronaghi AM, Taghavi SM, Jousset A (2016) Screening for the next generation heavy metal hyperaccumulators for dryland decontamination. J Environ Chem Eng 4:2350–2355. https://doi.org/10.1016/j.jece.2016.04.013

Romero MC, Cazau MC, Giorgieri S, Arambarri AM (1998) Phenanthrene degradation by microorganisms isolated from a contaminated stream. Environ Pollut 101:355–359. https://doi.org/10.1016/S0269-7491(98)00056-6

Saxena G, Bharagava RN (2015) Persistent organic pollutants and bacterial communities present during the treatment of tannery wastewater. In: Chandra R (ed) Environmental waste management, 1st edn. CRC Press, Taylor & Francis Group, Boca Raton, pp 217–247. https://doi.org/10.1201/b19243-10

Saxena G, Bharagava RN (2016) Ram Chandra: advances in biodegradation and bioremediation of industrial waste. Clean Techn Environ Policy 18:979–980. https://doi.org/10.1007/s10098-015-1084-9

Saxena G, Bharagava RN (2017) Organic and inorganic pollutants in industrial wastes, their ecotoxicological effects, health hazards and bioremediation approaches. In: Bharagava RN (ed) Environmental pollutants and their bioremediation approaches, 1st edn. CRC Press, Taylor & Francis Group, Boca Raton, pp 23–56. https://doi.org/10.1201/9781315173351-3

Saxena G, Bharagava RN, Kaithwas G, Raj A (2015) Microbial indicators, pathogens and methods for their monitoring in water environment. J Water Health 13:319–339. https://doi.org/10.2166/wh.2014.275

Saxena G, Chandra R, Bharagava RN (2016) Environmental pollution, toxicity profile and treatment approaches for tannery wastewater and its chemical pollutants. Rev Environ Contam Toxicol 240:31–69. https://doi.org/10.1007/398_2015_5009

Saxena G, Purchase D, Mulla SI, Saratale GD, Bharagava RN (2018) Phytoremediation of heavy metal-contaminated sites: environmental considerations, field studies, sustainability and future prospects. J Environ Manag 105:103–120

Segneanu AE, Orbeci C, Lazau C, Sfirloaga P, Vlazan P, Bandas C, Grozescu I (2013) Waste water treatment methods. In: Elshorbagy W (ed) Water treatment. ISBN: 978-953-51-0928-0, InTech. https://doi.org/10.5772/53755. Available from: http://www.intechopen.com/books/water-treatment/waste-water-treatment-methods

Shah AA, Fariha H, Abdul H, Safia A (2008) Biological degradation of plastics: a comprehensive review. Biotechnol Adv 26:2467–2650. https://doi.org/10.1016/j.biotechadv.2007.12.005

Shannon JM, Hauser LW, Liu X, Parkin GF, Mattes TE, Just GL (2015) Partial nitritation ANAMMOX in submerged attached growth bioreactors with smart aeration at 20 °C. Environ Sci Process Impacts 17:81–89. https://doi.org/10.1039/c4em00481g

Sheik S, Chandrashekar KR, Swaroop K, Somashekarappa HM (2015) Biodegradation of gamma irradiated low density polyethylene and polypropylene by endophytic fungi. Int Biodeterior Biodegrad 105:21–29. https://doi.org/10.1016/j.ibiod.2015.08.006

Shelton DR, Khader S, Karns JS, Pogell BM (1996) Metabolism of twelve herbicides by Streptomyces. Biodegradation 7:129–136. https://doi.org/10.1007/BF00114625

Silva IS, Grossman M, Durrant LR (2009) Degradation of polycyclic aromatic hydrocarbons (2-7 rings) under microaerobic and very-low-oxygen conditions by soil fungi. Int Biodeterior Biodegrad 63:224–229. https://doi.org/10.1016/j.ibiod.2008.09.008

Simpanen S, Mäkelä R, Mikola J, Romantschuk M (2016) Bioremediation of creosote contaminated soil in both laboratory and field scale: investigating the ability of methyl-β-cyclodextrin to enhance biostimulation. Int Biodeterior Biodegrad 106:117–126. https://doi.org/10.1016/j.ibiod.2015.10.013

Singh JS, Abhilash PC, Singh HB, Singh RP, Singh DP (2011) Genetically engineered bacteria: an emerging tool for environmental remediation and future research perspectives. Gene 480:1–9

Singh P, Saini HS, Raj M (2016) Rhamnolipid mediated enhanced degradation of chlorpyrifos by bacterial consortium in soil-water system. Ecotoxicol Environ Saf 134:156–162. https://doi.org/10.1016/j.ecoenv.2016.07.020

Sivan A (2011) New perspectives in plastic biodegradation. Curr Opin Biotechnol 22:422–426. https://doi.org/10.1016/j.copbio.2011.01.013

Somtrakoon K, Chouychai W, Lee H (2014) Phytoremediation of anthracene and fluoranthene contaminated soil by *Luffa acutangula*. Maejo Int J Sci Technol 8:221–231. https://doi.org/10.14456/mijst.2014

Throne-Holst M, Wentzel A, Ellingsen TE, Kotlar HK, Zotchev SB (2007) Identification of novel genes involved in long-chain n-alkane degradation by *Acinetobacter* sp. strain DSM 17874. Appl Environ Microbiol 73:3327–3332. https://doi.org/10.1128/aem.00064-07

Tong H, Liu C, Li F, Luo C, Chen M, Hu M (2015) The key microorganisms for anaerobic degradation of pentachlorophenol in paddy soil as revealed by stable isotope probing. J Hazard Mater 298:252–260. https://doi.org/10.1016/j.jhazmat.2015.05.049

Ueno R, Wada S, Ureno N (2008) Repeated batch culture of hydrocarbon-degrading, micro-algal strain Prototheca zopfii RND16 immobilized in polyurethane foam. Can J Microbiol 54:66–70. https://doi.org/10.1139/w07-112

Varjani SJ (2017) Microbial degradation of petroleum hydrocarbons. Bioresour Technol 223:277–286. https://doi.org/10.1016/j.biortech.2016.10.037

Varjani SJ, Upasani VN (2016) Biodegradation of petroleum hydrocarbons by oleophilic strain of *Pseudomonas aeruginosa* NCIM 5514. Bioresour Technol 222:195–201. https://doi.org/10.1016/j.biortech.2016.10.006

Varjani SJ, Upasani VN (2017) Critical review on biosurfactant analysis, purification and characterization using rhamnolipid as a model biosurfactant. Bioresour Technol 232:389–397. https://doi.org/10.1016/j.biortech.2017.02.047

Vergani L, Mapelli F, Zanardini E, Terzaghi E, Di Guardo A, Morosini C, Raspa G, Borin S (2016) Phyto-rhizoremediation of polychlorinated biphenyl contaminated soils: an outlook on plant-microbe beneficial interactions. Sci Total Environ 575:1395–1406. https://doi.org/10.1016/j.scitotenv.2016.09.218

Villaverde J, Rubio-Bellido M, Merchán F, Morillo E (2017) Bioremediation of diuron contaminated soils by a novel degrading microbial consortium. J Environ Manag 188:379–386. https://doi.org/10.1016/j.jenvman.2016.12.020

Warshawsky D, Cody T, Radike M, Reilman R, Schumann B, LaDow K, Schneider J (1995) Biotransformation of benzo[a]pyrene and other polycyclic aromatic hydrocarbons and heterocyclic analogs by several green algae and other algal species under gold and white light. Chem Biol Interact 97:131–148. https://doi.org/10.1016/0009-2797(95)03610-X

WerheniAmmeri R, MokniTlili S, Mehri I, Badi S, Hassen A (2016) Pentachlorophenol biodegradation by *Citrobacter freundii* isolated from forest contaminated soil. Water Air Soil Pollut. https://doi.org/10.1007/s11270-016-2959-z

Whyte LG, Bourbonnière L, Greer CW (1997) Biodegradation of petroleum hydrocarbons by psychrotrophic *Pseudomonas* strains possessing both alkanes (alk) and naphthalene (nah) catabolic pathways. Appl Environ Microbiol 63:3719–3723

Wilkes H, Buckel W, Golding BT, Rabus R (2016) Metabolism of hydrocarbons in n-Alkane utilizing anaerobic bacteria. J Mol Microbiol Biotechnol 26:138–151. https://doi.org/10.1159/000442160

Wu M, Li W, Dick WA, Ye X, Chen K, Kost D, Chen L (2017) Bioremediation of hydrocarbon degradation in a petroleum-contaminated soil and microbial population and activity determination. Chemosphere 169:124–130. https://doi.org/10.1016/j.chemosphere.2016.11.059

Xia Z-Y, Zhang L, Zhao Y, Yan X, Li S-P, Gu T, Jiang J-D (2017) Biodegradation of the herbicide 2,4-dichlorophenoxyacetic acid by a new isolated strain of *Achromobacter*sp. LZ35. Curr Microbiol 74:193–202. https://doi.org/10.1007/s00284-016-1173-y

Yang Y, Yang J, Wu WM, Zhao J, Song Y, Gao L, Yang R, Jiang L (2015a) Biodegradation and mineralization of polystyrene by plastic-eating mealworms: part 1. Chemical and physical characterization and isotopic tests. Environ Sci Technol 49:12080–12086. https://doi.org/10.1021/acs.est.5b02661

Yang Y, Yang J, Wu WM, Zhao J, Song Y, Gao L, Yang R, Jiang L (2015b) Biodegradation and mineralization of polystyrene by plastic-eating mealworms: part 2. Role of gut microorganisms. Environ Sci Technol 49:12087–12093. https://doi.org/10.1021/acs.est.5b02663

Yao Z, Li J, Xie H, Yu C (2012) Review on remediation technologies of soil contaminated by heavy metals. Procedia Environ Sci 16:722–729. https://doi.org/10.1016/j.proenv.2012.10.099

Zhang G, Ouyang X, Li H, Fu Z, Chen J (2016) Bioremoval of antimony from contaminated waters by a mixed batch culture of sulfate-reducing bacteria. Int Biodeterior Biodegrad 115:148–155. https://doi.org/10.1016/j.ibiod.2016.08.007

Zhao Y, Yao J, Yuan Z, Wang T, Zhang Y, Wang F (2016) Bioremediation of Cd by strain GZ-22 isolated from mine soil based on biosorption and microbially induced carbonate precipitation. Environ Sci Pollut Res:1–9. https://doi.org/10.1007/s11356-016-7810-y

Chapter 3
Application of Microbial Enzymes in Degradation and Detoxification of Organic and Inorganic Pollutants

Gaurav Saxena, Roop Kishor, and Ram Naresh Bharagava

Contents

Abstract Microbial enzymes have been reported to play a diverse role in various industrial applications. Microbial enzymes are also useful in bioremediation of environmental pollutants from industrial wastes due to their high specificity to a broad range of substrates (pollutants), use under extreme conditions that microbe cannot thrive, high effectiveness at low pollutant concentration, high activity in the presence of inhibitors of microbial metabolism, and high mobility (small size) than microorganisms. A variety of enzymes are produced by microorganisms that can be used in the degradation and detoxification of a wide range of organic and inorganic pollutants. This chapter provides an overview on the various microbial enzymes that can be used for the bioremediation of environmental pollutants. In addition, the prospects and challenges in applying microbial enzymes in bioremediation are also discussed in this chapter.

G. Saxena · R. Kishor · R. N. Bharagava (✉)
Laboratory of Bioremediation and Metagenomics Research (LBMR),
Department of Microbiology (DM), Babasaheb Bhimrao Ambedkar University
(A Central University), Lucknow, Uttar Pradesh, India
e-mail: ramnaresh_dem@bbau.ac.in

G. Saxena, R. N. Bharagava (eds.), *Bioremediation of Industrial Waste for Environmental Safety*, https://doi.org/10.1007/978-981-13-1891-7_3

Keywords Microbial enzymes · Biodegradation · Bioremediation · Biodetoxification · Environmental pollutants

3.1 Introduction

Environmental pollution is of serious global concern and the public outcry against it continually rising to ensure the safest and healthiest environment for living beings. Industries are the major players in the national economy of every country; however, these are also the major cause of serious environmental pollution. Industries use a variety of highly toxic chemicals in various operations to produce the good quality of products in short time and after processing, discharge these chemicals along with complex wastewaters responsible for toxicity in living beings (Arora et al. 2014, 2018; Saxena and Bharagava 2017). These highly toxic complex wastewaters are the responsible cause for serious soil and water pollution and severe toxic effects in living beings (Goutam et al. 2018; Bharagava et al. 2017a; Saxena and Bharagava 2015, 2017; Gautam et al. 2017; Saxena et al. 2015, 2016), and hence, their adequate treatment is necessary for environmental protection and public health safety.

Currently, physicochemical approaches are being applied to treat and manage the hazardous wastewaters. These approaches are environmentally destructive, cost high money to apply, and may cause secondary pollution by generating some toxic chemicals as a result of the remediation (Saxena and Bharagava 2017). Hence, these methods are not compatible with our natural environment and thus, lead to the need of some eco-friendly approaches. Bioremediation is an environmental friendly approach that can be the suitable alternative to effectively treat and manage the highly toxic complex industrial wastewater and its chemical pollutants. Bioremediation is the use of microbes and plants or their enzymes to degrade and detoxify the organic and inorganic pollutants present in the industrial wastewaters (Bharagava et al. 2017b, c, 2018; Saxena and Bharagava 2016, 2017; Chandra et al. 2015; Singh et al. 2011). But, currently, the use of microbial enzymes in the bioremediation of environmental pollutants is an emerging area of research. Microbial enzymes have diverse industrial applications and are also used in bioremediation of pollutants. The overview of microbial enzymes and their classification has been already reported in the previous literature; however, this chapter provides the application of various microbial enzymes in the degradation and detoxification of organic and inorganic pollutants.

3.2 Enzymes and Their Applications

Enzymes are biological catalysts that facilitate the conversion of substrates into products by providing favorable conditions that lower the activation energy of the reaction (Karigar and Rao 2011). An enzyme may be a protein or a glycoprotein and consists of at least one polypeptide moiety. The regions of the enzyme that are directly involved in the catalytic process are called the active sites. An enzyme may have one or more groups that are essential for catalytic activity associated with the active sites through either covalent or non-covalent bonds; the protein or glycoprotein moiety in such an enzyme is called the apoenzyme, while the nonprotein moiety is called the prosthetic group. The combination of the apoenzyme with the prosthetic group yields the holoenzyme.

The ultimate identification of a particular enzyme is possible through its enzyme commission (E.C.) number. The assignment of E.C. numbers is described in guidelines set out by the International Union of Biochemistry. All known enzymes fall into one of these six categories. The six main divisions are (Karigar and Rao 2011) as follows: (1) the oxidoreductases, (2) the transferases, (3) the hydrolases, (4) the lyases, (5) the isomerases, and (6) the ligases (synthetases). Oxidoreductases catalyze the transfer electrons and protons from a donor to an acceptor; transferases catalyze the transfer of a functional group from a donor to an acceptor; hydrolases facilitate the cleavage of C–C, C–O, C–N, and other bonds by water; lyases catalyze the cleavage of these same bonds by elimination, leaving double bonds (or, in the reverse mode, catalyze the addition of groups across double bonds); isomerases facilitate geometric or structural rearrangements or isomerizations; and ligases catalyze the joining of two molecules (Karigar and Rao 2011).

Industrial enzymes are enzymes that are commercially used in a variety of industries such as pharmaceuticals, chemical production, biofuels, food and beverage, and consumer products (Karigar and Rao 2011). Due to advancements in recent years, biocatalysis through isolated enzymes is considered more economical than use of whole cells.

Moreover, enzymes may present advantages over traditional technologies and also over microbial remediation. Indeed, enzymes are not inhibited by inhibitors of microbial metabolism, can be used under extreme conditions limiting microbial activity, highly effective at low pollutant concentrations and are active in the presence of microbial predators or antagonists, act against a given substrate (microorganisms may prefer more easily degradable compounds than the pollutant), and are more mobile than microorganisms because of their smaller size (Rao et al. 2010; Gianfreda and Bollag 2002). All these characteristics render enzymes useful in the environmental bioremediation.

Enzymes may act intracellularly, i.e., in the presence of or inside their originating cells; extracellularly, i.e., both in the presence and absence of their originating cells; free, i.e., soluble in solution and the catalysis will be homogenous; or immobilized, i.e., linked through different links to a solid matrix and the catalysis will be heterogeneous (Rao et al. 2010; Gianfreda and Rao 2004). The most representative enzymatic classes in the remediation of polluted environments are the following: hydrolases, dehalogenases, transferases, and oxidoreductases, and their main producers are bacteria; fungi, mainly white-rot fungi; plants; and microbe-plant associations (Rao et al. 2010). The microbial enzymes reported in the bioremediation of pollutants are presented in Table 3.1.

Table 3.1 Microbial enzymes reported in the bioremediation of organic and inorganic pollutants

Enzyme	Applications
Chrome reductase	A group of enzymes that catalyze the reduction of toxic and carcinogenic Cr(VI) to the less soluble and less toxic Cr(III)
Laccase	Decolorization of dyes and textile wastewater, degradation of distillery wastewater, detoxification of chlorophenols and phenols
Manganese peroxidase	Degradation of phenols, lignin, pentachlorophenol, and dyes
Lignin peroxidase	Degradation of aromatic compounds and phenolics. Remediation of kraft lignin wastewater
Monooxygenase	Dehalogenation, desulfurization, denitrification, and hydroxylation of aromatic compounds, bioremediation of pentachlorophenol, phenol, p-nitrophenol, 2, 4, 6-Trichlorophenol
Dioxygenase	Degradation of aromatic pollutants
Cellulase	Degradation of cellulase
Azoreductase	A commonly found enzyme that breaks down all azo dyes and azo compounds through reductive cleavage of azo bonds

Adapted from Karigar and Rao (2011) and Rao et al. (2010)

3.3 Microbial Enzymes in Degradation and Detoxification of Organic and Inorganic Pollutants

3.3.1 Laccase

Laccases [EC 1.10.3.2, para-bezenediol:dioxygen oxidoreductases] are multicopper proteins that use molecular oxygen to oxidize various aromatic and nonaromatic compounds by a radical-catalyzed reaction mechanism (Strong and Claus 2011). Laccase has been reported in plants, bacteria, and fungi and have diverse industrial applications as well as used in the bioremediation of various environmental pollutants (Strong and Claus 2011). The broad substrate range (which includes ortho (o)- and para (p)-diphenols, aromatic amines, polyphenols, and methoxy-substituted phenols), high catalytic constants, and its use of molecular oxygen give it a wide scope in bioremediation (Strong and Claus 2011; Chiacchierini et al. 2004). Laccase alone, or with redox mediators, has been reported to oxidize numerous dangerous compounds, which include phenols, halogenated phenols, aromatic amines, nitro-substituted compounds such as trinitrotoluene, hydroxy-indoles, the herbicide dymron, organophosphorous compounds, polycyclic aromatic hydrocarbons, chlorinated hydroxybiphenyls, bisphenol A and nonylphenol, hydroxyphenylureas, endocrine-disrupting chemicals, pesticides, and insecticides (Strong and Claus 2011; Durán and Esposito 2000). Laccase has been also reported in the treatment of distillery wastewater, pulp paper mill wastewater, municipal wastewater, olive oil mill wastewater, dye and printing wastewater, and detoxification of polluted soils (Strong and Claus 2011).

3.3.2 Chrome Reductase

Chromate reductases are a group of enzymes that catalyze the reduction of toxic and carcinogenic Cr(VI) to the less soluble and less toxic Cr(III) (Thatoi et al. 2014; Garcia-Arellano et al. 2004). These enzymes have recently raised enormous interest because of their central role in mediating chromium toxicity and their potential use in bioremediation and biocatalysis. A membrane-associated chromate reductase activity was first observed in *E. cloacae* HO1 where the insoluble form of reduced chromate precipitates was seen on the cell surface (Wang et al. 1989). The chromate reductase activity of *Shewanella putrefaciens* MR-1 was found to be associated with the cytoplasmic membrane of anaerobically grown cells where formate and NADH served as electron donors for the reductase (Myers et al. 2000). The anaerobic Cr(VI) reduction involving membrane-associated reductase has also been reported in some chromate-reducing bacteria which utilized H_2 as electron donor and Cr(VI) as an electron acceptor in the electron transport chain (Lovley and Phillips 1994; Quilntana et al. 2001). The type of bacterial chromate reductases depends on the nature of bacteria carrying out the reduction reaction, i.e., aerobic or anaerobic (Thatoi et al. 2014). Bacterial chromate reductases are either localized in the membrane fraction or in the cytosolic fractions of the chromate reducing bacteria (Thatoi et al. 2014). The enzymatic reduction of Cr(VI) in bacteria is accomplished in different ways. Soluble reductases can participate either in extracellular or intracellular reduction of Cr(VI) (Elangovan et al. 2010; Das et al. 2014), whereas membrane-bound reductase reduces by extracellular means (Wang et al. 1991). Cheung and Gu (2007) reported that extracellular enzymes (cytosolic proteins) are soluble chromate reductases such as flavin reductases, nitrate reductases, flavin proteins, and ferrireductases. *Streptomyces* sp. MC1, isolated from sugarcane, has been shown to reduce Cr(VI) intracellularly (Polti et al. 2011).

3.3.3 Ligninolytic Enzymes

Ligninolytic enzymes include lignin peroxidase (LiP) and manganese peroxidase (MnP). Lignin peroxidases (LiP) are the heme proteins secreted mainly by the white-rot fungus during secondary metabolism (Karigar and Rao 2011). In the presence of co-substrate, H_2O_2 and mediator like veratryl alcohol LiP degrade lignin and other phenolic compounds (Karigar and Rao 2011). Lignin peroxidase (LiP) also plays a central role in the biodegradation of the plant cell wall constituent lignin (Karigar and Rao 2011). LiP is able to oxidize aromatic compounds with redox potentials higher than 1.4 V (NHE) by single electron abstraction, but the exact redox mechanism is still poorly understood (Karigar and Rao 2011; Piontek et al. 2001). Manganese peroxidases (MnP) are an extracellular heme enzyme from the lignin-degrading basidiomycete's fungus that oxidizes Mn^{2+} to the oxidant Mn^{3+} in a multistep reaction, and Mn^{2+} stimulates the MnP production and functions as a substrate for MnP (Karigar and Rao 2011).

3.3.4 Azoreductase

Azoreductase is a commonly found enzyme that breaks down all azo dyes and azo compounds. Azo compounds are susceptible to biological degradation under both aerobic and anaerobic conditions (Chengalroyen and Dabbs 2013; Khehra et al. 2005). In general microbial degradation of azo dyes involves the reductive cleavage of azo bonds (–N55N–) with the help of an azoreductase enzyme under anaerobic conditions, and this involves a transfer of four electrons (reducing equivalents) (Saratale et al. 2011). This then proceeds through two stages at the azo linkage, and in each stage two electrons are transferred to the azo dye, which acts as a final electron acceptor, resulting in dye decolorization and the formation of colorless solutions (Saratale et al. 2011; Chang et al. 2000). The resulting intermediate metabolites (e.g., aromatic amines) are then further degraded aerobically or anaerobically (Saratale et al. 2011; Chang et al. 2004; Pandey et al. 2007). The optimal pH for many bacterial azoreductases seems to lie within a range of 6–8 with a temperature range between 25 and 45 °C. To name a few, previous reports include the 21.5 kDa reductase of *B. cereus* with an optimal activity at 40 °C and a pH of 7.0 (Pricelius et al. 2007), a *P. aeruginosa* reductase active at pH 7.0 and 35 °C (Nachiyar and Rajakumar 2005), and *A. hydrophila* with optimal activity at pH 5.0–10.0 and 25–37 °C (Chen et al. 2003).

3.3.5 Monooxygenase

Monooxygenases incorporate one atom of the oxygen molecule into the substrate. Monooxygenases are classified into two subclasses based on the presence cofactor: flavin-dependent monooxygenases and P450 monooxygenases (Karigar and Rao 2011). Flavin-dependent monooxygenases contain flavin as prosthetic group and require NADP or NADPH as coenzyme. P450 monooxygenases are heme-containing oxygenases that exist in both eukaryotic and prokaryotic organisms. The monooxygenases comprise a versatile superfamily of enzymes that catalyzes oxidative reactions of substrates ranging from alkanes to complex endogenous molecules such as steroids and fatty acids (Karigar and Rao 2011). Monooxygenases act as biocatalysts in bioremediation process and synthetic chemistry due to their highly regioselectivity and stereoselectivity on wide range of substrates (Karigar and Rao 2011). Majority of monooxygenase studied previously are having cofactor, but there are certain monooxygenases which function independent of a cofactor. These enzymes require only molecular oxygen for their activities and utilize the substrate as reducing agent (Karigar and Rao 2011; Arora et al. 2010). The desulfurization, dehalogenation, denitrification, ammonification, hydroxylation, biotransformation, and biodegradation of various aromatic and aliphatic compounds are catalyzed by monooxygenases (Karigar and Rao 2011). These properties have been explored in recent years for important application in biodegradation and biotransformation of aromatic compounds (Karigar and Rao 2011; Arora et al. 2010). Methane monooxygenase enzyme is the best characterized one, among monooxygenases. This enzyme is involved in the degradation of hydrocarbon such as substituted methanes, alkanes, cycloalkanes, alkenes, haloalkenes, ethers, and aromatic and heterocyclic hydrocarbons (Karigar and Rao 2011). Under oxygen-rich conditions, monooxygenase catalyzes oxidative dehalogenation reactions, whereas under low oxygen levels, reductive dechlorination takes place. Oxidation of substrate can lead to de-halogenation as a result of the formation of labile products that undergo subsequent chemical decomposition (Karigar and Rao 2011).

3.3.6 *Dioxygenase*

Dioxygenases are multicomponent enzyme systems that introduce molecular oxygen into their substrate. Aromatic hydrocarbon dioxygenases belong to a large family of Rieske nonheme iron oxygenases. These dioxygenases catalyze enantiospecifically the oxygenation of wide range of substrates. Dioxygenases primarily oxidize aromatic compounds and, therefore, have applications in environmental remediation (Karigar and Rao 2011). All members of this family have one or two electron transport proteins preceding their oxygenase components. The crystal structure of naphthalene dioxygenase has confirmed the presence of a Rieske (2Fe–2S) cluster and mononuclear iron in each alpha subunit (Karigar and Rao 2011; Dua et al. 2002). The catechol dioxygenases serve as part of nature's strategy for degrading aromatic molecules in the environment. They are found in the soil bacteria and involved in the transformation of aromatic precursors into aliphatic products (Karigar and Rao 2011).

3.3.7 *Cellulase*

Cellulases now promise the potential of converting waste cellulosic material into foods to meet burgeoning population and have been the subject of intense research (Karigar and Rao 2011; Bennet et al. 2002). Cellulases are usually a mixture of several enzymes. Three major groups of cellulases are involved in the hydrolysis process (Karigar and Rao 2011): (1) endoglucanase (EG, endo-1,4-Dglucanohydrolase) which attacks regions of low crystallinity in the cellulose fiber, creating free chain ends; (2) exoglucanase or cellobiohydrolase (CBH, 1,4-b-D-glucan cellobiohydrolase) which degrade the cellulose molecule further by removing cellobiose units from the free chain ends; (3) β-glucosidase which hydrolyzes cellobiose to glucose units. Cellulase enzymes are capable of degrading crystalline cellulose to glucose. Alkaline cellulases are produced by *Bacillus* strains and neutral and acidic cellulases by *Trichoderma* and *Humicola* fungi (Karigar and Rao 2011). In paper and pulp industry, cellulases have been employed for the removal of ink during recycling of paper (Karigar and Rao 2011).

3.4 Prospects and Challenges in Applying Enzymes in Bioremediation of Pollutants

Following concerns associated with the use of enzymes in bioremediation are the key challenges in the enzymatic bioremediation technology (Rao et al. 2010): (a) toxicity of the product of the enzyme-mediated reaction that should be less toxic than the substrate; (b) if enzymes require cofactors, their use may be problematic, unless a preparation containing both the enzyme and the respective cofactor is used; (c) use of enzymes to detoxify organic-polluted soil is given by the rapid degradation of the free enzyme by proteases released by soil microorganisms; (d) complexity of the pollution may entail possible negative or positive, synergistic effects on the enzyme efficiency; (e) reduction in the stability and reusability of enzymes in bioremediation, if the remediation is multitask; and (f) the cost of enzyme isolation and purification greatly hampers their practical application, mainly when their continuous feeding is needed. However, the possibilities to overcome several disadvantages linked to the use of free enzymes are to improve (Rao et al. 2010) (a) the functioning of the chosen enzyme by changing the conditions under which the enzyme displays its catalytic activity; (b) the enzyme itself by optimizing its intrinsic and inherent catalytic features by acting on its structure and catalytic function; (c) the performance of enzymes in the detoxification of pollutants through enzyme immobilization because immobilized enzymes have usually a long-term operational stability, being very stable toward physical, chemical, and biological denaturing agents; and (d) biomolecular engineering of enzymes by directed evolution or by site-specific mutagenesis to improve existing biodegradation pathways or to develop biocatalytic processes for the production of useful products.

However, the enzyme performance for the *in situ* bioremediation of contaminated groundwater and soils may be affected by many factors, which include (Rao et al. 2010) chemistry and toxicity of contamination, source and concentration of pollutants, solubility, transport, adsorption, dispersion of pollutant compounds, environmental conditions, nutrient sources and presence of electron acceptors, and mainly the biodegradability of contaminants. By overcoming above-discussed limitations, enzymatic remediation technology can provide the best remediation solution for contaminated environments.

3.5 Conclusion

Finding an eco-friendly solution for waste management is always a key requirement for sustainable development. Scientists around the globe are working hard to find the sustainable remediation technology for environmental waste management. Microbes are oftenly viewed as the eco-friendly tools for the treatment and management of industrial wastes containing highly toxic organic and inorganic pollutants to combat the environmental threats. However, the applicability of microbes in waste management can be limited under some conditions. Applying microbial enzymes in bioremediation can be the suitable alternative for waste management to protect the environment and public health. Further, research efforts are required to use the microbial enzymes in environmental remediation in a more sustainable and economically feasible manner.

Acknowledgments The financial support as "Major Research Projects" (Grant No.: EEQ/2017/000407) from "Science and Engineering Research Board" (SERB), Department of Science and Technology (DST), Government of India (GOI), New Delhi, India, and University Grant Commission (UGC) Fellowship received by Mr. Gaurav Saxena for doctoral studies is duly acknowledged.

References

Arora PK, Srivastava A, Singh VP (2010) Application of monooxygenases in dehalogenation, desulphurization, denitrification and hydroxylation of aromatic compounds. J Bioremed Biodegr 1:1–8

Arora PK, Srivastava A, Singh VP (2014) Bacterial degradation of nitrophenols and their derivatives. J Hazard Mater 266:42–59

Arora PK, Srivastava A, Garg SK, Singh VP (2018) Recent advances in degradation of chloronitrophenols. Bioresour Technol 250C:902–909

Bennet JW, Wunch KG, Faison BD (2002) Use of Fungi biodegradation. ASM Press, Washington, DC

Bharagava RN, Saxena G, Mulla SI, Patel DK (2017a) Characterization and identification of recalcitrant organic pollutants (ROPs) in tannery wastewater and its phytotoxicity evaluation for environmental safety. Arch Environ Contam Toxicol. https://doi.org/10.1007/s00244-017-0490-x

Bharagava RN, Chowdhary P, Saxena G (2017b) Bioremediation: an ecosustainable green technology: its applications and limitations. In: Bharagava RN (ed) Environmental pollutants and their bioremediation approaches, 1st edn. CRC Press, Taylor & Francis Group, Boca Raton, pp 1–22. https://doi.org/10.1201/9781315173351-2

Bharagava RN, Saxena G, Chowdhary P (2017c) Constructed wetlands: an emerging phytotechnology for degradation and detoxification of industrial wastewaters. In: Bharagava RN (ed) Environmental pollutants and their bioremediation approaches, 1st edn. CRC Press, Taylor & Francis Group, Boca Raton, pp 397–426. https://doi.org/10.1201/9781315173351-15

Bharagava RN, Purchase D, Saxena G, Mulla SI (2018) Applications of metagenomics in microbial bioremediation of pollutants: from genomics to environmental cleanup. In: Das S, Dash H (eds) Microbial diversity in the genomic era, 1st edn. Academic Press, Elsevier. https://doi.org/10.1016/B978-0-12-814849-5.00026-5

Chandra R, Saxena G, Kumar V (2015) Phytoremediation of environmental pollutants: an ecosustainable green technology to environmental management. In: Chandra R (ed) Advances in biodegradation and bioremediation of industrial waste, 1st edn. CRC Press, Taylor & Francis Group, Boca Raton, pp 1–30. https://doi.org/10.1201/b18218-2

Chang JS, Kuo TS, Chao YP, Ho JY, Lin PJ (2000) Azo dye decolorization with a mutant *Escherichia coli* strain. Biotechnol Lett 22:807

Chang JS, Chen BY, Lin YS (2004) Stimulation of bacterial decolorization of an azo dye by extracellular metabolites from *Escherichia coli* strain NO3. Bioresour Technol 91:243

Chen K-C, Wu J-Y, Liou D-J, Hwang S-CJ (2003) Decolorization of the textile dyes by newly isolated bacterial strains. J Biotechnol 101:57–68

Chengalroyen MD, Dabbs ER (2013) The microbial degradation of azo dyes: mini review. World J Microbiol Biotechnol 29:389–399

Cheung KH, Gu JD (2007) Mechanism of hexavalent chromium detoxification by microorganisms and bioremediation application potential: a review. Int Biodeterior Biodegrad 59(1):8–15

Chiacchierini E, Restuccia D, Vinci G (2004) Bioremediation of food effluents: recent applications of free and immobilized polyphenoloxidases. Food Sci Technol Int 10:373–382

Das S, Mishra J, Das SK, Pandey S, Rao DS, Chakraborty A, Sudarshan M, Das NN, Thatoi HN (2014) Investigation on mechanism of Cr(VI) reduction and removal by *Bacillus amyloliquefaciens*, a novel chromate tolerant bacterium isolated from chromite mine soil. Chemosphere 96:112–121

Dua M, Singh A, Sethunathan N, Johri A (2002) Biotechnology and bioremediation: successes and limitations. Appl Microbiol Biotechnol 59(2–3):143–152

Durán N, Esposito E (2000) Potential applications of oxidative enzymes and phenoloxidade-like compounds in wastewater and soil treatment: a review. Appl Catal B Environ 28:83–99

Elangovan R, Philip L, Chandraraj K (2010) Hexavalent chromium reduction by free and immobilized cell-free extract of Arthrobacter rhombi-RE. Appl Biochem Biotechnol 160(1):81–97

Garcia-Arellano H, Alcalde M, Ballesteros A (2004) Use and improvement of microbial redox enzymes for environmental purposes. Microb Cell Factories 3:10

Gautam S, Kaithwas G, Bharagava RN, Saxena G (2017) Pollutants in tannery wastewater, pharmacological effects and bioremediation approaches for human health protection and environmental safety. In: Bharagava RN (ed) Environmental pollutants and their bioremediation approaches, 1st edn. CRC Press, Taylor & Francis Group, Boca Raton, pp 369–396. https://doi.org/10.1201/9781315173351-14

Gianfreda L, Bollag J-M (2002) Isolated enzymes for the transformation and detoxification of organic pollutants. In: Burns RG, Dick R (eds) Enzymes in the environment: activity, ecology and applications. Marcel Dekker, New York, pp 491–538

Gianfreda L, Rao MA (2004) Potential of extra cellular enzymes in remediation of polluted soils: a review. Enzym Microb Technol 35:339–354

Goutam SP, Saxena G, Singh V, Yadav AK, Bharagava RN (2018) Green synthesis of TiO_2 nanoparticles using leaf extract of *Jatropha curcas* L. for photocatalytic degradation of tannery wastewater. Chem Eng J 336:386–396. https://doi.org/10.1016/j.cej.2017.12.029

Karigar CS, Rao SS (2011) Role of microbial enzymes in the bioremediation of pollutants: a review. Enzym Res 2011:805187

Khehra MS, Saini HS, Sharma DK, Chadha BS, Chimni SS (2005) Decolorization of various azo dyes by bacterial consortium. Dyes Pigments 67:55–61

Lovley DR, Phillips JPE (1994) Reduction of chromate by Desulfovibrio vulgaris and its C3 cytochrome. Appl Environ Microbiol 60(2):726–728

Myers CR, Carstens BP, Antholine WE, Myers JM (2000) Chromium (VI) reductase activity is associated with the cytoplasmic membrane of anaerobically grown Shewanella putrefaciens MR-1. J Appl Microbiol 88(1):98–106

Nachiyar CV, Rajakumar S (2005) Degradation of a tannery and textile dye, Navitan fast blue S5R by Pseudomonas aeruginosa. World J Microbiol Biotechnol 19:609–614

Pandey A, Singh P, Iyengar L (2007) Bacterial decolorization and degradation of azo dyes. Int Biodeter Biodegrad 59:73

Piontek K, Smith AT, Blodig W (2001) Lignin peroxidase structure and function. Biochem Soc Trans 29(2):111–116

Polti MA, Amoroso MJ, Abate CM (2011) Intracellular chromium accumulation by Streptomyces sp. MC1. Water Air Soil Pollut 214:49–57

Pricelius S, Held C, Murkovic M, Bozic M, Kokol V, Cavaco-Paulo A, Guebitz GM (2007) Enzymatic reduction of azo and indigoid compounds. Appl Microbiol Biotechnol 77:321–327

Quilntana M, Curutchet G, Donati E (2001) Factors affecting chromium (VI) reduction by *Thiobacillus ferrooxidans*. Biochem Eng J 9(1):11–15

Rao MA, Scelza R, Scotti R, Gianfreda L (2010) Role of enzymes in the remediation of polluted environments. J Soil Sci Plant Nutr 10(3):333–353

Saratale RG, Saratale GD, Chang JS, Govindwar SP (2011) Bacterial decolorization and degradation of azo dyes: a review. J Taiwan Inst Chem Eng 42:138–157

Saxena G, Bharagava RN (2015) Persistent organic pollutants and bacterial communities present during the treatment of tannery wastewater. In: Chandra R (ed) Environmental waste management, 1st edn. CRC Press, Taylor & Francis Group, Boca Raton, pp 217–247. https://doi.org/10.1201/b19243-10

Saxena G, Bharagava RN (2016) Ram Chandra: advances in biodegradation and bioremediation of industrial waste. Clean Techn Environ Policy 18:979–980. https://doi.org/10.1007/s10098-015-1084-9

Saxena G, Bharagava RN (2017) Organic and inorganic pollutants in industrial wastes, their ecotoxicological effects, health hazards and bioremediation approaches. In: Bharagava RN (ed) Environmental pollutants and their bioremediation approaches, 1st edn. CRC Press, Taylor & Francis Group, Boca Raton, pp 23–56. https://doi.org/10.1201/9781315173351-3

Saxena G, Bharagava RN, Kaithwas G, Raj A (2015) Microbial indicators, pathogens and methods for their monitoring in water environment. J Water Health 13:319–339. https://doi.org/10.2166/wh.2014.275

Saxena G, Chandra R, Bharagava RN (2016) Environmental pollution, toxicity profile and treatment approaches for tannery wastewater and its chemical pollutants. Rev Environ Contam Toxicol 240:31–69. https://doi.org/10.1007/398_2015_5009

Singh JS, Abhilash PC, Singh HB, Singh RP, Singh DP (2011) Genetically engineered bacteria: an emerging tool for environmental remediation and future research perspectives. Gene 480:1–9

Strong PJ, Claus H (2011) Laccase: a review of its past and its future in bioremediation. Crit Rev Environ Sci Technol 41(4):373–434

Thatoi H, Das S, Mishra J, Rath BP, Das N (2014) Bacterial chromate reductase, a potential enzyme for bioremediation of hexavalent chromium: a review. J Environ Manag 146:383–399

Wang PC, Mori T, Komori K, Sasatsu M, Toda K, Ohtake H (1989) Isolation and characterization of an *Enterobacter cloacae* strain that reduces hexavalent chromium under anaerobic conditions. Appl Environ Microbiol 55(7):1665–1669

Wang PC, Toda K, Ohtake H, Kusaka I, Yabe I (1991) Membrane-bound respiratory system of *Enterobacter cloacae* strain HO1 grown anaerobically with chromate. FEMS Microbiol Lett 78(1):11–15

Chapter 4
Persistent Organic Pollutants (POPs): Environmental Risks, Toxicological Effects, and Bioremediation for Environmental Safety and Challenges for Future Research

Ningombam Linthoingambi Devi

Contents

Abstract The persistent organic pollutants (POPs) such as organochlorine pesticides (OCPs), polychlorinated biphenyls (PCBs), and polycyclic aromatic hydrocarbons (PAHs) were detected in different media of environment such as air, water, and soil due to its persistence in nature. POPs have the efficacy to contaminate, accumulate, and bioaccumulate in any living organisms through the food

---

N. L. Devi (✉)
Centre for Environmental Science, School of Earth, Biological and Environmental Sciences, Central University of South Bihar, Patna, Bihar, India
e-mail: nldevi@cub.ac.in

G. Saxena, R. N. Bharagava (eds.), *Bioremediation of Industrial Waste for Environmental Safety*, https://doi.org/10.1007/978-981-13-1891-7_4

chain. Since these are lipophilic and semi-volatile in nature, can move or transport through air mass in the far distance such as Arctic region and mountainous region where the pollutants were never used before. Thus, POPs have the tendency to retain in cold climate for a longer time. Stockholm Convention is a focus to control or eliminate the production and implement the policy against the toxic POPs because these can lead to serious health effects including cancers, birth defects, and dysfunctional immune, nervous, and reproductive systems. The *Cucurbita* family of plant species including pumpkins, squash, and zucchini has been used widely to uptake the PCBs, DDT, and PAH from contaminated sites. Some of the bacterial group such as *Escherichia coli*, *Enterobacter aerogenes*, *Enterobacter cloacae*, *Pseudomonas aeruginosa*, *Pseudomonas putida*, and *Bacillus* were also reported for the degradation of DDT compounds. The gram-negative (*Pseudomonas*, *Alcaligenes*, *Achromobacter*, *Burkholderia* sp.) and gram-positive (*Rhodococcus*, *Corynebacterium*, and *Bacillus* sp.) bacteria have been reported for the degradation of PCB compounds, respectively. Many of the native microorganisms develop complex and significantly effective metabolic pathways that allow the biodegradation of toxic compounds released into the environment.

Keywords Persistent organic pollutant · Persistence · Toxicity · Health hazards · Bioremediation

4.1 Introduction

Persistent organic pollutants (POPs) are carbon-based organic chemical compounds and mixtures of industrial chemicals such as PCBs, agriculture pesticides like DDTs, and unwanted wastes such as dioxins (Saxena and Bharagava 2015). Industrial wastewaters are also responsible for releasing various persistent organic and inorganic pollutants (Goutam et al. 2018; Bharagava et al. 2017a; Gautam et al. 2017). POPs are primarily products and by-products of human industry that are of relatively fresh origin. POPs are a group of a toxic chemical that is persistent in the environment for a longer time and took longer time for degradation (UNEP 2006). Due to their persistence in nature, POPs can bioaccumulate in any living being and different environmental media with potentially significant impacts. Problems related to environmental pollution by POPs have been the rise and increasing attention of the world. Once POPs are released in any corner of the world, it can move or transfer far distances from their original places through a repeated process of evaporation and deposition. Therefore, it makes difficult to identify the original source point of POPs globally. Different international organizations have been trying to minimize the issues of POPs due to its lipophilic and accumulation in nature. Stockholm Convention also mainly focuses to implement and replace dirty and toxic POPs. Perhaps, POPs have the capability to move far distances where it was never been used before.

Persistent organic pollutants (POPs) mainly came into wide use after the Second World War. DDT is one of the famous and controversial persistent organic pesticides ever made. Since the 1940s, these inexpensive and toxic chemicals have been produced and applied worldwide. DDT was also used to control the vector-borne diseases. The heavy use of these persistent toxic pollutants is widespread in different components of the environment through different pathways. However, soon their negative impacts started showing. Initially apparent in fish, birds, mammals, and water bodies, further studies revealed their presence in the atmosphere and soil environment. Climate change and environmental contamination caused by POPs are a global concern. POPs accumulate in living organisms and subsequently in humans via food.

4.1.1 Persistent Organic Pollutants in Soil

The persistence of pesticides in soil depends upon a number of factors, such as the type of pesticide, soil type, organic matter content, clay content, the microflora, fauna present, cover crop, soil temperature, soil moisture, and cultivation of the soil. Different pesticides are taken up by crops from the soil even if they are not systemic. Some pesticides become absorbed into clay and organic matter fractions in soil. As a result, they are not readily taken into plants. In part, however, they may contaminate the groundwater supplies due to leaching (Zhao et al. 2006). POPs may be transported through far distance through long-range atmospheric transport and prior to deposition. Because of their persistence and semi-volatility, "old" previously emitted and deposited compound can undergo reemission from soils, vegetation, and water bodies and recycle through the environment (Wania and Mackay 1996; Lee et al. 1998). Studies reported that POPs can repeatedly undergo air-surface exchange, so that their ultimate sink may be far from source regions (Klecka et al. 2000). Recycling behavior of POPs is influenced by factors such as temperature, organic carbon content in soil, physicochemical properties of compounds, and air-surface partition coefficients. Soil receives a substantial proportion of POPs through atmospheric deposition and direct input from spillages or from landfilling (Sweetman and Jones 2000). Therefore, global surface soils are an important reservoir for persistent organic pollutants (POPs), and their burden is often used to interpret historical emissions and atmospheric deposition. Contamination of POPs in background surface soils is a function of cumulative atmospheric deposition, minus losses due to volatilization, biodegradation, and mixing/burial to depth. Atmospheric deposition fluxes are in turn influenced by different vegetation types such as canopy scavenging efficiency/temperature, etc. (Wania and McLachlan 2001).

4.1.2 Persistent Organic Pollutants in Air

POPs are known to undergo long-range atmospheric transport (LRAT), traveling according to their physical-chemical properties and to the characteristics of the environment that they encounter (soil properties, climate, wind direction and speed, and so forth), reaching remote regions where they have never been produced or used. This process can happen as a simple emission-transport-deposition event or following a series of "hops" by deposition onto a surface and subsequent emission (Wania and Mackay 1996). The number of air-surface exchange episodes depends on the type of surface (soil, water, vegetation, etc.) and on the physical-chemical properties of the given compounds. According to this principle, persistent chemicals with higher volatility will undergo LRAT followed by deposition in distant areas (e.g., Arctic regions), while those with lower vapor pressure will be deposited preferably in areas closer to the emission source.

Temperature is one of the main factors able to influence the global distribution of POPs, even if several other processes play an important role. Processes directly depending on temperature are emission rates from primary and secondary sources, the gas-particle partitioning of POPs in the air, reactions rates (biodegradation, photolysis, and oxidation in air), and the air-surface exchange (Halsall et al. 1999). In addition, important atmospheric depletion mechanisms for organic contaminants are photochemical degradation and the degradation initiated by the hydroxyl radical, whose concentration is related to solar activity and cloud cover (Mandalakis et al. 2003).

Polycyclic aromatic hydrocarbons (PAHs) belong to the group of persistent organic pollutants (POPs) of known chemical carcinogens. They are formed during the incomplete combustion and pyrolysis of organic matter such as coal, oil, wood, and fuels like diesel and petrol (Garban et al. 2002; Golomb et al. 2001; Halsall et al. 2001). PAHs are mainly concentrated in urban and industrial zones; they can also be found in rural areas (Gevao et al. 2000; Garban et al. 2002) or even in remote regions (Kallenborn et al. 1998; Migaszewske 1999) because of their persistence in the environment and their ability to be transported over long distances (Aamot et al. 1996; Halsall et al. 2001). Benzo(a)pyrene, benzo(b)fluoranthene, benzo(k)fluoranthene, and a number of other aromatic hydrocarbons are formed by the pyrolysis of simple hydrocarbons such as acetylene, butadiene, ethylbenzene, styrene, n-butylbenzene, and tetralin at 700 °C. Most of the lower molecular weight PAH compounds were detected at far from source point in gaseous foam; however, higher molecular weight PAH compounds were detected almost in particulate matter.

4.2 Physical and Chemical Properties of POPs

4.2.1 Organochlorine Pesticides (OCPs)

4.2.1.1 Dichlorodiphenyltrichloroethane (DDT)

Technical DDT is a white amorphous powder that melts over the range of 80–94 °C (Metcalf 1995). The technical grade DDT was used as an insecticide which is composed of active ingredient *p,p′*-DDT (65–80%), *o,p′*-DDT (15–21%), *p,p′*-DDD (4%), and 1.5% of 1-(p-chlorophenyl)-2,2,2-trichloroethanol (Metcalf 1995). DDT is highly insoluble in water and is soluble in most organic solvents. It is semi-volatile and can be expected to partition into the atmosphere as a result. Its presence is ubiquitous in the environment, and residues have even been detected in the Arctic. It is lipophilic, partitions readily into the fat of all living organisms, and has been demonstrated to bioconcentrate and biomagnified. The breakdown products of DDT, 1,1-dichloro-2,2-bis (4-chlorophenyl)ethane (DDD or DDE) and 1,1-dichloro-2,2-bis (4-chlorophenyl)ethylene (DDE) are also present virtually everywhere in the environment and are more persistent than the parent compound. DDT and its primary metabolites, DDE and DDD, are manufactured chemicals and are not known to occur naturally in the environment (WHO 1979). Historically, DDT was released to the environment during its production, formulation, and extensive use as a pesticide in agriculture and vector control applications. DDD was also used as a pesticide, but to a far lesser extent than was DDT. Although it was banned for use in the United States after 1972, DDT is still being used in some areas of the world. DDT and its metabolites are very persistent and bioaccumulate in the environment. This lipophilic property, combined with an extremely long half-life, is responsible for its high bioconcentration in aquatic organisms (i.e., levels in organisms exceed those levels occurring in the surrounding water).

4.2.1.2 Hexachlorohexanes (HCHs)

HCHs contain five isomers α-HCH, β-HCH, γ-HCH, δ-HCH, and ε-HCH, of which γ-HCH is the one having insecticide properties. Although it is toxic, its half-life in tissue, eggs, and other environmental compartments is relatively short in comparison to other pesticides. It breaks down readily into chlorobenzenes and chlorophenyls. The manufacture of HCH involves the photochlorination of benzene. Subsequent treatment with methanol or acetic acid followed by fractional recrystallization will concentrate the γ-HCH isomer to 99.9% (ATSDR 2000). These commercial mixtures typically contain 60–70% α, 5–12% β, 10–12% γ, 6–10% δ, and 3–4% ε (Kutz et al. 1991). HCH was first synthesized in 1825 by Michael Faraday, but the pesticidal properties were not identified until 1942 (Smith 1991). γ-HCH is used as an insecticide on fruits and vegetables, rice paddies, Christmas trees, and animals and as a seed treatment. The β-HCH has a much lower vapor pressure and a much higher melting point and bioconcentration factor (BCF) in human fat as compared to the α-isomer. These properties are largely dictated by the axial and equatorial positions of the chlorine atoms on each molecule. All of the chlorines on β-HCH are in the equatorial position, which seems to confer the greatest physical and metabolic stability to this isomer. This stability is reflected in the environmental and biological persistence of this isomer. For instance, the BCF in the human fat of β-HCH is nearly 30 times higher than that of γ-HCH. The γ-isomer has three chlorines in axial positions, creating two ways that HCl can be eliminated by anti-periplanar dehydrohalogenation and dehydrohalogenation, generating pentachloro-cyclohexene (PCCH) metabolites (Buser and Muller 1995).

4.2.1.3 Chlordanes (CHLs)

The physical and chemical properties of technical chlordane are difficult to specify since there are many components in the technical mixture. For example, technical chlordane is a viscous liquid made of a mixture of many compounds that are solids when pure (a eutectic mixture). The state of the technical product alone will affect the properties of the product. For example, the vapor pressure of the pure components of chlordane will be lower than the technical product because they are solids and have crystal lattice energies that reduce their vapor pressures relative to a liquid (Bidleman and Foreman 1987). The vapor pressure of the mixture will also change over time since the more volatile components will be removed faster, changing the composition of the mixture. Compositional changes with time may also result from different rates of degradation and transport among the constituents of the mixture.

The chemical name of chlordane is 1,2,3,4,5,6,7,8,8-octachloro-2,3,3a,4,7,7a-hexahydro-4,7-methano-1H-indene and the chemical formula is $C_{10}H_6Cl_8$ and molecular weight is 409.78. The basic properties of chlordane are as follows: melting point is <25 °C; boiling point is 165 °C at 2 mm Hg; KH is 4.8 × 10–5 atm m^3/mol at 25 °C; log K_{oc} is 4.58–5.57; log K_{ow} is 6.00; solubility in water is 56 ppb at 25 °C; vapor pressure is 10^{-6} mm Hg at 20 °C (Montgomery 1993).

Chlordane is a mixture of chlorinated hydrocarbons containing chlordane, heptachlor, non-chlordane, and related compounds, and technical chlordane typically contains 64–67% chlorine. Chlordane is highly insoluble in water and is soluble in organic solvents. It is semi-volatile and can be expected to partition into the atmosphere as a result. It binds readily to aquatic sediments and bioconcentrates in the fat of organisms as a result of its high partition coefficient (log K_{ow} = 6.00).

4.2.1.4 Endosulfans (Endos)

The common trade name of endosulfan is Thiodan. Technical-grade endosulfan is a mixture of two isomers, alpha-endosulfan (64–67%) and beta-endosulfan (29–32%) (GFEA-U 2007). The α- and β-isomers of endosulfan are present in the ratio of 7:3, respectively. Technical-grade endosulfan may also contain up to 2% endosulfan alcohol and 1% endosulfan ether. Endosulfan sulfate is usually included with the alpha and beta isomers of endosulfan as "total endosulfan" or "endosulfan (sum)" in the measurement of residues. And it is also found in the environment due to photolysis and in organisms as a result of oxidation by biotransformation (Coleman and Dolinger 1982). The endosulfan sulfate is regarded as being equally toxic and of increased persistence in comparison with the parent isomers (US EPA 2007).

4.2.1.5 Dieldrin and Endrin

Dieldrin is a colorless compound and remains a solid at an ambient temperature of melting point 175–176 °C, practically insoluble in water (0.186 mg/L at 20 °C) but partially soluble in aromatic hydrocarbons, halogenated hydrocarbons, ethers, esters, ketones, and alcohols (IPCS 1998). Endrin is white to light tan crystalline compound with a melting point of 226–230 °C and water solubility of 0.036 mPa and 0.230 mg/L at 25 °C. It can be dissolved in acetone, benzene, carbon tetrachloride, and xylene and moderately soluble in aliphatic hydrocarbons. Dieldrin and endrin have high octanol-water partition coefficient (log $K_{ow} = 6.2$; log $K_{ow} = 5.34$). Uses of dieldrin and endrin have been prohibited across the countries but still found in different environmental media such as soil, sediment, and groundwater. The high concentration of these two compounds has been detected in the agricultural soil and horticultural soils (Singh 2001; Wan et al. 2005; Hilber et al. 2008).

4.2.2 Polychlorinated Biphenyls (PCBs)

PCBs were found for the first time in the environment by Soren Jensen, who detected the contaminants in pike from Swedish waters. Within few years the adverse effects in organism were recognized. Then their use and production were reduced and finally banned in the late 1970s in most industrialized countries. The fact that there are only two chlorines and that both of the chlorine are in an ortho position would tend to make this congener of PCB less toxic than the case in which the congener contains six chlorines, but chlorines are lacking in all of the four ortho positions. A large number of chlorine present coupled with the fact that there are no ortho carbons make this congener a particularly toxic and bioaccumulative species of PCB.

PCBs are discharged into the environment by leaks from hydraulic or industrial systems; diffusion out of coatings, or other products, during production; and waste incineration or disposal or after accidents. The levels of individual congeners differ from location and environmental compartment. Atmospheric PCBs are present predominantly in the gas phase. However, the more chlorinated congeners are also present in the particulate phase (Pankow 1987). There are evidence that atmosphere PCBs are degraded by OH-radical reaction (Anderson and Hites 1996).

4.2.3 Polycyclic Aromatic Hydrocarbons (PAHs)

PAHs are compounds that consist of carbon and hydrogen atoms, grouped in the form of fused benzene rings. They are generally produced in all sorts of combustion encompassing from fossil fuels to organic materials, such as vegetation. PAH composition depends on the combusted organic material and the temperature at which combustion takes place (Howsam and Jones 1998). The higher molecular weight PAHs are easily generated at lower temperatures due to lower combustion efficiencies. Thus, combustion of diesel oil produces relatively higher molecular weight PAH than gasoline in vehicular engines (Khalili et al. 1995). In the late nineteenth century and early twentieth centuries, low-temperature coal combustion was the major source of PAHs. Nowadays fossil fuels, like gas, are used, which generates less PAHs. The occurrence of PAH in the urban air has caused particular concern because of the continuous nature of exposure and size of the population at risk. The urban atmosphere is a complex and dynamic system containing a large variety of interacting chemical species in both the gas and particulate phases. The higher molecular weight PAH compounds can be detected in solid foam but lower molecular weight PAH detected in gaseous foam. PAH compounds can reach human by four different modes of exposure: direct inhalation of polluted air, direct inhalation of tobacco smoke, ingestion of contaminated and processed food and water, and dermal contact.

4.3 Environmental Transport and Distribution of POPs

4.3.1 Soil-Air Exchange of Persistent Organic Pollutants (POPs)

Globally fate and transport of POPs are mainly affected by soil-air exchange processes in the environment. The soil is an important compartment or media with a large capacity for POP storage, and chemicals have long half-life in soils. Soil organic carbon plays an important role to hold the POPs in soil media, and it also reemitted in the atmosphere by air and soil exchange process. The atmospheric concentration of OCPs in background regions is found to be controlled by air and soil exchange mechanism (Cabrerizo et al. 2011). However, soil fugacity or exchange capacity depends on temperature, organic carbon present in the soil, vapor pressure, and enthalpy of the focus chemical. The air-soil exchange of any persistent organic pollutants is an important process due to their significant partition to the gas phase. Additionally, fugacity is the process to measure the chemical potentials in a particular medium that controls the transfer of chemical into one media to another media. The chemical partitioning between air and soil can be described by:

$$K_{\mathrm{SA}} = \frac{C_{\mathrm{S}}\rho_{\mathrm{S}}}{C_{\mathrm{A}}}$$

where C_{S} is the soil concentration (ng kg^{-1}, dry weight), ρ_{s} is the density of soil solids (kg m^{-3}), and C_{A} is the gas-phase air concentration (ng m^{-3}). K_{SA} is dependent on temperature, humidity, chemical, and soil properties (Meijer et al. 2003). Octanol air partition coefficient (K_{OA}) is a key parameter for chemical partitioning between the atmosphere and soil organic matter:

$$K_{\mathrm{SA}} = 0.411\rho_{\mathrm{S}}\varphi_{\mathrm{OC}}K_{\mathrm{OA}}$$

ρ_s is the density of soil solid (kgL^{-1}); ϕ_{oc} is the fraction of organic carbon. The factor 0.411 improves the correlation between the K_{SA} and K_{OA}. During the calculation of K_{SA}, it is assumed that the fugacity or exchange capacity of the soil entirely depends on organic carbon fraction. Thus, the net soil and air exchange flux of POPs is driven by the fugacity between air and surface soil:

$$\mathrm{ff} = f_{\mathrm{soil}} / \left(f_{\mathrm{soil}} + f_{\mathrm{air}} \right)$$

Generally, fugacity values can be calculated using the concentration of POP compounds in the two compartments such as soil and air. The equal fugacity values in the two compartments characterize the equilibrium (Wania 1993), and fugacity fraction (ff) for the chemical can be calculated using the above equation. If the 0.5 of fugacity fraction indicates the chemical is near soil-air equilibrium and fugacity fraction greater than 0.5 indicates that more than half of the chemical potential in the soil-air system is attributed toward soil, this means that net transfer occurs from soil to the air media. However, the fugacity fraction that is less than 0.5 could indicate that soil is a sink and net transfer occurs from air to the soil (Meijer et al. 2003).

4.3.2 Air-Water Gas Exchange of Persistent Organic Pollutants (POPs)

The gas flux between the air and water surface is a function of Henry law constant, concentration gradient, and overall mass transfer coefficient. The net flux can be calculated by the fugacity difference between air and surface water:

$$F_g = K_g \left(C_g - C_w H / RT \right)$$

where C_w and C_g are the water and air concentrations, H is the Henry law constant ($\mathrm{Pam}^{-3}\ \mathrm{mol}^{-1}$), R is the universal gas constant (8.314 Pa $\mathrm{m}^3\ \mathrm{mol}^{-1}\ \mathrm{K}^{-1}$), and T is the temperature at the air-water interface (K). Further, overall gas-phase mass transfer coefficient (K_g) is mostly related to individual mass transfer coefficient (K_g) for the liquid and gas films, K_w and K_a, as follows:

$$1 / K_g = \left(1 / K_g \right) + \left(H / \mathrm{RTK}_w \right)$$

Wind speed also played an important role for mass transfer coefficient of water vapor, i.e., oxygen and carbon dioxide. The below given equation can be used to estimate K_a and K_w for organic compounds (Schwarzenbach et al. 2003):

$$K_{\mathrm{a(compound)}} \left(\mathrm{cm}\, S^{-1} \right) = \left(0.2 U_{10} + 0.3 \right) \left[D_{\mathrm{a(compound)}} / D_{\mathrm{a(H2O)}} \right]^{0.67}$$

$$K_{\mathrm{a(compound)}} \left(\mathrm{cm}\, S^{-1} \right) = \left[\left(0.24 U_{10}^{\;2} + 0.061 U_{10} \right) / 3600 \right] \left[D_{\mathrm{w(compound)}} / D_{\mathrm{w(CO2)}} \right]^{0.5}$$

where D_a and D_w ($\mathrm{cm}^2\ \mathrm{s}^{-1}$) are the diffusivities in air and water and U_{10} is the wind speed 10 m above the water surface (ms^{-1}).

4.4 Long-Range Transport of POPs

Due to the specific characteristics such as volatility, resistance to photolytic, biological, and chemical degradation and capability of long-range atmospheric transport at far distances, POPs have been detected on every continent, countries, and even in remote places such as open oceans, deserts, and Arctic and Antarctic regions where no one used it (Iwata et al. 1994; Tanabe et al. 1982; Bidleman et al. 2002). Surprisingly, at the Arctic region, a high concentration of organochlorine was detected in the different environmental samples such as planktons and wild animals. Wania and Mackey (1993) explained that due to the global distillation or fractionation effect, POPs can migrate at polar regions. However, movement and migration of POPs also depend on physico-chemical properties of chemicals. The ambient temperature also strongly influences the volatility of POPs from the different environmental media. Especially, the migration and movement of POPs mainly take place from tropical and temperate regions of the globe and remain/accumulate in colder regions. Some of the highly volatile POPs will remain airborne and migrate faster toward polar regions through long-range atmospheric transport (LRAT). Thus, individual POPs separate out in the atmosphere according to this volatility; this process is known as global fractionation. Mostly POPs were emitted from different media of environment through exchange processes mainly from hot climatic areas and transport into remote sites where there is cold weather. However, the movement of POPs at a far distance depends on the physico-chemical properties of specific compounds such as HCH that can be detected at remote sites where there are no original sources. That's how POPs can move into the different corners of the globe through atmospheric transport, and there is no boundary for movement of air.

4.5 Toxicology of POPs

4.5.1 Environment

POPs can resist environmental degradation through chemical, biological, and photolytic process. The effect and toxicity of POPs have been the focus by Stockholm Convention to eliminate or restrict to use. The halogenated POPs including polychlorinated biphenyls (PCBs) and trichloroethylene are highly persistent in the different segments of the environment (soil, air, and water) (Petriello et al. 2014). Wang et al. (2014) study investigated that forest soil of Tibetan was contaminated by DDT, HCB, and PCB compounds, respectively. The Tibetan forest soil contains higher soil organic carbon (SOC), and it holds the POP compounds. Thus, the toxicity of soil environment takes place, and these compounds move through dietary intake of contaminated foods and through inhalation of airborne pollutants (Crinnion 2011). Many studies also investigated the contamination of organochlorine pesticides in agricultural soils, sediments, and surface water across the world (Wang et al. 2007; Chen et al. 2011). The shallow groundwater of Taihu Lake region, China, has been contaminated with OCPs in drinking water, mainly sources from historical and new inputs in that particular area. The health risk was noticed for local peoples who consumed the contaminated drinking water especially adult and childrens (Wu et al. 2014), respectively. Different POPs were estimated in groundwater and surface water of India which are mainly released from industrial waste and use of pesticides in agriculture, and more than 90% of water and fish samples of streams were contaminated by POPs (Singh et al. 2005). The Indian Himalayan region soil and air samples from India have been reported to be contaminated with PAH and OCP compounds due to past and present use (Devi et al. 2015, 2016).

4.5.2 Human Health

Bioaccumulation and biomagnification of POPs are detected to be significantly associated with fatty tissues of living organisms for long periods of time because of its high lipid solubility (Wania and Mackay 1996; Vallack et al. 1998). Most of the POPs are known to interfere the normal body functioning of the organs. Due to the global pollution of POPs in food stubs, it's riskier for consumers. Especially, fish are contaminated by these pollutants, once contaminated fish consumed into human body may cause adverse health effects (Borchers et al. 2010). And POPs respond to abnormal functioning of the reproductive system, pregnancy outcomes, as well as endometriosis. The cause of cancer and cardiovascular disease is mainly associated with the POP compounds (Ljunggren et al. 2014). Pregnant women are riskier because POPs may transfer into fetuses and to children through breast-feeding. However, the presence of POPs in the human body does not conclude that they are causing adverse health effects. But, it is also important to understand the exposure route of POPs through different purposes such as farmers applying pesticides without properly covering their body, farmers improperly handling pesticides, etc. Some of the studies have investigated that increasing the abnormal reproductive incidence such as testicular cancer in the male human could be due to exposure of POPs (Sharpe and Skakkebaek 1993; Thomas 1995). Therefore, it is an alarm to all the peoples living on this earth that are exploring the resources and using different chemicals for their productivity benefits which can directly or indirectly influence the human lifestyle. Many researchers already reported that POPs can affect the human being in significant impacts through different exposure routes.

4.6 International Conventions to Control POPs

The internal community has called for global action to reduce and eliminate the toxic chemicals released into the environment. Our earth has posed to risk due to the persistence and long-range atmospheric transportation of POPs in nature. The

Stockholm Convention is mainly focused to reduce or eliminate the POPs in an environment which came into force on 17 May 2004. The United Nation and Environment Programme (UNEP) also legally binds agreement to minimize the release of POPs in the different segments of the environment. The UNEP expanded its investigation on POPs in 1995 mainly focusing on "dirty dozen" compounds (aldrin, chlordane, DDT, dieldrin, dioxin, endrin, furans, heptachlor, hexachlorobenzene, mirex, PCBs, and toxaphene) and internally acted to eliminate and reduce the presence of POPs. However, many of these 12 compounds are no longer used for agricultural purposes, but some are used in developing countries. The Rotterdam Convention entered into force on 24 February 2004 which aims to promote and to share the responsibilities of handling hazardous chemicals and pesticides' importance and safety use. Convention on LongRange Transboundary Air Pollution (LRTAP), which entered into force on 23 October 2003, aims to reduce and prevent air pollution including long-range transboundary air pollution; the protocol is to control and reduce the emission of persistent organic pollutants. Hence, many international treaty organizations are coming together to avoid the use of such toxic POPs for any purposes in the environment.

4.7 Advance Bioremediation Techniques

4.7.1 Phytoremediation

Environmental pollution by persistent organic pollutants (POPs) is a worldwide problem. Remediation of POPs from soil environment is necessary for polluted arable land to reduce the health and environment risks. Phytoremediation technique is one of the advanced methods that uses plants to remove the organic and inorganic contaminants from the environment (Saxena et al. 2016; Saxena and Bharagava 2017; Bharagava et al. 2017b, c; Chandra et al. 2015). This technique is easily assessable for the environment. However, the most important aim of phytoremediation is to identify the accumulator plant species which can effectively uptake the targeted organic compounds.

Dieldrin and endrin pesticides are still detected in agricultural soils. The researcher has investigated that the members of family, Cucurbitaceae and Tiliaceae uptake a great level of dieldrin and endrin from soil media (Otani et al. 2007; Matsumoto et al. 2009). Johgenson (2001) investigated that concentration of dieldrin in soil was normally absorbed into the pulp of different vegetables such as cucumber, watermelons, and squash. The fact is the compounds having high octanol-water coefficient adsorb strongly to soil, and it's difficult to take up such compounds from the soil by plants. The *Cucurbita pepo* L. "Black Beauty" species can absorb the PCB (Arochlor 1268) compounds in roots, stems, leaves, and fruit parts (White et al. 2006). The *Cucurbita pepo* L. "Raven" and *Cucurbita pepo* L. "Baby Bear" also showed the uptake of *p,p'*-DDE metabolites in different parts of

Table 4.1 Uptake of different POPs from soil by plants

Common name	Scientific name	Targeted compound	References
White clove	*Trifolium repens*	Naphthalene	Smith et al. (2006)
Cocksfoot	*Dactylis glomerata*		
Tall fescue	*Festuca arundinacea*		
Red fescue	*Festuca rubra*		
Ryegrass	*Lolium perenne*		
Birds-foot trefoil	*Lotus corniculatus*		
Red clover	*Trifolium pratense*		
Cat claw	*Mimosa monancistra*	Benzo[a]pyrene	Bernel et al. (2007)
Switch grass	*Panicum virgatum*	PAH	Pradhan et al. (1998)
Little bluestem grass	*Schizachyrium scoparium*		
Alfalfa	*Medicago sativa* L.		
Zucchini	*Cucurbita pepo* L. "Black Tosca"	Dieldrin/endrin	Otani et al. (2007)
Pumpkin	*Cucurbita moschata Duch.* "Hayato"	Dieldrin/endrin	Otani et al. (2007)
Cucumber	*Cucumis sativus* L. "Straight Eight"	Dieldrin, heptachlor, heptachlor epoxide	Lichtenstein et al. (1965)
Zucchini	(*Cucurbita pepo* L. "Black Beauty")	PCB (Arochlor 1268)	White et al. (2006)
Zucchini pumpkin	*Cucurbita pepo* L. "Raven," *Cucurbita pepo L.* "Baby Bear"	*p,p'*-DDE	White (2001)
Zucchini pumpkin	*Cucurbita pepo* L. "Senator hybrid," *Cucurbita pepo* L. "Howden"	DDT	Lunney et al. (2004)
Zucchini	*Cucurbita pepo* L. "Black Beauty"	Chlordane	Mattina et al. (2004)

the plant (White 2001). Further Lunney et al. (2004) investigated the uptake of DDT and its metabolites by *Cucurbita pepo* L. "Senator hybrid" and *Cucurbita pepo* L. as well as in roots and shoot parts of the plant. Some other persistent organic pollutants like chlordane and heptachlor epoxide also taken up by *Cucurbita pepo* L. and *Lagenaria siceraria* species mostly in roots and aerial tissue (Mattina et al. 2004; Campbell et al. 2009) (Table 4.1). The uptake of POPs by plants may go through different pathways, for example, compounds may change the adsorbed state on a surface to a gaseous or liquid state that root uptakes from soil solution and transfers into areal parts and metabolizes in plants. That's how cucurbits absorb the compounds from soil and compounds were detected in xylem and aerial parts of the plant (Mattina et al. 2004; Lunney et al. 2004). The *Panicum virgatum* and *Schizachyrium* are also used for the uptake of PAH from polluted soil (Pradhan et al. 1998).

4.7.2 Microbial Degradation

Microorganism plays an important role in the degradation and detoxification of persistent organic and inorganic pollutants in the environment (Kishor et al. 2018; Bharagava et al. 2018; Arora et al. 2014, 2018; Bharagava et al. 2017c; Saxena and Bharagava 2016; Singh et al. 2011). Bioremediation has significant advantages on remediation methods for complete degradation of organic pollutants. Biochemical degradation by the soil organisms is perhaps the single most important method by which pesticides are removed from soils. The advance biotechnological technique used to clean or degrade the contaminated soil by bioremediation technique. Soil charcoal perfusion method is used to rapidly enrich and isolate the POP-degrading bacteria which can be adsorbent of organic chemicals (Takagi et al. 2011). Using this method can isolate the several bacteria which can be capable of degrading POPs.

DDT compounds can be degraded into reductively dechlorinated DDD using bacteria and fungi (Johnsen 1976; Lai and Saxena 1982; Kuhn and Sulflita 1986). The reductive dechlorination condition is the major mechanism for the microbial conversion of both the *o,p′*-DDT and *p,p′*-DDT metabolites into DDD. The pathway for bacterial degradation of DDT to DDD reaction involves the substitution of an aliphatic chlorine for a hydrogen atom. Some of the microbes have shown the degradation of DDT to DDD in the pure culture such as *Escherichia coli*, *Enterobacter aerogenes*, *Enterobacter cloacae*, *Klebsiella pneumoniae*, *Pseudomonas aeruginosa*, *Pseudomonas putida*, and *Bacillus* sp. and fungi *Saccharomyces cerevisiae*, *Phanerochaete chrysosporium*, and *Trichoderma viride* (Johnsen 1976; Lai and Saxena 1982; Subba Rao and Alexander 1985). DDT reductive microbes have been isolated from different habitats such as animal feces, soil, sewage, effluent sludge, and marine and freshwater sediments (Johnsen 1976; Lai and Saxena 1982). Under the anaerobic condition, changes of pH and temperature influence discrete enzymes involved in the metabolism of DDT by *E. aerogenes*. Figure 4.1 shows the pathways of degradation of DDT to yield DDNU in the anaerobic process. DDM is converted or degraded to DBP, which may undergo cleavage of one of the aromatic rings to form *p*-chlorophenylacetic acid. Perhaps, DBP could not be further metabolized under anaerobic conditions (Pfaender and Alexander 1972).

Bidlan and Manonmani (2002) investigate the degradation of DDT in aerobic condition, and microbes were isolated from DDT-contaminated soil which could degrade 5, 10, 15, and 25 ppm DDT by 48, 72, 96, and 144 h, respectively. The DDT metabolizer such as *Hydrogenomonas* sp. was isolated from sewage samples. Most of the gram-negative and gram-positive bacteria are investigated to have the capabilities to attack the DDT metabolites. The gram-positive bacteria could degrade DDT, DDD, and DDE in the presence of biphenyl (Awasthi et al. 2000). However, some of the environmental parameters such as pH, temperature, and other parameters may influence the growth of microorganisms and rate of degradation. Bidlan and Manonmani (2002) found that degradation of DDT metabolites was investigated at the increase in temperature as high as 50 °C.

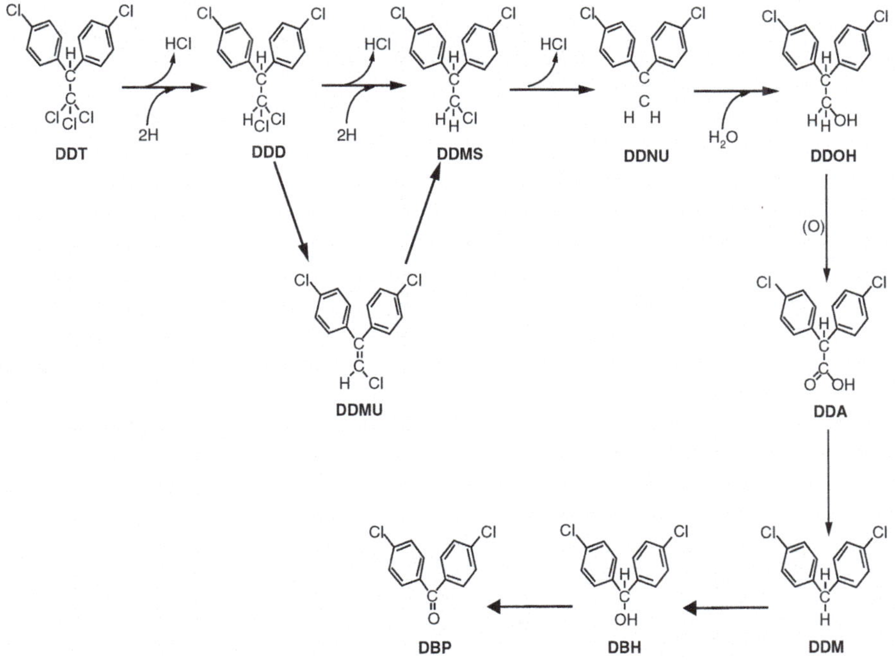

Fig. 4.1 Dechlorination of DDT into DDD (*DDT* 1,1,1-trichloro-2,2-bis(p-chloropenyl)ethane, *DDD* 1,l-dichloro-2,2-bis(p-chlorophenyl)ethane, *DDMU* l-chloro-2,2-bis(p-chlorophenyl)ethylene, *DDMS* 1-chloro-2,2-bis(p-chlorophenyl)ethane, *DDNU* 2,2-bis(p-chlorophenyl)ethylene, *DDOH* 2,2-bis(p-chlorophenyl)ethanol, *DDA* bis(f-chlorophenyl)-acetic acid, *DDM* bis(p-chlorophenyl)methane, *DBH* 4,4′-dichlorobenzhydrol, *DBP* 4,4′ dichlorobenzophenone, and *PCPA* *p*chlorophenylacetic acid)

PCB congeners can be degraded or dechlorinated by aerobic and anaerobic processes. Some of the gram-negative *(Pseudomonas*, *Alcaligenes*, *Achromobacter*, *Burkholderia*, *Comamonas*, *Sphingomonas*, *Ralstonia*, and *Acinetobacter)* and gram-positive (*Rhodococcus*, *Corynebacterium*, and *Bacillus*) bacteria can degrade the PCB compounds. The aerobic condition can cometabolize the lower-chlorinated PCB congeners and may also serve as a nutrient for growth; the higher-chlorinated PCB congeners undergo reductive dechlorination in anaerobic condition. Bacteria utilize the biphenyl as a nutrient source of carbon and energy and degraded the PCB compounds using biphenyl metabolic enzymes. However, PCB degradation decreases with increase in a number of chlorines, and depletion of PCB also depends on the bacterial strain. For the first time, the dehalogenation of PCB is reported in sediment, which could reduce the toxicity of PCBs in the environment (Brown et al. 1988). In the anaerobic condition, the degradation of three PCB congeners (2,3,5,6-CB, 2,3,4,5-CB, and 2,3,4,5,6-CB) has been investigated in sediment by Chang et al. (2001).

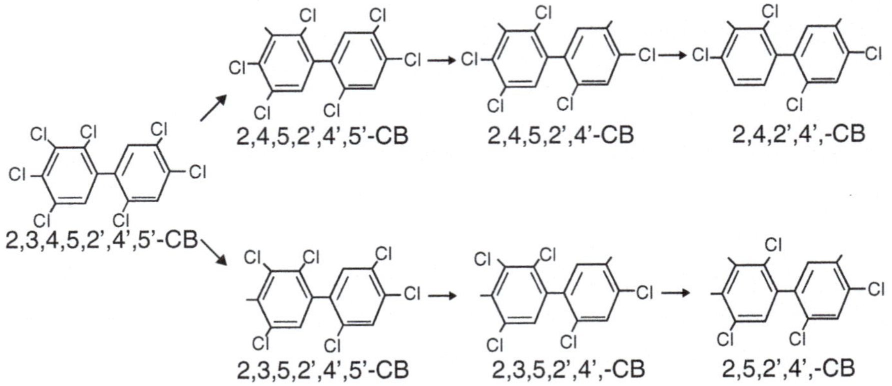

Fig. 4.2 Stepwise dechlorination of 2,3,4,5,2′,4′,5′-CB by *Dehalococcoides* bacteria. (Adapted from Furukawa and Fujihara 2008)

The Aroclor 1260 PCB has been dechlorinated by *Dehalococcoides* bacteria (Fig. 4.2). The PCB congeners of *meta*- and *para*-chlorines are capable to dechlorinate than *ortho* position chlorines in an anaerobic environment. Zhang et al. (2015) found that Anabaena PD-1 is capable of degrading Aroclor 1254 PCB compounds, and this bacteria is separated and isolated from contaminated agricultural soil. Approximately 85% of Aroclor can degrade in 25 days. Toxicity of most of the highly chlorinated PCB congeners converted or degraded in to lower types of chlorine (Mousa et al. 1996; Quensen et al. 1992, 1998). Phytoremediation has been investigated for the removal of PCBs from the soil, groundwater, air, sediment, and surface water (Russel 2005).

Bacteria also play an important role in the degradation of polycyclic aromatic hydrocarbons (PAH) compounds. Benzo[a]pyrene is the most carcinogenic isomer in PAH; 5% degradation of BaP was detected after observing 168 h during incubation with *Sphingomonas paucimobilis* strain EPA 505 (Ye et al. 1996). Some of the bacteria such as *Pseudomonas*, *Agrobacterium*, *Bacillus*, *Burkholderia*, and *Sphingomonas* species were isolated from contaminated soils these bacteria used for the degradation of BaP. It has been also noticed that *Rhodococcus* sp., *Mycobacterium*, and mixed culture of *Pseudomonas* and *Flavobacterium* species can degrade BaP compounds (Schneider et al. 1996; Trzesicka-Mlynarz and Ward 1995). From a heavily polluted petroleum refinery industry stream, the isolated P*seudomonas aeruginosa* and this species was actively grown in phenanthrene pollutants and showed a complete removal (100%) within 30 days as reported by Romero et al. (1998). The *Mycobacterium* spp. strain KR2 was isolated from contaminated soil of gas work plant degraded pyrene with different products such as cis-4,5-pyrene dihydrodiol, 4-5-phenanthrene dicarboxylic acid, 2-carboxybenzaldehyde, phthalic acid, and protocatechuic acid, respectively (Fig. 4.3).

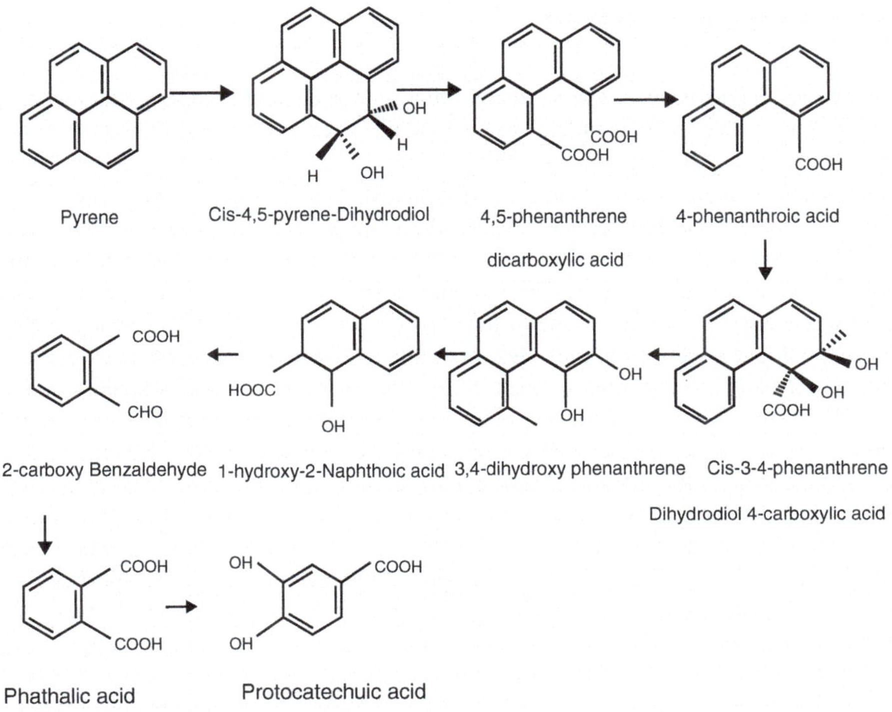

Fig. 4.3 Degradation of pyrene by *Mycobacterium* sp. strain KR2. (Adapted from Rehmann et al. 1998)

Degradation pathway of anthracene and phenanthrene of PAH isomer degradation was reported by Cajthmal et al. (2002). The ligninolytic fungus, *I. lacteus*, can degrade the anthracene and phenanthrene into anthraquinone and phenanthrene-9-10-dihydrodiol-9-10-dihydrodiol, respectively. However, degradation of PAH by fungal enzyme has different steps such as intracellular cytochrome P450 and extracellular lignin peroxidase, manganese peroxidase, and laccase. The lower molecular (two to three rings) PAH compounds can be degraded by soil fungi *Aspergillus* sp., *Trichocladium canadense*, and *Fusarium oxysporum*. The isolated fungi from sugarcane bagasse such as *A. niger*, *P. glabrum*, and *C. cladosporiosis* were able to reduce the phenanthrene by soil fermentation processes which could reduce 70%, 23%, and 54%, respectively (Cortes-Espinosa et al. 2006). The *Penicillium*, *Trichoderma*, and *Aspergillus* have the capability to convert the BaP into different forms under liquid state cultivation conditions within 5 days. Further, fungi have a greater tendency to reduce BaP than bacteria degradation (Machin-Ramirez et al. 2010). The *A. sclerotium* and *M. racemosus* marine fungi have the capability to convert BaP into hydroxybenzo[a]pyrene and benzo[a]pyrenyl sulfate metabolites in a highly available dextrose medium (Passarini et al. 2011).

4.8 Alternatives to POPs

Many researchers have investigated and found that persistent organic pollutants are one of the serious toxic chemicals to understand because of its fate in the environment. Shifting and removing the POPs chemical into nonchemical or nontoxic chemical in human and environment is needed. Many countries are facing the problems to identify and replacePOPs due to their low cost. Peoples need to have the background for education and training on the hazardous nature of POPs and should know what alternatives are available in each corner of the country. Integrated pest management (IPM) could be one of the eco-friendly methods for the management of pest without affecting the man and environment. Currently, some of the countries are moving toward the biopesticides to control the pest; the main objective is to avoid the side effects in any living beings. Some of the indigenous ideas may also apply to control the pest as biocontrol agent and as biopesticides. Swaminathan (1975) introduced the integrated pest management toward the cotton and rice crop which help to develop IPM at a particular location of two crops. Since 1981 integrated pest management was adopted in India for plant protection to reduce the chemical stress. The Stockholm Convention on POP meeting in Geneva approved to replace the endosulfan by nonchemical alternatives. The alternative chemical includes malathion, aldicarb, carbofuran, carbaryl, fipronil, methyl, parathion, and pyrethrin. Since 1993 the Commonwealth Agricultural Bureau International (CABI) of the United Kingdom actively supports the integrated pest management program for cotton plant. Then the Technology Mission on Cotton (TMC) was launched in 2000, and TMC was initiated with Central Institute of Cotton Research (CICR), Nagpur, India. The food and agricultural organization of United Nation also involved in reducing the use of chemical for the welfare of biodiversity. Mass education about the toxic chemical across the countries should be started, and stockholders should implement the policy regarding the persistent toxic chemicals. Individual peoples also must come forward with the replacement and shifting or removal of pesticides into non-chemical pest. Finding the alternatives of POPs is one of the major challenge for targeting to kill the insects. On the other hand, we must focus to have proper production from different productivity fields with practicing biopesticide applications to fulfill the demands.

4.9 Conclusions

Contaminations of persistent organic pollutants (POPs) in the environment have been reported from across the globe due to past and present users of these chemicals. The physicochemical properties and role of POPs in different media environment have been explained in this chapter. POPs have the tendency to be semi-volatile, lipophilic, and persistent in nature and can be found in far distances where they have never been used before. POPs can move to remote sites through long-range atmospheric transport (LRAT) at any corner of the earth. Therefore, it's a big alarm for everyone to come up together to control such toxic chemical through different international organization linked and may find alternative ways to stop using such chemicals.

Bioremediation is the advanced methods to transform the toxic compounds to less hazardous or nonhazardous forms with less affords. The advance bioremediation technique such as phytoremediation and microbial degradation can reduce the contamination level of POPs in the environment, respectively. Phytoremediation is also one of the methods to remove the POPs from contaminated sites. Different plant species are available for use to uptake the contaminant from the field such as DDTs, dieldrin, endrin, PCB, and PAH which can be easily uptake by the plant. However, selection of proper plant species for targeted compounds is also a difficult task. Microorganisms also play important role to degrade or remove the toxic compounds from the environment. Varieties of microbial groups are available for degradation of DDTs, PCBs, PAHs, etc. Most of the bacteria are isolated from contaminated sites and capable of degrading POPs. Furthermore, integrated pest management (IPM) could be one of the alternative methods to eliminate the use of POPs in the environment.

References

Aamot E, Steinnes E, Schmid R (1996) Polycyclic aromatic hydrocarbons in Norwegian forest soils; impact of long-range atmospheric transport. Environ Pollut 92:275–280

Anderson PN, Hites RA (1996) OH radical reactions: the major removal pathway for polychlorinated biphenyls from the atmosphere. Environ Sci Technol 30:1756–1763

Arora PK, Srivastava A, Singh VP (2014) Bacterial degradation of nitrophenols and their derivatives. J Hazard Mater 266:42–59

Arora PK, Srivastava A, Garg SK, Singh VP (2018) Recent advances in degradation of chloronitrophenols. Bioresour Technol 250C:902–909

Alvarez-Bernal D, Contreras-Ramos S, Marsch R, Dendooven L (2007) Influence of cat claw Mimosa monancistra on the dissipation of soil PAHs. Int J Phytoremediation 9:79–90

ATSDR (2000) Toxicological profile for endosulfan. Agency of Toxic Substances and Disease Registry, Atlanta http://www.atsdr.cdc.gov/toxprofiles/tp41.html

Awasthi N, Ahrya R, Kumar A (2000) Factors influencing the degradation of soil applied endosulfan isomers. Soil Biol Biochem 32:1697–1705

Bharagava RN, Saxena G, Mulla SI, Patel DK (2017a) Characterization and identification of recalcitrant organic pollutants (ROPs) in tannery wastewater and its phytotoxicity evaluation for environmental safety. Arch Environ Contam Toxicol. https://doi.org/10.1007/s00244-017-0490-x

Bharagava RN, Saxena G, Chowdhary P (2017b) Constructed wetlands: an emerging phytotechnology for degradation and detoxification of industrial wastewaters. In: Bharagava RN (ed) Environmental pollutants and their bioremediation approaches, 1st edn. CRC Press, Taylor & Francis Group, Boca Raton, pp 397–426. https://doi.org/10.1201/9781315173351-15

Bharagava RN, Chowdhary P, Saxena G (2017c) Bioremediation: an ecosustainable green technology: its applications and limitations. In: Bharagava RN (ed) Environmental pollutants and their bioremediation approaches, 1st edn. CRC Press, Taylor & Francis Group, Boca Raton, pp 1–22. https://doi.org/10.1201/9781315173351-2

Bharagava RN, Purchase D, Saxena G, Mulla SI (2018) Applications of metagenomics in microbial bioremediation of pollutants: from genomics to environmental cleanup. In: Das S, Dash H (eds) Microbial diversity in the genomic era, 1st edn. Academic Press, Elsevier. https://doi.org/10.1016/B978-0-12-814849-5.00026-5

Bidlan R, Manonmani HK (2002) Aerobic degradation of dichlorodiphenyltrichloroethane (DDT) by Serratia marcescens DT-1P. Process Biochem 38:49–56

Bidleman TF, Foreman WT (1987) Vapour-particle partitioning of semivolatile organic compounds. In: Hites RA, Eisenreich SJ (eds) Sources and fates of aquatic pollutants, American Chemical Society, Advances in Chemistry Series No. 216. American Chemical Society, New York, pp 27–56

Bidleman T, Jantunen L, Helm P, Brorstrom-Lunden E, Juntto S (2002) Chlordane enantiomers and temporal trends of chlordane isomers in Arctic air. Environ Sci Technol 36:539–544

Borchers A, Teuber SS, Keen CL, Gershwin ME (2010) Food safety. Clin Rev Allergy Immunol 39:95–141

Brown MB, Bush B, Rhee GY, Shane L (1988) PCB dechlorination in Hudson River sediment. Science 240:1674–1676

Buser HR, Muller MD (1995) Isomer and enantioselective degradation of hexachlorocyclohexane isomers in sewage-sludge under anaerobic conditions. Environ Sci Technol 29:664–672

Cabrerizo A, Dachs J, Moeckel C, Ojeda MJ, Caballero G, Barcelo D, Jone KV (2011) Factors influencing the soil-air partitioning and the strength of soils as a secondary source of polychlorinated biphenyls to the atmosphere. Environ Sci Technol 45(11):4785–4792

Cajthmal T, Moder M, Kacer P, Sasek V, Popp P (2002) Study of fungal degradation products of PAHs using gas chromatography with ion trap mass spectrometry detection. J Chromatogr A 974:213–222

Campbell S, Arakaki AS, Li QL (2009) Phytoremediation of heptachlor and heptachlor epoxide in soil by Cucurbitaceae. Int J Phytoremediation 11:28–38

Chandra R, Saxena G, Kumar V (2015) Phytoremediation of environmental pollutants: an eco-sustainable green technology to environmental management. In: Chandra R (ed) Advances in biodegradation and bioremediation of industrial waste, 1st edn. CRC Press, Taylor & Francis Group, Boca Raton, pp 1–30. https://doi.org/10.1201/b18218-2

Chang BV, Liu WG, Yuan SY (2001) Microbial dechlorination of three PCB congeners in river sediment. Chemosphere 45:849–856

Chen CH, Qian Y, Chen Q, Tao CH, Li CH, Li Y (2011) Evaluation of pesticide residues in fruits and vegetables from Xiamen, China. Food Control 22:1114–1120

Coleman PF, Dolinger PM (1982) Endosulfan monograph number four: environmental health evaluations of California restricted pesticides. Prepared by Peter M. Dolinger Associates, Menlo Park

Cortes-Espinosa DV, Fernandez-Perrino FJ, Arana-Cuenca A, Esparza-Garcıa F, Loera O, Rodrıguez-Vazquez R (2006) Selection and identification of fungi isolated from sugarcane bagasse and their application for phenanthrene removal from soil. J Environ Sci Health A Tox Hazard Subst Environ Eng 41:475–486

Crinnion WJ (2011) Polychlorinated biphenyls: persistent pollutants with immunological, neurological, and endocrinological consequences. Altern Med Rev 16:5–13

Devi NL, Yadav IC, Raha P, Shihua Q, Dan Y (2015) Spatial distribution source apportionment and ecological risk assessment of residual organochlorine pesticides in the Himalayas. Environ. Sci Pollut Res 22:20154–20166

Devi NL, Yadav IC, Shihua Q, Dan Y, Zhang G, Raha P (2016) Environmental carcinogenic polycyclic aromatic hydrocarbons in soil from Himalayas, India: implications for spatial distribution, sources apportionment and risk assessment. Chemosphere 144:493–502

Furukawa K, Fujihara H (2008) Microbial degradation of polychlorinated biphenyls: biochemical and molecular features. J Biosci Bioeng 105:433–449

Garban B, Blanchoud H, Motelav-Masei A, Ollivon D (2002) Atmospheric bulk precipitation of polycyclic aromatic hydrocarbons on to France at typical environmental sites. Atmos Environ 36:5395–5403

Gautam S, Kaithwas G, Bharagava RN, Saxena G (2017) Pollutants in tannery wastewater, pharmacological effects and bioremediation approaches for human health protection and environmental safety. In: Bharagava RN (ed) Environmental pollutants and their bioremediation approaches, 1st edn. CRC Press, Taylor & Francis Group, Boca Raton, pp 369–396. https://doi.org/10.1201/9781315173351-14

Gevao B, Hamilton-Taylor J, Jones KC (2000) Towards a complete mass balance model for PCB and PAH in a small rural lake, Cumbria, UK. Limnol Oceanogr 45:881–894

GFEA-U (2007) Endosulfan. Draft Dossier prepared in support of a proposal of endosulfan to be considered as a candidate for inclusion in the CLRTAP protocol on persistent organic pollutants. German Federal Environment Agency – Umweltbundesamt, Berlin

Golomb D, Barry E, Fisher G, Varanusupakul P, Koleda M, Rooney T (2001) Atmospheric deposition of polycyclic aromatic hydrocarbons near New England coastal waters. Atmos Environ 35:6245–6258

Goutam SP, Saxena G, Singh V, Yadav AK, Bharagava RN (2018) Green synthesis of TiO_2 nanoparticles using leaf extract of *Jatropha curcas* L. for photocatalytic degradation of tannery wastewater. Chem Eng J 336:386–396. https://doi.org/10.1016/j.cej.2017.12.029

Halsall CJ, Gevao B, Howsam M, Lee RGM, Ockenden WA, Jones KC (1999) Temperature dependence of PCBs in the UK atmosphere. Atmos Environ 33(4):541–552

Halsall CJ, Sweetman AJ, Barrie LA, Jones KC (2001) Modelling the behaviour of PAH during atmospheric transport from the UK to the Arctic. Atmos Environ 35:255–267

Hilber I, Mäder P, Schulin R, Wyss GS (2008) Survey of organochlorine pesticides in horticultural soils and there grown Cucurbitaceae. Chemosphere 73:954–961

Howsam M, Jones KC (1998) Sources of PAHs in the environment. Handbook of chemistry, vol 31. Springer, Berlin, pp 137–174

IPCS (1998) International chemical safety card—dieldrin, ICSC 0787. World Health Organization/ International Programme on Chemical Safety, Geneva

Iwata H, Tanabe S, Sakai N, Nishimura A, Tatsukawa R (1994) Geographical distribution of persistent organochlorines in air, water and sediments from Asia and Oceania, and their implications for global redistribution from lower latitudes. Environ Pollut 85:15–33

Johgenson JE (2001) Aldrin and dieldrin: a review of research on their production, environmental deposition and fate, bioaccumulation, toxicology, and epidemiology in the United States. Environ Health Perspect 109:113–139

Johnsen RE (1976) DDT metabolism in microbial systems. Pestic Rev 61:1–28

Kallenborn R, Oehme M, Wynn-Williams DD, Schlabach M, Harris J (1998) Ambient air levels and atmospheric long range transport of persistence organochlorines to Signy island, Antarctica. Sci Total Environ 220:167–180

Khalili NR, Scheff PA, Holsen TM (1995) PAH source fingerprints for coke ovens, diesel and gasoline engines, highway tunnels and wood combustion emissions. Atmos Environ 29:533–542

Kishor R, Bharagava RN, Saxena G (2018) Industrial wastewaters the major sources of dye contamination in the environment, ecotoxicological effects, and bioremediation approaches. In: Bharagava RN (ed) Recent advances in environmental management, 1st edn. CRC Press, Taylor & Francis, pp 1–25

Klecka G, Boethling B, Franklin J, Grady L, Graham D, Howard PH, Kannan K, Larson RJ, Mackay D, Muir D, van de Meent D (eds) (2000) Evaluation of persistence and long-range transport of organic chemicals in the environment. SETAC Press, Pensacola

Kuhn EP, Sulflita JM (1986) Dehalogenation of pesticides by anaerobic microorganisms in soils and groundwater-a review. In: Reactions and movement of organic chemicals in soil. SSSA special publication no. 22. Soil Science Society of America and American Society of Agronomy, Madison, pp 111–180

Kutz FW, Wood PH, Bottimore DP (1991) Organochlorine pesticides and polychlorinated biphenyls in human adipose tissue. Rev Environ Contam Toxicol 120:1–82
Lai R, Saxena DM (1982) Accumulation, metabolism, and effects of organochlorine insecticides on microorganisms. Microbiol Rev 46:95–127
Lee RGM, Hung H, Mackay D, Jones KC (1998) Measurement and modelling of the diurnal cycling of atmospheric PCBs and PAHs. Environ Sci Technol 32:2172–2179
Lichtenstein EP, Schulz KR, Skrentny RF, Stitt PA (1965) Insecticidal residues in cucumbers and alfalfa grown on aldrin- or heptachlor-treated soils. J Econ Entomol 58:742–746
Ljunggren SA, Helmfrid I, Salihovic S, van Bavel B, Wingren G, Lindahl M, Karlsson H (2014) Persistent organic pollutants distribution in lipoprotein fractions in relation to cardiovascular disease and cancer. Environ Int 65:93–99
Lunney AI, Zeeb BA, Reimer KJ (2004) Uptake of weathered DDT in vascular plants: potential for phytoremediation. Environ Sci Technol 38:6147–6154
Machin-Ramirez C, Morales D, Martinez-Morales F, Okoh AI, Trejo-Hernandez MR (2010) Benzo[a]pyrene removal by axenic- and co-cultures of some bacterial and fungal strains. Int Biodeterior Biodegrad 64:538–544
Mandalakis M, Berresheim H, Stephanou EG (2003) Direct evidence for destruction of polychlorobiphenyls by OH radicals in the subtropical troposphere. Environ Sci Technol 37:542–547
Matsumoto E, Kawanaka Y, Yun SJ, Oyaizu H (2009) Bioremediation of the organochlorine pesticides, dieldrin and endrin and their occurrence in the environment. Appl Microbiol Biotechnol 84:205–216
Mattina MJI, Eitzer BD, Iannucci-Berger W, Lee WY, White JC (2004) Plant uptake and translocation of highly weathered, soilbound technical chlordane residues: data from field and rhizotron studies. Environ Toxicol Chem 23:2756–2762
Meijer SN, Shoeib M, Jantunen LMM, Jones KC, Harner T (2003) Air-soil exchange of organochlorine pesticides in agricultural soils. 1. Field measurements using a novel in situ sampling device. Environ Sci Technol 37(7):1292–1299
Metcalf RL (1995) Insect control technology. In: Kroschwitz J, Howe-Grant M (eds) Kirk-Othmer encyclopedia of chemical technology, vol 14. Wiley, New York, pp 524–602
Migaszewske ZM (1999) Determining organic compound ratios in soils and vegetation of the Holy Cross Mts, Poland. Water Air Soil Pollut 111:123–138
Montgomery JH (1993) Agrochemicals desk reference. Environmental data. Lewis Publ, Chelsea, p 672
Mousa MA, Quensen JF, Chou K (1996) Microbial dechlorination alleviates inhibitory effects of PCBs on mouse gamete fertilization in vitro. Environ Sci Technol 30:2087–2092
Otani T, Seike N, Sakata Y (2007) Differential uptake of dieldrin and endrin from soil by several plant families and Cucurbita genera. Soil Sci Plant Nutr 53:86–94
Pankow JF (1987) Review and comparative analysis of the theories on partitioning between the gas and aerosol particulate phases in the atmosphere. Atmos Environ 21:2275–2283
Passarini MRZ, Rodrigues MVN, da Silva M, Sette LD (2011) Marine-derived filamentous fungi and their potential application for polycyclic aromatic hydrocarbon bioremediation. Mar Pollut Bull 62:364–370
Petriello MC, Newsome B, Hennig B (2014) Influence of nutrition in PCB-induced vascular inflammation. Environ Sci Pollut Res Int 21(10):6410–6418
Pfaender FK, Alexander M (1972) Extensive microbial degradation of DDT in vitro and DDT metabolism by natural communities. J Agric Food Chem 20:842–846
Pradhan SP, Conrad JR, Paterek JR, Srivastava VJ (1998) Potential of phytoremediation for treatment of PAHs in soil at MGP sites. Soil Sediment Contam 7(4):467–480
Quensen JF, Boyd SA, Tiedje JM (1992) Expected dioxin-like toxicity reduction as a result of the dechlorination of Aroclors. In: General electric company research and development program for the destruction of PCBs, eleventh progress report. General Electric Corporate Research and Development, Schenectady, pp 189–196.
Quensen JF, Mueller SA, Jain MK (1998) Reductive dechlorination of DDE to DDMU in marine sediment microcosms. Science 280:722–724

Rehman K, Noll HP, Steinberg CEW, Kettrup AA (1998) Pyrene metabolism by Mycobacterium sp. Strain KR2. Chemosphere 36:2977–2992

Romero MC, Cazau MC, Giorgieri S, Arambarri AM (1998) Phenanthrene degradation by microorganisms isolated from a contaminated stream. Environ Pollut 101:355–359

Russel K (2005) The use and effectiveness of phytoremediation to treat persistent organic pollutants. U.S. Environmental Protection Agency, Washington, DC

Saxena G, Bharagava RN (2015) Persistent organic pollutants and bacterial communities present during the treatment of tannery wastewater. In: Chandra R (ed) Environmental waste management, 1st edn. CRC Press, Taylor & Francis Group, Boca Raton, pp 217–247. https://doi.org/10.1201/b19243-10

Saxena G, Bharagava RN (2016) Ram Chandra: advances in biodegradation and bioremediation of industrial waste. Clean Techn Environ Policy 18:979–980. https://doi.org/10.1007/s10098-015-1084-9

Saxena G, Bharagava RN (2017) Organic and inorganic pollutants in industrial wastes, their ecotoxicological effects, health hazards and bioremediation approaches. In: Bharagava RN (ed) Environmental pollutants and their bioremediation approaches, 1st edn. CRC Press, Taylor & Francis Group, Boca Raton, pp 23–56. https://doi.org/10.1201/9781315173351-3

Saxena G, Chandra R, Bharagava RN (2016) Environmental pollution, toxicity profile and treatment approaches for tannery wastewater and its chemical pollutants. Rev Environ Contam Toxicol 240:31–69. https://doi.org/10.1007/398_2015_5009

Schneider J, Grosser R, Jayasimhulu K, Xue W, Warshawsky D (1996) Degradation of pyrene, benz[a]anthracene and benzo[a]pyrene by Mycobacterium sp. strain RJGII-135, isolated from a former coal gasification site. Appl Environ Microbiol 62:13–19

Schwarzenbach RP, Gschwend PM, Imboden DM (2003) Environmental organic chemistry, 2nd edn. Wiley, Hoboken

Sharpe RM, Skakkebaek NE (1993) Are oestrogens involved in falling sperm counts and disorders of the male reproductive tract? Lancet 341:1392–1395

Singh RP (2001) Comparison of organochlorine pesticide levels in soil and groundwater of Agra, India. Bull Environ Contam Toxicol 67:126–132

Singh KP, Malik A, Mohan D, Takroo R (2005) Distribution of persistent organochlo-rine pesticide residues in Gomti River, India. Bull Environ Contam Toxicol 74(1):146–154

Singh JS, Abhilash PC, Singh HB, Singh RP, Singh DP (2011) Genetically engineered bacteria: an emerging tool for environmental remediation and future research perspectives. Gene 480:1–9

Smith AG (1991) Chlorinated hydrocarbon insecticides. Classes of pesticides. In: Hayes WJ, Laws ER (eds) Handbook of pesticide toxicology. Academic, San Diego

Smith MJ, Flowers TH, Duncan HJ, Alder J (2006) Effects of polycyclic aromatic hydrocarbons on germination and subsequent growth of grasses and legumes in freshly contaminated soil and soil with aged PAHs residues. Environ Pollut 141:519–525

Subba Rao RV, Alexander M (1985) Bacterial and fungal co-metabolism of 1,1,1-trichloro-2,2-bis-(4-chlorophenyl) ethane (DDT) and its breakdown products. Appl Environ Microbiol 49:509–516

Swaminathan MS (1975) ICAR operational research projects, purpose and approach. Indian Farming

Sweetman AJ, Jones KC (2000) Declining PCB concentrations in the UK atmosphere: evidence and possible causes. Environ Sci Technol 34:863–869

Takagi K, Kataoka R, Yamazaki K (2011) Recent technology on bio-remediation of POPs and persistent pesticides. JARQ 45(2):129–136

Tanabe S, Tatsukawa R, Kawano M, Hidaka H (1982) Global distribution and atmospheric transport of chlorinated hydrocarbons: HCH (BHC) isomers and DDT compounds in the western Pacific, eastern Indian and Antarctic Oceans. J Ocean Turbul Jpn 38:137–148

Thomas JA (1995) Falling sperm counts. Lancet 346:636

Trzesicka-Mlynarz D, Ward OP (1995) Degradation of polycyclic aromatic hydrocarbons

UNEP (2006) The state of the marine environment: trends and processes. The Hague, The Netherlands
US-EPA (2007) Note to reader. Endosulfan Readers Guide. November 16. EPA-HQ-OPP-2002-0262-0057. http://www.regulations.gov.
Vallack HW, Bakker DJ, Brandt I, Broström-Ludén E, Brouwer A, Bull KR, Gough C, Guardans R, Holoubek I, Jansson B, Koch R, Kuylenstierna J, Lecloux A, Mackay D, McCutcheon P, Mocarelli P, Taalman RDF (1998) Controlling persistent organic pollutants – what next? Environ Toxicol Pharmacol 6:143–175
Wan MT, Kuo J, Pasternak J (2005) Residues of endosulfan and other selected organochlorine pesticides in farm areas of the lower Fraser valley, British Columbia, Canada. J Environ Qual 34:1186–1193
Wang Z, Chen J, Qiao X, Yang P, Tian F, Huang L (2007) Distribution and sources of polycyclic aromatic hydrocarbons from urban to rural soils: a case study in Dalian. Chin Chemosphere 68:965–971
Wang X, Xue Y, Gong P, Yao T (2014) Organochlorine pesticides and polychlorinated biphenyls in Tibetan forest soil: profile distribution and processes. Environ Sci Pollut Res Int 21(3):1897–1904
Wania FM (1993) Global fractionation and cold condensation of low volatility organochlorine compounds in polar regions. Ambio 22:10–18
Wania F, Mackay D (1993) Global fractionation and cold condensation of low volatility organochlorine compounds in polar regions. Ambio 22:10–18
Wania F, Mackay D (1996) Tracking the distribution of persistent organic pollutants. Environ Sci Technol 30(9):390A–396A
Wania F, McLachlan MS (2001) Estimating the influence of forests on the overall fate of semivolatile organic compounds using a multimedia fate model. Environ Sci Technol 35:582–590
White JC (2001) Plant-facilitated mobilization and translocation of weathered 2, 2-bis(p-chlorophenyl)-1, 1-dichloroethylene (*p, p'*- DDE) from an agricultural soil. Environ Toxicol Chem 20:2047–2052
White JC, Parrish ZD, Isleyen M, Gent MPN, Iannucci-Berger W, Eitzer BD, Kelsey JW, Mattina MI (2006) Influence of citric acid amendments on the availability of weathered PCBs to plant and earthworm species. Int J Phytoremediation 8:63–79
WHO (1979) DDT and its derivatives. Environmental health criteria. World Health Organization, Geneva, p 9
Wu C, Luo Y, Gui T, Huang Y (2014) Concentrations and potential health hazards of organochlorine pesticides in shallow groundwater of Taihu lake region, China. Sci Total Environ 470–471:1047–1055
Ye B, Siddiqi MA, Maccubbin AE, Kumar S, Sikka HC (1996) Degradation of polynuclear aromatic hydrocarbons by *Sphingomonas paucimobilis*. Environ Sci Technol 30:136–142
Zhang H, Jiang X, Lu L, Xiao W (2015) Biodegradation of polychlorinated biphenyls (PCBs) by the novel identified cyanobacterium Anabaena PD1. PLoS One. https://doi.org/10.1371/journal.pone.0131450
Zhao G, Xu Y, Han G, Ling B (2006) Biotransfer of persistent organic pollutants from a large site in China used for the disassembly of electronic and electrical waste. Environ Geochem Health 28:341–351

Chapter 5
Bioremediation of Distillery Effluent: Present Status and Future Prospects

Sushil Kumar Shukla, Vinod Kumar Tripathi, and Pradeep Kumar Mishra

Contents

S. K. Shukla (✉)
Centre for Environmental Sciences, Central University of Jharkhand, Ranchi, Jharkhand, India

V. K. Tripathi
Department of Farm Engineering, Institute of Agricultural Sciences, Banaras Hindu University, Varanasi, Uttar Pradesh, India

P. K. Mishra
Department of Chemical Engineering, Indian Institute of Technology (IIT), Banaras Hindu University, Varanasi, Uttar Pradesh, India
e-mail: pkmishra.che@iit.bhu.ac.in

G. Saxena, R. N. Bharagava (eds.), *Bioremediation of Industrial Waste for Environmental Safety*, https://doi.org/10.1007/978-981-13-1891-7_5

Abstract Environmental degradation due to industrial growth is putting pressure on the society and water resources near to the industry. In order to improve and protect the environment from pollution, sustainability between environment and development is vital. Environmental laws are given general applicability, and their enforcement has been increasingly stricter. So, in terms of health, environment, and economy, the fight against pollution has become a major issue. The pollution increase, industrialization, and rapid economic development impose severe risks to the availability and quality of water resources. Distilleries are considered as one of the most polluting and growth-oriented industries in the world. Distilleries consumed a huge amount of water in the manufacturing of alcohol and produce a large amount of wastewater which contains high organic load, has low pH, and is dark brown in color. This wastewater alters the physical, chemical, and biological characteristic of water and soil if thrown directly outside without treatment. In the present investigation, emphasis has been given to reduce the concentration of pollutants of distillery effluents through bioremediation techniques to meet the norms of environmental regulatory authorities. Biodegradation is one of the best techniques to reduce organic load from water bodies, but it has certain limitation. Taking this into consideration, developing an effective treatment plan for distilleries, i.e., coagulation followed by mixed culture aerobic treatment (fungal and algae), seems to be the viable cost-effective and eco-friendly technique.

Keywords Distillery effluent · Pollution · Anaerobically biodigested distillery effluent (ABDE) · Molasses · Melanoidin · Bioremediation

5.1 Introduction

Industries play an important role in the national economy of the country but also cause serious environmental pollution and toxicity in living beings (Bharagava et al. 2017a; Goutam et al. 2018; Gautam et al. 2017; Saxena et al. 2016; Saxena and Bharagava 2015). Among the industries, distilleries and sugar industries have a vital role in the agricultural economy. These are also considered as polluting and growth-oriented industries in the world. Distilleries are consuming a huge amount of water for manufacturing alcohol and producing by-product as wastewater. It contains high organic load, has low pH, and is dark brown in color. Wastewater alters the physical, chemical, and biological characteristic of soil and water system if thrown directly outside without appropriate treatment. Alcohol plays an important role in modern societies from ancient times, but nowadays there is an enhancement due to implantation of the rule of the Indian government to add 5% alcohol in gasoline. To throw the trated effluent effluents from industries require huge dilution, it limits the options for environmental engineers. Production of alcohol from agricultural raw materials displays a steady growth which is supported by its consumption in the form of industrial solvent and a beverage component as well as an alternative CO_2-neutral fuel. The latter usage attracts an increasing worldwide interest due to the shortage of nonrenewable energy resources and variability of oil and natural gas prices (Scott et al. 2010).

There are nearly 300 distilleries in India producing approximately 3×10^9 L of alcohol and generating 5×10^{10} L of wastewater each year (Uppal 2004). Over the years, as the sizes and number of distilleries have grown, larger and larger conventional biological treatment plants have been built to deal with increasing effluent volumes. Space and money to construct these installations are the biggest hindrances for such investments (Fumi et al. 1995). In India, there are a number of large-scale distilleries integrated with sugar mills. The waste products from sugar mill comprise bagasse, pressmud, and molasses. Bagasse is used in paper manufacturing and as fuel in boilers and molasses as a raw material in the distillery for alcohol production, while press mud has no direct industrial application (Nandy et al. 2002). The effluents from molasses-based distilleries contain large amounts of dark brown-colored spent wash (SW). In the distillation process, ethanol ranges from 5% to 12% by volume; hence it follows that the amount of waste varies from 88% to 95% by volume of the alcohol distilled. Thus, an average molasses-based distillery generates 12–15 L of spent wash per liter of alcohol produced (Beltran et al. 2001). Spent wash is one of the most difficult waste products for disposal because of low pH, high temperature, dark brown color, high ash content, and high percentage of dissolved organic and inorganic matter (Beltran et al. 1999). The biochemical oxygen demand (BOD) and chemical oxygen demand (COD) typically range between 35,000 and 50,000 and 100,000 and 150,000 mg/L, respectively (Nandy et al. 2002). However, the amount and the characteristics of the sugarcane molasses wastewater are highly variable and depend on the raw material used and on the ethanol production process (Pant and Adholeya 2007; Satyawali and Balakrishanan 2007). Washing water used to clean the fermenters, cooling water, and boiler water further contributes to its variability (Pant and Adholeya 2007). The main source of wastewater is the distillation step wherein large volumes of dark brown effluent (termed as spent wash, stillage, slop, or vinasse) are generated with a temperature range of 70–80 °C, acidic pH, and high concentration of organic materials and solids (Yeoh 1997; Nandy et al. 2002).

Worldwide environmental regulatory authorities are setting strict norms for discharge of wastewaters from industries. In India, the population equivalent of distillery wastewater based on BOD has been reported to be as high as 7.2 billion. The contribution of distillery waste to organic pollution is approximately seven times more than the entire municipal and domestic waste generated by the entire population of India.

In accordance with the Environment (Protection) Act, 1986, the Central Pollution Control Board (CPCB 2003) of India has promulgated the minimal national standards (MINAS) for the discharge of distillery wastewater. This MINAS requires that the DWW being released to the environment after treatment should have BOD <30 mg/L, COD <250 mg/L for disposal into inland surface water, and BOD <100 mg/L for disposal on land with pH 5.5–9.0 and TSS <100 mg/L. Color should be removed as per extent of possibility.

5.1.1 Wastewater Generation and Characteristics

Distilleries are 1 of the 17 most polluting industries listed by the Central Pollution Control Board (CPCB). The quantum and characteristics of wastewater generated at various stages in the manufacturing process are provided in Tables 5.1 and 5.2, respectively. Though wastewater is generated at various stages of alcohol production, wastewater from the fermenter sludge, spent wash, and spent lees are the main contributors to the pollution.

Molasses-based alcohol distilleries generate a large volume of dark-colored wastewater. The color is primarily attributed to a dark brown pigment, melanoidin, as well as to the presence of phenolic compounds and caramel (Kalavathi et al. 2001). Melanoidins are natural polymers known to be produced by the nonenzymatic

Table 5.1 Wastewater generations in different distillery operations

Sr. No.	Distillery operations	Average wastewater Generation (KLD/distillery)	Specific wastewater KL wastewater/ KL alcohol
1.	Spent wash (from distillation)	511.4	11.9
2.	Fermenter cleaning	108.2	2.5
3.	Fermenter cooling	307.7	7.2
4.	Condenser cooling	34.2	0.8
5.	Floor wash	47.6	1.1
6.	Bottling plant	126.9	3.0
7.	Others	33.3	0.8

Source: TERI (2003)

Table 5.2 Typical characteristics of various wastewater streams in distilleries

Sr. No.	Parameter	Spent wash	Fermenter cooling	Fermenter cleaning	Condenser cooling	Fermenter wash	Bottling plant
1.	Color	Dark brown	Colorless	Colorless	Colorless	Faint	Colorless
2.	pH	4.0–4.5	6.25	5.0–5.5	6.8–7.8	6.0	7.45
3.	Alkalinity (mg/L)	3500	300	Nil	–	40	80
4.	Total solids (mg/L)	100,000	1000–1300	1000–1500	700–900	550	400
5.	Suspended solids (mg/L)	10,000	220	400–600	180–200	300	100
6.	Biochemical oxygen demand (mg/L)	45,000–60,000	100–110	500–6000	70–80	15	5
7.	Chemical oxygen demand (mg/L)	80,000–120,000	500–1000	1200–1600	200–300	25	15

Source: TERI (2003)

browning reaction, the Maillard reaction between the amino and carbonyl groups in organic substances, and are closely related to the humic substances in the natural environment.

Maillard reaction precedes effectively at temperatures above 50 °C and pH 4.0–7.0 (Rivero-Perez et al. 2002). Spent wash contains 2% melanoidins, which has an empirical formula of $C_{17}H_{26-27}O_{10}N$ (Martins and Van Boekel 2004). Melanoidins have conjugated carbon-carbon double bonds –C=C– in their structure that are responsible for their brown color (Kim et al. 1997). The molecular weight distribution is between 5000 and 40,000. It consists of acidic, polymeric, and highly dispersed colloids, which are negatively charged due to the dissociation of carboxylic acids and phenolic groups. Melanoidins have recalcitrant compounds; thus the conventional treatment methods are not effective for complete color removal from this stream, and color can even be increased during anaerobic treatments, due to re-polymerization of compounds (Satyawali and Balakrishnan 2007; Singh 1999). Melanoidin has antioxidant properties causing toxicity to many microorganisms involved in wastewater treatment processes.

5.1.2 *Profile of Lords Industries Ltd. Nandganj, Ghazipur (Uttar Pradesh) India*

Wastewater samples were collected from Lords Industries Ltd., Nandganj Ghazipur, UP, India. The installed capacity of Lords Distillery Limited for manufacturing rectified spirit (RS) is 14,850 KL (94.5% alcohol) per annum. The molasses is converted into alcohol by fermentation using yeast through a batch fermentation process. The yeast culture used for fermentation converts the sugar present in molasses into alcohol. Alcohol is extracted from the fermenter wash by distillation. The final products are rectified spirit, country liquor, and Indian-made foreign liquor (IMFL). The capacity utilization of the plant in the study period has been almost 42%. The first stage of effluent treatment plant (ETP) is an anaerobic biological digester. The second stage comprises a primary clarifier, anaerobic digester with extended aeration system using diffused aeration mechanism, and a secondary clarifier. The third stage comprises anaerobic digester consisting of a hybrid aerobic biological reactor in which microorganisms supported on activated carbon particles are kept suspended within the reactor by means of air diffusion and a tertiary clarifier. In addition to the above treatment process, the company has developed a botanical treatment plant using the root zone process. Two lagoons of capacities 1435 m^3 and 890 m^3 have also been provided for storage of spent wash in case of emergencies. Effluent from the ETP is diluted if required and subsequently discharged onto neighboring agricultural land. The characteristics of spent wash and anaerobically digested distillery effluent (ABDE) of Lords Distillery Limited are given in Table 5.3, along with the detailed flowchart of effluent treatment plant used in Lords Ind. Ltd., Nandganj, which is presented in Fig. 5.1.

Table 5.3 Characteristics of distillery spent wash and anaerobically biodigested distillery effluent (ABDE) of Lords Distillery Ltd. Nandganj, Ghazipur

Sr. No.	Parameters	Distillery spent wash	Anaerobically digested distillery effluent (BDE)
1.	pH	4.5–5.4	7.5–8
2.	Biochemical oxygen demand (BOD_5) (mg/L)	55,000–65,000	5000–8000
3.	Chemical oxygen demand (COD) (mg/L)	110,000–130,000	40,000–52,000
4.	Total solid (TS) (mg/L)	130,000–160,000	70,000–75,000
5.	Total volatile solid (TVS) (mg/L)	60,000–75,000	48,000–60,000
6.	Total dissolved solids (TDS) (mg/L)	90,000–120,000	35,000–50,000
7.	Chlorides (mg/L)	6000–8500	5000–5500
8.	Phenols (mg/L)	8000–10,000	7000–8000
9.	Sulfate (mg/L)	7500–9000	3000–5000
10.	Phosphate (mg/L)	2500–2700	1500–1700
11.	Total nitrogen (mg/L)	5000–7000	4000–4200

5.1.3 Challenges and Research and Development Focus Areas in Distillery Effluent Treatment

A key challenge is to identify suitable secondary treatment steps for proper handling and disposal of the post-biomethanated effluent. Presently two-stage aeration process is used for COD reduction, but it is energy intensive and not efficient for color removal. Several physical, chemical, and biological treatment methods have been investigated (Satyawali and Balakrishnan 2007; Pant and Adholeya 2007). Physicochemical treatment methods such as adsorption, coagulation/flocculation, and oxidation processes have been examined in this context, but the cost of chemicals, as well as the associated sludge handling with disposal, is deterrent (Rajor et al. 2002). Thus, further work on low-cost adsorbents, including wastes generated in other industrial processes, is required. In spite of being a relatively expensive option, membrane treatment (RO) is being adapted to meet the zero-discharge norms. Although superiority of biodegradation processes for the treatment of distillery effluent cannot be questioned due to a large amount of wastewater being generated by these industries, efforts are required for an integrated approach to deal with the problem and minimize the wastage of freshwater which is currently used for dilution purposes.

5.1.4 Treatment Processes Employed in Distilleries

In distilleries, treatment of wastewater is done for removal of biochemical oxygen demand (BOD), chemical oxygen demand (COD), suspended solids, nutrients (nitrate, phosphate, and potash), colorants, pathogens, and toxic elements. BOD

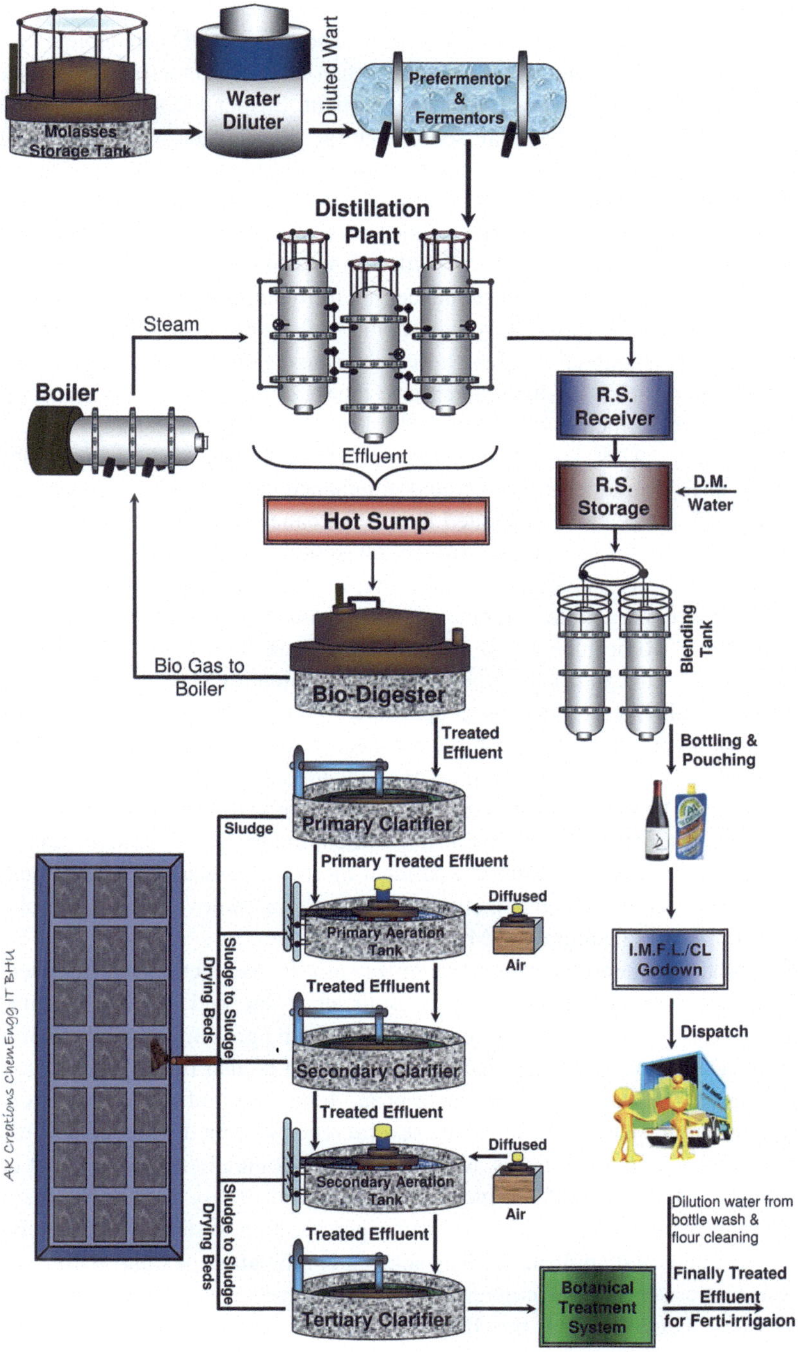

Fig. 5.1 Schematic process graphical flowchart of Lords Distillery Limited

exploits the ability of microorganisms to oxidize organic material to CO_2 and water using molecular oxygen as an oxidizing agent. Therefore, BOD can deplete the dissolved oxygen of receiving water leading to fish kills and anaerobiosis; hence its removal is a primary aim of wastewater treatment. Suspended solids are removed principally by physical/chemical sedimentation. Wastewater treatment systems are designed to remove nutrients, mainly dissolved nitrogen and phosphorus along with symbiotic associations between Fungi and Algae. It is an important step of treatment process. Discharge of these nutrients into sensitive water bodies leads to eutrophication by stimulating the growth of unwanted plants such as algae and aquatic macrophytes. Other consequences of nitrogen compounds in wastewater effluents are toxicity of non-ionized ammonia to fish and other aquatic organisms, interference with disinfection where a free chlorine residual is required, and methemoglobinemia in influents due to excessive nitrate concentrations (above 45 g/m^3) in drinking water (Lincolin and Earle 1990; Lincoln and Hill 1980).

Till the early 1970s, land disposal was practiced as one of the main treatment options in distilleries, since it was found to enhance the yield of certain crops. For example, in Brazil, vinasse generated from sugarcane juice fermentation is mainly used as a fertilizer due to its high nitrogen, phosphorus, and organic content. Its use is reported to increase sugarcane productivity; furthermore, under controlled conditions, the effluent is capable of replacing application of inorganic fertilizers (Cortez and Peréz 1997; Rodríguez 2000). Investigations have indicated that land disposal of distillery effluent can lead to groundwater contamination (Jain et al. 2002). Other disposal methods like evaporation of spent wash to produce animal feed and incineration of a spent wash for potash recovery have also been practiced. Wastewater treatment aims to remove unwanted compounds for safe discharge into the environment. This can be achieved by using physical, chemical, and biological methods either alone or in combination. Physical treatment methods such as screening, sedimentation, and skimming remove floating objects. Chemical treatment methods such as precipitation, pH adjustment, coagulation, etc. remove toxic materials and colloidal impurities (Stumm and Morgan 1962).

Biological treatments have been recognized as effective methods of treatment for highly polluted industrial wastewaters. Both anaerobic and aerobic systems are commonly used for the treatment of wastewater from agro-industrial plants including distilleries. In recent years, increasing attention is also being directed toward utilizing microbial activity of pure bacteria and fungi culture for decolorization and mineralization of spent wash. There are several reports citing the potential of microorganisms for use in this process. Moreover, the biologically treated effluent could be used safely and effectively to increase the soil productivity. Biological treatment system requires excessive dilution with proper aeration system which limits this process for industrial application. Nowadays a combination of fungi with microalgae (symbiotic associations) emerges as a new tool in organic-rich wastewater treatment system. Thus, solutions for effective management of molasses-based distillery wastewater are still evolving, and there is an urgent need to address the limitations in the existing methods and develop integrated treatment processes.

5.2 Bioremediation Techniques

Environmental biotechnology is not a new field; composting and wastewater treatments are familiar examples of old environmental biotechnologies. However, recent studies in molecular biology and ecology offer opportunities for more efficient biological processes. Notable accomplishments of these studies include the cleanup of polluted wastewater and land areas.

Bioremediation is the process in which organic wastes are biologically degraded under controlled conditions to an innocuous state or to levels below concentration limits established by regulatory authorities (Mueller et al. 1996). In other terms bioremediation is the use of living organisms, primarily microorganisms, to degrade the environmental contaminants into less toxic forms. It uses naturally occurring bacteria and fungi or plants to degrade or detoxify substances hazardous to human health and/or the environment (Saxena et al. 2018; Arora et al. 2018; Bharagava et al. 2017b, c; Saxena and Bharagava 2016, 2017; Chandra et al. 2015; Arora et al. 2014; Singh et al. 2011). The microorganisms may be indigenous to a contaminated area, or they may be isolated from elsewhere and brought to the contaminated site. Contaminant compounds are transformed by living organisms through reactions that take place as a part of their metabolic processes. Biodegradation of a compound is often a result of the actions of multiple organisms. Bioaugmentation is when microorganisms are imported to a contaminated site to enhance degradation. For bioremediation to be effective, microorganisms must enzymatically attack the pollutants and convert them to harmless products. As bioremediation can be effective only where environmental conditions permit microbial growth and activity, its application often involves the manipulation of environmental parameters to allow microbial growth and degradation to proceed at a faster rate. Like other technologies, bioremediation has its limitations. Some contaminants, such as chlorinated organic or high aromatic hydrocarbons, are resistant to microbial attack. They are degraded at a very slow rate or not degraded; hence it is not easy to predict the rates of cleanup for a bioremediation exercise; there are no rules to predict if a contaminant can be degraded. Bioremediation techniques are typically more economical than traditional methods such as incineration, and some pollutants can be treated on site, thus reducing exposure risks for cleanup personnel or potentially wider exposure as a result of transportation accidents. Since bioremediation is based on natural attenuation, the public considers it more acceptable than other technologies.

Microorganisms play a key role in bioremediation process and have been proven as an efficient, low-cost, and environmentally friendly alternative to physicochemical methods. Several microbial species including fungi, bacteria, and algae have been studied for their capacity to degrade and decolorize toxic chemical pollutants present in various industrial wastewater including distilleries. Free or immobilized cells have been studied widely for bioremediation of distillery wastewater. Immobilizing microorganism on inert support material including alginate, polyacrylamide, agar, polystyrene, and polyurethane is more advantageous compared to that of free cell. Some of the advantages include the compact physical structure of carrier pellets, high biomass retention, reusability of culture, and easier separation process (Sankaran et al. 2015; Raghukumar et al. 2004). The potential of microorganisms in distillery wastewater treatment is highly dependent on the type of chemical composition of wastewater, nutrient, pH, temperature, oxygen, and inoculum size (Surega and Krzywonos 2015; Gay et al. 1996).

In recent years, a number of studies have focused on microorganisms, which are able to biodegrade and biosorb coloring compounds present in wastewater. With an increasing consciousness about pollution control, biological decolorization is receiving prime attention for effluent treatment. With economic constraints on pollution control process, affordable and effective methods have become a necessity. Since spent wash is a high-strength wastewater, it is generally subjected to anaerobic digestion in huge methane reactors to convert organics into methane, which is used as a fuel. The anaerobically biodigested distillery effluent (ABDE) is, then, treated aerobically in activated sludge treatment plants to further reduce BOD/COD and used for composting pressmud, a by-product of sugar mills. The color of spent wash persists even after biomethanation.

5.2.1 Factors of Bioremediation

The control and optimization of bioremediation processes is a complex system of many factors. These factors include the existence of a microbial population capable of degrading the pollutants, the availability of contaminants to the microbial population, and the environment factors (a type of coloring material, temperature, pH, the presence of oxygen or other electron acceptors, and nutrients).

5.2.2 Microbial Populations for Bioremediation Process

Microorganisms can be isolated from almost any environmental conditions. Microbes will adapt and grow at subzero temperatures, as well as extreme heat, desert conditions, in water, with an excess of oxygen, and in anaerobic conditions, with the presence of hazardous compounds or on any waste stream. The main requirements are an energy source and a carbon source. Because of microbes and other biological systems, these can be used to degrade or remediate environmental hazards. We can subdivide these microorganisms into the following groups:

Aerobic (in the presence of oxygen): Examples of aerobic bacteria recognized for their degradative abilities are *Pseudomonas*, *Alcaligenes*, *Sphingomonas*, *Rhodococcus*, *Mycobacterium*, etc. There are several fungi also reported to degrade various hydrocarbon like *Phanerochaete chrysosporium*, *Aspergillus niger*, etc. These microbes have often been reported to degrade pesticides and hydrocarbons, both alkanes and polyaromatic compounds. Many of these bacteria and fungi use the contaminant as the sole source of carbon and energy.

Anaerobic (in the absence of oxygen): Anaerobic bacteria are not as frequently used as aerobic bacteria. There is an increasing interest in anaerobic bacteria used for bioremediation of polychlorinated biphenyls (PCBs) in river sediments and dechlorination of the solvent trichloroethylene (TCE) and chloroform.

Ligninolytic fungi: Fungi such as the white rot fungus *Phanerochaete chrysosporium* have the ability to degrade an extremely diverse range of persistent or toxic environmental pollutants. Common substrates used include straw, sawdust, or corncobs.

Methylotrophs (aerobic bacteria that grow utilizing methane for carbon and energy): The initial enzyme in the pathway for aerobic degradation, methane monooxygenase, has a broad substrate range and is active against a wide range of compounds, including the chlorinated aliphatic trichloroethylene and 1,2-dichloroethane.

5.3 Bioremediation of Anaerobically Biodigested Distillery Effluent (ABDE)

Biological treatments have been recognized as effective methods for treatment of highly polluted industrial wastewaters. Both anaerobic and aerobic systems are commonly used to treat the anaerobically biodigested distillery effluent (ABDE). In recent years, increasing attention is also being directed toward utilizing microbial activity (pure bacteria, fungi, algae, and mixed culture) for the decolorization and mineralization of ABDE. There are several reports citing the potential of microorganisms for use in this process. Moreover, the biologically treated effluent could be used safely and effectively to increase the soil productivity. This section is discussed in detail as anaerobic and aerobic treatments of anaerobically biodigested distillery effluent (ABDE):

5.3.1 Anaerobic Biodegradation

Anaerobic digestion is widely accepted as the first treatment step in distilleries. Wilkie et al. (2000) have reviewed the role of anaerobic digestion in spent wash treatment. Anaerobic digestion can convert a significant portion (>50%) of the COD to biogas, which may be used as fuel. At present, the anaerobic biodegradation of distillery effluents is widely practiced as an effective step in removing 90% of the COD in the effluent stream (Wolmarans and de Villiers 2002; Chandra and Singh 1999). During this stage, 80–90% BOD removal takes place and 85–90% biochemical energy is recovered as biogas. Common types of anaerobic reactors used for distillery effluent treatment is present in Table 5.4.

Table 5.4 Anaerobic method employed for distillery effluent treatment

Sr. No.	Reactor type	COD removal (%)	BOD removal (%)	References
1.	Downflow fixed-film reactor	60–73	85–97	Bories and Ranyal (1988)
2.	Granular-bed anaerobic baffled reactor (GRABR)	90–96	80–92	Akunna and Clark (2000)
3.	Hybrid anaerobic baffled reactor	70	–	Boopathy and Tilche (1991)
4.	Upflow anaerobic sludge blanket (UASB) reactor	39–67	80	Harada et al. (1996)
5.	Istanbul UASB reactor	90	–	Akarsubasi et al. (2006)
6.	Diphasic fixed-film reactor with granular activated carbon (GAC) as support media	67.1	–	Goyal et al. (1996)
7.	Anaerobic contact filter	73–98	–	Vijayaraghavan and Ramanujam (2000)

COD chemical oxygen demand, *BOD* biochemical oxygen demand

A number of environmental factors affect the activity of wastewater microbial populations and the rate of biochemical reactions. Of particular importance are temperature, pH, nutrients, and inhibiting toxic compounds. Due to acidic nature of spent wash, pH is one of the most relevant factors affecting the microbiological activity in the biological process. The pH of wastewater increases from 4.0 to 7.5 after anaerobic digestion due to the oxidation of organic acids to CO_2 and the reaction between the CO_2 and basic compounds to form carbonates and bicarbonates (Beltran et al. 1999). Because of its high organic load, the distillery wastewater is often diluted with tap water to simulate the concentration of a typical industrial effluent entering a wastewater treatment plant. However, spent wash even after anaerobic treatment does not meet the stringent effluent standards reported by CPCB, India, due to very high levels of BOD, COD, solids, etc. (Asthana et al. 2001).

Also, the secondary spent wash produced by the anaerobically biodigested distillery effluent (ABDE) is darker in color, needing a huge volume of water to dilute it, and is currently used as irrigation water causing gradual soil darkening. The effluent, therefore, is released after diluting with freshwater which is a very dear commodity to industries. Besides, sometimes failure of the anaerobic digestion threatens the discharge limits. To overcome this problem, either a large amount of water is used to dilute the wastewater prior to anaerobic digestion or chemical coagulants are added. These actions require expansion of the anaerobic digester volume, large amounts of water for dilution, and additional costs for coagulants (Kim et al. 1997). Hence aerobic treatment is necessary for anaerobically treated final effluent.

5.3.2 Aerobic Biodegradation

Anaerobically biodigested distillery effluent (ABDE) is subjected to aerobic treatment in activated sludge-type reactors or other advanced bioreactors using mixed or pure culture. Several workers have used pure aerobic bacterial cultures for the treatment of distillery effluent. In recent years, several basidiomycete- and ascomycete-type fungi have been used for the decolorization of natural and synthetic melanoidin of wastewaters from distilleries. The aim of fungal treatment is to purify the effluent by consumption of organic substances, thus, reducing its COD and BOD, and at the same time to obtain some valuable products, such as fungal biomass for protein-rich animal feed or some specific fungal metabolite. Filamentous fungi have lower sensitivity to variations in temperature, pH, nutrients, and aeration and have lower nucleic acid content in the biomass. Several fungi are used for fungal treatment process by various workers (Shukla et al. 2010; Pant and Adholeya 2009; Shayegan et al. 2004; Rajor et al. 2002; Singh and Dikshit 2010).

Bio-treatment with microalgae is particularly attractive because of their photosynthetic capabilities, converting solar energy into useful biomasses and incorporating nutrients such as nitrogen and phosphorus causing eutrophication. This fascinating idea launched some 55 years ago in the United States by Oswald and Gotaas (1957) has since been intensively tested in many countries (De Pauw and Van Vaerenbergh 1983).

Palmer (1974) surveyed microalgal genera from a wide distribution of waste stabilization ponds. In order of abundance and frequency of occurrence, the algae found were *Chlorella*, *Ankistrodesmus*, *Scenedesmus*, *Euglena*, *Chlamydomonas*, *Oscillatoria*, *Micractinium*, and *Golenkinia*. A survey of algal taxa in six-lagoon systems in Central Asia was completed by Erganshev and Tajiev (1986). Their analysis of long-term data revealed that the *Chlorophyta* was dominant both in variety and quantity followed by *Cyanophyta*, *Bacillariophyta*, and *Euglenophyta*. Palmer (1969) listed the algae in the order of their tolerance to organic pollutants as reported by 165 authors. The list was compiled for 60 genera and 80 species. The most tolerant eight genera were found to be *Euglena*, *Oscillatoria*, *Chlamydomonas*, *Scenedesmus*, *Chlorella*, *Nitzschia*, *Navicula*, and *Stigeoclonium*. Since the land-space requirements of microalgal wastewater treatment systems are substantial (De Pauw and Van Vaerenbergh 1983), efforts are being made to develop wastewater treatment systems based on the use of hyperconcentrated algal cultures. This proved to be highly efficient in removing N and P within very short periods of times, e.g., less than 1 h (De la Noue and De Pauw 1988). The algal systems can treat various wastewater like sewage and industrial wastes like distilleries (Kaplan et al. 1988). The bio-treatment of wastewater with algae to remove nutrients such as nitrogen and phosphorus and to provide oxygen for aerobic bacteria was proposed over 50 years ago by (Oswald and Gotaas 1957). Since then, there have been numerous laboratory and pilot studies of this process, and several sewage treatment plants using various versions of this systems have been constructed (Zou et al. 2008).

5.3.3 Challenges for Bioremediation of Anaerobically Biodigested Distillery Effluent (ABDE)

Microorganisms (bacteria/fungi/actinomycetes) due to their inherent capacity to metabolize a variety of complex compounds have been utilized since long back for biodegradation of complex, toxic, and recalcitrant compounds present in various industrial wastes for environmental safety. Microbial degradation and decolorization of industrial wastes are environment-friendly and cost-competitive alternatives to chemical decomposition process of waste minimization (Mohana et al. 2007; Pant and Adholeya 2007). Moreover, the utility of microbes in industrial waste treatment process largely depends on the enzymatic setup, nutrient requirement of microbes, as well as the nature and chemical structure of recalcitrant compounds and environmental conditions (Ohmomo et al. 1988).

The melanoidin and other color-causing compounds present in distillery wastewater, i.e., spent wash or ABDE, have antioxidant properties, which render the activity of conventional micro- and macroorganism and limit the treatment of ABDE by a biological organism. The greatest disadvantage associated with biological treatments, especially aerobic biodegradation, is the higher degree of dilution required to get the desired result.

Batch studies were performed using *Aspergillus niger* and *Phanerochaete chrysosporium* for biodegradation of anaerobically biodigested distillery effluent (ABDE) procured from Lord's Distillery Ltd., Nandganj, Ghazipur. The primary investigation revealed a requirement of excessive dilution resulting in additional wastage of water. The maximum percentage of decolorization thus obtained with *Phanerochaete chrysosporium* and *Aspergillus niger* were 84.0% and 63.5%, respectively, at an excessive dilution of 99% and 120 h of incubation. Similar results using other strains have been reported by several earlier workers (Shukla et al. 2010; Thakkar et al. 2006; Shayegan et al. 2004; Dahiya et al. 2001a, b; Fahy et al. 1997; Miranda et al. 1997). Critical examinations of available literature (Mohana et al. 2008; Mishra and Chaudhary 2007) in the field reveal the effectiveness of coagulation for minimizing the COD and BOD load of ABDE for making it suitable for aerobic biodegradation (APHA et al. 1989).

The ABDE was pretreated with different inorganic coagulant to reduce the BOD, COD, TDS, and color-causing compounds, and the filtrate was suitable for aerobic biodegradation at less dilution, i.e., reduced up to 90% dilution. Three inorganic coagulants were used for the treatment of the sample ABDE, it was observed that potash alum is the best coagulant among the three as shown in Table 5.5. The typical characteristics of ABDE before and after treatment by coagulants are shown in Table 5.6.

In aerobic biodegradation, coagulated ABDE filtrate was used. An experiment was performed at the optimized condition to find the maximum color removal. It was 63.5% after 72–96-h time intervals in aerobic degradation. The total decolorization obtained after complete treatment (fungal and coagulation) was 97.2%, and COD reduction was 91.7%, which indicates coagulation followed by aerobic degradation is a viable option for the treatment of anaerobically biodigested distillery effluent (ABDE) as shown in Table 5.7.

Table 5.5 Percent color, COD, and pH reduction of ABDE at optimum doses of coagulants

Sr. No	Particular's	Optimum dose (g/L)	% color removal	% COD reduction	pH reduction of ABDE
1.	Alum	80	92.45	78.5	4.8
2.	Aluminum chloride	35	74.6	66.8	3.9
3.	Ferric sulfate	30	67.0	59.4	4.2

Table 5.6 Characteristics of ABDE before and after coagulation

Sr. No.	Parameters	ABDE	Treated ABDE with		
			Alum	$AlCl_3$	$FeCl_3$
1.	Chemical oxygen demand	42,500.0	9450.0	14,110.0	17,425.0
2.	Biochemical oxygen demand	6200.0	1705.0	2790.0	3410.0
3.	Color	Blackish brown	Light yellow	Light brown	Greenish brown
4.	pH	7.6	4.8	3.9	4.2
5.	Ca	175.9	38.8	27.28	58.6
6.	K	930.0	860.0	86.0	820.0
7.	Total N	3900.0	1390	2118.0	2245.0
8.	SO_4	3200	1232.0	1650.0	1834.0
9.	Fe	18.0	67.86	8.92	1035.7
10.	PO_4	1650	1043.0	1254.0	1287

All value in mg/L except color and pH

Table 5.7 Typical characteristics of treated distillery effluent

Sr. No.	Parameters	ABDE	ABDE treated with alum	Biodegradation of coagulated ABDE
1.	COD	42,500.0	9450.0	3496
2.	BOD	6200.0	1705.0	494.0
3.	Color	Blackish brown	Light yellow	Light white
4.	pH	7.6	4.8	5.0
5.	Ca	175.9	38.8	18.9
6.	K	930.0	860.0	112.0
7.	Total N	3900.0	1390	635.0
8.	SO_4	3200	1232.0	765.0
9.	Fe	18.0	67.86	37.0
10.	PO_4	1650	1043.0	542.0

All value in mg/L except color and pH

From the above results, it is depicted that biodegradation alone is not satisfactory to treat ABDE to the desired prescribed level of MoEF and CPCB. Excessive dilution limits the scope of biodegradation. So, it is a challenge for environmental engineers to utilize biodegradation as a cost-effective and eco-friendly technique without excessive dilution.

5.4 Future Prospects of Bioremediation in the Context of Distillery Effluent

Combination treatment process (coagulation followed by aerobic biodegradation) is suitable for pretreated ABDE to reach the CPCB standards. Effluent contains large amount of organic and inorganic nutrients. So the addition of microalgae in the aerobic treatment process is a viable option to reduce the concentration at the optimum level prescribed by CPCB with some striking feature, which makes the process more economical and eco-friendly.

Microalgae have several peculiar properties. It is a vital tool in the field of wastewater treatment, particularly organic-rich effluents. Distillery industries employed various treatment techniques to treat effluent, but the most accepted path is anaerobic digestion followed by two-stage aerobic treatment by activated sludge processes. Aerobic treatment requires an aerator which maintains oxygen concentration in the reactor, which triggers the degradation processes by an anaerobic microorganism present in activated sludge. This process requires a huge amount of energy to maintain oxygen concentration. Finally, the effluent after aerobic treatment contains large amount of organic load along with higher nitrogen and phosphorus concentration, which produced as a stable product in the aerobic biodegrading process. It is not thrown outside as such because of strict environmental legislation.

So the application of microalgae in aerobic biodegradation process along with fungi seems to be a viable cost-effective and eco-friendly technique as compared to conventional treatment process because microalgae absorb CO_2 and water in the presence of sunlight and produce useful biomass with oxygen by the process of photosynthesis. These properties of microalgae reduced the cost of conventional treatment in which mechanical aerator is required to maintain oxygen concentration in the reactor. It is estimated that 1 kW of electricity is required to maintain oxygen concentration in the reactor which is sufficient to remove 1 kg f BOD from the reactor. The production of 1 kW of electricity is achieved by burning of fossil fuel which emits 1 kg of CO_2 in the atmosphere. So the conventional aerator system is not a cost-effective and eco-friendly process (Oswald 1988a, b). Nowadays there are a lot of new techniques employed in the industry to reduce and reuse sludge as a raw material in other industries. By this activity, industries tried to reduce the cost of treatment and the pressure on natural resources. Application of microalgae also reduced the sludge problem because it utilized the by-product of aerobic degradation in photosynthesis and other metabolic activities and produced biomass which is used to produce important useful product for humankind like biofuel, edible product, the raw material for industries, etc. Microalgae are considered as the largest carbon dioxide sequester because they absorb a huge amount of CO_2 and produce useful biomass. According to the National Renewable Energy Laboratory (NREL), under controlled environment, algae produced 40 times more biodiesel as compared to terrestrial oil plants (Sheehan et al. 1998; Oswald 1988a, b). Thus algal wastewater treatment is a cost-effective, eco-friendly, and sustainable technique of wastewater treatment along with useful by-products. Algal wastewater treatment also reduced the cost of disinfectant which is commonly used in conventional treatment to harmful pathogens. In carbon-limited condition, microalgae utilized carbonate as a carbon source and liberate hydroxide ions in the solution which increases the pH of the solution up to 9–10. At this pH, most of the pathogens are killed, so the water gets after this treatment pure as compared to conventional treatment.

5.5 Conclusion

Nowadays distilleries are considered as one of the most polluting and growth-oriented industries in the world. Distilleries consumed a huge amount of water in the manufacturing of alcohol and produce a large amount of wastewater which contains high organic load, has low pH, and is dark brown in color. This wastewater alters the physical, chemical, and biological characteristic of water and soil if thrown directly outside without treatment. Alcohol plays an important role in modern societies from ancient times, but now it boosts up by implanting the rule of the government to add 5% alcohol in gasoline. Huge dilution is required to thrown treated effluent from industries to outsides which limits the option for environmental engineers. Water is one of the important natural resources for the survival of the living being on this earth planet. Development of effective treatment plan for distilleries is one of the challenges to environmental engineers. Biodegradation is one of the best techniques to reduce organic load from water bodies, but it has certain limitation. Taking this into consideration, developing an effective treatment plant for distilleries, i.e., coagulation followed by mixed culture aerobic treatment (fungal and algae), seems to be the viable cost-effective and eco-friendly technique.

Acknowledgment The authors would like to acknowledge UGC for granting fund to carry out the research work. Combined laboratory facilities of School of Biochemical Engineering, Department of Chemical Engineering as well as the collaborative and continuous support of the Central University of Jharkhand made the work to reach its present shape.

References

Akarsubasi AT, Ince O, Oz NA, Kirdar B, Ince BK (2006) Evaluation of performance, acetoclastic methanogenic activity and archaeal composition of full-scale UASB reactors treating alcohol distillery wastewaters. Process Biochem 41:28–35

Akunna JC, Clark M (2000) Performance of a granular-bed anaerobic baffled reactor (GRABBR) treating whisky distillery wastewater. Bioresour Technol 74:257–261
APHA-AWWA-WPCF (1989) Standard methods for the examination of water and waste water, 17th edn. Washington, DC
Arora PK, Srivastava A, Singh VP (2014) Bacterial degradation of nitrophenols and their derivatives. J Hazard Mater 266:42–59
Arora PK, Srivastava A, Garg SK, Singh VP (2018) Recent advances in degradation of chloronitrophenols. Bioresour Technol 250C:902–909
Asthana A, Mishra SK, Chandra R, Guru R (2001) Treatment of colour and biochemical oxygen demand of anaerobically treated distillery effluent by aerobic bacterial strains. Indian J Environ Prot 21:1070–1072
Beltran FJ, Garcia-Araya JF, Alvarez PM (1999) Wine distillery wastewater degradation 2. Improvement of aerobic biodegradation by means of an integrated chemical (ozone)-biological treatment. J Agric Food Chem 47:3919–3924
Beltran FJ, Alvarez PM, Rodriguez EM, Garcia-Araya JF, Rivas J (2001) Treatment of high strength distillery wastewater (cherry stillage) by integrated aerobic biological oxidation and ozonation. Biotechnol Prog 17:462–467
Bharagava RN, Saxena G, Mulla SI, Patel DK (2017a) Characterization and identification of recalcitrant organic pollutants (ROPs) in tannery wastewater and its phytotoxicity evaluation for environmental safety. Arch Environ Contam Toxicol. https://doi.org/10.1007/s00244-017-0490-x
Bharagava RN, Saxena G, Chowdhary P (2017b) Constructed wetlands: an emerging phytotechnology for degradation and detoxification of industrial wastewaters. In: Bharagava RN (ed) Environmental pollutants and their bioremediation approaches, 1st edn. CRC Press, Taylor & Francis Group, Boca Raton, pp 397–426. https://doi.org/10.1201/9781315173351-15
Bharagava RN, Chowdhary P, Saxena G (2017c) Bioremediation: an ecosustainable green technology: its applications and limitations. In: Bharagava RN (ed) Environmental pollutants and their bioremediation approaches, 1st edn. CRC Press, Taylor & Francis Group, Boca Raton, pp 1–22. https://doi.org/10.1201/9781315173351-2
Boopathy R, Tilche A (1991) Anaerobic digestion of high strength molasses wastewater using hybrid anaerobic baffled reactor. Water Res 25:785–790
Bories A, Ranyal J (1988) Anaerobic digestion of high-strength distillery wastewater (cane molasses stillage) in a fixed-film reactor. Biol Wastes 23:251–267
Central Pollution Control Board (CPCB) (2003) Environmental management in selected industrial sectors status and needs, PROBES/97/2002-03. CPCB, Ministry of Environment and Forest, New Delhi
Chandra R, Singh H (1999) Chemical decolorization of anaerobically treated distillery effluent. Indian J Environ Protect 19(11):833–837
Chandra R, Saxena G, Kumar V (2015) Phytoremediation of environmental pollutants: an ecosustainable green technology to environmental management. In: Chandra R (ed) Advances in biodegradation and bioremediation of industrial waste, 1st edn. CRC Press, Taylor & Francis Group, Boca Raton, pp 1–30. https://doi.org/10.1201/b18218-2
Cortez LAB, Perez LEB (1997) Experiences on vinasse disposal: Part III: combustion of vinasse fuel oil emulsions. Braz J Chem Eng 14(1):71–77
Dahiya J, Singh D, Nigam P (2001a) Decolourisation of molasses wastewater by cells of *Pseudomonas fluorescens* immobilized on porous cellulose carrier. Bioresour Technol 78:111–114
Dahiya J, Singh D, Nigam P (2001b) Decolourisation of synthetic and spent wash melanoidins using the white-rot fungus Phanerochaete chrysosporium JAG-40. Bioresour Technol 78:95–98
De la Noue J, De Pauw N (1988) The potential of microalgal biotechnology. A review of production and uses of microalgae. Biotechnol Adv 6:725–770
De Pauw N, Van Vaerenbergh E (1983) Microalgal wastewater treatment systems: potentials and limits. In: Ghette PF (ed) Phytodepuration and the employment of the biomass produced. Centro Ric. Produz, Animali, Reggio Emilia, pp 211–287

Erganshev AE, Tajiev SH (1986) Seasonal variations of phytoplankton numbers. Acta Hydrochim Hydrobiol 14:613–625

Fahy VF j, Fitzgibbon G, Mcmullan D, Singh R (1997) Decolourization of molasses spent wash by *Phanerochaete chrysosporium*. Biotech Lett 16:97–99

Fumi MD, Parodi G, Parodi E, Silva A (1995) Optimization of long-term activated-sludge treatment of winery wastewater. Bioresour Technol 52:45–51

Gautam S, Kaithwas G, Bharagava RN, Saxena G (2017) Pollutants in tannery wastewater, pharmacological effects and bioremediation approaches for human health protection and environmental safety. In: Bharagava RN (ed) Environmental pollutants and their bioremediation approaches, 1st edn. CRC Press, Taylor & Francis Group, Boca Raton, pp 369–396. https://doi.org/10.1201/9781315173351-14

Gay M, Cerf O, Davey KR (1996) Significance of pre-incubation temperature and inoculum concentration on subsequent growth of *Listeria monocytogenes* at 14°C. J Appl Bacteriol 81:433–438

Goutam SP, Saxena G, Singh V, Yadav AK, Bharagava RN (2018) Green synthesis of TiO_2 nanoparticles using leaf extract of *Jatropha curcas* L. for photocatalytic degradation of tannery wastewater. Chem Eng J 336:386–396. https://doi.org/10.1016/j.cej.2017.12.029

Goyal SK, Seth R, Handa BK (1996) Diphasic fixed-film biomethanation of distillery spent wash. Bioresour Technol 56:239–244

Harada H, Uemura S, Chen AC, Jayadevan J (1996) Anaerobic treatment of a recalcitrant distillery wastewater by a thermophilic UASB reactor. Bioresour Technol 55:215–221

Jain N, Minocha AK, Verma CL (2002) Degradation of predigested distillery effluent by isolated bacterial strains. Ind J Exp Bot 40:101–105

Kalavathi DF, Uma L, Subramanian G (2001) Degradation and metabolization of the pigment-melanoidin in distillery effluent by the marine *Cyanobacterium oscillatoria boryana* BDU 92181. Enzym Microbiol 29:246–251

Kaplan D, Christiaen D, Arad S (1988) Binding of heavy metals by algal polysaccharides. In: Stadler T, Mollion J, Verdus MC, Karamanos Y, Morvan H, Christiaen D (eds) Algal biotechnology. Elsevier Applied Science, London, pp 179–187

Kim JS, Kim BG, Lee CH, Kim SW, Jee HS, Koh JH, Fane AG (1997) Development of clean technology in alcohol fermentation industry. J Clean Prod 5:263–267

Lincolin EP, Earle JFK (1990) Wastewater treatment with microalgae. In: Akatsuka I (ed) Introduction to applied phycology. SPB Academic Publishing by The Hague, The Netherlands, pp 429–446

Lincoln EP, Hill DT (1980) An integrated microalgae system. In: Shelef G, Soeder CJ (eds) Algae biomass. North Holland Biomedical Press, Amsterdam, pp 229–243

Martins SIFS, Van-Boekel MAJS (2004) A kinetic model for the glucose/glycine Maillard reaction pathways. Food Chem 90(1–2):257–269

Miranda M, Benito G, Cristobal N, Nieto H (1997) Color elimination from molasses wastewater by *Aspergillus niger*. Bioresour Technol 57:229–235

Mishra IM, Chaudhary PK (2007) Decolorization and removal of chemical oxygen demand (COD) with energy recovery: treatment of biodigester effluent of molasses-based alcohol distillery using inorganic coagulant. J Colloids Surf A Physicochem Eng Asp 296:238–247

Mohana S, Desai C, Madamwar D (2007) Biodegradation and decolourization of anaerobically treated distillery spent wash by a novel bacterial consortium. Bioresour Technol 98:333–339

Mohana S, Srivastava S, Divecha J, Madamwar D (2008) Response Surface Methodology for Optimization of Medium for Decolorization of Textile Dye Direct Black 22 by a Novel Bacterial Consortium. Bioresour Technol 99(3):562–569

Mueller JG, Cerniglia CE, Pritchard PH (1996) Bioremediation of environments contaminated by polycyclic aromatic hydrocarbons. In: Bioremediation: principles and applications. Cambridge University Press, Cambridge, pp 125–194

Nandy T, Shastry S, Kaul SN (2002) Wastewater management in a cane molasses distillery involving bioresource recovery. J Environ Manag 65:25–38

Ohmomo S, Daengsubha W, Yoshikawa H, Yui M, Nozaki K, Nakazima T, Nakamura I (1988) Screening of anaerobic bacteria with the ability to decolorize molasses melanoidin. Agric Biol Chem 52:2429–2435

Oswald WJ (1988a) Large-scale algal culture systems (engineering aspects). In: Borowitzka MA, Borowitzka LJ (eds) Micro- algal biotechnology. Cambridge Univ. Press, Cambridge, UK, pp 357–394

Oswald WJ (1988b) Micro-algae and wastewater treatment. In: Borowitzka MA, Borowitzka LJ (eds) Micro-algal biotechnology. Cambridge University Press, Cambridge, pp 305–328

Oswald WJ, Gotaas HB (1957) Photosynthesis in sewage treatment. Trans Am Soc Civil Eng 122:73–105

Palmer CM (1969) A composite rating of algae tolerating organic pollution. J Phycol 5:78–82

Palmer CM (1974) Algae in American sewage stabilization's ponds. Rev Microbiol (S-Paulo) 5:75–80

Pant D, Adholeya A (2007) Biological approaches for treatment of distillery wastewater: a review. Bioresour Technol 98:2321–2334

Pant D, Adholeya A (2009) Nitrogen removal from biomethanated spent wash using Hydroponic treatment followed by fungal decolourization. Environ Eng Sci 26:559–565

Raghukumar C, Mohandass C, Kamat S, Shailaja MS (2004) Simultaneous detoxification and decolorization of molasses spent wash by the immobilized white-rot fungus *Flavodon flavus* isolated from a marine habitat. Enzym Microb Technol 35:197–202

Rajor R, Singh R, Mathur RP (2002) Color removal of distillery waste by *Saccharomyces*. Indian J Environ Protect 22(12):1241–1252

Rivero-Pérez MD, Pérez-Magarin OS, Jose ML (2002) Role of melanoidins in sweet wines. Anal Chim Acta 458(1):169–175 and Technology 29: 246–251

Rodríguez JGO (2000) Effects of vinasse on sugarcane (saccharum officinarum) productivity. Rev Fac Agron Univ Zulia 17:318–326

Sankaran K, Pisharody L, Narayanan GS, Premalatha M (2015) Bacterial assisted treatment of anaerobically digested distillery wastewater. R Soc Chem Adv 5:70977–70984

Satyawali Y, Balakrishnan M (2007) Wastewater treatment in molasses-based alcohol distilleries for COD and colour removal: a review. J Environ Manag 86:481–497

Saxena G, Bharagava RN (2015) Persistent organic pollutants and bacterial communities present during the treatment of tannery wastewater. In: Chandra R (ed) Environmental waste management, 1st edn. CRC Press, Taylor & Francis Group, Boca Raton, pp 217–247. https://doi.org/10.1201/b19243-10

Saxena G, Bharagava RN (2016) Ram Chandra: advances in biodegradation and bioremediation of industrial waste. Clean Techn Environ Policy 18:979–980. https://doi.org/10.1007/s10098-015-1084-9

Saxena G, Bharagava RN (2017) Organic and inorganic pollutants in industrial wastes, their ecotoxicological effects, health hazards and bioremediation approaches. In: Bharagava RN (ed) Environmental pollutants and their bioremediation approaches, 1st edn. CRC Press, Taylor & Francis Group, Boca Raton, pp 23–56. https://doi.org/10.1201/9781315173351-3

Saxena G, Chandra R, Bharagava RN (2016) Environmental pollution, toxicity profile and treatment approaches for tannery wastewater and its chemical pollutants. Rev Environ Contam Toxicol 240:31–69. https://doi.org/10.1007/398_2015_5009

Saxena G, Purchase D, Mulla SI, Saratale GD, Bharagava RN (2018) Phytoremediation of heavy metal-contaminated sites: environmental considerations, field studies, sustainability and future prospects. J Environ Manag

Scott EL, Kootstra AMJ, Sanders JPM (2010) Perspectives on bioenergy and biofuels, O. Singh, S Harvey, Sustainable biotechnology, Springer, Heidelberg, 179–194

Shayegan J, Pazuki M, Afshari A (2004) Continuous decolorization of anaerobically digested distillery wastewater. Process Biochem 40:1323–1329

Sheehan J, Dunahay T, Benemann J, Roessler P (1998) A look back at the U.S. department of energy's aquatic species program- biodiesel from Algae, NERL/TP-580-24190. National Renewable Energy Laboratory, Golden, p 80401

Shukla SK, Mishra PK, Srivastava KK, Srivastava P (2010) Treatment of anaerobically digested distillery effluent by *Aspergillus niger*. IUP J Chem Eng 2:7–18
Singh R (1999) Chemical decolorization of anaerobically treated distillery effluent. Indian J Environ Protect 19:833–837
Singh SS, Dikshit AK (2010) Optimization of the parameters for decolourization by *Aspergillus niger* of anaerobically digested distillery spent wash pretreated with polyaluminium chloride. J Hazard Mater 176:864–869
Singh JS, Abhilash PC, Singh HB, Singh RP, Singh DP (2011) Genetically engineered bacteria: an emerging tool for environmental remediation and future research perspectives. Gene 480:1–9
Stumm W, Morgan JJ (1962) Chemical aspects of coagulation. J Am Water Assoc 54:971
Surega P, Krzywonos M (2015) Screening of medium components and process parameters for sugar beet molasses vinasse decolorization by *Lactobacillus plantarum* using Plackett-Burman experimental design. Pol J Environ Study 24:683–688
Thakkar AP, Dhamankar VS, Kapadnis BP (2006) Biocatalytic decolorization of molasses by *Phanerochaete chrysosporium*. Bioresour Technol 97:1377–1381
Uppal J (2004) Water utilization and effluent treatment in the Indian alcohol industry – an overview. In: Liquid assets, proceedings of Indo- EU workshop on promoting efficient water use in agro-based industries. TERI Press, New Delhi, pp 13–19
Vijayaraghavan K, Ramanujam TK (2000) Performance of anaerobic contact filters in series for treating distillery spent wash. Bioproc Eng 22:109–111
Wilkie AC, Riedesel KJ, Owens JM (2000) Stillage characterization and anaerobic treatment of ethanol stillage from conventional and cellulosic feedstocks. Biomass Bioenergy 19(2):63–102
Wolmarans B, de Villiers GH (2002) Start-up of a UASB effluent treatment plant on distillery wastewater. Water SA 28:63–68
Yeoh BG (1997) Two-phase anaerobic treatment of cane-molasses alcohol stillage. Water Sci Technol 36(6–7):441–448
Zou Y, Xiang C, Yang L, Sun LX, Xu F, Cao Z (2008) A mediatorless microbial fuel cell using polypyrrole coated carbon nanotubes composite as anode material. Int J Hydrogen Energy 33:4856–4862

Chapter 6
Plastic Waste: Environmental Hazards, Its Biodegradation, and Challenges

Kadapakkam Nandabalan Yogalakshmi and Sukhman Singh

Contents

Abstract Plastics have become an indispensable part of the society. Lightweight, easy handling, durability, flexibility, resistance to water, and other microbial attacks have made them ubiquitously popular. The extensive use of the long-lived polymer has confronted the environment with a challenging plastic pollution problem. Plastics are the product of coal, natural gas, crude oil, cellulose, and salt manufactured through energy-intensive technology. From cradle to grave, plastics affect the environment in a multifaceted way. The hazardous and ecologically (terrestrial and marine) damaging threats necessitate its removal from the environment. Incineration, landfilling, recycling, and degradation are the four most available options to manage

---

K. N. Yogalakshmi (✉) · S. Singh
Centre for Environmental Science and Technology, School of Environment and Earth Sciences, Central University of Punjab, Bathinda, Punjab, India

G. Saxena, R. N. Bharagava (eds.), *Bioremediation of Industrial Waste for Environmental Safety*, https://doi.org/10.1007/978-981-13-1891-7_6

the plastic waste. However, to avoid long-term environmental damage, degradation of plastic is the most preferred option among the management options. Plastic degradation is carried out by photodegradation, thermooxidative degradation, hydrolytic degradation, and biodegradation. Among them, microbial degradation using bacteria and fungi is an emerging strategy to manage plastic waste. Hence, this chapter highlights the benefits, concerns, and threats surrounding the use of plastics. The different perspectives covered in this chapter include plastic production and plastic waste generation, environmental and health effects of plastic pollution, plastic waste management options, biodegradation of plastic polymers and the mechanism involved, biodegradable plastics, and challenges and constraints of plastic waste biodegradation.

Keywords Plastics · Environmental effects · Polymers · Biodegradation · Biopolymers

6.1 Introduction

Plastic is one of the commonly used, versatile substances of the twentieth century. The term "plastic" was coined by Baekeland in 1909. It is derived from the Greek word where 'Plastikos' means fit for molding and 'Plastics' means molded. They are synthetic or semisynthetic amorphous substances prepared from fossil fuel resources. They resemble the natural resins obtained from the trees and plants. Since its invention, it has replaced wood, glass, and metal. Though the history of plastics dates back to 100 years, the development of modern polymers started 50 years back. Plastics have almost invaded all the industries, namely, food, clothing, construction, shelter, medical, transportation, sports, and many more. The wide application of the plastics was due to its multiple properties: durability, stability, low cost, lightweight, and unbreakable nature. Plastics are a long chain of monomers prepared by polymerization through addition and condensation process. The process is energy intensive and requires a huge amount of fossil fuels for production. The versatility of the polymer has resulted in the generation of huge amount of plastic waste. The buoyant nature of plastics spreads them to a huge area for a long distance. Moreover, the material from which they are made of resists them from degrading in the environment. Resistance and nonbiodegradability make plastic waste management very critical. The plastics remain in perpetuity when disposed of in landfills or open dumps.

Plastic waste has started attracting public attention due to the various environmental and health hazards. They are hazardous to terrestrial and aquatic life. The marine environment is largely affected by the plastics. The persistent organic pollutants that are coated on the plastics might enter into the food chain and cause harm to the animals and human beings. Animals also die due to indigestion and stomach bloating problems when plastic was consumed along with food. Marine animals die of suffocation when they get tangled with plastics. Dioxins released during the burning of plastics lead to cancer. The deterioration caused by the plastic pollution necessitates its removal from the environment. Microbial degradation can be one best option for its removal from the environment. Plastics are complex organic compounds and can be mineralized by microorganisms, unlike heavy metals. Henceforth, studies on plastic degradation using bacteria, fungi, and actinomycetes have generated a lot of interest among researchers.

Bacteria, fungi, and actinomycetes isolated from varied source, namely, soil, wastewater, sludge, mangrove soil, landfill soil, and plastic-accumulated soil, were efficient in degrading the natural and synthetic polymers. The microorganisms use intracellular and extracellular enzymes to cleave the polymer chain to oligomers, subsequent to which they are mineralized by aerobic and anaerobic metabolism. Weight loss, change in the structure of the polymers, and cracks in the polymers indicated the efficiency of microorganisms in degrading the polymers. However, the process of biodegradation was limited and challenged by factors such as polymer structure, composition, chemical bonding, stress, and environmental conditions (pH, temperature). Pretreatment of polymer with light, heat, oxygen, and UV enhances the biodegradation of polymers. This chapter will through light on plastic production, plastic waste generation, and its environmental and health effects. Biodegradation of plastic waste, factors influencing its degradation, and the various challenges associated are also discussed.

6.2 Plastic Production

Plastics are polymers that are manufactured chemically by polymerization of monomers of various synthetic and semisynthetic organic compounds. Monomers are a group of atoms that constitute unit cells. The monomers are made of elements such as carbon, hydrogen, sulfur, silicon, nitrogen, and chlorine. According to the type of polymers, the composition of monomers shall vary. For, example, polyethylene polymer consists of repeated units of carbon and hydrogen atoms. Other monomers include ethylene (ethene), propylene (propene), butene, and others. The monomers possess a double bond between carbon atoms such that it can react to form polymers. Polymerization of monomers occurs through two mechanisms: addition reaction and condensation. A catalyst is used to link one monomer to the other in an addition reaction, while in condensation reaction, the catalyst is used to fuse all monomers to adjacent monomers. This results in the formation of dimers and a by-product. The by-product has to be removed to avoid the release of environmental contaminants (American Chemistry Council 2005).

At the industrial level, the production of plastic began in the1940s. Presently, huge quantity of plastics is manufactured annually due to their application in the household, packing, agriculture, toys, and many more (Al-Salem et al. 2009). In the 1950s, the annual production of plastics was 1.5 million tones which have increased dramatically to 299 million tons in 2013 (Plastics Europe 2015). Natural products such as cellulose, natural gas, salt, coal, and petroleum derived from crude oil are used as polymers for the synthesis of plastics. Today, around 4% of crude oil is used as a feedstock for manufacturing plastics, and additional 4% of petrochemical products are used in energy-intensive process for plastic prototype or refined goods (Wu et al. 2013). During the distillation process of crude oil, different hydrocarbon products get separated into different fractions. One of such fraction includes naphtha which is a very crucial element for the production of plastics (Kalia et al. 2000). The most abundant plastic polymers produced include high-density polyethylene (HDPE), low-density polyethylene (LDPE), polystyrene (PS), expanded polystyrene (EPS), polypropylene (PP), and polyvinyl chloride (PVC) (Li et al. 2016). In order to improve the performance of plastics, various additives along with polymer resins are mixed during polymerization of plastic polymers. These include inorganic fillers, phthalate plasticizers, bisphenol A (BPA), thermal and UV stabilizers, colorings, carbon, silica, and flame retardants which provide various range of plastic products (Thompson et al. 2009). Generally, plastics are divided into two basic categories: (i) thermo-softening plastics and (ii) thermosetting plastics. Thermo-softening plastics are those which can be recycled and remolded into different shapes, whereas thermosetting plastics cannot be recycled and remolded (Singh et al. 2017). The polymers are processed into finished products through a different process. Extrusion, blow molding, film blowing, calendaring, expanded bead blowing, injection and rotational molding, casting, thermoforming, and compression folding are finishing steps in plastic production (American Chemistry Council 2005).

6.3 Plastic Waste Generation

Rapid growth and urbanization have led to the increased demand for plastics around the globe due to its multiple advantages such as lightweight, strength, durability, and low cost. Plastic waste generation depends on increasing population, lifestyle, socioeconomic background, and developmental activities. Since 1990 the demand for plastics has increased at an annual rate of 5% (Kunwar et al. 2016). Increase in plastic production leads to higher consumption which in turn results in more plastic waste generation. Plastic is used widely in various sectors such as household, agriculture, packaging, and industrial areas. So, the plastic waste generated can be categorized into municipal and industrial plastic waste. Industrial waste is homogenous which get generated during manufacturing of plastic products, packaging, and processing. Municipal waste is heterogeneous waste which includes plastic bags, wrappers, toys, wires, vending cups, CD, disposable cups, and few more. They are discarded as household waste (Panda et al. 2010). Some plastics products are prepared for single use which gets disposed off. According to Indian government's Plastic Waste Management Rules (2016), it is estimated that everyday approximately 15,000 tons of plastic waste is generated, out of which 60% is collected. Globally, around 32 billion tons of plastic waste is generated every day, out of which 65% of total plastic waste generated ends life in landfills. As mentioned earlier, plastics are categorized into two types: thermoset plastics and thermoplastics. Thermoplastics are recyclable plastics and they constitute 80% of plastic waste. The

Table 6.1 Resins used in manufacturing of plastics

Sr. no.	Thermoplastic	Thermoset plastic
1.	Polyethylene terephthalate (PET)	Bakelite
2.	Polypropylene (PP)	Epoxy
3.	Polyvinyl acetate (PVA)	Melamine
4.	Polyvinyl chloride (PVC)	Polyester
5.	Polystyrene (PS)	Polyurethane
6.	Low-density polyethylene (LDPE)	Urea-formaldehyde
7.	High-density polyethylene (HDPE)	

Source: CPCB (2011)

Table 6.2 Status of plastic waste generation

Sr. no.	Description	World	India
1.	Per capita per year consumption of plastic (kg)	24	6–7
2.	Recycling (%)	15–20	60
3.	Plastics in solid waste (%)	7	9

Source: Atulesh (2017)

resins produced for the preparation of varied types of plastics are summarized in Table 6.1. Plastic waste does not degrade unlike other municipal wastes such as paper, organic waste, etc. because it accumulates and remains there for longer periods of time (Sarker et al. 2011). Moreover, lack of awareness and unavailability of effective collection mechanism plastic waste generation are increasing day by day. In India, the per capita plastic waste generation is estimated as 5.7 kg/annum.

Recycling is the only means by which plastic waste generation is reduced or minimized. According to reports, the generation of plastic waste in India is comparatively lesser than the rest of the world (Table 6.2). Moreover, the recycling rate is also higher in India (60%) than the rest of the world (15–20%). The recycling is done only for a specific group of plastic. The mechanical recycling is done only for plastic bottles and PVCs. Polythene bag recycling is still facing issues. Segregation and cleaning are other factors that affect recycling.

6.4 Environmental Hazards of Plastic Waste

Plastics are extensively scattered in the general public and the environment due to their diverse chemical composition, properties, and applications. The higher the utilization, the more will be the generation of waste, and in case of plastics which are persistent in nature, the waste generation would be much higher (Thompson et al. 2009). Aesthetic problems get originated due to plastic debris. Plastics are very hazardous for both terrestrial and marine life. Conventional plastics are non-biodegradable and require very long time for degradation. They also show high resistance to aging, get converted to smaller debris, and accumulate in the environment (O'Brine and

Thompson 2010). Hazardous by-products or unbound monomers from plastics get released into the aquatic environment through the migration of plastic waste and landfill leachate to both freshwater and marine waters (Huang et al. 2012). In mid-oceans, plastic bags resemble prey to fishes. Fishes, mammals, turtles, and other sea animals consume these plastics along with their food, which gets accumulated in the gut of the sea animals. The plastics in the gut damage the inner lining of the stomach, block the digestive tract, and minimize the feed drive of animals (Laist 1987). Moore (2008) reported that around 267 marine species including mammals, fishes, turtles, seabirds, and invertebrates are affected by plastic waste. Marine animals get tangled in floating plastic debris which inhibits or limits their movement resulting in their drowning. Evidence of plastic debris in the guts of fishes have been reported by many researchers. Fragments of plastic enter the filter feeding organisms through which it enters the food chain and bioaccumulates in higher marine organisms. Plastics also destroy the habitat by acting as transportation agent for invasive species (Fergusson 1974). Microorganisms and other diatoms start growing over the plastics and create biofouling over the plastic material. Biofouling makes the plastic heavy and they eventually sink to the floor of the water body. Due to the buildup of plastic waste on the sea floor, hypoxia conditions may develop due to inhibition of gas exchange between pore waters of accumulated plastic debris and overlying waters. Hypoxia condition results in improper functioning of the normal marine ecosystem (Derraik 2002). Plastics floating in water suffocate marine animals to death. Plastic ingestion could increase the buoyancy of fish making it difficult for mesopelagic fish to return to deeper waters (Boerger et al. 2010).

A number of chemicals used for the manufacturing of plastics, unbounded monomers, additives, and their degradation products may get released into the environment during the lifecycle of plastic product which may have hazardous properties for biological life and environment. Bisphenol A (BPA), phthalates, polyhalogenated flame retardants, polyfluorinated compounds, nonylphenol, and antibacterial compounds such as triclosan pose serious threats to human health and other organisms (Halden 2010). Polybrominated diphenyl ethers (PBDEs) used in electrical and some non-electrical plastics as fire retardants are persistent in nature. In May 2009, Stockholm convention declared penta- and octa-BDE as persistent organic pollutants (POPs). The United States and the European Union have banned the use of these PBDEs (Tang et al. 2014). Bisphenol A and phthalates are of major concern due to their biological effects in humans as well as animals.

6.4.1 Bisphenol A (BPA)

Bisphenol A (BPA) is a monomer chemical that is used for the manufacturing of polycarbonate plastics and PVC and can leach into the environment (Schecter et al. 2010). During polymerization, some of the monomers remain unbound which get released and are commonly found in the leachate of the landfill where plastic waste is dumped (Halden 2010). It potentially results in ecosystem disruption of the region receiving leachates. Each year about 100 tons of BPA is released into the environment during synthesis alone (Vandenberg et al. 2009). BPA is of great concern because it poses some serious threats to humans. It has been recognized as potential endocrine disruptor which disrupts the normal hormonal functioning. Various other health effects associated with BPA are cardiovascular disease, heart attacks, male sexual dysfunction, and angina (Schecter et al. 2010).

6.4.2 Phthalates

Phthalates are diesters of phthalic acid used for flexibility of PVC, pliability, softening, and elasticity of rigid plastics. More than 25 different phthalate esters are used in plastic manufacturing. Phthalates do not bound covalently to polymer matrix due to which these are more prone to leaching into the environment (Halden 2010). Di(2-ethylhexyl) phthalate (DEHP) is associated with various health concerns for humans. On entering human body, DEHP is converted to mono-(2-ethylhexyl) phthalate (MEHP) by lipases and gets absorbed by the body (Singh and Li 2012). Dibutyl phthalate (DBP), benzyl butyl phthalate (BBP), di-isononyl phthalate (DINP), etc. are other important phthalate additives. Dibutyl phthalate and benzyl butyl phthalate produce toxic metabolite mono-butyl phthalate (Sathyanarayana 2008; Latini 2005). Adverse health effects of phthalates include hormone disruption, asthma, genital malformations, development of some cancers, and developmental and reproductive problems (Singh and Li 2012).

6.4.3 Land Pollution

The plastics when dumped in the open areas or landfills affect the aesthetics of the location. The wind carries the plastics and spreads them over a wide area. Improper segregation of plastics from other food wastes results in consumption of plastic by street animals. The animals die due to bloating of stomach and indigestion problems. In many places, especially in open dumpsites and landfills, the rag pickers set afire in the waste piles to segregate the materials. This results in burning of plastics along with other materials. According to NEERI (2010), burning of plastics results in the emission of 10,000 g of dioxins/furans every year to the atmosphere. Incomplete combustion of plastics results in the release of carbon monoxide (CO), dioxins, furans, polychlorinated biphenyls, soot, carbon black, aromatics (pyrene and chrysene), and heavy metals (chromium, copper, cobalt, selenium, lead, and cadmium). The burning of plastic waste causes headache, nausea, heart diseases, respiratory illness like asthma and emphysema, and neurological, respiratory, and reproductive diseases. It also damages the lungs, thyroid gland, eyes, and mucous membrane. Dioxins and its intermediate 2,3,7,8 tetrachlorodibenzo-*p*-dioxin (TCDD) are categorized under carcinogens and mutagens. Heavy metals released during burning interfere with plant process and accumulate in the tissues of organisms. Chlorine released from polyvinyl chlorides forms hydrochloric acid. Hydrochloric acid increases the acidity of the soil in addition to respiratory diseases (Verma et al. 2016).

6.5 Plastic Waste Management

Natural degradation or decomposition of plastics is very slow and takes about hundreds of years to get mineralized. Slow degradation has increased or necessitated the management of plastic waste throughout the world (Puncochar et al. 2012). Integrated waste management approaches should be followed effectively for better plastic waste management. Conventional approaches along with new technological methods are being used for better management of plastic waste which includes landfilling, incineration, recycling, and degradation (Panda et al. 2010). The conventional and most common method for managing waste is landfill. But there are certain long-term problems associated with this method of plastic waste management such as breakdown products of plastics, and leachate produced during degradation may pollute soil and groundwater (Hopewell et al. 2009). Plastic has high volume to weight ratio and, therefore, requires more space. Nowadays constructing a suitable landfill is costly, and space scarcity for landfill site is a major concern due to which other unconventional methods of plastic waste management are required to reduce the ultimate volume of plastic waste reaching landfill (Panda et al. 2010). In order to effectively manage or reduce plastic waste quantity, three Rs, reduce, reuse, and recycle, strategy should be opted (Hopewell et al. 2009). Reusing plastics is the best approach for managing plastic waste as it requires less energy. Plastic waste recycling is the process of recovering plastics. Recycling of plastic waste can be categorized into four major types: re-extrusion, mechanical recycling, chemical recycling, and energy recovery (Al-Salem et al. 2009). Re-extrusion or primary recycling is the use of plastic scrap for the manufacturing of new product both having similar features. Mechanical recycling is the process of recovering plastic waste that has been used in the manufacturing of new plastic products. Mechanical recycling involves various treatment and preparation steps such as shredding or cutting, floating, milling, washing, drying, etc. to yield high-quality end products (Singh et al. 2017). Heterogeneous plastic waste recycling is very difficult or cannot be performed in mechanical recycling which limits its usage to only homogenous or single polymer plastic waste. Chemical recycling also is known as tertiary recycling, where the plastics are depolymerized into smaller polymers due to alteration in the chemical structure of plastics. They are further used as feedstock for manufacturing of new plastics (Arena and Mastellone 1999). Chemical recycling is attained by three main approaches: depolymerization, partial oxidation, and cracking. Cracking is further categorized into thermal cracking, catalytic cracking, and hydrocracking (Panda et al. 2010). Energy recovery or quaternary recycling includes thermal decomposition, pyrolysis, gasification, and incineration of plastic waste.

Cost-effective and adequate infrastructure is required for the plastic waste collection, makeover to new product, and marketing of recycled goods for the success of plastic waste recycling (Rebeiz and Craft 1995). Incineration is another way of managing plastic waste which significantly reduces the need for landfilling of plastics. In this process the waste material is exposed to high temperatures (900–2000 °C) which leads to controlled and complete combustion of plastics and energy recovery, thereby reducing the volume by 80–90% (Sharholy et al. 2008). Generation of greenhouse gases and other toxic gases such as polychlorinated dibenzo-para-dioxins and polychlorinated dibenzofurans during incineration has led to its restriction in many countries especially for plastic waste (Panda et al. 2010).

Other than the conventional processes, some new technologies such as liquid fuel, plasma pyrolysis, coprocessing in cement kiln, and polymer-blended bitumen roads are being efficiently used in plastic waste management (Siddique et al. 2008). Thermal decomposition of plastics is done to convert plastics into liquid fuel produced through condensation of volatile liquid vapors, carbonized char, and synthetic gas. This thermal cracking or pyrolysis of plastic waste is done at 2700–3000 °C temperature by both non-catalytic and catalytic methods (Almeida and Marques 2016; Kumar and Singh 2013). Few studies have reported the thermal decomposition of polyethylene, polypropylene, polymethyl methacrylate, polyurethane, and polystyrene to liquid fuel. Gasoline-range liquid fuels having low octane value have been produced through thermal decomposition. The fuel produced requires further refining for usage which adds extra cost (Panda et al. 2010). Plasma pyrolysis technology is a fresh innovative technology that has been used to transform waste into valuable high calorific synthetic gas (syngas). In plasma pyrolysis, the high calorific plastic waste is exposed to very high temperature under oxygen-deficient conditions. High temperature decomposes plastics into syngas, which is used for energy production. Tar is produced along with syngas which is composed of CO, H_2, and low concentration of hydrocarbons, but the syngas produced in plasma pyrolysis has high heating value and low tar content in comparison with simple pyrolysis of plastic waste (Puncochar et al. 2012; Nema and Ganeshprasad 2002). Tang et al. (2003) used nitrogen plasma reactor for decomposition of polypropylene, which resulted in 96% of solid decomposition and 2160 mL/g of syngas production. Coprocessing means the use of waste material as a source of energy along with primary fuel. Plastic waste is used as alternative fuel in kilns to reduce the consumption of fossil fuel used in cement industry. Plastics used as fuel in coprocessing are known as an alternative fuel. Reduction in plastic waste volume along with energy recovery took place in coprocessing (De Queiroz Lamas et al. 2013). Another technology that has gained popularity in plastic waste management is polymer-based bitumen roads. In this, the plastic waste shredded into small pieces was blended with bitumen and used as alternative road material (Gawande et al. 2012). In Bangalore, India, the bitumen blended with shredded plastic waste was used for laying of the road (Khullar 2009).

Degradation is the best way of managing plastic waste as it provides a solution for longer periods of time. Any alteration in the chemical structure of a plastic polymer due to a physical, chemical, or biological constituent that leads to transformation is degradation of plastics. Physical, mechanical, and chemical properties of plastics get changed due to degradation (Pospisil and Nespurek 1997). Photo, thermal, and biological factors contribute to the degradation of plastics. Plastics absorb UV radiations from the sun which leads to photodegradation and further thermodegradation of plastics. Degradations begin with the integration of oxygen atom to plastic polymers. Both the processes make polymer delicate and break into smaller pieces until microbes start metabolizing. Microbes such as bacteria and fungi first transform low molecular weight polymers to monomers, and then microbial cells transform monomers to minerals. Four methods, photodegradation, thermooxidative degradation, hydrolytic degradation, and biodegradation by microorganisms, lead to complete degradation of plastics in sequence (Shah et al. 2008a, b; Singh and Sharma 2008). Biodegradation of plastics is far better than other plastic waste management processes as it is environment-friendly and completely mineralizes or transforms plastic waste, thus providing a sustainable solution to plastic waste.

6.6 Biodegradation of Natural and Synthetic Plastics

Bioremediation is the use of living organisms such as bacteria and plants or their enzymes to degrade and detoxify the environmental contaminants for environmental safety and public health protection (Saxena et al. 2018; Bharagava et al. 2017a, b; Saxena and Bharagava 2016, 2017; Chandra et al. 2015). In an environment, depending on the factors, plastic undergoes either abiotic or biotic degradation. Biotic degradation or biodegradation is a process by which the microorganisms bring about a change in the physical or chemical structure of the material (Paco et al. 2017). Bacteria, fungi, and actinomycetes are involved in the degradation of the polymers. Biodegradation results in the decomposition products such as carbon dioxide, methane, water, and inorganic compounds (nitrogen, sulfur, phosphorus). The microorganisms can perform degradation under both aerobic and anaerobic condition (Alshehrei 2017). The degree of degradation by the microorganisms depends on the molecular weight of the polymers. The higher the molecular weight, the lesser the solubility of the polymers. The microorganisms find it difficult to attack, break down, and metabolize the polymers of increased thickness. Precisely, the mechanism the microorganism follows to degrade the polymers includes the following:

1. The microorganisms secrete extracellular enzymes to breakdown the complex polymers into small oligomers, dimers, or monomers. This process is also termed as depolymerization. If the polymer is hydrophilic, then the microorganism attaches itself to the polymer, secretes the enzyme, and cleaves the polymers via hydrolysis (Schink et al. 1992). During hydrolysis, a nucleophilic attack occurs on the carbonyl carbon atom. The cleavage occurs at ester, peptide linkage, and glycosidic bonds.

2. The degraded substance is water soluble and can easily pass through the semi-permeable cell membrane.
3. The monomers on entering into the cell are used as a source of carbon and energy. The biodegradation process is favored by three processes such as hydrolysis, photooxidation, and physical disintegration. These processes weaken the surface of the polymer and decrease the molecular weight of the polymer (Muthukumar and Veerappapillai 2015).

6.6.1 Biodegradation of Plastics by Bacteria

Numerous bacteria have shown the ability to degrade natural and synthetic polymers (Table 6.3). In addition to degradation, bacteria also synthesize natural biopolymers. Polyhydroxyalkanoates (PHAs), a natural polymer, are synthesized during bacterial fermentation. Under starvation or nutrient-limited conditions, the PHAs are metabolized and degraded. Another important natural polymer that is produced from the low-cost feedstock is polyhydroxybutyrate (PHB). Aerobic and anaerobic bacteria isolated from different sources, namely, activated sludge, soil, anaerobic sludge, and seawater, were used for natural polymer degradation. *Pseudomonas lemoignei*, *Comamonas* spp., *Variovorax paradoxus*, *Streptoverticillium kashmeriense* AF1, *Bacillus megaterium* AF3, and *Ilyobacter delafieldii* are some of the strains reported for PHA and PHB degradation (Lee 1996a, b; Shah et al. 2007). Molitoris et al. (1996) and Sang et al. (2002) observed grooves and erosion pits on the polymer subjected to bacterial degradation.

Biodegradation of polyethylene glycol (PEG) has been widely studied due to multiple reasons such as huge application, long biodegradation time, fear of entry into the environment, water solubility, and availability in different molecular weights (200–50000 Åg/mol). Aerobic and anaerobic bacteria have been used for PEG degradation. Inoculum such as aerobic and anaerobic sludge, seawater, municipal sewage, and wetland sediments was used for PEG degradation. However, seawater showed just 40% PEG degradation. The degradation pathway followed by the bacteria to degrade PEG differed with the environment. *Desulfovibrio desulfuricans DG2*, *Bacteroides* sp., *Pseudonocardia* sp. strain K1, *Sphingomonas*, *Acetobacterium*, and *Pseudomonas* sp. were among those bacteria reported for PEG degradation (Kawai et al. 1978; Dwyer and Tiedje 1983; Zgola-Grzeskowiak et al. 2006). According to Kawai et al. (1978), PEG degradation was better when the synergistic mixed culture of *Flavobacterium* and *Pseudomonas* sp. was used. PEGs pretreated with ozone showed accelerated degradation (Suzuki et al. 1978).

Recently, a novel bacteria *Ideonella sakaiensis* 201-F6 isolated from the debris of PET bottle recycling site were used to degrade poly(ethylene terephthalate) (PET) polymer popularly used in the manufacturing of bottles. The bacteria degraded 75% of PET film at 30 °C within 6 weeks of incubation. Carbon dioxide evolution was observed during degradation which indicated the mineralization of the polymer (Yoshida et al. 2016). Polythene bags and plastic cups were buried in

Table 6.3 Bacteria degrading different types of plastics

S. No.	Polymers		Microorganisms	References
1.	Polyethylene	LDPE	*Rhodococcus ruber* C208	Chandra and Rustgi (1997)
			Pseudomonas stutzeri	Sharma and Sharma (2004)
			Brevibacillus borstelensis	Hadad et al. (2005)
			Rhodococcus ruber C208	Sivan et al. (2006) and Gilan et al. (2004)
			Bacillus sp., *Micrococcus* sp., *Listeria* sp. and *Vibrio* sp.	Kumar et al. (2007)
			Rhodococcus rhodochrous ATCC 29672	Fontanella et al. (2009)
			Staphylococcus epidermidis	Chatterjee et al. (2010)
			Bacillus cereus CI	Balasubramanian et al. (2011)
			Bacillus cereus, *Bacillus megaterium*, *Bacillus subtilis* and *Brevibacillus borstelensis*	Abrusci et al. (2011)
			Pseudomonas aeruginosa PAO1 (ATCC 15729), *Pseudomonas aeruginosa* (ATCC 15692), *Pseudomonas putida* (KT2440 ATCC 47054) and *Pseudomonas syringae* (DC3000 ATCC 10862)	Kyaw et al. (2012)
			Stenotrophomonas pavanii	Mehmood et al. (2016)
			Aneurinibacillus aneurinilyticus btDSCE01, *Brevibacillus agri* btDSCE02, *Brevibacillus* sp. btDSCE03 and *Brevibacillus brevis* btDSCE04	Skariyachan et al. (2018)
		LDPE powder	*Pseudomonas* sp., *Staphylococcus* sp. and *Bacillus* sp.	Usha et al. (2011)
		LLDPE	*Rhodococcus rhodochrous* ATCC 29672	Fontanella et al. (2009)
			Bacillus cereus, *Bacillus megaterium*, *Bacillus subtilis* and *Brevibacillus borstelensis*	Abrusci et al. (2011)
			Serratia marcescens	Odusanya et al. (2013)
		HDPE	*Bacillus* sp., *Micrococcus* sp., *Listeria* sp. and *Vibrio* sp.	Kumar et al. (2007)

			Rhodococcus rhodochrous ATCC 29672	Fontanella et al. (2009)
			Arthrobacter sp. and *Pseudomonas* sp.	Balasubramanian et al. (2010)
			Aneurinibacillus aneurinilyticus btDSCE01, *Brevibacillus agri* btDSCE02, *Brevibacillus* sp. btDSCE03 and *Brevibacillus brevis* btDSCE04	Skariyachan et al. (2018)
		Degradable polyethylene	*Rhodococcus rhodochrous* ATCC 29672 and *Nocardia steroids* GK911	Bonhomme et al. (2003)
			Bacillus mycoides	Seneviratne et al. (2006)
		Polyethylene bags and plastic cups	*Streptococcus* sp., *Staphylococcus* sp., *Micrococcus* sp., *Moraxella* sp. and *Pseudomonas* sp.	Kathiresan (2003)
		Polyethylene carry bags	*Bacillus cereus*, *Pseudomonas* sp.	Aswale and Ade (2008)
			Serratia marcescens	Aswale and Ade (2009)
			Serratia marcescens 724, *Bacillus cereus*, *Pseudomonas aeruginosa*, *Streptococcus aureus* (B-324) and *Micrococcus lylae* (B-429), *Arthrobacter* sp. and *Pseudomonas* sp.	Aswale (2010)
			Pseudomonas aeruginosa, *Pseudomonas putida* and *Bacillus subtilis*	Obidi et al. (2010)
		Polyethylene and cups	*Bacillus* sp., *Staphylococcus* sp., *Streptococcus* sp., *Diplococcus*., *Micrococcus* sp., *Pseudomonas* sp. and *Moraxella* sp.	Reddy (2008)
		Natural and synthetic polyethylene	*Pseudomonas* sp. (P1, P2 and P3)	Nanda et al. (2010)
		Polyethylene	*Brevibacillus*, *Pseudomonas* and *Rhodococcus* spp.	Nanda and Sahu (2010)
			Bacillus subtilis	Vimala and Mathew (2016)
		Polyethylene terephthalate (PET)	*Serratia plymuthica*	Janczak et al. (2018)

(continued)

Table 6.3 (continued)

S. No.	Polymers		Microorganisms	References
2.	Polythene		*Pseudomonas* sp.	Sinosh et al. (2015)
			Staphylococcus sp., *Bacillus* sp., *Pseudomonas* sp.	Singh et al. (2016)
3.	Polyurethane		*Aeromonas salmonicida*; *Alcaligenes denitrificans*	Kay et al. (1991) and Akutsu et al. (1998)
			Arthrobacter globiformis	
			Comamonas acidovorans TB-35	
			Corynebacterium sp.; *Enterobacter agglomerans*; *Pseudomonas aeruginosa*; *Serratia rubidaea*	
			Comamonas acidovorans	Nakajima-Kambe et al. (1995)
			Acinetobacter calcoaceticus; *Arthrobacter globiformis*; *Pseudomonas cepacia*; *Pseudomonas* sp.	Halim et al. (1996)
			Methanotrix sp.	Varesche et al. (1997)
			Bacillus sp.	Blake and Howard (1998)
			Pseudomonas fluorescens	Howard and Blake (1999)
			Pseudomonas chlororaphis	Howard and Hilliard (1999)
			Pseudomonas chlororaphis	Zheng et al. (2005)
			Pseudomonas sp.	
			Rhodococcus equi	Akutsu-Shigeno et al. (2006)
			Alicycliphilus spp.; *Alicycliphilus denitrificans*	Oceguera-Cervantes et al. (2007)
			Bacillus pumilus	Nair and Kumar (2007)
			Corynebacterium sp.	Shah et al. (2008a)
			Micrococcus sp.	Shah et al. (2008b)
			Acinetobacter gerneri P7	Howard et al. (2012)
			P. Aeruginosa strain MZA-85	Shah et al. (2013)
			Pseudomonas putida	Peng et al. (2014)
			Bacillus subtilis	Nakkabi et al. (2015)
		Polyurethane film	*Bacillus* sp. AF8, *Pseudomonas* sp. AF9, *Micrococcus* sp. 10, *Arthrobacter* sp. AF11 and *Corynebacterium* sp. AF12	Shah et al. (2008a)

4.	Polypropylene	Isotactic polypropylene	*Pseudomonas chlororaphis*, *Pseudomonas stulzeri* and *Vibrio* sp.	Cacciari et al. (1993)
		Polypropylene films and pellets	*Aneurinibacillus aneurinilyticus* btDSCE01, *Brevibacillus agri* btDSCE02, *Brevibacillus* sp. btDSCE03 and *Brevibacillus brevis* btDSCE04	Skariyachan et al. (2018)
5.	Polyvinyl chloride	PVC powder	*Pseudomonas aeruginosa*	Peciulyte (2002)
		PVC	*Pseudomonas fluorescens* B-22, *P. putida* AJ, *P. chlororaphis*, *Ochrobactrum* TD, *Aspergillus niger*	Dussud and Ghiglione (2014) and Shah et al. (2008a)
6.	Polylactic acid		*Penicillium roqueforti*, *Amycolatopsis* sp., *Bacillus brevis*, *Rhizopus delemar*	Shah et al. (2008a)
			Serratia plymuthica	Janczak et al. (2018)
7.	Polystyrene		*Rhodococcus zopfii* stoecker, *Enterococcus faecalis*, *Pseudomonas putida* and *Salmonella*	Savoldelli et al. (2017)
8.	Polyesteramides		*Pseudomonas aeruginosa*	Novotný et al. (2015)
9.	Plastic bags		*Aneurinibacillus migulanus*	Chaisu (2016)
10.	Degradable plastic		*Pseudomonas* sp., *Bacillus subtilis*, *Staphylococcus aureus*, *Streptococcus lactis*, *Proteus vulgaris* and *Micrococcus luteus*	Priyanka and Archana (2011)
11.	Milk covers		*Pseudomonas putida*	Saminathan et al. (2014)
12.	UV-treated microplastics from polyethylene (PE), polyethylene terephthalate (PET), polypropylene (PP) and polystyrene (PS)		*Bacillus cereus* and *Bacillus gottheilii*	Auta et al. (2017)

Source: Adapted and updated from Devi et al. (2016); Pathak and Navneet (2017) mangrove soil to assess in situ degradation. It was observed that in a span of 9 months, polythene bags and plastic cups showed 4.21% and 0.25% degradation. The inoculum showed the presence of gram-positive and gram-negative bacteria. The species that were predominant were *Streptococcus*, *Staphylococcus*, *Micrococcus* (Gram-positive), *Moraxella*, and *Pseudomonas* (Gram-negative). *Bacillus*, *Listeria*, and *Vibrio* are bacteria observed in mangrove soil of Fiji in addition to the abovementioned bacteria. The bacteria generated hydrolytic enzymes such as amylases, lipases, and gelatinases for high-density and low-density polythene (Kumar et al. 2007). Garbage and garden soil, when used as an inoculum, showed 13.6% weight loss of polythene. *Staphylococcus arlettae* that was responsible for the weight loss altered the chemical properties of the polythene (Divyalakshmi and Subhashini 2016). *Staphylococcus aureus* and *Pseudomonas aeruginosa* also showed polythene and plastic degradation when incubated for a period of 1 month (Kannahi and Sudha 2013).

Synthetic polymers such as polyester, polycaprolactone, and polyurethane were also degraded by bacteria isolated from soil, activated sludge, and other sources. The bacteria utilized polymer as carbon and nitrogen source. *Comamonas acidovorans* and *Alcaligenes faecalis* degraded polyester polyurethane and polycaprolactone (Nakajima-Kambe et al. 1999; Oda et al. 1997). *Alcaligenes faecalis* produced polycaprolactone depolymerase enzyme to degrade polycaprolactone.

6.6.2 Biodegradation of Plastics by Fungi

Fungi like other microorganism have gained huge attention among researchers due to its biodegradation potential. The degradation ability of fungi is attributed to the enzymes they produce. Moreover, the hyphal structure of fungi through which it can penetrate into the contaminated sites has made it more advantageous for bacteria. Fungi are explored for the degradation of different types of natural and synthetic polymers such as low-density polyethylene, high-density polyethylene, polyvinyl chloride, polyurethane, polyolefins, poly(vinyl alcohol), and polyhydroxyalkanoates (Table 6.4). Irrespective of the type of polymer, genera *Aspergillus*, *Penicillium*, and *Phanerochaete* are popularly reported for plastic degradation. Polyvinyl chloride is a strong plastic used in pipes and fittings, flooring material, synthetic leathers, insulation wires, textiles, shoes, etc. It has low moisture and resists biodegradation. However, thermal and photodegradation are effective for PVC degradation. In fact, they serve as pretreatment, thereby facilitating better biodegradation by fungi. White rot fungi can degrade low molecular weight PVCs. According to Kirbas et al. (1999), Kitamoto (2008), and Ali et al. (2014), *Phanerochaete chrysosporium*, *Lentinus* spp., *Aspergillus niger*, *Penicillium janthinellum*, and *Doratomyces* spp. can degrade PVC. Many yeast strains, namely, *Rhodotorula aurantiaca* and *Kluyveromyces* spp., and other fungi *Aureobasidium pullulans* also degraded PVCs (Webb et al. 2000).

Table 6.4 Fungi involved in plastic degradation

S. No.	Polymers		Microorganisms	References
1.	Polyethylene	Polyethylene	*Penicillium simplicissimum*	Yamada et al. (2001)
			Brevibacillus borstelensis and *Rhodococcus ruber*	Hadad et al. (2005)
			Trichoderma harzianum	Sowmya et al. (2014)
		Polyethylene terephthalate (PET)	*Laccaria laccata*	Janczak et al. (2018)
		Polyethylene carry bags	*Aspergillus niger*	Obidi et al. (2010)
			Phanerochaete chrysosporium, *Pleurotus ostreatus*, *Aspergillus niger* and *Aspergillus glaucus*	Aswale (2010)
			Aspergillus niger	Aswale and Ade (2008)
		Polyethylene bags and plastic cups	*Aspergillus niger* and *Aspergillus glaucus*	Kathiresan (2003)
			Aspergillus niger, *Aspergillus ornatus*, *Aspergillus nidulans*, *Aspergillus cremeus*, *Aspergillus flavus*, *Aspergillus candidus* and *Aspergillus glaucus*	Reddy (2008)
		LDPE	*Penicillium pinophilum* and *Aspergillus niger*	Volke-Sepulveda et al. (2002)
			Aspergillus niger, *Penicillium funiculosum*, *Chaetomium globosum*, *Gliocladium virens* and *Pullularia pullulans*	Gilan et al. (2004)
			Fusarium sp. AF4	Shah et al. (2009)
			Aspergillus oryzae	Konduri et al. (2011)
			Mucor circinelloides and Aspergillus flavus	Pramila and Vijaya Ramesh (2011)
			Aspergillus sp.	Raaman et al. (2012)
			Aspergillus spp.	Kumar et al. (2013)
			Rhizopus oryzae NS 5	Awasthi et al. (2017)
			Penicillium oxalicum NS4 (KU559906) and *Penicillium chrysogenum* NS10	Ojha et al. (2017)

(continued)

Table 6.4 (continued)

S. No.	Polymers		Microorganisms	References
		LDPE powder	*Aspergillus nidulans* and *Aspergillus flavus*	Usha et al. (2011)
			Aspergillus versicolor and *Aspergillus* sp.	Pramila and Vijaya Ramesh (2011)
		HDPE	*Phanerochaete chrysosporium* ME-446 and *Trametes versicolor* (IFO 7043, IFO 15413)	Liyoshi et al. (1998)
			Aspergillus niger, *Aspergillus oryzae* and *Aspergillus flavus*	Konduri et al. (2010)
			Aspergillus terreus MF12	Balasubramanian et al. (2014)
			Penicillium oxalicum NS4 (KU559906) *and Penicillium chrysogenum* NS10	Ojha et al. (2017)
		Degradable polyethylene	*Cladosporium cladosporioides*	Bonhomme et al. (2003)
			Penicillium frequentans	Seneviratne et al. (2006)
2.	Polyurethane		*Chaetomium globosum* and *Aspergillus terreus*	Boubendir (1993)
			Curvularia senegalensis, *Fusarium solani*, *Aureobasidium pullulans* and *Cladosporium* sp.	Crabbe et al. (1994)
3.	PVC		*Polyporus versicolor*, *Pleurotus sajor-caju*, *Phanerochaete chrysosporium* ME 446, *Pleurotus ostreatus*, *Pleurotus sapidus*, *Pleurotus eryngii*, *Pleurotus florida*	Kirbas et al. (1999)
		PVC powder	*Aureobasidium pullulans*, *Geotrichum candidum*, *Alternaria alternaria*, *Cladosporium* sp., *Aspergillus* sp., *Penicillium* sp. and *Ulocladium atrum*	Peciulyte (2002)
		PVC films	*Phanerochaete chrysosporium* PV1, *Lentinus tigrinus* PV2, *Aspergillus niger* PV3 and *Aspergillus sydowii* PV4	Ali et al. (2014)
		Plasticized PVC	*Aureobasidium pullulans*, *Rhodotorula aurantiaca* and *Kluyveromyces* spp.	Webb et al. (2000)
			Aspergillus versicolor, *Aspergillus niger*, *Aspergillus flavus*, *Chrysonilia sitophila* and *Penicillin* sp.	Sachin and Mishra (2013)

4.	Polyesteramides		Yeast *Candida guilliermondii* and micromycete *Aspergillus fumigatus*	Novotný et al. (2015)
5.	Disposable plastic films		*Aspergillus flavus* and *Mucor rouxii* NRRL 1835	El-Shafei et al. (1998)
6.	Degradable plastic	Degradable plastic	*Phanerochaete chrysosporium*	Lee et al. (1991)
			Aspergillus niger, *Aspergillus nidulans*, *Aspergillus lavus*, *Aspergillus glaucus* and *Penicillium* sp.	Priyanka and Archana (2011)

Source: Adapted and updated from Devi et al. (2016); Pathak and Navneet (2017)

Four fungal species, namely, *Curvularia senegalensis*, *Fusarium solani*, *Aureobasidium pullulans*, and *Cladosporium* sp., are soil fungi with potential to degrade ester-based polyurethane. These fungi produce extracellular enzymes such as ureases, proteases, and esterases in the presence of polyurethane. The release of ammonia is indicated in the degradation of polyurethane (Pathirana and Seal 1985; Howard et al. 2012). According to Cosgrove et al. (2007), *Geomyces pannorum* and *Phoma* species dominated the soil involved in polyurethane degradation.

Polyethylene is hydrophobic high molecular weight polymers which are difficult to degrade by fungi. The low-density polyethylene is degraded by genus *Penicillium*, *Aspergillus*, *Chaetomium*, *Pullularia*, and *Gliocladium*. Likewise, the high-density polymers are degraded by *Trametes versicolor* and *Phanerochaete chrysosporium* using lignocellulosic enzymes (Iiyoshi et al. (1998). Around 0.2% loss in weight of polymers was observed in 10 years. To facilitate better degradation, the high molecular weight polyethylene is pretreated with UV light and nitric acid. Pretreatment weakens the structure of polymers making it easy for fungi to act on the polymers. The biodegradation of polymers can be confirmed by the presence of oxidation products. The structural changes can be confirmed with an electron microscope and FTIR (Hasan et al. 2007). According to Friedrich et al. (2007), polyamide-6 also termed as nylon-6 is also degraded by white rot fungi and other species (*Fusarium* spp., *Phanerochaete chrysosporium*, *Aspergillus* spp., *Penicillium* spp., *Cladosporium* spp., *Ulocladium* spp., *Trichoderma* spp., *Gliocladium roseum*, *Pithomyces chartarum*, *Trichothecium roseum*, and *Mucor hiemalis*). Aliphatic polyester resin Bionolle and aliphatic-aromatic co-polyester Ecoflex (BASF) degradation were also assessed. No significant weight loss was observed in the polymer except for structural changes (279). *Phanerochaete chrysosporium* and *Talaromyces wortmannii* are also reported to degrade natural polymer poly(hydroxybutyrate-hydroxyvalerate) which is a biodegradable and biocompatible thermoplastic (Coelho et al. 2008). Polyhydroxyalkanoates are another natural polymer that showed 93% weight loss when inoculated with *Aspergillus fumigatus* (Bhatt et al. 2008). Other natural polymers that were degraded by fungi strains include poly(caprolactone) (PCL), poly(butylene succinate) (PBS), poly(butylene succinate and adipate) (PBSA), and poly(lactide) (PLA), poly(butylene succinate), and poly(alkaline succinate). The hydrolytic degradation products obtained during polymer degradation did not affect the diversity of other organisms in the soil. Incidence of marine and mangrove bacteria accumulating polyhydroxyalkanoates on the midwest coast of India has been reported by Rawte et al. (2002).

6.6.3 Biodegradation of Plastics by Actinomycetes

Actinomycetes are organisms that possess characteristics of a bacteria and fungi. These unicellular organisms with nonseptate mycelium possess the ability to degrade plastics. Table 6.5 summarizes the various actinomycetes species involved in plastic biodegradation. Species of genera *Streptomyces* and *Acinetobacter* are

Table 6.5 Actinomycetes and plastic degradation

S. No.	Types of plastic used		Microorganisms	References
1.	Polyethylene	Polyethylene	Streptomyces sp.	Méndez et al. (2007)
			Streptomyces sp., *Pseudonocardia*, *Actinoplanes*, *Sporichthya*	Sathya et al. (2012)
		Starch polyethylene-prooxidant degradable polyethylene	*Streptomyces badius* 252	Lee et al. (1991)
			Streptomyces setonii 75Vi2 and *Streptomyces viridosporus* T7A	Pometto et al. (1992)
		LPDE powder	*Streptomyces* KU8, *Streptomyces* KU5, *Streptomyces* KU1 and *Streptomyces* KU6	Usha et al. (2011)
			Streptomyces sp.	Madhuri (2015)
2.	Polyurethane		*Acinetobacter calcoaceticus*	Halim et al. (1996)
			Acinetobacter gerneri P7	Howard et al. (2012)
3.	Disposable plastic films		*Streptomyces* spp. (8 strains)	El-Shafei et al. (1998)
4.	Poly(tetramethylene succinate) (PTMS)		*Microbispora rosea*, *Excellospora japonica* and *E. viridilutea*	Jarerat and Tokiwa (2001)

Source: Adapted and modified from Devi et al. (2016)

popularly reported to degrade polyethylene, polyurethane, plastic films, and LPDE powder. Other genera such as *Actinoplanes*, *Pseudonocardia*, and *Sporichthya* are also used in plastic degradation. Similar to bacteria, the actinomycetes too attacked the polymers using enzymes. The change in the polymer structure and weight confirms the polymer degradation by actinomycetes.

6.6.4 Factors Affecting Biodegradation of Plastics

The process of biodegradation is associated to certain drawbacks and is influenced by various factors. Some of them are discussed below.

6.6.4.1 Anaerobic/Aerobic Condition

The process of biodegradation is greatly influenced by the presence or absence of oxygen. Aerobic condition is more favorable and efficient than anaerobic condition. The biodegradation under aerobic condition was found to be faster than anaerobic condition. Moreover, the aerobic process released only CO_2, while anaerobic process produced more harmful CH_4 and CO_2. Microbial degradation of plastics is carried anaerobically in the landfills and results in the generation of methane having global warming potential of 56 in 20 years. According to Kumar et al. (2011), aerobic condition is elemental for the biodegradation of plastics by fungi.

6.6.4.2 Chemical Composition of the Plastics

The chemical composition of plastics is crucial in deciding the fate of plastic degradation. The microbes decomposing the plastics feed and thrive on it by consuming it as nutrient source; therefore, starch-based polymers are preferred by the microbes as the hydrolytic enzymes released by them act on the polymer matrix and hence reduce plastic into simpler form. The polymers derived from flax fibers or starch is inclined to relatively easy biodegradation. Polymers based on petroleum such as polyolefins are photodegraded by microbes (Sen and Raut 2015). Molecular weight of the polymers is one of the important factors which determine the biodegradation. Low molecular weight is suitable for biodegradation as the rate of enzymatic hydrolysis is rapid in case of smaller molecular weight (Tokiwa and Suzuki 1977). The melting point of the polymer too influences the enzymatic biodegradability. A polymer with higher melting temperature tends to have lower biodegradability. The enzymatic activity is also observed to decrease with time.

6.6.4.3 Pretreatment

Nowadays, during manufacturing or processing of polymers/biopolymers, some additives such as benzophenone are used to avoid or minimize the photodegradation of the plastics. These additives affect the thermal sensitivity and UV absorption capacity of the polymer. It is observed that chemically/thermally sensitive polymers have better biodegradation rate. The thermal exposure prior to subjecting microbes to the plastics leads to the breakdown of the polymer into simpler form on which bacteria can easily attack. An alkaline substance known as nodax is also added to alter the chemical structure of the polymer to make it available for the microbes (Augusta et al. 1993). According to Moore (2008), the natural conditions such as the hydrolytic properties of sea water, atmosphere's oxidative properties and solar radiation help in breaking polymers into fragile and simpler substrate for the microbes.

6.6.4.4 Intrinsic Parameters

Factors such as temperature, pH, humidity, sunlight, water stress, culture conditions, and sunlight predominantly affect microbial community and enzymatic activity for polymer degradation (Gu 2003). Soil bacteria are known to reduce plastics in the presence of optimal water and temperature. Fungus when grown at lowest pH showed maximum CO_2 evolution and efficient lignolytic activity (Glass and Swift 1990).

6.6.4.5 Microbial Population

The quality and quantity of microbial population determine the efficiency of biodegradation. Irrespective of the source, many bacteria have shown their potential in plastic degradation. Bacteria, algae, and fungi isolated from various resources such as plastic and soil at the dumping sites, rhizosphere soil of mangroves, marine water, polythene buried in the soil have been reported to degrade plastics (Rutkowska et al. 2002). Low density polythene was found to be effectively degraded using *Mucor circinelloides* and *Aspergillus flavus* isolated from municipal landfill area (Pramila and Vijaya Ramesh 2011). The microbial population use the plastics as a source of energy and carbon. Fuhs in 1961 reported several microorganisms that fed on paraffin as a carbon source. Jen-hou and Schwartz (1961) showed bacteria consumed polyethylene as carbon source.

6.7 Alternate to Plastics

6.7.1 Biodegradable Plastics

The invention of plastic has been a boon to the population to lead a sophisticated life. Excessive usage, improper disposal, and poor management practices have necessitated the synthesis of alternate, eco-friendly biodegradable or bio-based plastics (BDPs) that should possess properties of conventional plastic with a high degree of biodegradability. Such plastics are known as bioplastics which are synthesized from the polymers derived from renewable carbon sources. According to the literature, bioplastics consist of either biodegradable plastics (i.e., synthesized using fossil materials) or bio-based plastics (i.e., plastics synthesized from renewable resources or biomass) (Tokiwa et al. 2009). Biodegradable plastics or bioplastics are nonpersistent in the environment and get degraded naturally on exposure to sunlight, air, and microorganisms. Under aerobic conditions, they are degraded to CO_2, inorganic compounds, and H_2O and under anaerobic conditions to CH_4, CO_2, and H_2O (Lee 1996a, b). The principal mechanism involved in degradation of bioplastics includes the enzymatic action of microorganisms. Renewable raw materials such as agricultural/animal resources like cellulose, starch, collagen, casein, soy protein polyesters, and triglycerides are used for the production of bioplastics. The composition of the final product and the chemical structure of the material used decide the biodegradability of plastics. The bioplastics are manufactured using natural or synthetic resins. Natural biodegradable plastics can be naturally created or either synthesized from renewable resources. Synthetic biodegradable plastics are a petroleum-based nonrenewable resource. Naturally biodegradable plastics are unified with synthetic polymers to produce plastics to meet the properties of the conventional plastics (Demirbas 2007).

The different types of biodegradable or bioplastics are listed below. Some of the polymers used as bioplastics include polylactic acid (PLA), poly-β-hydroxybutyric acid (PHB), and polyhydroxyalkanoate (PHA). They are produced using natural polymers, polymers derived from microbes, and conventional chemical or fermentation process for bio-based monomer polymerization (Harding et al. 2007; Kalia et al. 2000).

Plant-based bioplastics: Bioplastics are derived from plant raw materials such as corn, soy, and potato starch. They are degraded by composting and more than 60% of it gets degraded within 180 days. Biodegradation of these plastics requires water, air, and sunlight. These plastics meet the American society for testing and material standards. The biggest disadvantage of the plant-based bioplastics is that they are not economical and cannot be recycled. Spudware is one example of plant-based bioplastics prepared from potato starch and used in the manufacturing of spoons, forks, and plates.

Oil-based bioplastics: They are manufactured from nonrenewable sources such as crude oil. Oil-based bioplastics are produced using energy-intensive techniques.

Thermal-based film: These are bioplastics with a specific additive which helps the plastic to degrade when exposed to high temperature. They include two types of film plastics, namely, oxo- and hydro-biodegradable. The medium of destruction differentiates oxo- and hydro-biodegradable plastics. The former requires oxygen for degradation and the latter requires moisture. Hydro-biodegradable plastics degrade much faster than oxo-degradable plastics. Oxo-biodegradable plastics are less expensive. The major disadvantage of these plastics is they start to degrade when stored in hot place.

Additive-based film combo: It is another type of film with additives. During degradation, they convert into humus and then to carbon dioxide and water. The advantage of these types of plastic is they can degrade with and without oxygen, can be recycled, and possess a long expiration date, for example, green film.

BAK is a 100% biodegradable polyester amide plastic that is developed by Bayers. They are recyclable and possess excellent properties. According to reports, these degrade at a temperature as low as 0.5 °C. Moisture is another important factor for degradation of BAK. Preservation is required for BAK which is the major drawback of the product.

Degradable polyethylene: These plastics are manufactured with the purpose of degrading them within 60 days to 5 years of incubation according to the requirement. They are degraded by light and moisture. Within 55 days of incubation, it gets converted to mulch. The additive used in the plastic is less toxic and can be safely disposed off in landfills.

The waste management option for the bioplastics or biodegradable plastics is domestic and municipal composting (Davis and Song 2006). Anaerobic digestion is also popularly used in bioplastic degradation (Murphy and Bartle 2004). Under mesophilic conditions, the material gets degraded within a maximum period of 24 months (Briassoulis and Dejean 2010). The advantage of biodegradable plastics over conventional plastics is that carbon emissions are eight times lesser for biodegradable plastics. Conventional plastic manufacturing process requires more energy in comparison with bioplastics. Bioplastics can be decomposed by composting on a regular basis, and the compost can be used as valuable soil additive (Song et al. 2009).

6.7.2 *Compostable Plastics*

Compostable plastics are new generation plastics that are degraded by composting. Cornstarch, potato starch, cellulose, and soy resin are the polymers used for the production of compostable plastics. They possess most of the properties of conventional plastics such as tensile strength, oxygen barrier, and resistance to heat and impact or shock. Compostable plastics are degraded in a composting facility within 90–180 days. However, the thickness and composition of the compostable plastics determine the degradation efficiency and time. A compostable plastic should possess three characteristics, namely, biodegradability, disintegration, and toxicity. The compostable plastic should not produce any toxic intermediates.

6.8 Challenges in Plastic Degradation

Most of the polymers are biologically inert and recalcitrant due to high molecular weight, hydrophobic nature, and absence of functional groups which are compatible with the microbial enzymatic system. Degradation of the polymer is characterized by discoloration, phase separation, cracking, erosion, and delamination. One of the biggest challenges remains high molecular weight of the synthetic polymers as the rate of biodegradability is directly dependent on the molecular weight of the polymer. The other limiting factors which need to be explored are breakage of bonds, transformation due to chemicals, and synthesis of new functional groups which are responsible for the variation in the structure of the polymer to make it easy for degradation. Moreover, since biodegradation involves microorganisms, the type and nature of organisms, availability, size of the inoculum, and environmental conditions are also important challenges that need more attention in polymer biodegradation. The challenges of the polymer biodegradation are discussed below in detail:

(a) Chemical composition: The chemical composition of the polymers plays a crucial role in polymer degradation. The long carbon chain in the thermoplastic makes the polymers resistant to biodegradation. The strength of the polymer chain is affected by assimilation with heterogroups such as O_2. Unsaturated portions and amorphous regions are susceptible to oxidation. The rate of oxidation depends on the reactivity of the peroxy radicals formed and on the dissociation energies of available carbon-hydrogen bonds in the polymer substrate.
(b) Molecular weight: Molecular weight determines the rate of degradation of plastics. The higher the molecular weight, the lower the rate of degradation. It has been reported that some microorganisms utilize polyolefins with low molecular weight faster as compared to high molecular weight polyolefins. Lower molecular weight polymers support the growth of microorganisms.
(c) Hydrophobicity: Hydrophobicity interferes with the formation of microbial biofilm, and hence degradation gets limited.

(d) Size of the molecules: Degradation (mechanical, thermal, and biological) is influenced by the size of the molecules. The smaller the size of the molecule, the higher the rate of degradation.
(e) Introduction of functionality: Addition of carbonyl groups or chromophores to the polymers increases its photodegradation capacity. This might be due to the availability of more sites to absorb photons to initiate degradation. The carbonyl chromophore absorbs near-UV radiation and forms radicals by the Norrish Types I and II and H-abstraction processes for photochemical degradation. Reagents based on sulfuric acid help in slow sulfonation and oxidative degradation of polyolefins. In certain cases, starch is used to modify the polymer. Addition of starch helps in bacterial and fungal degradations in the soil.
(f) Additives: Addition of fillers, pigments, photostabilizers, and other polymerization catalysts makes the polymer susceptible to degradation. Lignocellulosic fillers and metals decrease the thermal stability of the polymer. Heating in the presence of oxygen changes the properties of the polymer via oxidation. The extent of oxidation determines the structure and chemical properties of the polymer.
(g) Chemical bonding: Linkage between the monomer units determines the resistance of the plastics. During polymerization, the head-to-head addition of monomer units and the tail-to-tail addition of monomer units create weaker portions of the polymer favoring degradation. Likewise branching increases degradation, and cross-linking reduces the rate of degradation. Cross-linking locks the polymer structure and prevents lamellar unfolding.
(h) Methods of synthesis: Polymer preparation method too determines the rate of biodegradation.
(i) Effect of substituents: The addition of substituents in the polymer backbone influences the stability of the plastics. For example, hydrogen, phenyl group, and hydrochloric acid affect the thermal stability, while fluorine and aromatic compounds increase the stability of polymers.
(j) Effect of stress: Polymers face two types of stress, namely, tensile and compressional stress. The former favors degradation and the latter retards it. A structural or morphological change is observed in the polymers exposed to stress. Stressed polymers are more susceptible to degradation. During stress, polymer strains are straightened, stretched, and cleaved by light.
(k) Environmental conditions: Polymer biodegradation depends upon environmental conditions such as moisture, temperature, oxygen, and a suitable population of microorganisms. In warm climates when the relative humidity exceeds above 70%, the rate of polymer degradation by the microorganisms increases. High temperature and high humidity enhance hydrolytic degradation of the polymers. Moisture and temperature do some primary degradation and weaken the polymer, thereby facilitating the microbial attack. Briefly, the photostabilizers become active when exposed to light and leach out of the polymer matrix in the presence of moisture. After a preliminary breakage in the polymer chain, it is easy for the microorganisms to attack the polymer chain. Tropical regions are more favorable for degradation. It is a general rule that at an increase of temperature by 10 °C, the rate of the reaction doubles. Oxygen in the air due to its scavenging property also facilitates polymer attack at ambient temperature.

6.9 Conclusion

Plastics are ubiquitous due to their multiple advantages. They are part of our day today's life. Improper management and disposal practices have ended up in the accumulation of plastic waste in the terrestrial and aquatic environment. The harmful effects caused by plastic waste accretion are drastically increasing day by day. Henceforth, their removal from the environment is very vital. Among thermal, photooxidative, ozone-induced, mechanochemical, and catalytic degradation methods, biodegradation is considered as the best option for plastic waste degradation due to low cost and eco-friendly nature. Also, they do not release any secondary pollutants. The microorganisms (bacteria, fungi, and actinomycetes) degrade the polymers through secretion of extracellular and intracellular enzymes. The complexity of the polymer limited the degradation by the microorganisms. Polymer pretreatment enhanced the biodegradation efficiency of the microorganisms. Henceforth, plastic biodegradation can be successful when biodegradation is combined with pretreatment. Further, bioprospecting of the isolated microorganism is essential to obtain efficient plastic-degrading bacteria. Although biodegradation can be one of the solutions, still more research is required to develop biodegradable alternate plastics that can degrade naturally without causing any harm to the environment.

6.10 Future Outlook

As mentioned earlier, plastics have become an indispensable part of our society. It is highly impossible to ban it from the society. Henceforth, to minimize the problems of conventional plastics, it is vital for us to focus on the preparation of polymers from natural sources. Further, the modifications or additives used in synthetic polymers to strengthen them can be minimized or replaced with alternate sources that can cause minimum damage to the environment. In recent times, the focus is also shifted to the preparation of bioplastics from waste materials such as organic wastes, agricultural wastes, paper wastes, and many more. Reusability can be another best option for minimizing plastic pollution. Hence, more research should be focused on the preparation of durable materials that can be reused again and again.

References

Abrusci C, Pablos JL, Corrales T, López-Marín J, Marín I, Catalina F (2011) Biodegradation of photo-degraded mulching films based on polyethylenes and Â stearates of calcium and iron as pro-oxidant additives. Int Biodeter Biodegr 65(3):451–459

Akutsu Y, Nakajima-Kambe T, Nomura N, Nakahara T (1998) Purification and properties of a polyester polyurethane-degrading enzyme from *Comamonas acidovorans* TB-35. Appl Environ Microbiol 64:62–67

Akutsu-Shigeno Y, Adachi Y, Yamada C, Toyoshima K, Nomura N, Uchiyama H, Nakajima-Kambe T (2006) Isolation of a bacterium that degrades urethane compounds and characterization of its urethane hydrolase. Appl Microbiol Biotechnol 70(4):422–429

Ali MI, Ahmed S, Robson G, Javed I, Ali N, Atiq N, Hameed A (2014) Isolation and molecular characterization of polyvinyl chloride (PVC) plastic degrading fungal isolates. J Basic Microbiol 54(1):18–27

Almeida D, Marques MDF (2016) Thermal and catalytic pyrolysis of plastic waste. Polímeros 26(1):44–51

Alshehrei F (2017) Biodegradation of synthetic and natural plastic by microorganisms. Journal of Applied and Environmental Microbiology 5:8–19

Al-Salem SM, Lettieri P, Baeyens J (2009) Recycling and recovery routes of plastic solid waste (PSW): a review. Waste Manag 29(10):2625–2643

American Chemistry Council (2005) Plastics industry producer statistics group

Arena U, Mastellone ML (1999) Particle agglomeration during energy recovery from plastic wastes by means of fluidized bed reactors (No. CONF-990534--). Univ. Federico II of Naples, Napoli

Ashwin Kumar A, Karthick K, Arumugam KP (2011) Biodegradable polymers and its applications. Int J Biosci Biochem Bioinformatics 1(3):173–176

Association of Plastic Manufacturers Europe (2015) An analysis of European plastics production, demand and waste data. European Association of Plastics Recycling and Recovery Organisations, Belgium, pp 1–32

Aswale P (2010) Studies on bio-degradation of polythene. PhD thesis, Dr. Babasaheb Ambedkar Marathwada University, Aurangabad, India

Aswale P, Ade A (2008) Assessment of the biodegradation of polythene. Bioinfolet 5:239–245

Aswale PN, Ade AB (2009) Effect of pH on biodegradation of polythene by Serretia marscence. The Ecotech 1:152–153

Atulesh (2017) http://www.wealthywaste.com/plastic-consumption-in-india

Augusta J, Müller R-J, Widdecke H (1993) A rapid evaluation plate-test for the biodegradability of plastics. Appl Microbiol Biotechnol 39(4–5):673–678

Auta HS, Emenike CU, Fauziah SH (2017) Screening of Bacillus strains isolated from mangrove ecosystems in peninsular Malaysia for microplastic degradation. Environ Pollut 231:1552–1559

Awasthi S, Srivastava N, Singh T, Tiwary D, Mishra PK (2017) Biodegradation of thermally treated low density polyethylene by fungus *Rhizopus oryzae* NS 5. 3 Biotech 7(1)

Balasubramanian V, Natarajan K, Hemambika B, Ramesh N, Sumathi CS, Kottaimuthu R, Rajesh Kannan V (2010) High-density polyethylene (HDPE)-degrading potential bacteria from marine ecosystem of gulf of Mannar, India. Lett Appl Microbiol 51:205–211

Balasubramanian S, Maruthamuthu S, Alika K, Palanisamy N, Muralidharan VS, Ragunathan R, Kannan M, Navaneetha Pandiyaraj K (2011) Influence of thermal oxidation on surface and thermo-mechanical properties of polyethylene. J Polym Res 18(6):2175–2184

Balasubramanian V, Natarajan K, Rajesh Kannan V, Perumal P (2014) Enhancement of in vitro high-density polyethylene (HDPE) degradation by physical, chemical and biological treatments. Environ Sci Pollut Res 21:12,549–12,562

Bharagava RN, Saxena G, Chowdhary P (2017a) Constructed wetlands: an emerging phytotechnology for degradation and detoxification of industrial wastewaters. In: Bharagava RN (ed) Environmental pollutants and their bioremediation approaches, 1st edn. CRC Press, Taylor & Francis Group, Boca Raton, pp 397–426. https://doi.org/10.1201/9781315173351-15

Bharagava RN, Chowdhary P, Saxena G (2017b) Bioremediation: an ecosustainable green technology: its applications and limitations. In: Bharagava RN (ed) Environmental pollutants and their bioremediation approaches, 1st edn. CRC Press, Taylor & Francis Group, Boca Raton, pp 1–22. https://doi.org/10.1201/9781315173351-2

Bhatt R, Panchal B, Patel K, Sinha VK, Trivedi U (2008) Synthesis, characterization, and biodegradation of carboxymethylchitosan-g-medium chain length polyhydroxyalkanoates. J Appl Polym Sci 110(2):975–982

Blake RC, Howard GT (1998) Adhesion and growth of a Bacillus sp. on a polyesterurethane. Int Biodeterior Biodegrad 42:63–73
Boerger CM, Lattin GL, Moore SL, Moore CJ (2010) Plastic ingestion by planktivorous fishes in the North Pacific Central Gyre. Mar Pollut Bull 60(12):2275–2278
Bonhomme S, Cuer A, Delort A-M, Lemaire J, Sancelme M, Scott G (2003) Environmental biodegradation of polyethylene. Polym Degrad Stab 81(3):441–452
Boubendir A (1993) Purification and biochemical evaluation of polyurethane degrading enzymes of fungal origin. Dissertation Abstract International 53:4632
Briassoulis D, Dejean C (2010) Critical review of norms and standards for biodegradable agricultural plastics part I biodegradation in soil. J Polym Environ 18:384–400
Cacciari I, Quatrini P, Zirletta G, Mincione E, Vinciguerra V, Lupatelli P, Sermanni GG (1993) Isotactic polypropylene biodegradation by a microbial community: physicochemical characterization of metabolites produced. Appl Environ Microbiol 59(11):3695–3700
Central Pollution Control Board (2011) Annual report on plastic waste management. http://cpcb.nic.in/displaypdf
Chandra R, Rustgi R (1997) Biodegradation of maleated linear low-density polyethylene and starch blends. Polym Degrad Stab 56(2):185–202
Chandra R, Saxena G, Kumar V (2015) Phytoremediation of environmental pollutants: an eco-sustainable green technology to environmental management. In: Chandra R (ed) Advances in biodegradation and bioremediation of industrial waste, 1st edn. CRC Press, Taylor & Francis Group, Boca Raton, pp 1–30. https://doi.org/10.1201/b18218-2
Chatterjee S, Roy B, Roy D, Banerjee R (2010) Enzyme-mediated biodegradation of heat treated commercial polyethylene by staphylococcal species. Polym Degrad Stab 95(2):195–200
Coelho NS, Almeida Y, Vinhas GM (2008) The biodegradation of polyhydroxybutyrate-co-valerate/amphiprotic starch in the presence of microorganisms. Polímeros 18(3):270–276
Cosgrove L, McGeechan PL, Robson GD, Handley PS (2007) Fungal communities associated with degradation of polyester polyurethane in soil. Appl Environ Microbiol 73(18):5817–5824
Crabbe JR, Campbell JR, Thompson L, Walz SL, Schultz WW (1994) Biodegradation of a colloidal ester-based polyurethane by soil fungi. Int Biodeter Biodegr 33(2):103–113
Davis G, Song JH (2006) Biodegradable packaging based on raw materials from crops and their impact on waste management. Ind Crop Prod 23(2):147–161
De Queiroz Lamas W, Palau JCF, de Camargo JR (2013) Waste materials co-processing in cement industry: ecological efficiency of waste reuse. Renew Sust Energ Rev 19:200–207
Demirbas A (2007) Biodegradable plastics from renewable resources. Energy Sources, Part A 29(5):419–424
Derraik JG (2002) The pollution of the marine environment by plastic debris: a review. Mar Pollut Bull 44(9):842–852
Devi RS, Kannan VR, Natarajan K, Nivas D, Kannan K, Chandru S, Antony AR (2016) The role of microbes in plastic degradation. In: Environmental waste management. Taylor and francis publication, Boca Raton
Divyalakshmi S, Subhashini A (2016) Screening and isolation of polyethylene degrading bacteria from various soil environments. IOSR J Environ Sci Toxicol Food Technol 10(12):1–7
Dussud C, Ghiglione JF (2014) Bacterial degradation of synthetic plastics. In CIESM Workshop Monogr (No. 46)
Dwyer DF, Tiedje JM (1983) Degradation of ethylene glycol and polyethylene glycols by methanogenic consortia. Appl Environ Microbiol 46(1):185–190
El-Shafei HA, Abd El-Nasser NH, Kansoh AL, Ali AM (1998) Biodegradation of disposable polyethylene by fungi and Streptomyces species. Polym Degrad Stab 62(2):361–365
Fergusson WC (1974) Summary. In: Staudinger JJP (ed) Plastics and the environment. Hutchinson and Co, London, p 2
Fontanella S, Bonhomme S, Koutny M, Husarova L, Brusson JM, Courdavault JP, Pitteri S et al (2009) Comparison of the biodegradability of various polyethylene films containing pro-oxidant additives. Polym Degrad Stab 95:1011–1021

Friedrich J, Zalar P, Mohorčič M, Klun U, Kržan A (2007) Ability of fungi to degrade synthetic polymer nylon-6. Chemosphere 67(10):2089–2095

Gawande A, Zamare G, Renge VC, Tayde S, Bharsakale G (2012) An overview on waste plastic utilization in asphalting of roads. J Eng Res Stud 3(2):1–5

Gilan I, Hadar Y, Sivan A (2004) Colonization, biofilm formation and biodegradation of polyethylene by a strain of *Rhodococcus ruber*. Appl Microbiol Biotechnol 65:97–104

Glass JE, Swift G (1990) Agricultural and Synthetic Polymers, Biodegradation and Utilization, ACS Symposium Series No. 433. ACS, Washington, pp. xi + 323

Hadad D, Geresh S, Sivan A (2005) Biodegradation of polyethylene by the thermophilic bacterium Brevibacillus borstelensis. J Appl Microbiol 98(5):1093–1100

Halden RU (2010) Plastics and health risks. Annu Rev Public Health 31:179–194

Halim A, El-Sayed MM, Mahmoud WM, Davis EM, Coughlin RW (1996) Biodegradation of polyurethane coatings by hydrocarbondegrading bacteria. Int Biodeter Biodegr 37(1–2):69–79

Harding KG, Dennis JS, Von Blottnitz H, Harrison STL (2007) Environmental analysis of plastic production processes: comparing petroleum-based polypropylene and polyethylene with biologically-based poly-β-hydroxybutyric acid using life cycle analysis. J Biotechnol 130(1):57–66

Hasan F, Shah AA, Hameed A, Ahmed S (2007) Synergistic effect of photo and chemical treatment on the rate of biodegradation of low density polyethylene by fusarium sp. AF4. J Appl Polym Sci 105(3):1466–1470

Hopewell J, Dvorak R, Kosior E (2009) Plastics recycling: challenges and opportunities. Philos Trans R Soc Lond B: Biol Sci 364(1526):2115–2126

Howard GT, Blake RC (1999) Growth of Pseudomonas fluorescens on a polyester-polyurethane and the purification and characterization of a polyurethanase-protease enzyme. Int Biodeterior Biodegrad 42:213–220

Howard GT, Hilliard NP (1999) Use of Coomassie blue-polyurethane interaction inscreening of polyurethanase proteins and polyurethanolytic bacteria. Int Biodeter Biodegr 43(1–2):23–30

Howard GT, Norton WN, Burks T (2012) Growth of Acinetobacter gerneri P7 on polyurethane and the purification and characterization of a polyurethanase enzyme. Biodegradation 23(4):561–573

Huang YQ, Wong CKC, Zheng JS, Bouwman H, Barra R, Wahlström B, Wong MH (2012) Bisphenol A (BPA) in China: a review of sources, environmental levels, and potential human health impacts. Environ Int 42:91–99

Iiyoshi Y, Tsutsumi Y, Nishida T (1998) Polyethylene degradation by lignin-degrading fungi and manganese peroxidase. J Wood Sci 44(3):222–229

Janczak K, Hrynkiewicz K, Znajewska Z, Dąbrowska G (2018) Use of rhizosphere microorganisms in the biodegradation of PLA and PET polymers in compost soil. Int Biodeter Biodegr 130:65–75

Jarerat A, Tokiwa Y (2001) Degradation of poly(tetramethylene succinate) by thermophilic actinomycetes. Biotechnol Lett 23(8):647–651

Jen-hou L, Schwartz A (1961) Zum Verhalten von bakteriengemischen gegentiber polyfithylen verschiedenen mittleren Molekulargewichts. Kunststoffe 51:317–320

Ji-Dong G (2003) Microbiological deterioration and degradation of synthetic polymeric materials: recent research advances. Int Biodeter Biodegr 52(2):69–91

Kalia VC, Raizada N, Sonakya V (2000) Bioplastics. J Sci Ind Res 59:433–445

Kannahi M, Sudha P (2013) Screening of polythene and plastic degrading microbes from Muthupet mangrove soil. J Chem Pharm Res 5(8):122–127

Kathiresan K (2003) Polythene and plastics-degrading microbes from the mangrove soil. Rev Biol Trop 51(3–4):629–633

Kawai F, Kimura T, Fukaya M, Tani Y, Ogata K, Ueno T, Fukami H (1978) Bacterial oxidation of polyethylene glycol. Appl Environ Microbiol 35(4):679–684

Kay MJ, Morton LHG, Prince EL (1991) Bacterial degradation of polyester polyurethane. Int Biodeterior 27(2):205–222

Khullar M (2009) Plastic roads offer greener way to travel in India. The New York Times 13
Kırbaş Z, Keskin N, Güner A (1999) Biodegradation of polyvinylchloride (PVC) by white rot fungi. Bull Environ Contam Toxicol 63(3):335–342
Kitamoto H (2008) Where are biodegradable plastic-degrading microorganisms? BRAIN Technol News 129:1–6
Konduri MKR, Anupam KS, Vivek JS, Kumar RDB, Narasu ML (2010) Synergistic effect of chemical and photo treatment on the rate of biodegradation of high density polyethylene by indigenous fungal isolates. International Journal of Biotechnology and Biochemistry 6:157–174
Konduri MKR, Koteswarareddy G, Kumar DBR, Reddy BV, Narasu ML (2011) Effect of pro-oxidants on biodegradation of polyethylene (LDPE) by indigenous fungal isolate, *Aspergillus oryzae*. J Appl Polym Sci 120:3536–3545
Korawit Chaisu (2016). Study of plastic degrading bacteria by *Aneurinibacillus migulanus*, Management and Innovation Technology International Conference (MITicon), MIT-209 - MIT-212, IEEE Conferences
Kumar S, Singh RK (2013) Thermolysis of high-density polyethylene to petroleum products. J Pet Eng 2013:987568
Kumar S, Hatha AAM, Christi KS (2007) Diversity and effectiveness of tropical mangrove soil microflora on the degradation of polythene carry bags. Rev Biol Trop 55(3–4):777–786
Kumar S, Das M, Jeyanthi Rebecca L, Sharmila S (2013) Isolation and identification of LDPE degrading fungi from municipal solid waste. J Chem Pharm Res 5(3):78–81
Kunwar B, Cheng HN, Chandrashekaran SR, Sharma BK (2016) Plastics to fuel: a review. Renew Sust Energ Rev 54:421–428
Kyaw BM, Champakalakshmi R, Sakharkar MK, Lim CS, Sakharkar KR (2012) Biodegradation of low density polythene (LDPE) by Pseudomonas species. Indian J Microbiol 52(3):411–419
Laist DW (1987) Overview of the biological effects of lost and discarded plastic debris in the marine environment. Mar Pollut Bull 18(6):319–326
Latini G (2005) Monitoring phthalate exposure in humans. Clin Chim Acta 361(1):20–29
Lee SY (1996a) Bacterial polyhydroxyalkanoates. Biotechnol Bioeng 49(1):1–14
Lee SY (1996b) Plastic bacteria? Progress and prospects for polyhydroxyalkanoate production in bacteria. Trends Biotechnol 14(11):431–438
Lee B, Pometto AL, Fratzke A, Bailey TB (1991) Biodegradation of degradable plastic polyethylene by Phanerochaete and Streptomyces species. Appl Environ Microbiol 57:678–685
Li WC, Tse HF, Fok L (2016) Plastic waste in the marine environment: a review of sources, occurrence and effects. Sci Total Environ 566:333–349
Liyoshi Y, Tsutsumi Y, Nishida T (1998) Polyethylene degradation by lignin-degrading fungi and manganese peroxidase. J Wood Sci 44:222–229
Madhuri JR (2015) Biodegradation of low density polyethylene by micro-organisms from garbage soil. J Exp Biol Agric Sci 3(1):15–21
Mehmood CT, Qazi IA, Hashmi I, Bhargava S, Deepa S (2016) Biodegradation of low density polyethylene (LDPE) modified with dye sensitized titania and starch blend using *Stenotrophomonas pavanii*. Int Biodeter Biodegr 113:276–286
Méndez CR, Vergaray G, Vilma R, Karina B, Cárdenas J (2007) Isolation and characterization of polyethylene-biodegrading mycromycetes. Rev Peru Biol 13(3):203–205
Molitoris HP, Moss ST, De Koning GJM, Jendrossek D (1996) Scanning electron microscopy of polyhydroxyalkanoate degradation by bacteria. Appl Microbiol Biotechnol 46(5):570–579
Moore CJ (2008) Synthetic polymers in the marine environment: a rapidly increasing, long-term threat. Environ Res 108(2):131–139
Murphy, R, Bartle, I (2004) Summary Report, Biodegradable Polymers and Sustainability: Insight from Life-cycle Assessment. National Non Food Crops Centre, UK. (PDF) Compostable packaging based on fibrous natural materials and bioplastics. Preliminary results. Available from: https://www.researchgate.net/publication/318361044_Compostable_packaging_based_on_fibrous_natural_materials_and_bioplastics_Preliminary_results. Accessed Aug 27 2018

Muthukumar A, Veerappapillai S (2015) Biodegradation of plastics – a brief review. Int J Pharm Sci Res 3(2):204–209

Nair S, Kumar P (2007) Molecular characterization of a lipase-producing *Bacillus pumilus* strain (NMSN1d) utilizing colloidal water-dispersible polyurethane. World J Microbiol Biotechnol 23:1441–1449

Nakajima-Kambe T, Onuma F, Kimpara N, Nakahara T (1995) Isolation and characterization of a bacterium which utilizes polyester polyurethane as a sole carbon and nitrogen source. FEMS Microbiol Lett 129(1):39–42

Nakajima-Kambe T, Shigeno-Akutsu Y, Nomura N, Onuma F, Nakahara T (1999) Microbial degradation of polyurethane, polyester polyurethanes and polyether polyurethanes. Appl Microbiol Biotechnol 51(2):134–140

Nakkabi A, Sadiki M, Fahim M, Ittobane N, Ibnsouda Koraichi S, Barkai H, El abed S (2015) Biodegradation of poly(ester urethane)s by *Bacillus subtilis*. Int J Environ Res 9(1):157–162

Nanda S, Sahu SS (2010) Biodegradability of polyethylene by Brevibacillus, Pseudomonas, and Rhodococcus spp. N Y Sci J 3(7):95–98

Nanda S, Sahu S, Abraham J (2010) Studies on the biodegradation of natural and synthetic polyethylene by Pseudomonas spp. J Appl Sci Environ Manag 14(2)

National Environmental Engineering Research Institute, NEERI (2010) Air quality assessment. Emissions Inventory and Source Apportionment Studies, Mumbai

Nema SK, Ganeshprasad KS (2002) Plasma pyrolysis of medical waste. Curr Sci 83:271–278

Novotný Č, Erbanová P, Sezimová H, Malachová K, Rybková Z, Malinová L, Brožek J (2015) Biodegradation of aromatic-aliphatic copolyesters and polyesteramides by esterase activity-producing microorganisms. Int Biodeter Biodegr 97:25–30

Obidi OF, Nwachukwu SCU, Aboaba OO (2010) Possible prevention of environmental health Hazard of Lead-based paints. Res J Microbiol 5(7):595–606

Oceguera-Cervantes A, Carrillo-García A, López N, Bolaños-Nuñez S, Cruz-Gómez MJ, Wacher C, Loza-Tavera H (2007) Characterization of the polyurethanolytic activity of two *Alicycliphilus* sp. strains able to degrade polyurethane and N-methylpyrrolidone. Appl Environ Microbiol 73(19):6214–6223

Odusanya SA, Nkwogu JV, Alu N, Etuk Udo GA, Ajao JA, Osinkolu GA, Uzomah AC (2013) Preliminary studies on microbial degradation of plastics used in packaging potable water in Nigeria. Nigerian Food Journal 31(2):63–72

O'Brine T, Thompson RC (2010) Degradation of plastic carrier bags in the marine environment. Mar Pollut Bull 60(12):2279–2283

Oda Y, Oida N, Urakami T, Tonomura K (1997) Polycaprolactone depolymerase produced by the bacterium *Alcaligenes faecalis*. FEMS Microbiol Lett 152(2):339–343

Ojha N, Pradhan N, Singh S, Barla A, Shrivastava A, Khatua P, Rai V, Bose S (2017) Evaluation of HDPE and LDPE degradation by fungus, implemented by statistical optimization. Sci Rep 7(1)

Paço A, Duarte K, da Costa JP, Santos PSM, Pereira R, Pereira ME, Freitas AC, Duarte AC, Rocha-Santos TAP (2017) Biodegradation of polyethylene microplastics by the marine fungus *Zalerion maritimum*. Sci Total Environ 586:10–15

Panda AK, Singh RK, Mishra DK (2010) Thermolysis of waste plastics to liquid fuel: a suitable method for plastic waste management and manufacture of value added products-a world prospective. Renew Sust Energ Rev 14(1):233–248

Vinay Mohan Pathak, Navneet (2017) Review on the current status of polymer degradation: a microbial approach. Bioresources and Bioprocessing 4 (1)

Pathirana RA, Seal KJ (1985) Studies on polyurethane deteriorating fungi: IV: a note on the spectro-chemical changes during fungal deterioration. Int Biodeterior 21(2):123–125

Peciulyte D (2002) Microbial colonization and biodeterioration of plasticized polyvinyl chloride plastics. Ekologija 4:7–15

Peng Y-H, Shih Y-h, Lai Y-C, Liu Y-Z, Liu Y-T, Lin N-C (2014) Degradation of polyurethane by bacterium isolated from soil and assessment of polyurethanolytic activity of a *Pseudomonas putida* strain. Environ Sci Pollut Res 21(16):9529–9537

Plastic Waste Management Rules (2016), Ministry of Environment forest and climate change notification http://www.moef.gov.in/sites/default/files/PWM%20Rules,%202016_0.pdf

Pometto AL III, Lee BT, Johnson KE (1992) Production of an extracellular polyethylene degrading enzyme(s) by Streptomyces species. Appl Environ Microbiol 58:731–733

Pospíšil J, Nešpůrek S (1997) Highlights in chemistry and physics of polymer stabilization. Macromol Symp 115(1):143–163

Pramila R, Vijaya Ramesh K (2011) Biodegradation of low density polyethylene (LDPE) by fungi isolated from municipal landfill area. J Microbiol Biotech Res 1(4):131–136

Priyanka N, Archana T (2011) Biodegradability of polythene and plastic by the help of microorganism: a way for brighter future. J Environ Anal Toxicol 01(02)

Punčochář M, Ruj B, Chatterj PK (2012) Development of process for disposal of plastic waste using plasma pyrolysis technology and option for energy recovery. Procedia Eng 42:420–430

Raaman N, Rajitha N, Jayshree A, Jegadeesh R (2012) Biodegradation of plastic by aspergillus spp. isolated from polythene polluted sites around Chennai. Journal of Academia and Industrial Research 1(6):313–316

Rawte T, Padte M, Mavinkurve S (2002) Incidence of marine and mangrove bacteria accumulating polyhydroxyalkanoates on the mid-west coast of India. World J Microbiol Biotechnol 18:655–659

Rebeiz KS, Craft AP (1995) Plastic waste management in construction: technological and institutional issues. Resour Conserv Recycl 15(3–4):245–257

Reddy RM (2008) Impact of soil composting using municipal solid waste on biodegradation of plastics. Indian J Biotechnol 7:235–239

Rutkowska M, Heimowska A, Krasowska K, Janik H (2002) Biodegradability of polyethylene starch blends in sea water. Pol J Environ Stud 11:267–274

Sachin SS, Mishra RL (2013) Screening and identification of soil fungi with potential of plastic degrading ability. Indian Journal of Applied Research 3(11):34–36

Saminathan P, Sripriya A, Nalini K, Sivakumar T, Thangapandian V (2014) Biodegradation of plastics by Pseudomonas putida isolated from garden soil samples. J of Advanced Botany and Zoology Vol 1(3):1–4

Sang BI, Hori K, Tanji Y, Unno H (2002) Fungal contribution to in situ biodegradation of poly (3-hydroxybutyrate-co-3-hydroxyvalerate) film in soil. Appl Microbiol Biotechnol 58(2):241–247

Sarker M, Rashid MM, Molla M (2011) Waste plastic conversion into chemical product like naphtha. J Fundam Renew Energy Appl 1:110101

Sathya R, Ushadevi T, Panneerselvam A (2012) Plastic degrading actinomycetes isolated from mangrove sediments. International Journal of Current Research 4(10):001–003

Sathyanarayana S (2008) Phthalates and children's health. Curr Probl Pediatr Adolesc Health Car 38(2):34–49

Savoldelli J, Tomback D, Savoldelli H (2017) Breaking down polystyrene through the application of a two-step thermal degradation and bacterial method to produce usable byproducts. Waste Manag 60:123–126

Saxena G, Bharagava RN (2016) Ram Chandra: advances in biodegradation and bioremediation of industrial waste. Clean Techn Environ Policy 18:979–980. https://doi.org/10.1007/s10098-015-1084-9

Saxena G, Bharagava RN (2017) Organic and inorganic pollutants in industrial wastes, their ecotoxicological effects, health hazards and bioremediation approaches. In: Bharagava RN (ed) Environmental pollutants and their bioremediation approaches, 1st edn. CRC Press, Taylor & Francis Group, Boca Raton, pp 23–56. https://doi.org/10.1201/9781315173351-3

Saxena G, Purchase D, Mulla SI, Saratale GD, Bharagava RN (2018) Phytoremediation of heavy metal-contaminated sites: environmental considerations, field studies, sustainability and future prospects. J Environ Manag

Schecter A, Malik N, Haffner D, Smith S, Harris TR, Paepke O, Birnbaum L (2010) Bisphenol a (BPA) in US food. Environ Sci Technol 44(24):9425–9430

Schink B, Janssen PH, Frings J (1992) Microbial degradation of natural and of new synthetic polymers. FEMS Microbiol Lett 103(2–4):311–316
Sen SK, Raut S (2015) Microbial degradation of low density polyethylene (LDPE): a review. J Environ Chem Eng 3(1):462–473
Seneviratne G, Tennkoon NS, Weerasekara MLMAW, Nandasena KA (2006) Polythene biodegradation by a developed Penicillium- Bacillus biofilm. Curr Sci 90:20–21
Shah AA, Hasan F, Hameed A, Ahmed S (2007) Isolation and characterization of poly (3-hydrox ybutyrate-co-3-hydroxyvalerate) degrading bacteria and purification of PHBV depolymerase from newly isolated Bacillus sp. AF3. Int Biodeter Biodegr 60(2):109–115
Shah AA, Hasan F, Akhter JI, Hameed A, Ahmed S (2008a) Degradation of polyurethane by novel bacterial consortium isolated from soil. Ann Microbiol 58(3):381–386
Shah AA, Hasan F, Hameed A, Ahmed S (2008b) Biological degradation of plastics: a comprehensive review. Biotechnol Adv 26(3):246–265
Shah AA, Hasan F, Akhter J, Hameed A, Ahmed S (2009) Isolation of fusarium sp. AF4 from sewage sludge, with the ability to adhere the surface of polyethylene. Afr J Microbiol Res 3:658–663
Shah Z, Hasan F, Lee K, Atkas D, Shah AA (2013) Degradation of polyester polyurethane by newly isolated *Pseudomonas aeruginosa* strain MZA-85 and analysis of degradation products by GC-MS. Int Biodeter Biodegr 79:105
Sharholy M, Ahmad K, Mahmood G, Trivedi RC (2008) Municipal solid waste management in Indian cities–a review. Waste Manag 28(2):459–467
Sharma A, Sharma A (2004) Degradation assessment of low density polythene (LDPE) and polythene (PP) by an indigenous isolate of *Pseudomonas stutzeri*. J Sci Ind Res 63:293–296
Siddique R, Khatib J, Kaur I (2008) Use of recycled plastic in concrete: a review. Waste Manag 28(10):1835–1852
Singh S, Li SSL (2012) Epigenetic effects of environmental chemicals bisphenol a and phthalates. Int J Mol Sci 13(8):10143–10153
Singh B, Sharma N (2008) Mechanistic implications of plastic degradation. Polym Degrad Stab 93(3):561–584
Singh G, Singh AK, Bhatt K (2016) BIODEGRADATION OF POLYTHENES BY BACTERIA ISOLATED FROM SOIL. Int J Res Dev Pharm Life Sci 5(2):2056–2062
Singh N, Hui D, Singh R, Ahuja IPS, Feo L, Fraternali F (2017) Recycling of plastic solid waste: a state of art review and future applications. Compos Part B 115:409–422
Sinosh S, Megha M, Kini MN, Mukund KM, Rizvi A, Vasist K (2015) Selection and screening of microbial consortia for efficient and ecofriendly degradation of plastic garbage collected from urban and rural areas of Bangalore, India. Environ Monit Assess 187(1)
Sivan A, Szanto M, Pavlov V (2006) Biofilm development of the polyethylene-degrading bacterium *Rhodococcus ruber*. Appl Microbiol Biotechnol 72(2):346–352
Skariyachan S, Patil AA, Shankar A, Manjunath M, Bachappanavar N, Kiran S (2018) Enhanced polymer degradation of polyethylene and polypropylene by novel thermophilic consortia of Brevibacillus sps. And Aneurinibacillus sp. screened from waste management landfills and sewage treatment plants. Polym Degrad Stab 149:52–68
Song JH, Murphy RJ, Narayan R, Davies GBH (2009) Biodegradable and compostable alternatives to conventional plastics. Philos Trans R Soc B 364(1526):2127–2139
Sowmya HV, Ramalingappa Krishnappa M (2014) Environ Monit Assess 186:65–77
Suzuki J, Hukushima K, Suzuki S (1978) Effect of ozone treatment upon biodegradability of water-soluble polymers. Environ Sci Technol 12(10):1180–1183
Tang L, Huang H, Zhao Z, Wu CZ, Chen Y (2003) Pyrolysis of polypropylene in a nitrogen plasma reactor. Ind Eng Chem Res 42(6):1145–1150
Tang Z, Huang Q, Cheng J, Yang Y, Yang J, Guo W, Jin L (2014) Polybrominated diphenyl ethers in soils, sediments, and human hair in a plastic waste recycling area: a neglected heavily polluted area. Environ Sci Technol 48(3):1508–1516

Thompson RC, Moore CJ, Vom Saal FS, Swan SH (2009) Plastics, the environment and human health: current consensus and future trends. Philos Trans R Soc Lond B: Biol Sci 364(1526):2153–2166

Tokiwa Y, Suzuki T (1977) Hydrolysis of polyesters by lipases. Nature 270(5632):76–78

Tokiwa Y, Calabia B, Ugwu C, Aiba S (2009) Biodegradability of plastics. Int J Mol Sci 10(9):3722–3742

Usha R, Sangeetha T, Palaniswamy M (2011) Screening of polyethylene degrading microorganisms from garbage soil. Libyan Agricultural Research Center Journal International 2:200–204

Vandenberg LN, Maffini MV, Sonnenschein C, Rubin BS, Soto AM (2009) Bisphenol-A and the great divide: a review of controversies in the field of endocrine disruption. Endocr Rev 30(1):75–95

Varesche MB, Zaiat M, Vieira LGT, Vazoller RF, Foresti E (1997) Microbial colonization of polyurethane foam matrices in horizontal-flow anaerobic immobilized-sludge reactor. Appl Microbiol Biotechnol 48(4):534–538

Verma R, Vinoda KS, Papireddy M, Gowda ANS (2016) Toxic pollutants from plastic waste-a review. Procedia Environ Sci 35:701–708

Vimala PP, Mathew L (2016) Biodegradation of polyethylene using *Bacillus Subtilis*. Procedia Technology 24:232–239

Volke-Sepulveda T, Saucedo-Castaneda G, Gutierrez-Rojas M, Manzur A, Favela-Torres E (2002) Thermally treated low density polyethylene biodegradation by *Penicillium pinophilum* and *Aspergillus niger*. J Appl Polym Sci 83:305–314

Webb JS, Nixon M, Eastwood IM, Greenhalgh M, Robson GD, Handley PS (2000) Fungal colonization and biodeterioration of plasticized polyvinyl chloride. Appl Environ Microbiol 66(8):3194–3200

Wu G, Li J, Xu Z (2013) Triboelectrostatic separation for granular plastic waste recycling: a review. Waste Manag 33(3):585–597

Yamada-Onodera K, Mukumoto H, Katsuyaya Y, Saiganji A, Tani Y (2001) Degradation of polyethylene by a fungus, Penicillium simplicissimum YK. Polym Degrad Stab 72(2):323–327

Yoshida S, Hiraga K, Takehana T, Taniguchi I, Yamaji H, Maeda Y, Oda K (2016) A bacterium that degrades and assimilates poly (ethylene terephthalate). Science 351(6278):1196–1199

Zgoła-Grześkowiak A, Grześkowiak T, Zembrzuska J, Łukaszewski Z (2006) Comparison of biodegradation of poly(ethylene glycol)s and poly(propylene glycol)s. Chemosphere 64(5):803–809

Zheng Y, Yanful EK, Bassi AS (2005) A review of plastic waste biodegradation. Crit Rev Biotechnol 25:243–250

Chapter 7
Textile Industry Wastewaters as Major Sources of Environmental Contamination: Bioremediation Approaches for Its Degradation and Detoxification

Rijuta Ganesh Saratale, J. Rajesh Banu, Han-Seung Shin, Ram Naresh Bharagava, and Ganesh Dattatraya Saratale

Contents

R. G. Saratale
Research Institute of Biotechnology and Medical Converged Science, Dongguk University-Seoul, Goyang-si, Gyeonggi-do, Republic of Korea

J. Rajesh Banu
Department of Civil Engineering, Regional Center of Anna University, Tirunelveli, Tamilnadu, India

H.-S. Shin
Department of Food Science and Biotechnology, Dongguk University-Seoul, Goyang-si, Gyeonggi-do, Republic of Korea

R. N. Bharagava
Laboratory of Bioremediation and Metagenomics Research (LBMR), Department of Microbiology (DM), Babasaheb Bhimrao Ambedkar University (A Central University), Lucknow, Uttar Pradesh, India

G. D. Saratale (✉)
Research Institute of Biotechnology and Medical Converged Science, Dongguk University-Seoul, Goyang-si, Gyeonggi-do, Republic of Korea

Department of Food Science and Biotechnology, Dongguk University-Seoul, Goyang-si, Gyeonggi-do, Republic of Korea
e-mail: gdsaratale@dongguk.edu

G. Saxena, R. N. Bharagava (eds.), *Bioremediation of Industrial Waste for Environmental Safety*, https://doi.org/10.1007/978-981-13-1891-7_7

Abstract In the past few decades, rapid industrialization led to increasing the demand for textile products which leads to increase in water pollution. Effluent released by the textile industry pose a threat to environmental safety throughout the world. Bioremediation approaches could be considered as an efficient and effective way to treat textile effluent relative to existing physical and chemical methods. This book chapter describes various methods used to treat textile industry wastewater. Bioremediation approaches using live cells system, enzymes, and phytoremediation approaches and their detailed mechanisms have also been discussed. Further discussion on the various types of the bioreactors employed to treat textile industry wastewater at large scale is summarized. The performance of the system and their key challenges and future technological aspects are also briefly discussed. The aim of this chapter is to provide an overview of bioremediation approaches to resolve issues related to textile industry wastewater and to minimize pollution and control their associated effects on the environment.

Keywords Textile wastewater · Pollution · Toxicity · Oxidoreductive enzymes · Fixed-bed bioreactor · Bioremediation

7.1 Introduction

Industrialization is one of the major aspects of economic growth of the nation; however, industrial wastewaters are also responsible for pollution and toxicity in living beings (Goutam et al. 2018; Bharagava et al. 2017a, b; Saxena and Bharagava 2015, 2017; Saxena et al. 2016; Gautam et al. 2017). Textile industries have huge financial impact and employment generation. However a well-flourished textile industrial development is also responsible for environmental pollution, a number of pollutants generated get discharged directly into the nearby surroundings without any additional treatment (Pandey et al. 2005). It was reported that during textile processing,

10–15% of the total dyes are discharged into the water stream causing severe water pollution; however, it also depends on processing stages, types of synthetic dyes, fabric material, processing operational conditions employed during the conversion of fiber to textile fabric, and concentration of chemical agents (Banat et al. 1996; Kalyani et al. 2008; Robinson et al. 2001). Textile industrial effluents can be considered as one of the higher generators of liquid effluent pollutants, and literature showed that about 280,000 tons of textile dyes are discharged in such industrial effluents every year worldwide (Jin et al. 2007; Saratale et al. 2017a). Textile effluents have been a major environmental concern for a long time since it affects the aquatic ecosystem, the fertility of soil, plant growth and productivity of plants, and plant susceptibility toward pathogens (Saratale et al. 2009b, c, 2011c). In 1974, ETAD (Ecological and Toxicological Association of the Dyestuff Manufacturing Industry) was established with the aim of controlling environment spoiling, protecting users and consumers, and assisting entirely with public anxiety about the harmful impact of their products with respect to toxicity.

The problem related to textile effluent is not the presence of dyes but they're recalcitrant to degradation and stability (Jadhav et al. 2008). Research in the field of textile wastewater treatment is increasing day by day because of strict government legislation regarding the release of contaminated effluent (Saratale et al. 2009a). In this book chapter, we have described various bioremediation approaches including live cell system, enzymes, and phytoremediation approaches used to treat textile wastewater, and their detailed mechanisms have also been discussed. Special emphasis is devoted to the various types of bioreactors employed to treat textile wastewater, and their key challenges and future technological aspects are also briefly discussed.

7.2 Dye Composition and Classification

Textile dyes are colored compounds which constitute chromophore and auxochrome groups and having the ability to absorb electromagnetic energy in the visible range (~350 to 700 nm). The colored compound group is called chromophore whereas, auxochrome group increases color intensity of the chromophore group by shifting the overall energy of the electron system (Forgacs et al. 2004). The usual chromophores are N=N, C=N, C=C, $-NO_2$, C=O, and quinoid rings, and auxochromes are $-NH_3$, $-SO_3H$, –COOH, and –OH (Zollinger 1987). On the basis of chemical structure or chromophore group, dyes are classified as azo (monoazo, disazo, triazo, polyazo), phthalocyanine, anthraquinone, triarylmethane, indigoid, diarylmethane, oxazine, azine, thiazine, xanthene, nitro, nitroso, methine, indamine, thiazole, indophenol, aminoketone, lactone, hydroxy ketone stilbene, and sulfur dyes. In 1924, each dye was given a C.I. generic name on the basis of its color, application, and characteristics by the Society of Dyers and Colorists, and the American Association of Textile Chemists and Colorists. Dyes can be classified into natural dyes and synthetic dyes based on their origin. Natural dyes are a group of dyes that are extracted from biological material like plant and animal residues.

According to color index, dyes are classified as reactive dyes, acid dyes, metal complex dyes, direct dyes, basic dyes, mordant dyes, disperse dyes, pigment dyes, vat dyes, anionic and ingrain dyes, sulfur dyes, solvent dyes, fluorescent dyes, and other dyes including food and natural dyes. Among the various classes of dyes, reactive dyes and azo dyes used exhaustively in commercial applications like textile, dyeing, and paper printing.

7.3 Treatment Methodologies for Textile Industry Wastewater

The textile industrial development also increases awareness toward the environmental problems arising due to effluent discharge, and with strict government legislation, both color creating and color using industries are constrained to search for novel physicochemical treatments and technologies for the treatment of textile industrial wastewater. The methods that have now been employed to treat textile wastewater are physicochemical and biological methods (Vandevivere et al. 1998; Swaminathan et al. 2003; Golab et al. 2005; Saratale et al. 2011a, 2016a, b). It was observed that each methods result and its technical and practical applicability depend on various factors including the chemical composition of wastewater, types of dyes present, operational cost of instrument used, handling costs, and environmental fate of generated by-products. In the below section, we have discussed the various physicochemical and biological methods used to treat dye-containing wastewater.

7.3.1 Physical Methods

Dye decolorization using adsorption technique is considered as an effective method in which rapid separation of adsorbate from liquid/gaseous phase on to solid phase takes place and could be a useful textile dye industrial wastewater treatment (Saratale et al. 2016a, b). However, the process performance can be influenced by physical-chemical factors such as dye/sorbent interaction, particle size, the surface area of sorbent, contact time, pH, and temperature (Anjaneyulu et al. 2005). In general, activated carbon (AC) is widely used as an adsorbent and found very effective to the various types of dyes, but due to their high cost, it is not used conventionally (Robinson et al. 2001). In adsorption treatment, various types of adsorbents are required, and also their regeneration or disposal makes the process economically unfeasible (Karcher et al. 2001). Nowadays to reduce the process cost, many investigators used low-cost adsorbents from waste materials, for example, radix *Angelicae dahuricae* (Baizhi) residue (Sun et al. 2018), wood chips (Nigam et al. 2000), waste slurry of rice husk after hydrolytic enzymes production (Saratale et al. 2017a), water hyacinth (Mahamadi and Epias 2014), and peach stone (Saratale et al. 2016b).

Membrane filtration technique is used for clarification, concentration, and separation of dye from the effluent. In the current scenario, pressure-driven membrane processes such as nanofiltration, ultrafiltration, and reverse osmosis are more popular and applicable techniques in the area of wastewater treatment (Kook et al. 2017). Membrane filtration is well suitable for effluents containing low molecular weight dyes at low concentrations, but it is ineffective for a reduction in dissolved solids, due to which water reuse becomes a difficult task and high investment costs, membrane fouling, produces secondary waste streams which need further treatment (Saratale et al. 2011c; Kook et al. 2017).

Colored wastewater is passed through ion exchange resin until the saturation of available exchange sites. This method is used to remove both cationic and anionic dyes from textile wastewater. Using this method, removal of soluble dyes, solvent reclamation after use is possible, and also adsorbent loss is avoided. However, the major disadvantage of this method is expensive organic solvents and the ion exchange method is not very useful for disperse dyes.

7.3.2 Chemical Methods for Textile Wastewater Treatment

The chemical oxidation process can be considered as an easy procedure where removal of dye from effluent takes place by cleavaging aromatic ring present in dye structure. Oxidizing agents, for example, Fenton's reagents, chlorine, UV/peroxide, ozone, and UV/ozone, are used individually or in combination for color removal. Chlorine is the most popular oxidizing agent due to its low operational and capital costs, but the possibility of formation of undesirable by-products has limited the suitability of this treatment method. Various types of an oxidizing agent such as ozone (O_3), hydrogen peroxide (H_2O_2), and permanganate (MnO_4) have been used in chemical oxidation methods. Moreover in advanced oxidation processes (photochemical and photocatalytic), oxidizing agents such as O_3 and H_2O_2 or with heterogeneous photocatalysts are used with catalysts such as TiO_2, ZnO_2, Mn, CuO and Fe, in the presence or absence of an irradiation source are found effective by generating (OH) radical for the destruction of hazardous dye pollutants (Saratale et al. 2018; Forgacs et al. 2004). This method is found better in terms of higher COD removal and decolorization efficiencies for both soluble and insoluble dyes; however, high sludge generation due to the flocculation of reagents and dye molecules still limits this process (Van der Zee 2002).

Electrochemical treatment of textile wastewater is an effective method because of its color removal efficiency. This method is generally found more advantageous because it can run under lower temperatures relative to other non-electrochemical treatment methods, with no requirement of other chemicals and nonhazardous products (Saratale et al. 2015). The main oxidizing agent for electrochemical oxidation is hypochlorite ion or hypochlorous acid produced from the naturally occurring chloride ions; however the high cost of electricity limits the process (Lopez-Grimau and Gutierrez 2006; Robinson et al. 2001).

Coagulation-flocculation is most widely used the method in textile wastewater treatment plants for the removal of main sulfur and disperse dyes (Vandevivere et al. 1998). Either individually or in combination, aluminum sulfate, ferric chloride, ferrous and ferric sulfate, copper sulfate, and calcium chloride are used for color removal from textile wastewater (Golab et al. 2005). The pH adjustment, dewatering problems, costly chemicals for precipitation, highly concentrated residual cations in the supernatant, and disposal problems of sludge generated after coagulation are some of the limitations of this method.

7.3.3 Biological Methods for Textile Wastewater Treatment

Textile wastewater treatment is mainly done by physical and chemical treatment methods; however these techniques have major limitations including high energy requirements, high cost, generation of huge amount of sludge, and possibility of formation of harmful by-products (Maier et al. 2004; Daeshwar et al. 2007; Saratale et al. 2009b, 2013, 2017a). Bioremediation is defined as a spontaneous or managed biological method of eliminating or remediating environmental pollutants using catalytic reactions of living organisms especially of microorganisms and plants (Saxena et al. 2018; Arora et al. 2014, 2018; Bharagava et al. 2018; Chandra et al. 2015; Singh et al. 2011; Saratale et al. 2006). By applying advances in biotechnology in the bioremediation research in which microorganisms showed better ability to utilize variously concentrated and toxic pollutants such as petroleum products, pesticides, textile dyes, aliphatic, and aromatic hydrocarbons thus became one of the most fast-growing fields of environmental biotechnology (Saratale et al. 2011a, b).

Microorganisms are ubiquitous in nature considered as makes them capable for adapting themselves to make use of xenobiotic or toxic compounds as a source of energy and growth the best candidates (Saratale et al. 2006, 2007a, b; Bharagava et al. 2017c). Microorganisms act as natural recyclers by converting toxic compounds to harmless by-products, often water, and carbon dioxide. To increase microbial activity, the main goal should be an exploration of microbial diversity in contaminated areas to investigate microorganisms capable of degrading a wide range of pollutants (Saratale et al. 2015; Bedekar et al. 2014a; Sahasrabudhe et al. 2014). Literature survey suggests that microbial systems for decolorization of textile wastewater are superior in decolorizing textile dyestuff with complete decolorization in just a few hours (Saratale et al. 2015; Bedekar et al. 2014b; Sahasrabudhe et al. 2014). The use of microbial or enzymatic treatment for the complete decolorization and degradation of industrial dyes from textile effluent possesses considerable advantages; being eco-friendly and cost-competitive and having less sludge-producing properties, end products with complete mineralization and non-toxic products could help to reduce the enormous water consumption compared to physicochemical methods (Saratale et al. 2009c, 2010; Bedekar et al. 2014a). Plants also have great potential to degrade individual dyes as well as dye mixtures (Kagalkar et al. 2009).

7.3.3.1 Bacteria

Bacteria are known as faster degrader and mineralizer of many reactive azo dyes than other microorganisms. Bacterial oxygenase and hydroxylase are able to degrade the intermediate products synthesized during dye decolorization (Saratale et al. 2009b, c, 2011c; Banat et al. 1996). Mostly, potent dye degrading bacteria are isolated from contaminated soil of textile industry by using enrichment techniques. 16s rRNA sequencing is performed to identify bacterial species. Literature survey also suggests that there are some bacterial species which showed excellent dye decolorization capacity even though they are not isolated from dye-contaminated soil. *Enterobacter* sp., *Pseudomonas* sp., and *Morganella* sp. isolated from powdered activated carbon process showed good textile dye decolorization capacity (Barragan et al. 2007). Decolorization of textile dyes under anoxic conditions was reported by Chen et al. (1999) and Yu et al. (2001) using *Aeromonas hydrophila*, *Pseudomonas luteola*, and *Proteus mirabilis*. Nigam et al. (1996) and Khehra et al. (2005) have reported anoxic dye decolorization using facultative anaerobic and mixed aerobic microbial consortia. Available literature suggests that effective degradation of azo dyes usually takes place under anaerobic conditions. The main drawbacks of anaerobic degradation are the formation of toxic aromatic amines due to the transfer of reducing equivalents at end of the process as well as the requirement of complex organic sources, including peptone, yeast extract, or combination of carbohydrate and complex organic sources (Khehra et al. 2005). Hence the combination of anaerobic and aerobic treatment for textile dye degradation that gives fast results with the formation of nontoxic end products is advantageous (Banat et al. 1996; Bharagava et al. 2018). Bacteria have been preferred over fungi for the treatment of textile wastewaters as the rates of decolorization and mineralization of the dyes present in the effluents are higher. In addition to this major advantage, the use of bacteria for treatment leads to less sludge generation, and the process is cost-effective as well. Reductions in COD and color by 66.7% and 96.9%, respectively, were observed when *Bacillus* MK-8 strain was used in the form of granules (Kolekar and Kodam 2012). In our research group, we have explored the decolorization ability of various dyes as well as textile wastewater by isolated as well as commercially available bacterial strains such as *Enterococcus faecalis* YZ 66, *Lysinibacillus* sp. RGS, *Proteus vulgaris* NCIM-2027, *Micrococcus glutamicus* NCIM-2168, *Rhizobium radiobacter* MTCC 8161, and *Kocuria rosea* MTCC-1532 (Sahastrabuddhe et al. 2014; Saratale et al. 2009b, 2010, 2013; Parshetti et al. 2006, 2009).

7.3.3.2 Fungi

Lignolytic and nonligninolytic fungi are mostly used for the decolorization of dye in wastewater. *Phanerochaete chrysosporium*, *Trichophyton rubrum* LSK-27, *Ganoderma* sp. WR-1, *Trametes versicolor*, *Funalia trogii*, *Irpex lacteus*, etc. are widely used for textile dye decolorization and degradation (Machado et al. 2006; Robinson et al. 2001; Rodrı'guez et al. 1999; Saratale et al. 2011c). Moreover non-white rot fungi such as *Aspergillus niger*, *Pleurotus ostreatus*, and *Cunninghamella elegans* (Ambrosio and Campos-Takaki 2004) showed ability of textile dye decolorization (Wesenberg et al. 2003; Christian et al. 2005). White rot fungi, in particular, have been found very effective in the degradation of various dyes and other xenobiotics (Saratale et al. 2006, 2007a, b; Bhosale et al. 2006; Humnabadkar et al. 2008). Lignolytic enzymes, such as laccase, manganese peroxidase, and lignin peroxidase, are required for biodegradation of dyes using white rot fungi. While in case of dye degradation using nonligninolytic fungi, they require laccase, bilirubin oxidase, and azoreductase (Lu et al. 2007). A mixed fungal culture of *Pleurotus ostreatus* and *Coriolus versicolor* reduced the COD and BOD of wastewater along with achieving its decolorization. Moreover requirement of nitrogen-limiting conditions and long growth cycle leads to prolonged hydraulic retention times, and preserving the dominance of fungal cultures in the reactor system is the major challenge that prevents the use of fungi for full-scale applications (Acikgoz et al. 2016; Saratale et al. 2011b, c).

7.3.3.3 Yeast

Interest in textile wastewater treatment using yeasts is due to the ability of the yeast biomass to absorb and accumulate toxic chromophores as well as to degrade them into simpler compounds. Decolorization of azo dyes using yeast *Candida zeylanoides* is reported (Ramalho et al. 2002). Ramalho et al. (2004) further described the enzymatic characterization in *Issatchenkia occidentalis*. Enzymatic system involved in *Saccharomyces cerevisiae* for dye decolorization was also reported. Aksu (2003) reported bioaccumulation of various reactive dyes, viz., Remazol Blue, Remazol Red RB, and Remazol Black B, during growth of *S. cerevisiae* on molasses. Dead biomass of yeast has been utilized as a biosorbent for the biosorption of dyes, for example, modified baker's yeast has been used for biosorption of acridine orange, crystal violet, aniline blue, safranin, and malachite green (Šafaříková et al. 2005). Yeasts also possess an enzyme system that can degrade dyes present in the textile wastewater. In our study, we have proved the ability of *Trichosporon beigelii* in the decolorization of Navy Blue HER and also determined the enzyme system responsible for dye decolorization (Saratale et al. 2009c).

7.3.3.4 Algae

Microalgae are microscopic, unicellular photosynthetic plants which have the ability to convert solar energy with CO_2 and H_2O by using nutrients to produce biomass (Saratale et al. 2017b). The general mechanism behind algae decolorization where firstly chromophores group is utilized for production of algal biomass, carbon dioxide, and water or transformation of colored to noncolored compound or by adsorption mechanism. *Oscillatoria* (Jinqi and Houtian 1992), *Chlorella* (Acuner and Dilek 2004), and *Spirogyra* (Mohan et al. 2002) species have been reported for their azo dye decolorization ability to degrade azo dyes to aromatic amines to simple compounds and subsequently to CO_2. Several researchers have reported the potential of various algal strains to treat textile wastewater owing to their ability to degrade the azo bonds. *Scenedesmus bijugatus* showed 68% decolorization of azo dyes after 6 days of incubation and *Enteromorpha* sp. showed 83.5% decolorization of Basic Red 46 within 5 h of incubation (Jianwei et al. 1997; Daneshwar et al. 2007).

7.3.3.5 Actinomycetes

Actinomycetes are a group of abundant and metabolically diverse bacteria in soil (Saratale et al. 2012). *Streptomyces rochei* A10, *Streptomyces chromofuscus* 11, *Streptomyces diastaticus* A12, *Streptomyces diastaticus* A13, and *Streptomyces rochei* A14 were able to degrade various textile dyes through their extracellular oxidative enzyme system which includes lignin peroxidase and manganese peroxidase (Saratale et al. 2011c).

7.3.3.6 Plants

In phytoremediation plants that degrade the dye molecules into nontoxic products are preferable. In this treatment method, selection of the plant will depend upon the genetic makeup of the plant in terms of varied enzyme activities and differential absorptive capacities resulting into variable patterns for the removal of dyes. In addition to this selected plants should be in the same stage of growth and should have almost equivalent dry weights and similar root and shoot lengths to accomplish reproducibility of results. Moreover it is preferable that selected plants should be from the same area, and other influencing factors such as plant age, soil conditions, nutrient status, and light availability should be optimized to get significant results. It was recommended to use flowering plants for these treatment methods to give esthetically appealing systems or other plants that could be useful for bioenergy production (Peuke and Rennenberg 2005). Prof. Govindwar's research group widely studied many plants for the degradation of dyes as well as wastewater even they have proposed wetland using wild plants for actual textile effluents and field trial (Kagalkar et al. 2009; Saratale et al. 2011c).

7.3.3.7 Microbial Consortia

Use of mixed microbial population for textile wastewater treatment is necessary to achieve higher rate of decolorization, degradation, and mineralization. Synergistic metabolic activities of microbial community during degradation of synthetic dyes have considerable advantage over the use of pure culture (Khehra et al. 2005; Junnarkar et al. 2006). Microbial consortium is widely used in the field of textile wastewater treatment. Bacteria-bacteria consortia (bacteria-fungi consortia, and bacteria-yeast consortia have been reported for decolorization and degradation of textile dyes (Saratale et al. 2011c). Inhabitant microbial communities play an important role in the detoxification of contaminated sites. These natural degraders are acclimatized to xenobiotic compounds via various genetic mechanisms such as transposition, genetic recombination, gene transfer, and mutational drift.

7.4 Dye Decolorization Mechanism

7.4.1 Role of Enzymes in Dye Degradation

The enzymatic approach is also could be considered as an effective alternative approach to treat dye-containing wastewater for the conventional physicochemical treatments. Microbial enzymes deal with the dye molecules in cleavage and successive degradation of the dye efficiently. Highly reactive free radicals generated by oxidative and reductive enzymes present in microorganisms undergo spontaneous cleavage reactions serially, and thus enzymatic treatment could be considered as most effective when target pollutants are in highest concentration and other contaminants are less in concentration. The oxidoreductive enzymes, mainly lignin peroxidase, laccases, tyrosinase, azoreductase, riboflavin reductase, and NADH-DCIP reductase, are involved in microbial decolorization and degradation of dyes and are described in this section.

7.4.2 Reductive Enzymes

Azoreductases catalyze reductive cleavage of azo bond (N=N) in the presence of FADH or NADH or NADPH as an electron donor (Russ et al. 2000). Azoreductases are flavoproteins in nature and are classified on a structural basis as flavin-dependant azoreductases (Nakanishi et al. 2001; Chen et al. 2005) and flavin-independent azoreductases. The flavin-dependent azoreductases are again classified into three groups, NADPH only, NADH only, or both, which act as electron donors (Chen et al. 2005; Nakanishi et al. 2001). Bacterial cell membrane either intracellularly or extracellularly carries azoreductases. The enzymatic action of azoreductases many times results in the formation of colorless aromatic amines (Stolz 2001). There are few reports where purification and characterization of azoreductase enzymes from *Xenophilus azovorans* KF46F, *Pigmentiphaga kullae* K24, *Enterococcus faecalis*, *Staphylococcus aureus*, *Escherichia coli*, *Bacillus* sp. OY1-2, and *Rhodobacter sphaeroides* have been reported (Chen et al. 2005; Suzuki et al. 2001). In our studies we have reported the involvement of intracellular azoreductase in bacterial decolorization of dyes (Saratale et al. 2009a, 2010, 2013). It was also reported that due to high polarities and complex structure, it is difficult to diffuse azo dyes through the cell membrane; thus cytoplasmic as well as intracellular azoreductase enzyme activity is involved in azo dyes decolorization. Another marker enzyme, NADH-DCIP reductase, has been widely studied in the decolorization and detoxification of xenobiotics by fungi and bacteria (Bhosale et al. 2006; Saratale et al. 2007a, b). In addition to this, riboflavin reductase is also studied for the decolorization of dyes. The basic enzymatic mechanism where the reduction of various flavins such as; riboflavin, flavin mononucleotide (FMN), and flavin adenine dinucleotide (FAD), at the expense of reduced pyridine nucleotides takes place (Ingelman et al. 1999). The involvement of this enzyme in the degradation of Scarlet R by Consortium-GR (consisting of *Proteus vulgaris* and *Micrococcus glutamicus*) and Reactive Green 19A by *M. glutamicus* NCIM 2168 has been reported earlier (Saratale et al. 2009b).

7.4.3 Oxidative Enzymes

There are some reports where oxidative enzymes are also involved in the decolorization and degradation of textile dyes. Laccases (E.C.1.10.3.2) are copper-containing enzymes that catalyze the oxidation of several aromatic and inorganic substances with the concomitant reduction of oxygen to water the detailed mechanism shown below (Majcherczyk et al. 1998). The molecular structure of laccases shows four neighboring copper atoms which are distributed among different binding sites and having molecular weight in the range of 60–390 kDa (Kalme et al. 2009).

Laccases have been widely studied focusing their industrial applications such as degradation of various recalcitrant compounds such as polyaromatic hydrocarbons, chlorophenols, phenols, structures related to lignin, aromatic dyes, and organophosphorus compounds (Kalyani et al. 2008; Saratale et al. 2009c). Very few bacteria, for example, *Micrococcus glutamicus* NCIM 2168, *Pseudomonas desmolyticum* NCIM 2112, and *Bacillus* sp. ADR, exhibited laccase activity that catalyzes the oxidation of azo dyes (Kalme et al. 2009; Saratale et al. 2009c). Kalme et al. (2009) reported the purification and characterization of an extracellular laccase from a *Pseudomonas* sp. LBC1 and tested it against various azo dyes.

Lignin peroxidases (EC 1.11.1.14; LiP) are glycoproteins having around 38–46 kd molecular weight. Lignin peroxidases are reported for their recalcitrant aromatic compounds mineralization capacity such as dyes, polychlorinated biphenyls, PAHs (Krcmar and Ulrich 1998). Oxidation of nonphenolic aromatic lignin moieties and similar substances is catalyzed by lignin peroxidase. Lignin peroxidase catalyzes several oxidation reactions in the lignin side chains and related compounds (Tien and Kirk 1983), and one electron is abstracted to form reactive radicals of L form. The mechanism of azo dye oxidation by peroxidases such as LiP probably involves the oxidation of the phenolic group to produce a radical at the carbon bearing the azo linkages. Some investigators reported the purification and characterization of the lignin peroxidase activity from the *Aeromonas hydrophila* for the degradation of Crystal violet (Bharagava et al. 2018), from *Micrococcus glutamicus* for Green HE4BD (Saratale et al. 2009b) and *Lysinibacillus* sp RGS for Reactive Orange 4 (Saratale et al. 2015) was studied.

Tyrosinase (EC1.14.18.1), also known as monophenol monooxygenase, catalyzes the oxidation of phenol. Moreover, the involvement of tyrosinase in the degradation of Direct Blue-6 by *Pseudomonas desmolyticum* NCIM 2112 (Kalme et al. 2007) and disperse dye brown 3REL by a microbial consortium consisting of *Galactomyces geotrichum* MTCC 1360 and *Bacillus* sp. VUS (Jadhav et al. 2008) has been demonstrated. Substrate breakdown by desulfonation and oxidative cleavage is carried out in the presence of veratryl alcohol oxidase enzyme. Induction of veratryl alcohol oxidase activity during decolorization and degradation of Blue HERD was reported by Bedekar et al. (2014a). Decolorization of textile dyes, Direct Blue GLL, Red HE7B, methyl orange, Remazol black, and Red HE3B, using a purified form of this enzyme is also reported (Jadhav et al. 2008). Polyphenol oxidases oxidize single phenolic ring of amino acid tyrosine to form *o*-quinone hence referred as tyrosinases. Polyphenol oxidase acts as a catalyst for oxidation of phenolic compounds to form quinones in the presence of molecular oxygen. Polyphenol oxidases have been reported in *Streptomyces glaucescens*, *Streptomyces antibioticus*, *Bacillus licheniformis*, *Bacillus natto*, and *Bacillus sphaericus* (Echiqo and Ristuko 2001) (Figs. 7.1 and 7.2).

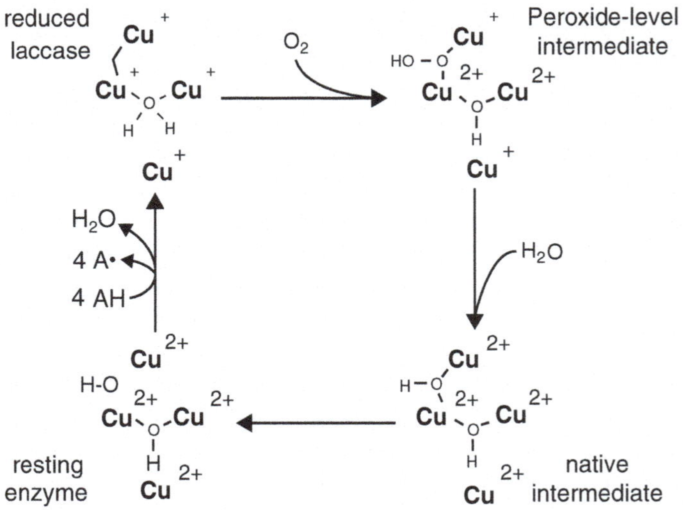

Fig. 7.1 The catalytic cycle of laccases. (Adapted from Wesenberg et al. 2003)

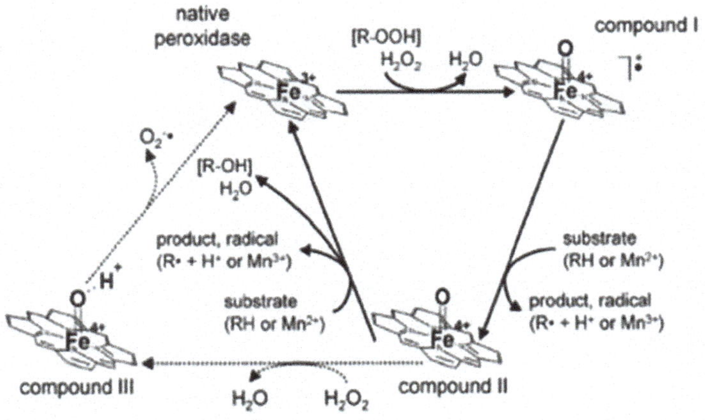

Fig. 7.2 Catalytic cycle of peroxidases. (Adapted from Wesenberg et al. 2003)

7.5 Bioreactors Studies for the Treatment of Dye-Containing Wastewater

7.5.1 Overview

Wastewater generated from textile industries is highly colored in nature due to the presence of dye. Dye-containing wastewater consists of many organic and inorganic substances. These substances are very complex in nature. The main contaminants present in the dye-containing wastewater include recalcitrant compounds, residuals of used dyes, color dyes, surfactants, chlorine compounds, alkalis, acids, amines, etc. This dye-containing wastewater has a higher level of biological oxygen demand (BOD), chemical oxygen demand (COD), suspended solids, dissolved solids, alkalinity, and pH and is strongly odorous in nature. The presence of refractory organics such as dyes, sizing agents, and dyeing aids provides recalcitrant and less biodegradable nature to these wastewaters. Various bioreactors (both aerobic and anaerobic) have been configured and designed to treat the dye wastewaters. In this section, the bioreactors employed to treat the dye wastewaters both aerobically and anaerobically are dealt in detail.

7.5.2 Bioreactors to Treat Dye Wastewaters

Conventional biological treatment systems are not efficiently designed to degrade these types of wastewaters since the color and chemicals present in these wastewaters are not efficiently removed. Once they enter the water bodies, they pollute them.

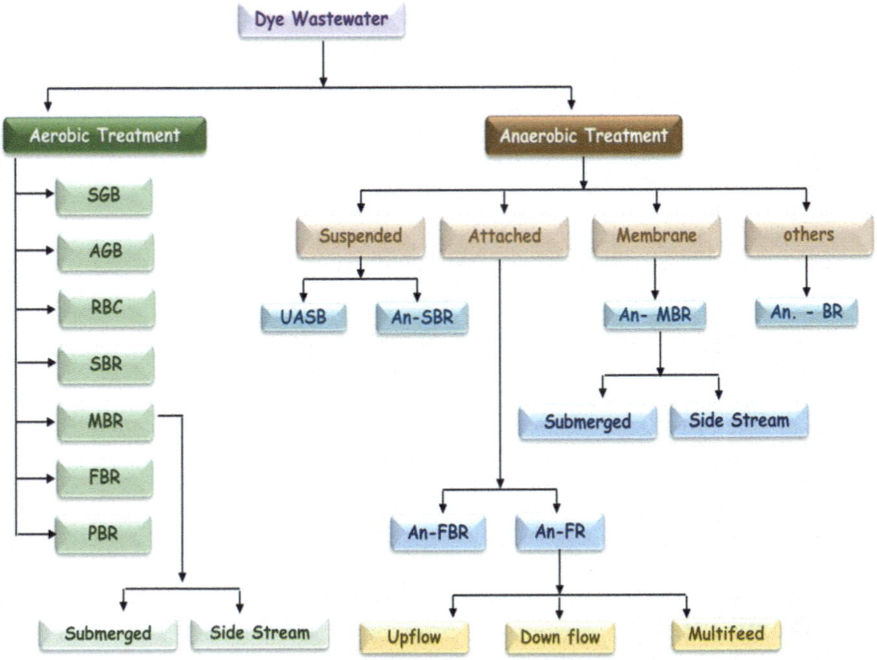

Fig. 7.3 Flow diagram of various types of bioreactors

The biological treatment process for dye wastewaters includes both aerobic and anaerobic process. A suitable bioreactor is essential to treat these wastewaters. The bioreactors are commonly categorized into two types, namely, aerobic and anaerobic reactors. Figure 7.3 shows the various types of aerobic and anaerobic reactors and their subdivisions employed to treat dye wastewaters.

7.5.3 Aerobic Bioreactors

Aerobic bioreactors are mainly employed to treat soluble organic matter and nitrogenous compounds of dye wastewaters at a high rate. Commercially existing aerobic reactors are mainly employed for decolorization, reduction of suspended solids, BOD, COD, and other pollutants. In small-scale industries, the aerobic reactors are operated as batch, stable volume, and completely mixing bioreactors, since the availability of influent is not continuous. To improve the contact of dissolved oxygen, microbial population, and wastewater, a complete mixing is provided in aeration compartment. The aerobic bioreactors are mainly classified into the following types: suspended growth bioreactors, attached growth bioreactors, rotating biological contactors, sequencing batch bioreactor, membrane bioreactors, fluidized-bed bioreactors, and packed bed bioreactors.

Suspended growth bioreactors (SGB): The suspended growth bioreactors are otherwise called scaled down bioreactors. These reactors have two ports such as an inlet for influent access and an outlet for the effluent exit. An aerator is suspended within the reactor to provide aeration and a baffle for complete mixing. In these types of reactors, the organics are utilized by the bacteria present in the effluent (Farhadian et al. 2008). The dispersed bacterial community that is present in the biological aeration chambers is utilized by the protozoa and rotifers. The protozoa and rotifers serve as predators of bacterial community, and thus the dispersed bacteria are segregated. The biomass is completely mixed with the easily biodegradable portion of organics. The microbial community agglomerates with each other and forms a mass-like community structure called biological flocs. The mass of microbial community along with the wastewater is called as mixed liquor.

Attached growth bioreactors (AGB): Attached growth bioreactors are also called as fixed-film bioreactors. These types of reactors have an inert carrier medium for the microbial attachment. The wastewater follows a cross flow through the medium, and this flow assists the biofilm (biological flocs) to assimilate the fine and colloidal soluble organics. The wastewaters and the dissolved oxygen contact with this attached biofilm by two ways: by pumping the wastewaters over the inert carrier medium and by moving the inert carrier through the wastewaters. In these types of reactors, attached growth is provided as suspended growth in the same chamber. In case of turbulent flow, channels having higher diameter are used. The attached growth biofilm is submerged, and it flows above the surfaces. The submerged growth biofilm will develop in the channels. Aeration is supplied by two means: (1) direct air injection and (2) by the flow of water into the air-liquid interface. The excess attached growth will be settled down, and this must be removed regularly for proper maintenance. These reactors have the following advantages: (1) strengthened microbial biofilm and (2) the treated wastewaters contain less amount of residual solids and BOD.

Rotating biological contactor (RBC): These bioreactors are configured as the merge of both suspended and attached growth reactors. In these types of reactors, many circular disks are positioned closely and are fixed on a shaft. These disks are rolled slowly. Approximately, 35–40% of the disk is suspended in the wastewater, and the shaft is positioned above the water surface. This configuration facilitates the contact of air and dye wastewaters regularly with that of disk surface during its rotation. Materials such as polystyrene or polyvinyl chloride are often employed as a disk. The biofilm develops on the surface of the disk. At the time of the suspension, the microbes come in contact with the dye wastewaters. If the disks are configured in sequence and sometimes in stages, then an elevated amount of organics can be removed (Kapdan and Kargi 2002). Therefore, the organic loading rate decreases based on the rate of the reactor. The fixed biofilm utilizes oxygen once it is exposed to the air during the rotation of the disk. If an excess of oxygen is absorbed, then it is mixed with the wastewater. During this time, the surface of the contactor moves backward via the wastewater. Once the size of the disk attached biofilm is increased, then some part of excess growth is settled. This excess biomass must be removed for proper maintenance. The reactor tanks are normally sealed to prevent odor problems. The main advantages of RBC are as follows: (1) the rotating disks provide more interfacial space to facilitate the contact between the biofilm and the pollutants present in the wastewater, and (2) the presence of mixed microbial population aids in the complete biodegradation of dye wastewaters.

7.5.4 Sequencing Batch Reactor (SBR)

These types of reactors are also called as suspended growth activated sludge reactors (Vives et al. 2003). In SBR, sequential processes such as equalization of flow, aeration, clearance, and wastage of microbial community are performed in the same tank. SBRs are mainly used to treat the dye and recalcitrant compounds present in the water. These reactors provide selective pressures which select specific microbes that are capable to biodegrade the specific pollutant. Thus, these reactors can be used to treat dye wastewaters which are recalcitrant in nature. To prolong the durability of SBR functioning, the reactor tank should not be entirely drained out; instead, some amount of suspended solids should be retained in the bioreactor. The solids that are settled can be employed as inoculum for the subsequent rotation. The aerobic SBRs are employed for color removal and biodegradation of dye wastewaters such as azo dye-containing wastewater (Sandhya et al. 2005), azo dye, acid red (Buitron et al. 2004), and diazo dyes (Lourenco et al. 2001). One cycle of SBR involves the following four steps:

(a) Filling – during this stage, 25% of the total cycle time may be utilized. The bioreactor is filled with dye wastewater from the primary unit of treatment.
(b) Reacting – 35% of the total time is utilized for this phase. For complete biodegradation of organics and nitrogen-containing compounds, aeration is provided.
(c) Settling – 20% of the total time is utilized for this phase. This is the clarification phase during which separation of liquid-solid takes place. The introduced air is removed to maintain anoxic condition.
(d) Decanting – 20% of the total time is utilized for this phase. During this phase, the treated wastewater (clear supernatant) is removed with the help of weirs. The solids which settled in the bottom are segregated.

7.5.5 Membrane Bioreactors (MBR)

Membrane bioreactor (MBR) is the biological reactor that consists of membranes which unite both clarification and filtration of activated sludge process into a one-stage process. The separation of sludge can be done easily with MBR as the surface of the membrane serves as a barrier. During operation, the colloidal particles, bacterial community, and other biomasses retain in the membrane, whereas the treated wastewater moves through the membrane. As a result, the volume of the reactor and the sludge generation are highly reduced. These bioreactors can operate sludge concentration up to 12,000 mg/L. These reactors are of two types, namely, submerged and side stream. In submerged MBR, the reactor vessel has membrane units, air diffusers, and an outlet for sludge exit. The reactor is added with activated sludge. The permeable substances can be removed with the help of a pump. The membrane unit generates transmembrane pressure. This pressure can be monitored using a sensor. The two commonly employed membrane modules in submerged bioreactors are flat sheet and hollow fiber membrane module. Compared to flat sheet membranes, hollow fiber membranes have less space between them. As a result, a greater density of growth can be obtained. On the other hand, it causes clogging problems and makes the cleaning difficult (Le-Clech et al. 2005). The major advantages of membrane bioreactors are (1) achievement of greater solid-liquid separation, (2) less energy input, (3) ease of operation, (4) pollution-free, (5) solids biodegraded entirely, (6) reduced volume of the reactor which increases the loading rate and allows the biomass to develop inside the reactor, and (7) decoupling of solid retention time from hydraulic retention time due to retention of biofilm inside the bioreactor.

7.5.6 Fluidized-Bed Reactors (FBR)

These bioreactors can treat larger volumes of dye wastewaters. Wastewaters can be fluidized only when the smaller solid particles are suspended in an upward path. In FBRs, the microbial community is suspended in the aqueous phase. A permeable disk is located at the base of the reactor. This assists uniform flow of wastewaters. Aeration is maintained by adequate supply of air at less flow rate below the distributor plate. When the bed is fluidized, a decrement in pressure is achieved across the bed. At the same time, the height of the bed increases with increase in flow. These reactors are functioned at high speed with very less biomass loss. The advantages of fluidized-bed reactors are as follows: (1) no temperature gradients in the bed as there is complete mixing of solids; (2) solids can be separated easily; and (3) greater heat transfer rate can be achieved due to vigorous movement of solids.

7.5.7 Packed-Bed Reactors (PBR)

In these reactors, the dye wastewater is introduced at the apex of the clean packed bed which serves as an adsorbent. Packed-bed reactors employ single or mixed microbial population of bacteria and fungi for decolorization of dye wastewaters. In addition to color removal, these reactors are also employed for higher reduction of BOD and COD. Solids are initially removed from the adsorbent located at the bottom of the reactor. This zone begins to expand till the tip of the zone touches the bottom of reactor. During this breakthrough time, the effluent increases hastily. The breakthrough time is defined as the adsorption time at which the effluent concentration is 10% of the influent concentration (Nidheesh et al. 2012). Intensity of zone, rate of adsorption, and capability of adsorption are some of the major factors to be taken into account while designing packed-bed reactor (Suresh and Babu 2010). In these reactors, transport of solute within the reactor is the critical rate determining factor. Various models such as Weber and Morris, Boyd, and Urano-Tachkawa have been proposed to determine the adsorption process of these processes (Shaverdi 2012). In our studies we have studied *Proteus vulgaris* NCIM-2027 cells immobilized on different polymeric immobilized matrix and *Luffa cylindrica* (Loofa) on the decolorization of individual dye C.I. Reactive Blue 172 and mixture of textile dyes and also mineralization performance in terms of total organic carbon (TOC) and chemical oxygen demand (COD) reduction at 37 °C and pH 8.0 under static incubation (Saratale et al. 2012). The results are depicted in Fig. 7.4.

Packed bed reactor with *Proteus vulgaris* cells immobilized (A) calcium alginate and (B) on Loofa

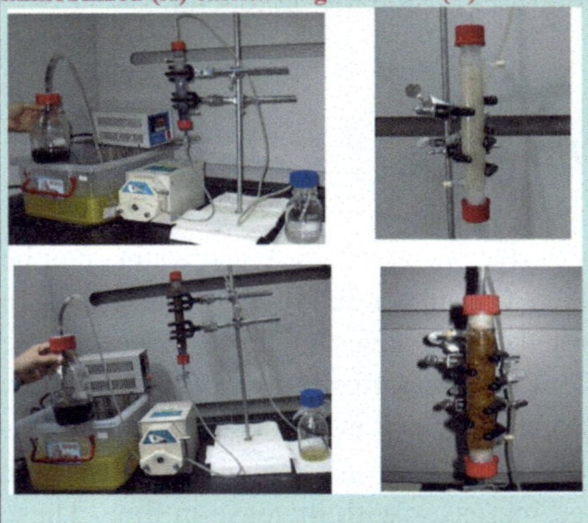

Dye degradation pathway using enzymatic and analytical studies

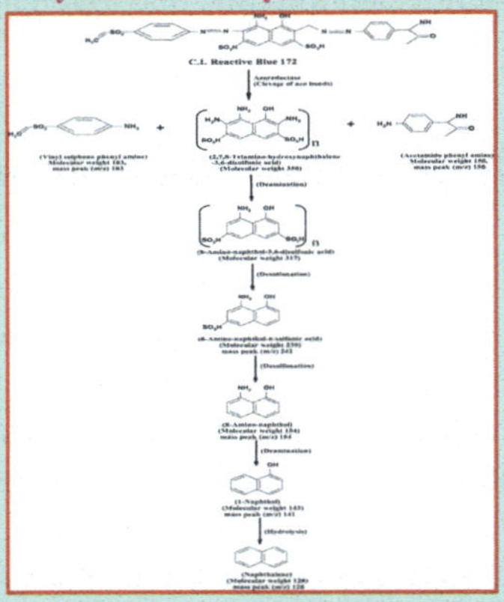

The decolorization performance, TOC and COD reduction of C.I. Reactive Blue 172 and mixture of nine reactive dyes (each at 30 mg L^{-1}) in fixed bed decolorizer containing *P. vulgaris* cells immobilized on Loofa at different volumetric feeding rate and feeding dye concentration

Dye concentration	Parameters	Volumetric feeding rate ($mL\ h^{-1}$)		
		10	20	30
50 mg L^{-1}	Specific decolorization rate	1.24	2.75	3.65
	TOC reduction (%)	82	78	75
	COD reduction (%)	82	76	74
100 mg L^{-1}	Specific decolorization rate	2.21	5.08	5.98
	TOC reduction (%)	78	72	70
	COD reduction (%)	75	69	68
150 mg L^{-1}	Specific decolorization rate	3.20	6.75	7.55
	TOC reduction (%)	72	65	60
	COD reduction (%)	68	62	58
200 mg L^{-1}	Specific decolorization rate	4.04	8.66	6.51
	TOC reduction (%)	58	46	36
	COD reduction (%)	54	46	42
250 mg L^{-1}	Specific decolorization rate	3.62	6.66	4.02
	TOC reduction (%)	38	28	20
	COD reduction (%)	35	22	20
Mixture of reactive dyes (each 50 mg L^{-1})	Specific decolorization rate	6.66	14.12	7.49
	TOC reduction (%)	55	28	ND
	COD reduction (%)	49	22	ND

Phytotoxicity studies of dye and degraded metabolites.

Fig. 7.4 Flow diagram of various types of bioreactors

7.5.8 Anaerobic Bioreactors

Compared to aerobic treatment, anaerobic treatment of dye wastewater is beneficial due to the following advantages: (1) recalcitrant organics and any undesirable substances can be easily biodegraded, (2) biodegradability of wastewater is greatly improved, (3) the removal efficiency of solids is higher, and (4) no requirement of oxygen which reduces the cost and energy consumption. The high rate bioreactors are mainly employed for treating dye wastewaters. These bioreactors have many advantages such as high organic loading rate (OLR), low hydraulic retention time (less than 10 h), and requirement of less space. The high-rate anaerobic bioreactors are of four types, namely, suspended growth, attached growth, membrane bioreactors, and other bioreactors.

Suspended growth bioreactors: In these types of bioreactors, the anaerobic processes are mediated by microbial aggregates (microbial flocs or immobilized granules). These microbial aggregates treat the dye wastewaters and are retained in suspension through suitable mixing processes. These reactors are of two types, namely, upflow anaerobic sludge blanket (UASB) bioreactors and anaerobic sequencing batch (An-SBR) bioreactors.

Upflow anaerobic sludge blanket (UASB) bioreactors: In UASB bioreactor, OLR and HRT are regulated to assist the generation of anaerobic granules (Khanal 2008). These bioreactors have a reactor tank with separate phases for solid, liquid, and gas on its upper portion (Rajesh Banu et al. 2006, 2007a). The substrate (dye wastewaters) is introduced and circulated evenly from the base of the bioreactor through the influent inlet. These wastewaters are then transported via the reaction phase which contains anaerobic granules of about 0.5–2 mm in dia. The granules contain numerous microbial populations which include acidogenic, acetogenic, and methanogenic microbes (Rajesh Banu and Kaliappa 2007; Rajesh Banu et al. 2007b). These granules enhance the catabolic reactions among the microbes and in addition protect the perceptive microbes (methanogenic microbes) from the harmful environment. These granules are shaken regularly in the bed of sludge through superficial flow velocity, and as a result, bubble of biogas generates. Once the introduced feed wastewater contacts the granules, anaerobic biodegradation of the organics takes place and resulted in the generation of biogas. The generated gas bubbles facilitate sufficient mixing by hydraulic instability. This biogas is separated through the gas liquid-solid separator. In the apex of the bioreactor, the biogas and the treated wastewaters are collected in a separate manner through various outlets. The fine-sized particles are settled in the reactor back. The main advantages of these reactors are the following: (1) they do not need mechanical mixers, and (2) the simplicity of the bioreactor configuration makes the process beneficial and cost-effective.

Anaerobic sequencing batch reactors (An-SBR): In these types of bioreactors, the operation takes place in four steps. The first step is called filling step. During this step, the dyeing wastewater is introduced into the bioreactor. The second step is called reaction step. At this step, the organic matter is removed from the wastewater. The third step is called settle step. During this third step, the microbial aggregates are detached from the wastewater. The final step is called decant step during which the treated wastewater is collected with the maintained biomass in the bioreactor. The parameters that influence the performance of the bioreactor include reactor configuration, temperature, mixing mode, suspended solids concentration, hydraulic retention time, and organic loading rate. The advantages of these bioreactors include (1) excellent flocculation, (2) enhanced solids separation, (3) extended retention time, and (4) proficient utilization of organic compounds for methane generation. In comparison with upflow anaerobic sludge blanket reactors, these An-SBR do not need substrate dispersion and gas-solid-liquid separation; hence it is simpler in configuration. In case of lower and medium organic loading rates, the operation mode of UASB and An- SBR is similar. In case of higher organic loading rates, UASB is more efficient than ASBR on the basis of COD removal, biogas generation, and VFA generation.

Attached growth bioreactors: The attached growth bioreactors are also called anaerobic biofilm reactors. In attached growth bioreactors, the microorganisms which mediate the treatment process developed by attaching to an inert medium such as pebbles, synthetic fillings, etc. The removal solids from substrates and transport of electrons take place among the microbial biofilm structure. The substrate reduction rates are mediated by diffusion. These reactors are of two types based on the wastewater flow direction (Korsak 2008). They are anaerobic filter bioreactors and anaerobic fluidized-bed bioreactors.

Anaerobic filter bioreactors (An-FR): Anaerobic filter bioreactors are packed bed bioreactors in which the wastewaters to be treated flow via the fixed-packed biofilm. Anaerobic filter bioreactors work under strict anaerobic and swamped conditions in which the air won't contact with the biomass. In these reactors, the dye wastewaters flow via a column packed with an inert medium for the biofilm development. These reactors efficiently treat wastewater with reduced sludge generation. Based on feeding mode, these reactors are of three types, upflow, downflow, and multi-feed filters. In upflow anaerobic filter bioreactors, the wastewater is introduced in an upward direction from the bottom of the bioreactor (Rajkumar et al. 2011). In downflow anaerobic filter bioreactor, the wastewater is introduced in a downward direction from the top of the bioreactor through a packed bed of biofilm adhered to an inert medium. In multi-feed anaerobic filter bioreactors, the wastewater is introduced through various feed systems of the bioreactors. Ten percent (v/v) of the inoculum is usually employed for these reactors to achieve an improved outcome. The OLR should be in the range of 1–10 kg COD/m^3.d to achieve 75–85% organic removal. The HRT should be kept in the range of 18–24 h.

Anaerobic fluidized-bed reactors (An-FBR): In these types of bioreactors, the aggregates attach to the carrier inert medium. In these reactors, both fluidization and mixing occur. These bioreactors work under high flow velocity of about 10–25 m/h. The biofilm bed expansion ranges from 25% to 30% of the volume of settled bed.

This reduces the clogging problems and short circuits, upholds more biomass load, and works under long sludge retention time. The high upflow velocities make the carrier material containing biomass load to move liberally in the bioreactor. An-FBR has the following advantage: Good mixing avoids short circuit and development of dead zones; therefore it requires minimal space.

7.5.9 *Membrane Processes*

In membrane processes, the membrane serves as a barrier, and it selectively allows only some compounds to move through it. By retaining other materials, the biological system performance can be protected. The membrane process avoids biomass loss and decreases the concentration of pollutants by pollutant separation.

Anaerobic membrane bioreactor (An-MBR): In anaerobic membrane bioreactors, the anaerobic treatment combined with membrane separation for efficient treatment of dye wastewaters. These bioreactors are affected by various parameters such as shear force, water flux, mode of operation, temperature, sludge retention time, and hydraulic retention time (Rajesh Banu et al. 2011). The other factors that should be taken into account are physiochemical characteristics of the sludge and also design configuration of membrane modules. The main advantages of anaerobic bioreactors are (1) improved biomass retention, (2) improved effluent characteristics, and (3) less sludge generation. These reactors are of two types, namely, sidestream and submerged membrane bioreactors (Uan et al. 2009, 2012, 2013). In sidestream reactors, the membranes are located in an external loop, whereas in the submerged system, the membranes are immersed within the bioreactor. The main advantages of the submerged bioreactor are minimal space and energy requirement; when compared to sidestream, it does not need energy input for continuous flow through the membrane. The main disadvantage of the submerged system is membrane fouling (Judd 2010).

Anaerobic baffled reactor (An-BR): In these types of bioreactors, a sequence of perpendicular baffles is fitted in both upflow and downflow compartment. These reactors give reliable treatment efficiency for dye wastewaters containing particulate organics. The wastewater to be treated flows through compartments to obtain phase separation and improved biodegradation (Manariotis and Grigoropoulos 2002). The main advantages of these bioreactors are (1) excellent retention of solids and (2) steady performance of bioreactor by enhancing the growth conditions of the diverse microbial population (both acidogenic and methanogenic microbes). These advantages can be exploited for dye wastewater treatment under harsh environmental circumstances which are rigorous hydraulic and organic load shock, discontinuous feed, temperature alterations and due to recalcitrant compounds.

7.6 Toxicity of Dyes and Their Studies

Variety of synthetic dyestuff released by the textile industries poses a threat to the environmental safety. It is estimated that 10–15% of the dyes are lost in the effluent during the dyeing process (Robinson et al. 2001; Saratale et al. 2011c), and in the case of reactive azo dyes, up to 50–75% of the dye is lost during the dyeing process which shows negative esthetic effect on the wastewater (Saratale et al. 2010). The release of textile and dyehouse effluent may cause abnormal coloration of surface waters that creates the problems for both the public and the environment. There are not only artistic problems; the greatest environmental concern with the azo dyes is their absorption, water quality, and penetration of sunlight which directly affects the aquatic flora and fauna. In addition, the human health impact of azo dyes has caused concern for a number of years. Recently, some research has been carried out to study the effects of dyestuffs and dye-containing effluents on the activity of both aerobic and anaerobic bacteria in wastewater treatment systems (Pandey et al. 2005). Furthermore, the occupational exposure to dyestuffs of human workers in dye manufacturing and dye utilizing industries has received attention. It was observed that purified forms of azo dyes are directly mutagenic and carcinogenic except for some azo dyes. The bacterial degradation of azo dyes is carried out with three mechanisms: (i) azo dyes that are toxic only after reduction and cleavage of the azo linkage to give aromatic amines, mostly via intestinal anaerobic bacteria (the aromatic amines are metabolically oxidized to reactive electrophilic species that covalently bind DNA), (ii) azo dyes with structures containing free aromatic amine groups that can be metabolically oxidized without azo reduction, and (iii) azo dyes that may be activated via direct oxidation of the azo linkage to highly reactive electrophilic diazonium salts. Each mechanism may be compound specific; thus azo toxicity is probably caused by more than one mechanism. It was observed that the transformed intermediates of azo dyes are highly toxic and mutagenic in nature. Obviously, lack of adequate toxicity figures of contaminants to cell populations makes bioremediation (e.g., decolorization or degradation) unpredictable and unreliable for on-site operation. Toxicity of azo dyes and their metabolic intermediates has been investigated by many researchers. This toxicity can be studied by assessing genotoxicity, cytotoxicity, phytotoxicity, and microbial toxicity of the sample of before and after bacterial treatment.

7.7 Current Bottlenecks and Challenges for Biological Treatment

7.7.1 Removal of Nonbiodegradable Compounds and Micropollutants

Biological treatment methods are widely used effective methods for treatment of dye wastewaters. Even though these methods are extensively used, they have some limitations, bottlenecks, and challenges which have to be addressed. The major problem of dye wastewaters is color and nonbiodegradable substances. There is a practical limit to remove COD and BOD removal in biological treatment methods. But in case of nonbiodegradable substances, the biological treatment lacks the ability to treat them. In those circumstances, the biological methods cannot be used as a single method. It can be combined with other methods to treat the dye-containing wastewater for color removal and removal of nonbiodegradable substances.

Presence of organic micropollutants such as aromatic amines in dye wastewater is also a drawback which affects the quality of treated effluent. Removal of micropollutants is also a major problem in biological treatment methods. Even though the biological treatment methods are less expensive, current methods are not capable of degrading the wide range of micropollutants. Since the recent biological treatment methods are developed to treat the pollutants at mg/L, the presence of pollutants at ng/L can be treated by designing and configuring new treatment strategies. These micropollutants are very difficult to degrade by biological treatment since their chemical nature is so diverse. Biological treatment methods are not able to remove these pollutants, and as a result, they are detected in treated wastewater. No particular approaches have been effectively designed to improve removal of micropollutant through biological treatment systems. The next phase in the design of biological treatment systems should focus on designing and implementing new and novel metabolic qualities of microbes or by stimulating the production of enzymes capable of removing micropollutants.

7.7.2 *Production of Excess Sludge*

Production of excess sludge is an important drawback in biological treatment systems. It is a rising challenge that has to be addressed in an efficient way. Benefits of excess sludge reduction can be understood from many strategies such as lysis cryptic growth, uncoupling metabolic traits, and maintenance of metabolic reactions, but the cost incurred by these strategies must be taken into account. These are the major bottlenecks that still have to be overcome. The issues associated with these strategies have to be solved for their practical implementation such as acclimatization of chemical uncouplers and enzyme-secreting microbial population. The ecological effect (odor problems, the presence of toxic chemical uncouplers in the treated wastewaters) stimulated by the strategies of sludge reduction should be considered.

7.7.3 *Microbial Cultures for Dye Decolorization and Biomass Loss*

Reactive azo dyes are the main constituents in dye wastewaters. They are mainly utilized for dyeing cotton. Even though these reactive dyes are easily hydrolyzable and get converted into more fractions of unfixed reactive dyes. A 50% of these dyes are present in the treated wastewater. Treatment of these dyes by aerobic treatment is possible by using pure cultures of bacteria and fungi. But the process is not economically feasible since the isolated culture is dye specific (Sen and Demirer 2003). The anaerobic treatment for dye decolorization is considered to be efficient, but its main drawback is the addition of external carbon such as glucose to get efficient color removal. It is the main issue that should be taken into account while analyzing the practical applicability of this process. Anaerobic treatment of dye wastewaters needs the united action of the diverse microbial population which possesses different metabolic traits. To achieve an effective treatment process, all the sequential phases concerned with the decomposition of organics and the microbes mediating these processes must work in a coordinated way. Methanogenic microbial populations have lengthy duplication period up to 30 days. These microbes are recognized to be the very sensitive ones, and it mediates process effectiveness. Therefore it is essential to protect this microbial population from being washed out during operation and treatment. Even though many discoveries have been developed to overcome these issues, still many studies regarding removal of toxic compounds, a combination of biological treatments with membrane separation has to be improved more. A better understanding of the metabolic reaction of microbial population and the interaction of different microbial population is still needed.

7.7.4 Bioreactor Configuration

Even though the anaerobic treatment process is considered to be very effective due to its less energy input and bioenergy generation, its application is limited in treating dye wastewaters because of less substrate concentration which needs further investigations and developments. The budding MBR technique with less energy generation and improved quality effluent is a promising method to treat dye wastewaters. In these types of bioreactors, membrane fouling is a serious problem to be addressed. To overcome this, anaerobic fluidized membrane bioreactor with granular activated carbon (GAC) carrier would be efficient to prevent membrane fouling and obtaining high biomass retention and removal organics. On the other hand, the higher operational and capital cost associated with GAC addition and fluidization of these reactors is still need to be considered. Many of the anaerobic treatment systems are unable to meet the treated effluent release standard. To overcome this limitation, hybrid reactors are essential which unite two anaerobic treatment systems or combine anaerobic with aerobic treatment systems (Rajesh Banu et al. 2014). These hybrid reactors gained much attention due to its simplicity and reduced sludge generation. However, scaling of these treatment systems is lacking. Further research and development are required for evaluating their operation on large scales and the examples are presented in (Table 7.1).

Table 7.1 Different type of reactors studied for the decolorization textile dyes and textile wastewater

Wastewater	Type of bioreactor	Color and COD removal efficiency (%)	Concentration	HRT	Other operating conditions	References
Reactive azo dyes	Fixed-film reactor	Color: 99.5 COD: 97.5	Initial conc.: 300 mg/L	HRT: 24 h	COD: 7200 mg/L	Balapure et al. (2015)
Synthetic reactive azo dyes	Continuous biofilm reactor	Color: 70–80	Initial conc.: 30 mg/L	HRT: 12 h	COD: 500–900 mg/L	Yang et al. (2009)
Textile wastewater	Continuous biofilm reactor	Color: 89	Initial conc.: 200–320 dilutions	HRT: 12 h	COD: 750–1175 mg/L	Deowan et al. (2013)
Textile wastewater	Submerged membrane bioreactor	Color: 20–70 COD: ~90	COD: 2450 mg/L	HRT: 40–80 h	Permeate flux, 4 L/m^2 h; TMP, 50 mbar; MLSS, 12 g/L; pH, 8.2–10.5; temperature, 18.2 °C	Blanquez et al. (2008)
Textile wastewater	Air-pulsed bioreactor using *Trametes versicolor*	Color: >90	Initial conc.: 150 mg/L	HRT: 48 h	–	Selvakumar et al. (2010)
Procion Blue 2G	Membrane bioreactor biochemical oxidation using *Pseudomonas aeruginosa*	Color: 100 COD: 90	COD: 2579 mg/L	–	–	Konsowa et al. (2011)
Azo dye Acid Orange 7	Membrane bioreactor	Color: ~100 COD: 60–80	Initial conc.: 50–400 mg/L	HRT: 4–24 h	COD: 95–550 mg/L	Ramavandi et al. (2014)
Textile wastewater	Packed-bed column	Color: 93–100 COD: ~50	–	Flow rate: 0.5 ml mL/min	Packing, *Candida tropicalis* within sodium alginate matrix; initial pH, 3.0–6.0; bed height, 5–15 cm	Duarte et al. (2013)

(continued)

Table 7.1 (continued)

Wastewater	Type of bioreactor	Color and COD removal efficiency (%)	Concentration	HRT	Other operating conditions	References
Reactive Blue 172 and mixture of dyes	Fixed-bed column containing *Proteus vulgaris* NCIM-2027 immobilized on *Luffa cylindrica* sponge	COD: 35–82	Initial concentration 50–250 mg/L; mixture of reactive dyes (each 50 mg L^{-1})	20 mL h^{-1}, rather than at 30 mL h^{-1}	Glass columns (ca. 2.2 cm in diameter) with the fixed-bed length of 15 cm	Ramavandi et al. (2014)
		COD: 49				
Anionic dye Orange II	Packed-bed column	Color: 74–94	Concentration: 50–200 mg/L	Flow rate: 4–10 mL/min	Packing, bimetallic chitosan; initial pH, 3.0–6.0; bed height, 3e7 cm	Saratale et al. (2012)
Synthetic textile effluent	Fluidized-bed reactor	Color: 76	Initial COD: 2000 mg/L		Temperature, 50 °C; Cr(VI), 5–45 mg/L	Cirik et al. (2013)
		COD: 60				
Reactive Black 5, Reactive Orange 16, and Reactive Blue 2	Fluidized-bed reactor	Sulfate: 50	–	HRT: 100 min	–	Su et al. (2011)
		chromium: 93				
		Color: 82–100				
		COD: 57–91				
Reactive Blue 13	Fluidized-bed reactor	Color: 83	–	HRT: 70 h	pH, 7.0; residence time, 70 h; glucose conc., 2 g/L	Lin et al. (2010)
		COD: 91				
Azo dye (Red RBN)	Fluidized-bed reactor	Color: 90	Initial dye conc.: <2200 mg/L	HRT: 8 h	Polyvinyl alcohol-immobilized cell beads	Wu et al. (2005)

Acknowledgments This research was completely supported by Dongguk University, Seoul, South Korea, under research fund 2018-2020.

References

Acikgoz C, Ülküye DG, Özan K, Borazan AA (2016) Degradation of Reactive Blue by the mixed culture of *Aspergillus* versicolor and Rhizopus arrhizus in membrane bioreactor (MBR) system. Desalin Water Treat 57:3750–3756

Acuner E, Dilek FB (2004) Treatment of tectilon yellow 2G by *Chlorella vulgaris*. Process Biochem 39:623–631

Anjaneyulu Y, Sreedhara Chary N, Raj DSS (2005) Decolourization of industrial effluents-available methods and emerging technologies-a review. Rev Environ Sci Biotechnol 4:245–273

Arora PK, Srivastava A, Singh VP (2014) Bacterial degradation of nitrophenols and their derivatives. J Hazard Mater 266:42–59

Arora PK, Srivastava A, Garg SK, Singh VP (2018) Recent advances in degradation of chloronitrophenols. Bioresour Technol 250C:902–909

Ambrósio ST, Campos-Takaki GM (2004) Decolorization of reactive azo dyes by Cunninghamella elegans UCP 542 under co-metabolic conditions. Bioresour Technol 91:69–75

Aksu Z (2003) Reactive dye bioaccumulation by saccharomyces cerevisae. Process Biochem 38:1437–1444

Balapure K, Bhatt N, Madamwar D (2015) Mineralization of reactive azo dyes present in simulated textile wastewater using down flow microaerophilic fixed film bioreactor. Bioresour Technol 175:1–7

Banat IM, Nigam P, Singh D, Marchant R (1996) Microbial decolorization of textile-dye-containing effluents: a review. Bioresour Technol 58:217–227

Barragán BE, Costa C, Márquez MC (2007) Biodegradation of azo dyes by bacteria inoculated on solid media. Dyes Pigments 75:73–81

Bedekar PA, Saratale GD, Saratale RG, Govindwar SP (2014a) Development of low cost UCB reactor for degradation and detoxification of Blue HERD and textile effluent by *Lysinibacillus* sp. RGS immobilized on Loofa. Int Biodeterior Biodegrad 96:112–120

Bedekar PA, Saratale RG, Saratale GD, Govindwar SP (2014b) Oxidative stress response in dye degrading bacterium *Lysinibacillus* sp. RGS exposed to Reactive Orange 16, degradation of RO16 and evaluation of toxicity. Environ Sci Pollut Res 21:11075–11085

Bharagava RN, Saxena G, Mulla SI, Patel DK (2017a) Characterization and identification of recalcitrant organic pollutants (ROPs) in tannery wastewater and its phytotoxicity evaluation for environmental safety. Arch Environ Contam Toxicol. https://doi.org/10.1007/s00244-017-0490-x

Bharagava RN, Saxena G, Chowdhary P (2017b) Constructed wetlands: an emerging phytotechnology for degradation and detoxification of industrial wastewaters. In: Bharagava RN (ed) Environmental pollutants and their bioremediation approaches, 1st edn. CRC Press, Taylor & Francis Group, Boca Raton, pp 397–426. https://doi.org/10.1201/9781315173351-15

Bharagava RN, Chowdhary P, Saxena G (2017c) Bioremediation: an ecosustainable green technology: its applications and limitations. In: Bharagava RN (ed) Environmental pollutants and their bioremediation approaches, 1st edn. CRC Press, Taylor & Francis Group, Boca Raton, pp 1–22. https://doi.org/10.1201/9781315173351-2

Bharagava RN, Mani S, Mulla SI, Saratale GD (2018) Degradation and decolourization potential of an ligninolytic enzyme producing Aeromonas hydrophila for crystal violet dye and its phytotoxicity evaluation. Ecotox Environ Safety 156:166–175

Bhosale SK, Saratale GD, Govindwar SP (2006) Biotransformation enzymes in *Cunninghamella blakesleeana* (NCIM-687). J Basic Microbiol 46:444–448

Blanquez P, Sarra M, Vicent T (2008) Development of a continuous process to adapt the textile wastewater treatment by fungi to industrial conditions. Process Biochem 43:1–7

Buitron G, Quezada M, Moreno G (2004) Aerobic degradation of the azo dye acid red 151 in a sequencing batch biofilter. Bioresour Technol 92:143–149

Chandra R, Saxena G, Kumar V (2015) Phytoremediation of environmental pollutants: an eco-sustainable green technology to environmental management. In: Chandra R (ed) Advances in biodegradation and bioremediation of industrial waste, 1st edn. CRC Press, Taylor & Francis Group, Boca Raton, pp 1–30. https://doi.org/10.1201/b18218-2

Chen KC, Huang WT, Wu JY, Houng JY (1999) Microbial decolorization of azo dyes by *Proteus mirabilis*. J Ind Microbiol Biotechnol 23:686–690

Chen H, Hopper SL, Cerniglia CE (2005) Biochemical and molecular characterization of an azoreductase from *Staphylococcus aureus*, a tetrameric NADPH-dependent flavoprotein. Microbiology 151:1433–1441

Christian V, Shrivastava R, Shukla D, Modi HA, Vyas BRM (2005) Degradation of xenobiotic compounds by lignin-degrading white-rot fungi: enzymology and mechanism involved. Indian J Exp Biol 43:301–312

Cirik K, Dursun N, Sahinkaya E, Cinar O (2013) Effect of electron donor source on the treatment of Cr (VI) containing textile wastewater using sulfate-reducing fluidized bed reactors (FBRs). Bioresour Technol 133:414–420

Daeshwar N, Ayazloo M, Khataee AR, Pourhassan M (2007) Biological decolorization of dye solution containing Malachite Green by microalgae *Cosmarium* sp. Bioresour Technol 98:1176–1182

Deowan SA, Galiano F, Hoinkis J, Figoli A, Driolic E (2013) Submerged membrane bioreactor (SMBR) for treatment of textile dye wastewater towards developing novel MBR process. APCBEE Procedia 5:259–264

Duarte F, Morais V, Maldonado-Ho'dar FJ, Madeira LM (2013) Treatment of textile effluents by the heterogeneous Fenton process in a continuous packed-bed reactor using Fe/activated carbon as catalyst. Chem Eng J 232:34–41

Echiqo O, Ritsuko T (2001) Bacterial polyphenol oxidase from bacillus for use in oxidation of colored substances. US Patent No. 6184014

Farhadian M, Duchez D, Vachelard C, Larroche C (2008) Monoaromatics removal from polluted water through bioreactors a review. Water Res 42:1325–1341

Forgacs E, Cserhati T, Oros G (2004) Removal of synthetic dyes from wastewaters: a review. Environ Int 30:953–971

Gautam S, Kaithwas G, Bharagava RN, Saxena G (2017) Pollutants in tannery wastewater, pharmacological effects and bioremediation approaches for human health protection and environmental safety. In: Bharagava RN (ed) Environmental pollutants and their bioremediation approaches, 1st edn. CRC Press, Taylor & Francis Group, Boca Raton, pp 369–396. https://doi.org/10.1201/9781315173351-14

Golab V, Vinder A, Simonic M (2005) Efficiency of the coagulation/flocculation method for the treatment of dyebath effluent. Dyes Pigments 67:93–97

Goutam SP, Saxena G, Singh V, Yadav AK, Bharagava RN (2018) Green synthesis of TiO_2 nanoparticles using leaf extract of *Jatropha curcas* L. for photocatalytic degradation of tannery wastewater. Chem Eng J 336:386–396. https://doi.org/10.1016/j.cej.2017.12.029

Humnabadkar RP, Saratale GD, Govindwar SP (2008) Decolorization of purple 2 R by *Aspergillus ochraceus* (NCIM-1146). Asian J Microbiol Biotechnol Environ Sci 10:693–697

Ingelman M, Ramaswamy S, Niviere V, Fontecave M, Eklund H (1999) Crystal structure of NAD(P)H: flavin oxidoreductase from *Escherichia coli*. Biochemist 38:7040–7049

Jadhav SU, Jadhav UU, Dawkar VV, Govindwar SP (2008) Biodegradation of disperse dye Brown 3REL by microbial consortium of *Galactomyces geotrichum* MTCC 1360 and *Bacillus* sp. VUS. Biotechnol Bioprocess Eng 13:232–239

Jianwei M, Xingdong Y, Gulan Z (1997) Studies on the decolorization of azo dyes in aqueous system using metal ions together with algae (*Sargassum*). Huanj Huax Chin 16:238–240

Jin X, Liu G, Xu Z, Tao W (2007) Decolourisation of a dye industry effluent by *Aspergillus fumigatus* XC6. Appl Microbiol Biotechnol 74:239–243

Jinqi L, Houtian L (1992) Degradation of azo dyes by algae. Environ Pollut 75:273–227

Judd S (2010) The MBR book, principles and applications of membrane bioreactors for water and wastewater treatment. Elsevier, Oxford

Junnarkar N, Murty DS, Bhatt NS, Madamwar D (2006) Decolorization of diazo dye Direct red 81 by a novel bacterial consortium. World J Microbiol Biotechnol 22:163–168

Kagalkar AN, Jagtap UB, Jadhav JP, Bapat VA, Govindwar SP (2009) Biotechnological strategies for phytoremediation of the sulfonated azo dye Direct Red 5B using *Blumea malcolmii* Hook. Bioresour Technol 100:4104–4110

Kalme SD, Parshetti GK, Jadhav SU, Govindwar SP (2007) Biodegradation of benzidine based dye Direct Blue-6 by *Pseudomonas desmolyticum* NCIM 2112. Bioresour Technol 98:1405–1410

Kalme S, Jadhav S, Jadhav M, Govindwar S (2009) Textile dye degrading laccase from *Pseudomonas* desmolyticum NCIM 2112. Enzym Microb Technol 44:65

Kalyani DC, Telke AA, Dhanve RS, Jadhav JP (2008) Ecofriendly biodegradation and detoxification of Reactive Red 2 textile dye by newly isolated *Pseudomonas* sp. SUK1. J Hazard Mater 163:735–742

Kapdan IK, Kargi F (2002) Biological decolorization of textile dyestuff containing wastewater by *Coriolus versicolor* in a rotating biological contactor. Enzym Microb Technol 30:195–199

Karcher S, Kornmuller A, Jekel M (2001) Screening of commercial sorbents for the removal of reactive dyes. DyesPigments 51:111–125

Khanal SK (2008) Anaerobic biotechnology for bioenergy production. Wiley-Blackwell, Iowa, p 179

Khehra MS, Saini HS, Sharma DK, Chadha BS, Chimn SS (2005) Decolorization of various azo dyes by bacterial consortium. Dyes Pigments 67:55–61

Kolekar YM1, Kodam KM (2012) Decolorization of textile dyes by *Alishewanella* sp. KMK6. Appl Microbiol Biotechnol 95(2):521–529

Konsowa AH, El-Rahman HA, Moustafa MA (2011) Removal of azo dye acid orange 7 using aerobic membrane bioreactor. Alex Eng J 50:117–125

Koók L, Nemestóthy N, Bakonyi P, Zhen G, Kumar G, Lu X, Su L, Saratale GD, Kim SH, Gubicza L (2017) Performance evaluation of microbial electrochemical systems operated with Nafion and supported ionic liquid membranes. Chemosphere 175:350–355

Korsak L (2008) Anaerobic treatment of wastewater in a UASB reactor. Department of Chemical Engineering and Technology, Royal Institute of Technology, Stockholm

Krcmar P, Ulrich R (1998) Degradation of polychlorinated biphenyl mixtures by the lignin-degrading fungus *Phanerochaete chrysosporium*. Folia Microbiol 43:79–84

Le-Clech P, Jefferson B, Judd SJ (2005) A comparison of submerged and side stream tubular membrane bioreactor configurations. Desalination 173:113–122

Lin J, Zhang X, Li Z, Lei L (2010) Biodegradation of reactive blue 13 in a two-stage anaerobic/aerobic fluidized beds system with a Pseudomonas sp. isolate. Bioresour Technol 101:34–40

Lopez-Grimau V, Gutierrez MC (2006) Decolorization of simulated reactive dyebath effluents by electrochemical oxidation assisted by UV light. Chemosphere 62:106–112

Lourenco ND, Novais JM, Pinheiro HM (2001) Effect of some operational parameters on textile dye biodegradation in a sequential batch reactor. J Biotechnol 89:163–174

Lu L, Zhao M, Zhang B-B, Yu S-Y, Bian X-J, Wang W, Wang Y (2007) Purification and characterization of laccase from Pycnoporus sanguineus and decolorization of an anthraquinone dye by the enzyme. Appl Microbiol Biotechnol 74:1232–1239

Machado KMG, Compart LCA, Morais RO, Rosa LH, Santos MH (2006) Biodegradation of reactive textile dyes by basidiomycetous fungi from Brazilian ecosystems. Braz J Microbiol 37:481–487

Maier J, Kandelbauer A, Erlacher A, Cavaco-Paulo A, Gubitz GM (2004) A new alkali-thermostable azoreductase from *Bacillus* sp. strain SF. Appl Environ Microbiol 70:837–844

Majcherczyk A, Johannes C, Hu¨ttermann A (1998) Oxidation of polycyclic aromatic hydrocarbons (PAH) by laccase of *Trametes versicolor*. Enzym Microb Technol 22:335–341

Manariotis ID, Grigoropoulos SG (2002) Low-strength wastewater treatment using an anaerobic baffled reactor. Water Environ Res 74:170–176

Martins MAM, Cardoso MH, Queiroz MJ, Ramalho MT, Campos AMO (1999) Biodegradation of azo dyes by the yeast *Candida zeylanoides* in batch aerated cultures. Chemosphere 38:2455–2460

Mohan SV, Rao CN, Prasad KK, Karthikeyan J (2002) Treatment of simulated Reactive Yellow 22 (Azo) dye effluents using *Spirogyra* species. Waste Manag 22:575–582

Mahamadi C, Epias M (2014) High adsorption of dyes by water hyacinth fixed on alginate. Environ. Chem. Lett. 12:313–320.

Nakanishi M, Yatome C, Ishida N, Kitade Y (2001) Putative ACP phosphodiesterase gene (acpD) encodes an azoreductase. J Biol Chem 276:46394–46399

Nidheesh PV, Gandhimathi R, Ramesh ST, Singh TSA (2012) Adsorption and desorption characteristics of crystal violet in bottom ash column. J Urban Environ Eng 6:18–29

Nigam P, Banat IM, Singh D, Marchant R (1996) Microbial process for the decolorization of textile effluent containing azo, diazo and reactive dyes. Process Biochem 31:435–442

Nigam P, Armour G, Banat IM, Singh D, Marchant R (2000) Physical removal of textile dyes from effluents and solid-state fermentation of dye-adsorbed agricultural residues. Bioresour Technol 72:219–226

Pandey A, Singh P, Iyengar L (2005) Bacterial decolorization and degradation of azo dyes. Int Biodeterior Biodegrad 59:73–84

Parshetti GK, Kalme SD, Saratale GD, Govindwar SP (2006) Biodegradation of Malachite green by *Kocuria rosea* MTCC-1532. Acta Chim Slov 53:492–498

Parshetti GK, Saratale GD, Govindwar SP (2009) Biodegradation of hazardous triphenylmethane dye methyl violet by *Rhizobium radiobacter* (MTCC 8161). J Basic Microbiol 49:1–7

Peuke AD, Rennenberg H (2005) Phytoremediation: molecular biology, requirements for application, environmental protection, public attention and feasibility. EMBO Rep 6:497–501

Rajakumar R, Meenambal T, Rajesh Banu J, Yeom IT (2011) Treatment of poultry slaughterhouse wastewater in upflow anaerobic filter under low upflow velocity. Int J Environ Sci Technol 8:149–158

Rajesh Banu J, Kaliappan S (2007) Treatment of tannery wastewater using hybrid upflow anaerobic sludge blanket reactor. J Environ Eng Sci 6:415–421

Rajesh Banu J, Kaliappan S, Beck D (2006) High rate anaerobic treatment of Sago wastewater using HUASB with PUF as carrier. Int J Environ Sci Technol 3:69–77

Rajesh Banu J, Kaliappan S, Yeom IT (2007a) Two-stage anaerobic treatment of dairy wastewater using HUASB with PUF and PVC carrier. Biotechnol Bioprocess Eng 12:257–264

Rajesh Banu J, Kaliappan S, Yeom IT (2007b) Treatment of domestic wastewater using upflow anaerobic sludge blanket reactor. Int J Environ Sci Technol 4:363–370

Rajesh Banu J, Kaliappan S, Adish Kumar S, Yeom IT, Uan DK (2011) Effect of low temperature thermochemical pretreatment on sludge reduction potential of membrane bioreactor treating primary treated dairy wastewater. Water Qual Res J Can 46:12–320

Rajesh Banu J, Arulazhagan P, Adish Kumar S, Kaliappan S, Lakshmi AM (2014) Anaerobic co-digestion of chemical- and ozone-pretreated sludge in hybrid upflow anaerobic sludge blanket reactor. Desalin Water Treat 54:3269–3278

Ramalho PA, Scholze H, Cardoso MH, Ramalho MT, Oliveira-Campos AM (2002) Improved conditions for the aerobic reductive decolourisation of azo dyes by *Candida zeylanoides*. Enzym Microb Technol 31:848–854

Ramalho PA, Cardoso MH, Cavaco-Paulo A, Ramalho MT (2004) Characterization of azo reduction activity in a novel ascomycete yeast strain. Appl Environ Microbiol 70:2279–2288

Ramavandi B, Farjadfard S, Ardjmand M (2014) Mitigation of orange II dye from simulated and actual wastewater using bimetallic chitosan particles: continuous flow fixed-bed reactor. J Environ Chem Eng 2:1776–1784

Robinson T, McMullan G, Marchant R, Nigam P (2001) Remediation of dyes in textile effluent: a critical review on current treatment technologies with a proposed alternative. Bioresour Technol 77:247–255

Rodrı'guez E, Pickard MA, Vazquez-Duhalt R (1999) Industrial dye decolorization by laccases from ligninolytic fungi. Curr Microbiol 38:27–32

Russ R, Rau J, Stolz A (2000) The function of cytoplasmic flavin reductases in the reduction of azo dyes by bacteria. Appl Environ Microbiol 66:1429–1434

Sahasrabudhe M, Saratale RG, Saratale GD, Pathade GR (2014) Decolorization and detoxification of sulfonated toxic diazo dye C.I. Direct Red 81 by isolated *Enterococcus faecalis* YZ 66. J Environ Health Sci Eng 12:151–163

Sandhya S, Padmavathy S, Swaminathan K, Subrahmanyam YV, Kaul SN (2005) Microaerophilic aerobic sequential batch reactor for treatment of azo dyes containing simulated wastewater. Process Biochem 40:885–890

Saratale GD, Kalme SD, Govindwar SP (2006) Decolorization of textile dyes by *Aspergillus ochraceus* (NCIM-1146). Indian J Biotechnol 5:407–410

Saratale GD, Bhosale SK, Kalme SD, Govindwar SP (2007a) Biodegradation of kerosene in *Aspergillus ochraceus* (NCIM-1146). J Basic Microbiol 47:400–405

Saratale GD, Humnabadkar RP, Govindwar SP (2007b) Study of mixed function oxidase system in *Aspergillus ochraceus* (NCIM-1146). Indian J Microbiol 47:304–309

Saratale RG, Saratale GD, Chang JS, Govindwar SP (2009a) Decolorization and biodegradation of textile dye navy blue HER by *Trichosporon beigelii* (NCIM- 3326). J Hazard Mater 166:1421–1428

Saratale RG, Saratale GD, Chang JS, Govindwar SP (2009b) Ecofriendly decolorization and degradation of Reactive Green 19A using *Micrococcus glutamicus* NCIM-2168. Bioresour Technol 100:3897–3905

Saratale RG, Saratale GD, Kalyani DC, Chang JS, Govindwar SP (2009c) Enhanced decolorization and biodegradation of textile azo dye scarlet R by using developed microbial consortium-GR. Bioresour Technol 100:2493–2500

Saratale RG, Saratale GD, Chang JS, Govindwar SP (2010) Decolorization and biodegradation of reactive dyes and dye wastewater by a developed bacterial consortium. Biodegradation 21:999–1015

Saratale GD, Saratale RG, Chang JS, Govindwar SP (2011a) Fixed-bed decolorization of Reactive Blue 172 by *Proteus vulgaris* NCIM-2027 immobilized on *Luffa cylindrica* sponge. Int Biodeterior Biodegrad 65:494–503

Saratale RG, Saratale GD, Chang JS, Govindwar SP (2011b) Decolorization and degradation of reactive azo dyes by fixed-bed bioreactors containing immobilized cells of *Proteus vulgaris* NCIM-2027. Biotechnol Bioprocess Eng 14:830–842

Saratale RG, Saratale GD, Chang JS, Govindwar SP (2011c) Outlook of bacterial decolorization and degradation of azo dyes: a review. J Taiwan Inst Chem Eng 42:138–157

Saratale RG, Purankar M, Gandhi S, Kurade M, Oh SE, Govindwar SP, Saratale GD (2013) Decolorization and degradation of C.I. Remazol Red and textile effluent by isolated *Lysinbacillus fusiformis* strain IGI. J Biosci Bioeng 115:658–667

Saratale RG, Saratale GD, Govindwar SP, Kim DS (2015) Exploiting the efficacy of *Lysinibacillus* sp. RGS for decolorization and detoxification of industrial dyes, textile effluent and bioreactor studies. J Environ Sci Health A 50:176–192

Saratale RG, Hwang KJ, Song JY, Saratale GD, Kim DS (2016a) Electrochemical oxidation of phenol for wastewater treatment using Ti/PbO_2 electrode *ASCE*. J Environ Eng 142(2):04015064

Saratale RG, Sivapathan SS, Jeong W, Kim H-Y, Saratale GD, Kim DS (2016b) Preparation of activated carbons from peach stone by $H_4P_2O_7$ activation and its adsorption behavior for Acid Red 18 and dye containing wastewater. J Environ Sci Health A 51:1–10

Saratale GD, Saratale RG, Ghodake GS, Jiang YY, Chang JS, Shin HS (2017a) Solid state fermentative lignocellulolytic enzymes production, characterization and its application in the saccharification of rice waste biomass for ethanol production: an integrated biotechnological approach. J Taiwan Inst Chem Eng 76:51–58

Saratale RG, Kuppam C, Mudhoo A, Saratale GD, Periyasamy S, Zhen G, Koók L, Bakonyi P, Nemestóthy N, Kumar G (2017b) Bioelectrochemical systems using microalgae – a concise research update. Chemosphere 177:35–43

Saxena G, Bharagava RN (2015) Persistent organic pollutants and bacterial communities present during the treatment of tannery wastewater. In: Chandra R (ed) Environmental waste management, 1st edn. CRC Press, Taylor & Francis Group, Boca Raton, pp 217–247. https://doi.org/10.1201/b19243-10

Saxena G, Bharagava RN (2017) Organic and inorganic pollutants in industrial wastes, their ecotoxicological effects, health hazards and bioremediation approaches. In: Bharagava RN (ed) Environmental pollutants and their bioremediation approaches, 1st edn. CRC Press, Taylor & Francis Group, Boca Raton, pp 23–56. https://doi.org/10.1201/9781315173351-3

Saxena G, Chandra R, Bharagava RN (2016) Environmental pollution, toxicity profile and treatment approaches for tannery wastewater and its chemical pollutants. Rev Environ Contam Toxicol 240:31–69. https://doi.org/10.1007/398_2015_5009

Saxena G, Purchase D, Mulla SI, Saratale GD, Bharagava RN (2018) Phytoremediation of heavy metal-contaminated sites: environmental considerations, field studies, sustainability and future prospects. J Environ Manag 105:103–120

Selvakumar KV, Basha CA, Prabhu HJ, Kalaichelvi P, Nelliyan S (2010) The potential of free cells of Pseudomonas aeruginosa on textile dye degradation. Bioresour Technol 101:2678–2684

Sen S, Demirer GN (2003) Anaerobic treatment of real textile wastewater with a fluidized bedreactor. Water Res 37:1868–1878

Shaverdi G (2012) Developing a model for mass transfer in adsorption packed-bed filters (Ph.D. thesis), Concordia University, Montreal, Quebec, Canada

Singh JS, Abhilash PC, Singh HB, Singh RP, Singh DP (2011) Genetically engineered bacteria: an emerging tool for environmental remediation and future research perspectives. Gene 480:1–9

Stolz A (2001) Basic and applied aspects in the microbial degradation of azo dyes. Appl Microbiol Biotechnol 56:69–80

Su CC, Pukdee-Asa M, Ratanatamskul C, Lu MC (2011) Effect of operating parameters on decolorization and COD removal of three reactive dyes by Fenton's reagent using fluidized-bed reactor. Desalination 278:211–218

Suresh G, Babu BV (2010) Experimental investigations and theoretical modeling aspects in column studies for removal of Cr (VI) from aqueous solutions using activated tamarind seeds. J Water Resour Protect 2:706–716

Suzuki Y, Yoda T, Ruhul A, Sugiura W (2001) Molecular cloning and characterization of the gene coding for azoreductase from *Bacillus* sp. OY1-2 isolated from soil. J Biol Chem 276:9059–9065

Swaminathan KS, Sandhya A, Carmalin S, Pachhade K, Subrahmanyam K (2003) Decolorization and degradation of H-acid and other dyes using ferrous-hydrogen peroxide system. Chemosphere 50:619–625

Sun Q, Saratale RG, Saratale GD, Kim DS (2018) Pristine and modified radix Angelicae dahuricae (Baizhi) residue for the adsorption of methylene blue from aqueous solution: a comparative study. J Mol Liq 265:36–45

Saratale RG, Ghodake GS, Shinde SK, Cho SK, Saratale GD, Pugazhendhi A, Bharagava RN (2018) Photocatalytic activity of CuO/cu(OH)2 nanostructures in the degradation of reactive green 19A and textile effluent, phytotoxicity studies and their biogenic properties (antibacterial and anticancer). J Environ Manag 223:1086–1097

Saratale GD, Saratale RG, Oh SE (2012) Production and characterization of multiple cellulolytic enzymes by isolated Streptomyces sp. MDS. Biomass Bioenergy 47:302–315

Šafaříková M, Ptáčková L, Kibriková I, Šafařík I (2005) Biosorption of water-soluble dyes on magnetically modified *Saccharomyces cerevisiae* sub sp. *uvarum* cells. Chemosphere 59:831–835

Tien M, Kirk TK (1983) Lignin-degrading enzyme from the hymenomycete *Phanerochaete chrysosporium* burd. Science 221:661–663

Uan DK, Rajesh Banu J, Chung IJ, Yeom IT (2009) Effect of thermochemical sludge pretreatment on sludge reduction and on performances of anoxic-aerobic membrane bioreactor treating low strength domestic wastewater. J Chem Technol Biotechnol 84:1350–1355

Uan DK, Banu RJ, Son DH, Yeom IT (2012) Influence of ferrous sulfate on thermochemical sludge disintegration and on performances of wastewater treatment in a new process: anoxic-oxic membrane bioreactor coupled with sludge disintegration step. Biochem Eng J 66:20–26

Uan DK, Yeom IT, Arulazhagan P, Rajesh Banu J (2013) Effects of sludge pretreatment on sludge reduction in a lab-scale anaerobic/anoxic/oxic system treating domestic wastewater. Int J Environ Sci Technol 10(495):502

Van der Zee F (2002) Anaerobic azo dye reduction. Environ Technol. Wageningen University, Wageningen, The Netherlands

Vandevivere PC, Bianchi R, Verstraete W (1998) Treatment and reuse of wastewater from the textile wet-processing industry: review of emerging technologies. J Chem Technol Biotechnol 72:289–302

Vives MT, Balaguer MD, Garcıa S, Garcıa R, Colprim J (2003) Textile dyeing wastewater treatment in a sequencing batch reactor system. J Environ Sci Health Part A 38:2089–2099

Wesenberg D, Kyriakides I, Agathos SN (2003) White-rot fungi and their enzymes for the treatment of industrial dye effluents. Biotechnol Adv 22:161–187

Wu JY, Hwang SCJ, Chen CT, Chen KC (2005) Decolorization of azo dye in a FBR reactor using immobilized bacteria. Enzym Microb Technol 37:102–112

Yang Q, Li C, Li H, Li Y, Yu N (2009) Degradation of synthetic reactive azo dyes and treatment of textile wastewater by a fungi consortium reactor. Biochem Eng J 43:225–230

Yu J, Wang X, Yue P (2001) Optimal decolorization and kinetic modeling of synthetic dyes by *Pseudomonas* strains. Water Res 35:3579–3586

Zollinger H (1987) Colour chemistry-synthesis, properties of organic dyes and pigments. VCH Publishers, New York, pp 92–100

Chapter 8
Management of Petroleum Industry Waste Through Biosurfactant-Producing Bacteria: A Step Toward Sustainable Environment

Amar Jyoti Das, Shweta Ambust, and Rajesh Kumar

Contents

Abstract Petroleum oil and its product contain various toxic compounds that can be regarded as toxic for environment and health. Terrestrial ecosystems contaminated by petroleum industry waste are considered to be hazardous due to hydrocarbon toxicity. Oil-contaminated soil causes toxicity in microorganism, plants, and animals. Hence, various approaches (biological, chemical, and physical technologies) have been adopted for remediation petroleum-contaminated terrestrial environment. Biosurfactants are amphiphilic, heterogeneous metabolites secreted by microorganisms consisting of hydrophilic and hydrophobic moieties, having the ability to manage petroleum oil and its product at various environmental conditions. They are

A. J. Das (✉) · S. Ambust · R. Kumar
Rhizospheric Biology Laboratory, Department of Environmental Microbiology, School for Environmental Sciences, Babasaheb Bhimrao Ambedkar University (A Central University), Lucknow, Uttar Pradesh, India
e-mail: Rajesh_dem@bbau.ac.in

G. Saxena, R. N. Bharagava (eds.), *Bioremediation of Industrial Waste for Environmental Safety*, https://doi.org/10.1007/978-981-13-1891-7_8

highly degradable in soil and water and are low toxic in nature. Hence, the present chapter discusses details about biosurfactant and biosurfactant-producing bacteria and their role in management of petroleum-contaminated terrestrial ecosystem.

Keywords Terrestrial ecosystems · Petroleum industry waste · Biosurfactant · Management · Hazardous

8.1 Introduction

Petroleum industries and refineries convert crude oil into a variety of petroleum product through various physical and chemical processes. During these processes, they release contaminants that can cause air, water, and soil pollution. Hence, the petroleum industries and refineries are regulated based on environmental laws (Environmental Update 2003). Petroleum oil and its product contain various toxic compounds that can be regarded as hazardous for the environment and health (Salanitro et al. 1997; Ekperusi and Aigbodion 2015). Crude oil, diesel, and gasoline termed as total petroleum hydrocarbons (TPH) are persistent in nature (Huang et al. 2004, 2005; Atlas and Brag 2009; Hussain et al. 2018). Accidental release of oil spills in environment and their associated pollution have been reported worldwide (Van Hamme et al. 2003; Atlas and Hazen 2011). Soil pollution by petroleum oil or TPH creates hydrophobic conditions in soil, such as loss of vegetation, infertile soils, and presence of microbial population (Hutchinson et al. 2001; Ramadass et al. 2015; Liu et al. 2016; Hussain et al. 2018). Various physical, chemical, and biological methods have been used for the management of petroleum-contaminated sites, but still there is a search for suitable cost-effective technique (Hussain et al. 2018). Hence, present chapter demonstrated the strategy of biosurfactant-producing bacteria for the management of petroleum industry waste.

8.2 Biological Techniques for the Management of Petroleum-Contaminated Soil

8.2.1 Bioventing

In the bioventing process, air is injected into the contaminated media at a rate designed to maximize in situ biodegradation and minimize the off-gassing of volatilized contaminants to the atmosphere (Khan et al. 2004). Baker and Moore (2000) investigated the optimized performance and effectiveness of in situ bioventing. Diele et al. (2002) and Mihopoulos et al. (2002) have reported the numerical models and their applications in bioventing system design and their operation. This method is more efficient on midrange petroleum products such as diesel, because lighter products tend to volatilize quickly and the heavier products generally take long time to biodegrade (Khan et al. 2004; Kumar et al. 2015).

8.2.2 Biopiles

Biopiles are a soil treatment technique employed for petroleum-contaminated soil remediation (Khan et al. 2004). This treatment involves the heaping of contaminated soils (petroleum) into piles and then stimulating the aerobic microbial activity by aeration and addition of required amount of moisture, nutrients, and minerals (Filler et al. 2001; Khan et al. 2004). Biopile has been proven effective in reducing the amount of contaminants (petroleum) found at underground storage tank through which air passes and has a covering to prevent runoff, evaporation, and volatilization (Khan et al. 2004; Kumar et al. 2015). This treatment can be accomplished within a short span of time.

8.2.3 Land Farming

Land farming is the prominent method for remediation of petroleum-contaminated sites (FRTR 1999; Khan et al. 2004). This technique is employed to reduce the concentration of petroleum constituents present in soil through process accomplished with biological-based remediation (Khan et al. 2004). This technology usually involves the spreading of contaminated soil in a thin layer on the uppermost layer of the land up to 1.5 m which to be treated and microorganism along with proper aeration, moisture, and nutrient. Microbes are used for the breakdown of petroleum hydrocarbons to increase the efficiency of remediation rate of petroleum-contaminated products (Riser-Roberts 1998; USEPA 1998; Hejazi 2002; Khan et al. 2004; Kumar et al. 2015).

8.2.4 Enzymatic Degradation

Heme-thiolate monooxygenase is a potent degradative enzyme of cytochrome P-450 alkane hydroxylases superfamily which specifically degrades petroleum hydrocarbon and other related compounds with the help of microorganism (Zimmer et al. 1996; Das and Chandran 2011). Further, biodegradation of petroleum hydrocarbon necessarily requires enzyme system which incorporates oxygen in substrate complex, and the enzyme system basically focuses on the carbon content and chain length of hydrocarbon pollutant (Das and Chandran 2011). In higher eukaryotes, there is diversity of cytochrome P450 and their isoforms specialized for desired substrate conversion. Therefore, they have the ability to use n-alkanes and other aliphatic hydrocarbon as a source of carbon and energy that regulated by various microsomal cytochrome P-450 (Zimmer et al. 1996; Scheuer et al. 1998; Das and Chandran 2011).

8.2.5 *Phytoremediation*

Phytoremediation is a biological method for rehabilitation of environment by employing plants (Raskin 1996; Ndimele 2010). Phytoremediation is a broad term that has been use to describe the use of plants to reduce the concentration of contaminants in soil and groundwater (McCutcheon and Schnoor 2003; Borjana et al. 2001; Kumar et al. 2015). The phytoremediation method adopts various mechanisms to clean up the contaminants. Such mechanisms include (a) phytoextraction, (b) phytovolatilization, (c) phytodegradation, (d) rhizodegradation, (e) rhizofiltration, and (f) phytostabilization (Fig. 8.1). Each of these mechanisms will have an effect on the volume, mobility, or toxicity of contaminants, as the application of phytoremediation is intended to do (EPA 2000).

8.3 Management of Petroleum-Contaminated Site by Biosurfactant-Producing Bacteria

8.3.1 *Biosurfactant*

Biosurfactants are amphiphilic, heterogeneous, and secondary metabolites produced by cell surface of microorganism consisting of hydrophilic and hydrophobic moieties and having the capacity to decrease the surface tensions and interfacial tension by gathering two immiscible fluids like water and oil interface (Ilori et al. 2005; Mahmound et al. 2008; Muthusamy et al. 2008; Kiran et al. 2009; Obayori et al. 2009; Das and Chandran 2011). The biosurfactant consists of hydrophobic moiety of saturated or unsaturated hydrocarbons or fatty acids and hydrophilic moiety along with an acid, peptide cations or anions, and mono-, di-, or polysaccharides. On the basis of molecular weight, they are classified into two parts: one is low molecular weight biosurfactant, and other one is high molecular weight which is referred as bioemulsifiers (Karanth et al. 1999; Mulligan 2005; Kumar et al. 2015; Das et al. 2016, 2017).

8.3.2 *Types of Biosurfactant*

Biosurfactants are categorized into high and low molecular weight. The low molecular weight surfactants include lipopeptides, glycolipids, and phospholipids, while polymeric and particulate surfactants are included in high molecular weight category (Kappeli and Finnerty 1979; Nitschke and Coast 2007) as mentioned below.

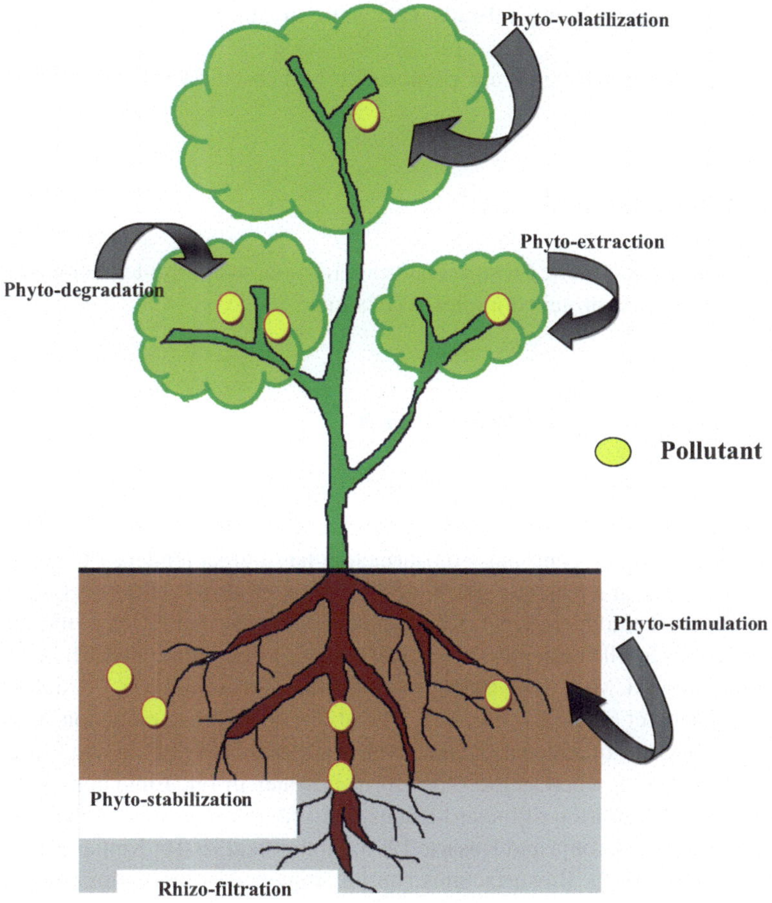

Fig. 8.1 Various mechanisms behind remediation of petroleum hydrocarbons by plants

8.3.2.1 Glycolipids

Glycolipids are widely known biosurfactants due to their properties and application. They consist of carbohydrate moieties linked to fatty acid moieties. Cellobiose lipids, mannosylerythritol lipids, rhamnolipids, sophorolipids, and trehalose lipids are the most popular glycolipids (Alahmad 2015; Mnif and Ghribi 2016).

8.3.2.2 Lipopeptides

Lipopeptides are low molecular weight biosurfactants that consist of a lipid attached to a polypeptide chain. Surfactin, lichenysin, iturin, and fengycin are some common types of lipopeptides (Wang et al. 2008; Shaligram and Singhal 2010; Meena and Kanwar 2015; Kumar et al. 2015).

8.3.2.3 Fatty Acids, Phospholipids, and Neutral Lipids

These types of biosurfactants are produce by both bacteria and yeast (Alahmad 2015).

8.3.2.4 Polymeric Surfactants

The most studied polymeric biosurfactants are liposan, mannoprotein, emulsan, and other polysaccharide-protein complexes (Alahmad 2015).

8.3.3 Properties of a Biosurfactant

The properties of biosurfactants are highly influenced by their molecular mass. Biosurfactants with high molecular mass have the ability to stabilize oil water emulsions, while a low molecular mass biosurfactant has a great tendency to reduce the interfacial and surface tensions (Pacwa-Płociniczak et al. 2011; Geys et al. 2014). The biosurfactant activity is generally dependent on the concentration of the surface-active compounds till it attends the critical micelle concentration (CMC). In concentration above CMC; biosurfactant molecules form bilayers, vesicles, and micelles (Whang et al. 2008) (Fig. 8.2). The ability to stabilize emulsion is one the important properties of a biosurfactant. Emulsion generally exists in two forms; one form is oil in water, whereas the other form is water in oil. Emulsions have less stability, but with addition of biosurfactants, the stability of emulsion increases for months or years (Velikonja and Kosaric 1993; Santos et al. 2016; Kumar et al. 2015; Das et al. 2016, 2017). Biosurfactants can be employed at high temperatures and various pH and salt concentrations. They are highly degradable by microorganisms in soil and water and are low toxic in nature (Santos et al. 2016).

8.3.4 Management of Petroleum-Contaminated Site by Rhamnolipid Biosurfactant-Producing Bacteria

8.3.4.1 Mechanism Behind Ability of the Biosurfactant-Producing Bacteria to Remediate Petroleum Hydrocarbon

Nowadays, biosurfactant-assisted remediation of petroleum-contaminated technology is gaining interest due to their effective and eco-friendly nature (Pacwa-Płociniczak et al. 2011). Petroleum-contaminated site management efficiency of biosurfactant-producing bacteria is generally based on two different mechanisms:

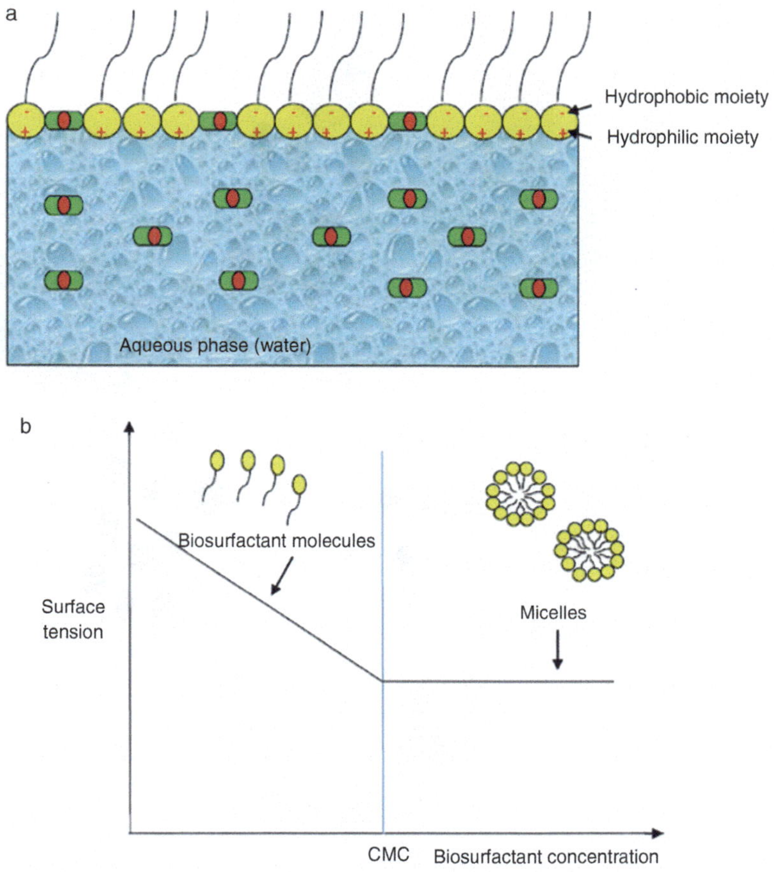

Fig. 8.2 The relationship between concentration of biosurfactant, surface tension, and formation of micelles. (Source: Das et al. 2017)

firstly, increase of bioavailability of substrate for microorganisms and secondly cell surface interaction that enhances the hydrophobicity of the cell surface to get associated with the hydrophobic substrates (Mulligan and Gibbs 2004; Pacwa-Płociniczak et al. 2011). Cameotra and Singh (2009) demonstrated the role of rhamnolipid biosurfactant synthesized by *Pseudomonas aeruginosa* in the uptake of n-alkane. They explored an exciting mechanism for hydrocarbon uptake through hydrocarbons internalization inside the cell. Their proposed mechanism involves dispersion of hydrocarbons by biosurfactant into microdroplets and enhances the hydrocarbon bioavailability to the bacterial cell surface. Thenceforth, biosurfactant-coated hydrocarbon droplets are up taken by the bacterial cell. This process is quite similar to pinocytosis process (Fig. 8.3).

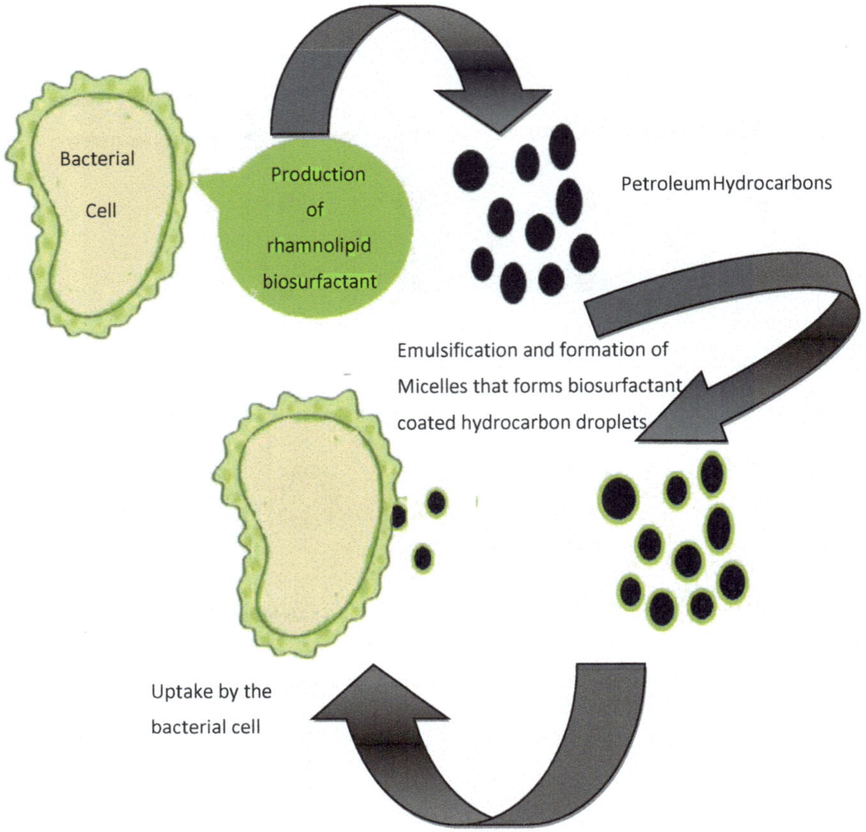

Fig. 8.3 Uptake of petroleum hydrocarbon by *Pseudomonas* sp. with the help of rhamnolipid biosurfactant

8.3.4.2 Rhamnolipid Biosurfactant-Facilitated Remediation of Petroleum Hydrocarbon

There are overwhelming reports that state the benefits of employing rhamnolipid biosurfactant or rhamnolipid biosurfactant producers in remediation of petroleum hydrocarbon (Table 8.1).

Table 8.1 Recent studies on rhamnolipids-assisted bioremediation by increasing bioavailability via solubilization

Pollutants	Bioremediators	Maximum removal (%)	References
Polycyclic aromatic hydrocarbons	*Sphingobacterium* sp. QPH-3	75.6	Zhu et al. (2013)
Polycyclic aromatic hydrocarbons	*P. chrysosporium*	71.5	Wang et al. (2014)
Petroleum hydrocarbons	*Burkholderia multivorans* NG1	87	Mohanty et al. (2013)
Polyaromatic hydrocarbons	Indigenous microbes	95	Nikolopoulou et al. (2013)
Polyaromatic hydrocarbons	Indigenous microbes	90	Sponza and Gök (2010)
Crude oil	*P. aeruginosa* DN1	90.52	Ma et al. (2016)
n-alkanes	Indigenous microbes	97	Nikolopoulou et al. (2013)
Eicosane, pristane, and fluoranthene	*P. aeruginosa* DSVP20	97, 75, 47	Sharma et al. (2015)
Diesel oil	Indigenous microbes	100	Wang et al. (2008)
Phenanthrene	*P. putida* ATCC 17484	91	Gottfried et al. (2010)
Phenanthrene	*P. aeruginosa* BP9	92	Bezza and Chirwa (2014)
Pyrene	Pyrene-degrading bacteria	86.4	Jorfi et al. (2014)
Pyrene	*Mycobacterium gilvum* VM552	70	Congiu and Ortega-Calvo (2014)
Atrazine	Acinetobacter A6	80	Singh and Cameotra (2013)
Anthracene	*Sphingomonas* sp. 12A	52	Cui et al. (2008)
Anthracene	*Sphingomonas* sp. 12B	39	Cui et al. (2008)
Chlorpyrifos	Indigenous microbes	82.3	Singh et al (2016)
Hexachlorocyclohexane	*Sphingomonas* sp. NM05	95	Manickam et al. (2012)
Organochlorine pesticides	*P. chrysosporium*	62.5	Wang et al. (2014)
Triclosan	Indigenous microbes	94	Qian et al. (2016)

Adapted and modified from Liu et al. (2017)

8.4 Conclusion

The environment pollution due to petroleum oil contamination is increasing day-by-day. Various methods are applied, but still there is a requirement of some new novel technology. Hence, implication of biosurfactant or biosurfactant production could open up a new vista and diminish the ecotoxicological effects due to petroleum hydrocarbon.

Acknowledgement The research support received from DST Inspire program (INSPIRE Code: IF140569) to the author Amar Jyoti Das is duly acknowledged.

References

Alahmad K (2015) The definition, preparation and application of Rhamnolipids as biosurfactants. Int J Nutr Food Sci 4(6):613–623

Atlas R, Bragg J (2009) Bioremediation of marine oil spills: when and when not – the Exxon Valdez experience. Microb Biotechnol 2:213–221

Atlas RM, Hazen TC (2011) Oil biodegradation and bioremediation: a tale of the two worst spills in U.S. history. Environ Sci Technol 45:6709–6715

Baker RS, Moore AT (2000) Optimizing the effectiveness of in situ bioventing. Pollut Eng 32(7):44–47

Bezza FA, Chirwa EMN (2014) Optimization strategy of polycyclic aromatic hydrocarbon contaminated media bioremediation through biosurfactant addition. Chem Eng Trans 39:1597–1602

Borjana KT, George RI, Nelly EC (2001) Biosurfactant production by a new Pseudomonas putida strain. Z Naturforsch 57:356–360

Cameotra SS, Singh P (2009) Synthesis of rhamnolipid biosurfactant and mode of hexadecane uptake by Pseudomonas species. Microb Cell Factories 8:1–7

Cui CZ, Zeng C, Wan X, Chen D, Zhang JY, Shen P (2008) Effect of rhamnolipids on degradation of anthracene by two newly isolated strains, Sphingomonas sp. 12A and Pseudomonas sp. 12B. J Microbiol Biotechnol 18(1):63–66

Das N, Chandran P (2011) Microbial degradation of petroleum hydrocarbon contaminants: an overview. Biotechnol Res Int., Article ID 941810, 13p. https://doi.org/10.4061/2011/941810

Das AJ, Shastri B, Lal S, Kumar R (2016) Bioremediation of petroleum hydrocarbons and heavy metal contaminated sites by biosurfactants: an eco-friendly and sustainable technology bioremediation of industrial pollutants 1, 91–118. In: RN Bhargava, G Saxena (ed) Book

Das AJ, Lal S, Kumar R, Verma C (2017) Bacterial biosurfactants can be an ecofriendly and advanced technology for remediation of heavy metals and co-contaminated soil. Int J Environ Sci Technol 14(6):1343–1354

Diele F, Notarnicola F, Sgura I (2002) Uniform air velocity field for a bioventing system design: some numerical results. Int J Eng Sci 40(11):1199–1210

Ekperusi OA, Aigbodion FI (2015) Bioremediation of petroleum hydrocarbons from crude oil contaminated soil with the earthworm: Hyperiodrilus africanus. Biotech 5:957–965. https://doi.org/10.1007/s13205-015-0298-1

Environmental Update#12 (2003) Published by the hazardous substance research centers/South & Southwest outreach program. https://cfpub.epa.gov/ncer_abstracts/index.cfm/fuseaction/display.files/fileID/14522

EPA (2000) A citizen's guide to phytoremediation. EPA 542-F-98-011. United States Environmental Protection Agency, p 6. Available at: http://www.bugsatwork.com/XYCLONYX/EPA_GUIDES/PHYTO.PDF

Filler DM, Lindstrom JE, Braddock JF, Johnson RA, Nickalaski R (2001) Integral biopile components for successful bioremediation in the Arctic. Cold Reg Sci Technol 32(2–3):143–156

FRTR (1999) Thermal desorption. Federal remediation technologies roundtable. USEPA, Washington, DC http://www.frtr.gov/matrix2/section4/4_29.html

Geys R, Soetaert W, Bogaert IV (2014) Biotechnological opportunities in biosurfactant production. Curr Opin Biotechnol 30:66–72

Gottfried A, Singhal N, Elliot R, Swift S (2010) The role of salicylate and biosurfactant in inducing phenanthrene degradation in batch soil slurries. Appl Microbiol Biotechnol 86(86):1563–1571

Hejazi RZ (2002) Oily sludge degradation study under arid conditions using a combination of landfarm and bioreactor technologies. PhD thesis, Faculty of Engineering and Applied Science, Memorial University of Newfoundland, St John's, Canada

Huang XD, El-Alawi Y, Penrose DM, Glick BR, Greenberg BM (2004) A multiprocess phytoremediation system for removal of polycyclic aromatic hydrocarbons from contaminated soils. Environ Pollut 130:465–476

Huang XD, El-Alawi Y, Gurska J, Glick BR, Greenberg BM (2005) A multi-process phytoremediation system for decontamination of persistent total petroleum hydrocarbons (TPHs) from soils. Microchem J 81:139–147

Hussain I, Puschenreiter M, Gerhard S, Schöftner P, Yousaf S, Wang A, Syed HJ, Reichenauer GT (2018) Rhizoremediation of petroleum hydrocarbon-contaminated soils: improvement opportunities and field applications. Environ Exp Bot 147(2018):202–219

Hutchinson SL, Banks MK, Schwab AP (2001) Phytoremediation of aged petroleum sludge. J Environ Qual 30:395–403

Ilori MO, Amobi CJ, Odocha AC (2005) Factors affecting biosurfactant production by oil degrading Aeromonas spp. isolated from a tropical environment. Chemosphere 61(7):985992

Jorfi S, Rezaee A, Jaafarzadeh NA, Esrafili A, Akbari H, Ali GAM (2014) Bioremediation of pyrenecontaminated soils using biosurfactant. Jentashapir J Health Res 5(5):e23228

Kappeli O, Finnerty WR (1979) Partition of alkane by an extracellular vesicle derived from hexadecane grown Acinetobacter. J Bacteriol 140:707–712

Karanth NGK, Deo PG, Veenanadig NK (1999) Microbial production of biosurfactants and their importance. Curr Sci 77:116–123

Khan FI, Husain T, Hejazi R (2004) An overview and analysis of site remediation technologies. J Environ Manag 71:95–122

Kiran GS, Hema TA, Gandhimathi R, Selvin J, Thomas TA, Rajeetha Ravji T, Natarajaseenivasan K (2009) Optimization and production of a biosurfactant from the sponge-associated marine fungus Aspergillus ustus MSF3. Colloids Surf B 73(2):250–256

Kumar R, Das AJ, Lal S (2015) Petroleum hydrocarbon stress management in soil using microorganisms and their products, Waste Management Edited by Environmental Ram Chandra. CRC Press Taylor and Francis, Boca Raton ISBN

Liu G, Niu J, Guo W, An X, Zhao L (2016) Ecological and health risk-based characterization of agricultural soils contaminated with polycyclic aromatic hydrocarbons in the vicinity of a chemical plant in China. Chemosphere 163:461–470

Liu G, Zhong H, Yang X, Liu Y, Shao B, Liu Z (2017) Advances in applications of rhamnolipids biosurfactant in environmental remediation: a review. Biotechnol Bioeng

Mahmound A, Aziza Y, Abdeltif A, Rachida M (2008) Biosurfactant production by Bacillus strain injected in the petroleum reservoirs. J Ind Microbiol Biotechnol 35:1303–1306

Ma KY, Sun MY, Dong W, He CQ, Chen FL, Ma YL (2016) Effects of nutrition optimization strategy on rhamnolipid production in a Pseudomonas aeruginosa strain DN1 for bioremediation of crude oil. Biocatal Agric Biotechnol 6:144–151

Manickam N, Bajaj A, Saini HS, Shanker R (2012) Surfactant mediated enhanced biodegradation of hexachlorocyclohexane (HCH) isomers by Sphingomonas sp. NM05. Biodegradation 23(5):673–682

McCutcheon SC, Schnoor JL (2003) Phytoremediation: transformation and control of contaminants. Wiley, Hoboken ISBN: 0-471-39435-1. 987 pp

Meena KR, Kanwar SS (2015) Lipopeptides as the antifungal and antibacterial agents: applications in food safety and therapeutics. Biomed Res Int 2015:473050

Mihopoulos PG, Suidan MT, Sayles GD, Kaskassian S (2002) Numerical modeling of oxygen exclusion experiments of anaerobic bioventing. J Contam Hydrol 58(3–4):209–220

Mnif I, Ghribi D (2016) Glycolipid biosurfactants: main properties and potential applications in agriculture and food industry. J Sci Food Agric 96(13):4310–4320

Mohanty S, Jasmine J, Mukherji S (2013) Practical considerations and challenges involved in surfactant enhanced bioremediation of oil. Biomed Res Int 2013:1–16

Mulligan CN (2005) Environmental applications for biosurfactants. Environ Pollut 133(2):183–198

Mulligan CN, Gibbs BF (2004) Types, production and applications of biosurfactants. Proc Indian Natl Sci Acad 1:31–55

Muthusamy K, Gopalakrishnan S, Ravi TK, Sivachidambaram P (2008) Biosurfactants: properties, commercial production and application. Curr Sci 94(6):736–747

Ndimele PE (2010) A review on the phytoremediation of petroleum hydrocarbon. Pak J Biol Sci 13(15):715–722

Nikolopoulou M, Pasadakis N, Norf H, Kalogerakis N (2013) Enhanced ex situ bioremediation of crude oil contaminated beach sand by supplementation with nutrients and rhamnolipids. Mar Pollut Bull 77(1):37–44
Nitschke M, Coast SG (2007) Biosurfactants in food industry. Trends Food Sci Technol 18:252–259
Obayori OS, Ilori MO, Adebusoye SA, Oyetibo GO, Omotayo AE, Amund OO (2009) Degradation of hydrocarbons and biosurfactant production by Pseudomonas sp. strain LP1. World J Microbiol Biotechnol 25:1615–1623
Pacwa-Płociniczak M, Płaza GA, Piotrowska-Seget Z, Cameotra SS (2011) Environmental applications of biosurfactants: recent advances. Int J Mol Sci 12:633–654
Qian G, Jia Y, Wen J, Hu Y, Chen Y, Wu W (2016) Rhamnolipidenhanced aerobic biodegradation of triclosan (TCS) by indigenous microorganisms in water-sediment systems. Sci Total Environ 571:1304–1311
Ramadass K, Megharaj M, Venkateswarlu K, Naidu R (2015) Ecological implications of motor oil pollution: earthworm survival and soil health. Soil Biol Biochem 85:72–81
Raskin I (1996) Plants genetic engineering may help with environmental clean-up. Proc Nat Acad Sci USA 93:3164–3166
Riser-Roberts E (1998) Remediation of petroleum contaminated soil: biological, physical, and chemical processes. Lewis Publishers, Boca Raton
Salanitro JP, Dorn PB, Huesemann MH, Moore KO, Rhodes IA, Jackson LMR, Vipond TE, Western MM, Wisniewski HL (1997) Crude oil hydrocarbon bioremediation and soil ecotoxicity assessment. Environ Sci Technol 31(6):1769–1776. https://doi.org/10.1021/es960793i
Santos DKF, Rufino RD, Luna JM, Santos VA, Sarubbo LA (2016) Biosurfactants: multifunctional biomolecules of the 21st century. Int J Mol Sci 17(3):401
Scheuer U, Zimmer T, Becher D, Schauer F, Schunck WH (1998) Oxygenation cascade in conversion of n-alkanes to α, ω-dioic acids catalyzed by cytochrome P450 52A3. J Biol Chem 273(49):32528–32534
Singh AK, Cameotra SS (2013) Rhamnolipids production by multimetal-resistant and plant-growthpromoting phizobacteria. Appl Biochem Biotechnol 170(5):1038–1056
Singh P, Saini HS, Raj M (2016) Rhamnolipid mediated enhanced degradation of chlorpyrifos by bacterial consortium in soil-water system. Ecotoxicol Environ Saf 134:156–162
Shaligram NS, Singhal RS (2010) Surfactin–a review on biosynthesis, fermentation, purification and applications. Food Technol Biotechnol 48(2):119–134
Sharma D, Ansari MJ, Al-Ghamdi A, Adgaba N, Khan KA, Pruthi V, Al-Waili N (2015) Biosurfactant production by Pseudomonas aeruginosa DSVP20 isolated from petroleum hydrocarbon-contaminated soil and its physicochemical characterization. Environ Sci Pollut Res 22(22):1–8
Sponza DT, Gök O (2010) Effect of rhamnolipid on the aerobic removal of polyaromatic hydrocarbons (PAHs) and COD components from petrochemical wastewater. Bioresour Technol 101(3):914
USEPA (1998) Landfarming. Office of the underground storage tank, US Environmental Protection Agency, Publication # EPA 510-B-95-007. Available at: http://www.epa.gov/swerust/cat/landfarm
Van Hamme JD, Singh A, Ward OP (2003) Recent advances in petroleum microbiology. Microbiol Mol Biol Rev 67:503–549
Velikonja J, Kosaric N (1993) Biosurfactants in food applications. In: Kosaric N, Sukan FV (eds) Biosurfactants: production, properties, applications. CRC Press, New York, pp 419–448
Wang CP, Li J, Jiang Y, Zhang ZY (2014) Enhanced bioremediation of field agricultural soils contaminated with PAHs and OCPs. Int J Environ Res 8(4):1271–1278
Wang D, Liu Y, Lin Z, Yang Z, Hao C (2008) Isolation and identification of surfactin producing Bacillus subtilis strain and its effect of surfactin on crude oil. Wei Sheng Wu Xue Bao 48(3):304–311
Zhu, S., Liang, S., & You, X. (2013, January). Effect of rhamnolipid biosurfactant on solubilization and biodegradation of polycyclic aromatic hydrocarbons. In Intelligent System Design and Engineering Applications (ISDEA), 2013 Third International Conference on (pp. 651–654). IEEE
Zimmer T, Ohkuma M, Ohta A, Takagi M, Schunck WH (1996) The CYP52 multigene family of *Candida maltosa* encodes functionally diverse n-alkane-inducible cytochromes p450. Biochem Biophys Res Commun 224(3):784–789

Chapter 9
Environmental Contamination, Toxicity Profile, and Bioremediation Approaches for Detoxification of Paper Mill Wastewater

Shiv Shankar, Shikha, Arpna Ratnakar, Shailja Singh, and Shalu Rawat

Contents

Abstract Pulp and paper mills are considered as core sector industries and are the fifth largest contributor to industrial water pollution globally. During the processing of lignocellulosic material for paper production, a large quantity of brown/black effluent is generated with high BOD, COD, potentially toxic chlorinated compounds, adsorbable organic halides (AOXs), suspended solids, tannins, resin acids, sulfur compounds, and lignins. Lignin and chlorinated derivatives (e.g., ligno-sulfonic acid, resins, phenols, and hydrocarbons) released during pulping and bleaching operations

S. Shankar · Shikha (✉) · A. Ratnakar · S. Singh · S. Rawat
Department of Environmental Science, School for Environmental Sciences, Babasaheb Bhimrao Ambedkar University (A Central University), Lucknow, Uttar Pradesh, India

G. Saxena, R. N. Bharagava (eds.), *Bioremediation of Industrial Waste for Environmental Safety*, https://doi.org/10.1007/978-981-13-1891-7_9

impart dark color and toxicity to the effluent. The discharge of untreated/partially treated effluent into the receiving water bodies results in an increase in BOD/COD, slime growth, thermal problems, scum formation, discoloration, loss of aesthetic quality, and toxicity to the aquatic life. The toxic chlorinated organic compounds and hazardous metals present in wastewater pose serious human health risks through long-term exposure via drinking water and/or through consumption of fish that can bioaccumulate certain toxic metals from the food chain. Hence, several attempts have been made to degrade/detoxify pulp and paper mill wastewater. Keeping in view several disadvantages associated with conventional physicochemical treatment methods, biological methods have drawn the significant attention of the researchers as safer and cost-effective technologies for the treatment of pulp and paper mill effluent. In aforesaid context, present chapter encompasses the problem of environmental contamination due to pulp and paper mill effluent and its adverse effects on the environment and human population. It also highlights novel, energy, and cost-effective environment-friendly technologies to treat pulp and paper mill effluent.

Keywords Pulp paper mill wastewater · Lignin · Environmental contamination · Toxicity profile · Bioremediation · Detoxification

9.1 Introduction

Industries are important for the economic growth of the country; however, these are also the major pollution causing commercial establishment as these discharge a highly toxic complex wastewater containing organic and inorganic pollutants responsible for pollution and toxicity in living beings (Goutam et al. 2018; Bharagava et al. 2017a; Gautam et al. 2017; Saxena et al. 2016; Saxena and Bharagava 2015). Among the industries, pulp and paper industries are the important commercial sectors. Globally, pulp and paper industries are engaged in the production of pulp, paper, paper boards, and different cellulose-based products by using wood as a raw material. China, the United States of America, Canada, Finland, Sweden, Russia, Japan, and South Korea are leading producers of the pulp and paper products. For the production of pulp and paper products, pulp and paper industries use wood, fossil fuel, electricity, and water at massive scale. In addition, wide spectrums of pollutants are released during manufacturing of pulp and paper products causing adverse effects on the natural environment.

Pulp and paper industry releases different solid, liquid, and gaseous pollutants into the environment and has been placed under world's top 6 environment-polluting industries (after oil, cement, leather, textile, and steel industries). In Canada, pulp and paper is the third largest industrial polluter to water, air, and land, sixth largest in the United States. In Canada, pulp and paper industry generated 174,000 tonnes of water, air, and land pollutants (5.3%) out of 3.3 million tonnes of pollutants released by other industries.

In the United States of America, pulp and paper industry contributed 79,000 tonnes of pollutants which were 5% of pollutants released by other industries in 2015. Out of total pollutants released by pulp and paper industry in the United States of America, 66% pollutants are released into the air, 24% onto the land, and 10% into the aqueous streams.

Pulp and paper production involves three steps: pulping, bleaching, and paper-making. The wastewater produced during pulping operations is known as black liquor which is dark brown in color due to the presence of dissolved lignin and its degradation products, resins acids, hemicelluloses, and phenols (Berryman et al. 2004). During bleaching of the pulp, highly toxic compounds such as chlorinated organic compounds, derivatives of lignin, and hemicellulose and adsorbable organic halides (AOXs) are generated. The color of the effluent is persistent dark due to the presence of lignin and its derivatives, such as chlorolignin (Prasongsuk et al. 2009). Lignin, a complex heteropolymer, consists of phenylpropanoid aryl-C3 units connected with each other via diverse ether and C–C bonds. Lignin binds cellulose fiber and covalently linked to hemicelluloses. The ether and C–C bonding present in lignin make it unsusceptible toward hydrolytic attack, and therefore, lignin is highly resistant to breakdown (Bugg et al. 2011). A large quantity of wastewater is released by pulp and paper industries which is a major cause of concern to the scientists (Tewari et al. 2009; Saxena and Bhargava 2016, 2017).

9.2 Pulp and Paper Industry: General Overview

In pulp and paper industry, dark-colored effluent is generated during the process of pulping and bleaching. The wastewater released from pulp and paper industry adversely affects water bodies as it has high chemical oxygen demand (COD), biochemical oxygen demand (BOD), chlorinated compounds as adsorbable organic halides (AOXs), fatty acids, suspended solids, tannins, lignin and its derivatives like chlorolignins, resin acids, and compounds of sulfur (Chandra et al. 2007). It has been well documented that many of such pollutants are acute and chronic toxicants. Genotoxicants like furans and dioxins released from pulp and paper industry contaminate aqueous streams and exert aquatic biota and public health negatively (Chandra et al. 2009). Stringent environmental regulations, the environmental toxicity of pollutants, and increased public awareness have forced pulp and paper industry to treat wastewater to the desired limit before its discharge into the environment (D'Souza et al. 2006). Hence, treatment of pulp and paper mill wastewater is required to reduce adverse environmental impacts of the pollutants. For detoxification and decolorization, different physicochemical and biotechnological methods have been practiced. Physicochemical methods are not environment-friendly and economical. Processes like coagulation and precipitation generate a huge amount of sludge, which is difficult to dewater. Generally, an extreme pH range is employed for optimum treatment, and the pH is required to be readjusted to neutral before discharge. Oxidation processes using ozone and hydrogen peroxide are costly and un-ecofriendly as they result information of secondary pollutants such as chlorinated organics. The membrane filtration techniques generally require pretreatment and a large capital investment. Membrane fouling is another drawback of the membrane filtration techniques. Keeping in view several disadvantages associated with conventional physicochemical treatment methods, biological methods have drawn the significant attention of the researchers as safer and cost-effective technologies for the treatment of pulp and paper mill effluent (Gupta et al. 2015).

In aforesaid context, present chapter encompasses the problem of environmental contamination due to pulp and paper mill effluent and its adverse effects on the environment and human population. It also highlights novel, energy, and cost-effective environment-friendly technologies to treat pulp and paper mill effluent.

9.3 Pulp and Paper Production: Description of the Process

The paper manufacturing process involves three major steps: pulping, bleaching, and finally papermaking (Fig. 9.1).

9.3.1 Different Methods of Pulp Production

9.3.1.1 Mechanical Pulping

In mechanical pulping, the pulp is produced using grindstones embedded with aluminum oxide or silicon carbide. The wood is converted into stone pulp using grindstones. The pulp is called pressure ground pulp if it is exposed to steam before grinding.

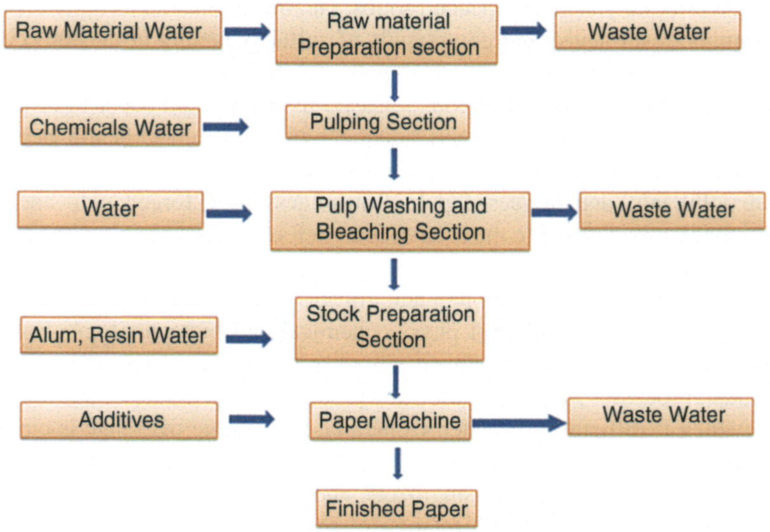

Fig. 9.1 Typical process of paper manufacturing

Manufactured grindstones with embedded silicon carbide or aluminum oxide can be used to grind small wood logs called "bolts" to make stone pulp (SGW). If the wood is steamed prior to grinding, it is known as pressure groundwood pulp (PGW). Mechanical pulps are produced for manufacturing of paper products requiring less strength, e.g., paper boards and newsprint.

9.3.1.2 Thermomechanical Pulping

Thermochemical pulping involves processing of chips of the wood by employing heat and mechanical refining. In this process bark is removed from the wood and wood is converted into small chips. Then, a mechanical force is applied to grind the chips for separation of fibers. During this process heat and vapors are generated which soften lignin component. The pulp generated in the process is cleaned and reprocessed. The pulp obtained using thermochemical pulping is rigid and hard.

9.3.1.3 Chemi-thermomechanical Pulping

In this process, wood chips are pretreated with chemicals like sodium hydroxide, sodium carbonate, and sodium sulfite before refining under moderate temperature and pH. Pulps produced in this process are called chemo-thermomechanical pulps (CTMP).

9.3.1.4 Chemical Pulping

In chemical pulping, the pulp is produced by digesting wood chips in large vessels by employing chemicals and heat. During the process of digestion of wood, heat and the chemicals dissolve lignin, which connects cellulose fibers with each other. The kraft process is a potential chemical pulping method, followed by pulping.

9.3.1.5 Recycled Pulping

Recycled pulp is produced by removing printing inks from the waste paper during recycling of waste paper in a process called deinking. The process of deinking is performed using chemicals like sodium sulfate, sodium silicate, and hydrogen peroxide. Different types of paper like newsprint, facial tissue, and toilet papers are produced using the recycled pulp.

9.3.1.6 Organosolv Pulping

In organosolv pulping, lignin and hemicelluloses are converted into soluble components using organic solvents at a temperature more than 140 °C. The pulping liquid is harvested by distillation. Ethanol, methanol, and formic acids are generally employed for organosolv pulping.

9.3.1.7 Bleaching

Pulp obtained after pulping of wood chips is bleached in order to produce bright paper. Chlorine dioxide, hydrogen peroxide, oxygen, and ozone are used in bleaching operations.

9.3.1.8 Papermaking

In papermaking step, the wood pulp is diluted with 100 times of its weight. The fiber liquid is then run through the machine. In the paper machine, water is removed, and the papers' properties are developed. In finishing and coating stages, the paper is coated, depending on the desired end product.

9.4 Pulp and Paper Mill Wastewater: Generation

Wastewater is generated in several processes in pulp and paper industry. These processes include debarking of the wood/making of the chip, pulping and bleaching operations, manufacturing of the paper, and deinking of wastewater. In all the processes, different types of pollutants are generated (Tables 9.1 and 9.2). The

Table 9.1 Major pollutants released from paper and pulp making process

Process stages	Wastewater (v)	Pollution load	Pollutants
Raw material preparation	Low	Low	Suspended solids including bark particles, fiber pigments, dirt, grit, BOD and COD
Pulping	Low	High	Color, bark particles, soluble wood materials, resin acids, fatty acids, AOX, VOCs, BOD, COD and dissolved inorganics
Bleaching	High	High	Dissolved lignin, color, COD, carbohydrate, inorganic chlorines, AOX, EOX, VOCs, chlorophenols, and halogenated hydrocarbons
Papermaking	Depends on the extent	Low	Particulate wastes, organic and inorganic compounds, COD and BOD

Adapted from Ali and Sreekrishnan (2001) and Hossain and Ismail (2015)

Table 9.2 Pollutants generated during different stages of pulp and paper production

Unit operations	pH	COD (mg/L)	BOD_5 (mg/L)	TSS (mg/L)	References
Recycled paper mill	6.2–7.8	3380–4930	1650–2565	1900–3138	Zwain et al. (2013)
Integrated pulp and paper mill	6.5	3791	1197	1241	Avsar and Demirer (2008)
Paper machine	6.5	1116	641	645	Avsar and Demirer (2008)
Bleaching[a]	8.2	3680	352	950	Kansal et al. (2008)
Pulping process operations[b]	5.5	9065	2440	1309	Avsar and Demirer (2008)
Kraft cooking section	13.5	1669.7	460	40	Wang and Ma (2007)
Chemical thermomechanical pulping[c]	7.43	7521	3000	350	Liu et al. (2011)
Wood yard and chipping[d]	7	1275	556	7150	Avsar and Demirer (2008)
Thermomechanical pulping	4.0–4.2	3343–4250	–	330–510	Qu et al. (2012)

[a]A combination of chlorination and alkaline extraction stages
[b]Pulping, pulp screening, pulp washing and thickening, bleaching, and kraft repulping
[c]Alkaline peroxide mechanical pulping (APMP)
[d]Pulpwood storage, debarking, and chipping

characteristics and volume of the wastewater generated from pulp and paper industry depend on the scale of production, feedstock employed, and technology of paper production (Tiku et al. 2007). Soil, dirt, and bark are removed from the raw material during preparation of wood which results in the generation of colored wastewater containing tannin and resin acids. Cooking of feedstock for production of cellulose pulp in kraft pulping processes results in the generation of a large amount of colored wastewater containing lignin. Wood pulping operations result in the generation of a wide spectrum of toxic pollutants mainly resin acids (namely, sandacopimaric, abietic, isopimaric, dehydroabietic, and neoabietic), unsaturated fatty acids (linoleic acid, oleic acid and linolenic acid), chlorinated resin acids, juvaniones, and diterpene alcohols (Ali and Sreekrishnan 2001; Vepsäläinen et al. 2011). Lignin generated during pulping imparts dark color and high COD to the wastewater. Effluent generated during pulping also contains chemicals such as soluble silicates ($3SiO_2$ Na_2O), known as water glass (Huuha et al. 2010). Besides aforesaid compounds, black liquors released during pulping may contain sodium hydroxide and sodium sulfide, which are used to dissolve lignin (Betancur et al. 2009).

The pulp obtained from pulping of the wood is dark colored due to the presence of residual lignin which is subjected to bleaching for improvement of its brightness using chemicals like chlorine dioxide, hydrogen peroxide, oxygen, and ozone. Use of chlorine compounds in bleaching operations results in serious environmental concerns in the forms of ozone depletion. Pollutants like adsorbable organic halogens

(AOXs), chlorophenols, extractable organic halogens (EOXs), and traces of toxic substances such as polychlorinated biphenyls, DDT, and polychlorinated dibenzodioxins (Karrasch et al. 2006) are present in bleach plant effluent. In addition, chemicals like chlorinated ligno-sulfonic acids, chlorinated resin acids, dioxins, and furans are generated during bleaching operations (Ali and Sreekrishnan 2001). Keeping in view toxicity and environmental concerns, it is imperative to develop efficient and eco-friendly technologies for the pulping and bleaching of the pulp.

9.5 Environmental Contamination and Toxicity

For the production of 1-ton paper, pulp and paper industry produces 150–200 m^3 wastewater characterized by 500–1100 kg chemical oxygen demand (COD)/ton paper, 85–370 kg biochemical oxygen demand (BOD)/ton paper, and 90–240 kg suspended solids/ton paper (Kumar et al. 2015a, b). Pollutants present in pulp and paper mill wastewater have adverse effects on the aquatic and terrestrial environment. Lignin present in pulp and paper mill wastewater is very difficult to degrade due to its peculiar structure (Shankar and Shikha 2012). Lignosulfonates and chlorolignins present in wastewater have been reported to inhibit functions of enzymes, toxins, and antibiotics in abiological system (Farstad and Naess 1977; Sponza 2003; Kumar et al. 2010). The color of the pulp and paper mill effluent is dark brown which deteriorates the aesthetic value of receiving water bodies and reduces penetration of solar radiation in water, thereby affecting the productivity of aquatic plants. Chlorinated lignins present in sludge adversely affect the growth of the plant. Chlorinated lignins have been reported to inhibit the germination of the wheat.

Chlorophenolics present in pulp and paper mill wastewater have been reported to have toxic, carcinogenic, mutagenic, and estrogenic effects (Michałowicz and Duda 2007). Chlorophenolics interrupt synthesis of ATP and oxidative phosphorylation during respiration in aquatic plants and animals. The chlorophenols result in the formation of extremely toxic dioxins and furans upon incineration (Veningerová et al. 1994; Criado et al. 2004). Chlorophenolics tend to accumulate into sludge and soil sediments (de Morais et al. 2012). In addition, they show toxic effects on fish (Oikari et al. 1988). In addition, chloroguaiacols show more toxicity in fresh water than in marine conditions or in brackish water systems (Renberg et al. 1980). Chlorinated anisoles and veratroles are toxic and cause off flavors in receiving water fish and tend to bioaccumulate (Brownlee et al. 1993). Wood extractives have been reported to cause adverse effects on aquatic biota (Ali and Sreekrishnan 2001; Ratia et al. 2013). Wood sterols like stigmastanol, β-sitosterol, stigmasterol, and campesterol (Ayebo et al. 2006) may result in hormonal disruptions in fish. Stigmasterol and β-sitosterol have been found to bind with estrogen receptor and disturb the reproductive functions in aquatic fish (Tremblay and van der Kraak 1999). Wood sterols cause risk to bottom feeding fish as they are hydrophobic and easily adsorbed on soil sediments (Rämänen et al. 2010). Fatty acids present in the pulp and paper mill wastewater cause toxicity to water salmonoids (Voss and Rapsomatiotis 1985).

The growth of the methanogenic bacteria like acetoclastic is inhibited by long-chain fatty acids (Ali and Sreekrishnan 2001). Resin acids present in pulp and paper mill wastewater may have detrimental effects on plants, humans, and aquatic animals. They exert toxicity to fish at a concentration of 200–800 μg/L (Ali and Sreekrishnan 2001). Resin acids are more toxic at low pH since their water solubility increases with increase in the pH. Resin acids result in impairment in liver functions in invertebrates and fish (Ratia et al. 2013). Resin acids bioaccumulate in the bile and liver of fish resulting in intense decomposition of red cells and impairment in the metabolism of iron (Kostamo et al. 2004). Resin acids reduce the secretion of reproductive hormones and disrupt cell division (Christianson-Heiska et al. 2008).

9.6 Treatment of Pulp and Paper Mill Effluent Wastewater Treatment

9.6.1 Physicochemical Methods

Keeping in view quantum of fresh water used in paper industry, environmental, and economic concerns and strict environmental regulations, the treatment of pulp and paper mill wastewater using safer methods has gained significant importance. Generally, physicochemical processes are employed for the treatment of pulp and paper mill wastewater. These processes are used during preliminary, primary, or tertiary stages of effluent treatment. Physicochemical treatment processes like flotation, sedimentation, adsorption, screening, oxidation, ozonation, coagulation, reverse osmosis, electrolysis, the replacement of chlorine by hypochlorite, ultrafiltration and nano-filtration technologies, activated carbon, and all phenolic compounds and sorption on hypo and alum sludge are used for the removal of colloidal particles, suspended solids, colors, floating matters, and toxic compounds (Dey et al. 2013; Sharma et al. 2007; Singh et al. 2011a, b) from pulp and paper mill wastewater. Physicochemical processes used for the treatment of pulp and paper mill wastewater are not commercially viable and eco-incentive. In addition, recalcitrant pollutants like lignin and its derivatives are not completely eliminated and persist in water in several chemical forms (Singh and Kaur 2015). Coagulation and precipitation methods of treatment of wastewater generate voluminous sludge, which is very difficult to dewater. Usually, an extreme pH range is needed for optimum treatment of the sludge produced. The pH of the sludge requires being neutral before its discharge into the environment. Employment of ozone and hydrogen peroxide in processes like oxidation is costly and results in formation of secondary pollutants like chlorinated organics. Membrane filtration techniques demand pretreatment and huge capital input. In addition, the problem of membrane fouling is routine problem encountered during pulp and paper mill effluent treatment. Biotechnological methods of treatment of pulp and paper mill wastewater are cheap and efficient alternatives to the physicochemical methods.

9.7 Biological Treatment of Pulp and Paper Mill Wastewater (Microbe-Based Bioremediation)

Biological methods of treatment of pulp and paper mill effluent employ microorganisms like bacteria, fungi, algae, and their enzyme system. Biological processes are considered more eco-friendly and eco-incentive than physicochemical methods. Biological processes significantly reduce toxicity and chemical load of the wastewater. Bioremediation is an effective waste management technique which employs biological organisms such as bacteria and plants to detoxify/transform a wide spectrum of organic and inorganic pollutants in less toxic and less harmful compounds (Saxena et al. 2018; Arora et al. 2014, 2018; Bharagava et al. 2017b, c; Saxena and Bharagava 2015; Chandra et al. 2015; Singh et al. 2011a, b). It is therefore, bioremediation is suggested as a greener approach to eliminate pollutants from effluent of pulp and paper industry (Abd El-Rahim and Zaki 2005; Shankar and Shikha, 2011).

9.7.1 Treatment of Pulp and Paper Mill Wastewater Using Bacteria

Bacteria have been utilized for the treatment of pulp and paper mill wastewater under anaerobic and anaerobic conditions. For the detoxification of biodegradable organic pollutants, a combination of aerobic and anaerobic treatment methods is more effective (Pokhrel and Viraraghavan 2004).

A large number of bacteria have been explored for the treatment of pulp and paper mill wastewater, and some promising bacterial isolates have been successfully used at commercial scale (Table 9.3). A lignin peroxidases-producing bacterium *Serratia liquefaciens* has been employed for the treatment of pulp and paper industry effluent. The bacterium was found to reduce 84% chemical oxygen demand (COD), 72% color, 61% lignin, and 95% phenolic content (Haq 2017). Abhishek et al. (2017) studied bacterial degradation and detoxification of kraft lignin using two bacteria *Citrobacter freundii* and *Serratia marcescens*. The mixture of bacteria resulted in 87% decolorization of kraft lignin (2000 mg/L). The axenic culture of *Citrobacter freundii* and *Serratia marcescens* decolorized 64, 60% model, and 50, 55% kraft lignin, respectively, at 34 °C, pH 8.2. The bacterial culture was found to reduce 76, 61% total organic content; 80, 67% chemical oxygen demand; and 87, 65% lignin from model and kraft lignin, respectively (Abhishek et al. 2017). Due to high chemical load and color contributing substances, pulp and paper mill effluents result serious aquatic and soil pollution. A lignin-degrading bacterium, *Serratia liquefaciens*, has been reported to significantly reduce 72% color, 58% lignin %, 85% COD, and 95% phenol during the incubation period of 144 h at 30 °C, pH 7.6 and 120 rpm. The outcomes of the study suggested that the bacterium *S. liquefaciens* is a promising candidate for bioremediation of pulp and paper mill effluent (Haq 2017). In pulp and paper mill effluent, lignin is the main pollutant. Bacterial

Table 9.3 Pulp and paper mill wastewater detoxifying microorganisms

Sr. no.	Bacteria	References
1.	*Serratia liquefaciens*	Haq (2017)
2.	*Citrobacter freundii*, *Serratia marcescens*	Abhishek et al. (2017)
3.	*Brevibacillus sagri*	Hooda et al. (2015)
4.	*Bacillus subtilis and Micrococcus luteus*	Tyagi et al. (2014)
5.	*Alcaligenes* sp., *Klebsiella* sp., and *Cronobacter* sp.	Kumar et al. (2014)
6.	*Streptomycetes badius* and *S. Viridosporous*	Chandra et al. (2011)
7.	*Pseudomonas putida* and *Acinetobacter calcoaceticus*	Murugesan (2003)
8.	*Ancylobacter*, *Methylobacterium*	Keharia and Madamwar (2003)
9.	*Bacillus* sp.	Mishra and Thakur (2010)
10.	*Pseudomonas stutzeri* CL7A	Karn et al. (2010)
11.	*Bacillus* sp. and *Serratia marcescens*	
12.	*Bacillus cereus* (ITRC-S6) and *Serratia marcescens* (ITRC-S7)	Chandra et al. (2009)
Fungi		
1.	*Phanerochaete chrysosporium*, *Tinctoria borbonica*, and *Trametes versicolor*	Fukuzumi et al. (1977)
2.	*Aspergillus niger*	Dashtban et al. (2010)
3.	*Trichoderma* spp.	Kamali and Khodaparast (2015)
4.	*Schizophyllum* sp.	Dashtban et al. (2010)
5.	*Tinctoria borbonica*	Abd El-Rahim and Zaki (2005)
6.	*P. chrysosporium*	Singhal et al. (2005)
7.	*P. chrysosporium*, *T. versicolor*, *P. ostreatus* and *L. edodes*	Wu et al. (2005)
8.	*Emericella nidulans* and *Aspergillus nidulans*	Singhal and Thakur (2009)
9.	*P. ostreatus* and *Pleurotus sajor-caju*	Belém et al. (2008)
10.	*T. versicolor*	Pendroza et al. (2007)
11.	*Pleurotus sajorcaju*, *Trametes versicolor* and *Phanerochaete chrysosporium*, *Rhizopus oryzae*	Freitas (2009)
12.	*Phanerochaete chrysosporium*	Sharari et al. (2011)
13.	*Gliocladium virens*	Kamali and Khodaparast (2015)
14.	*Fibrodontia* sp. RCK783S	Kreetachat et al. (2016)
15.	*Paecilomyces* sp.	Chuphal et al. (2005)
16.	*Pleurotus sajorcaju*, *Trametes versicolor*, *Phanerochaete chrysosporium*, *Rhizopus oryzae*	Rocha-Santos et al. (2010)
17.	*Planococcus* sp. *TRC1*	Priyadarshinee et al. (2017)
18.	*Pleurotus ostreatus*	Kannan et al. (1990)
19.	*Peniophora sp. (NFCCI-2131)*	Shankar and Shikha (2012)
20.	*Heterobasidium* sp.	Daniel et al. (1998)
21.	*Fusarium* sp.	Abedin and Taha (2008)
Algae		
1.	*Chlorella* sp.	Min et al. (2011)
2.	*Scenedesmus* sp.	

(continued)

Table 9.3 (continued)

Sr. no.	Bacteria	References
3.	*Chlorococcum* sp.	Sethunathan et al. (2004)
4.	*Selenastrum capricornutum*	Arbib et al. (2014)
5.	*Microcystis* spp.	Iyovo et al. (2010) and Sharma et al. (2014)
6.	*Chlamydomonas* sp.	Sharma et al. (2014) and Chandra and Singh (2012)

enzymes, due to their stability at extreme environmental conditions, play a very crucial role in the degradation of lignin. The treatment of pulp and paper mill effluent by a bacterium, *Brevibacillus agri*, isolated from sludge has been carried out under batch and semicontinuous conditions (Hooda et al. 2015). Results revealed that the bacterium reduces 69% COD, 37% lignin and adsorbable organic halides, and 19% color under batch study. The bacterium was found to reduce 40% AOX, 30% lignin, 37% color, and 62% COD under semicontinuous reactor study. The results of the study substantiated that the bacterium can significantly reduce the chemical load of the pulp and paper mill wastewater.

Bacteria like *Bacillus subtilis* and *Micrococcus luteus* have been found to reduce BOD up to 87% and COD up to 94.7% and lignin up to 97% during the incubation period of 9 days. Both the strains were found to reduce pH of the effluent up to 7 and improve dissolved oxygen from 0.8 to 6.8 mg/l (Tyagi et al. 2014). Bacterial consortium (*Alcaligenes* sp., *Klebsiella* sp., and *Cronobacter* sp.) was employed for the treatment of pulp and paper mill wastewater in a sequential batch reactor by Kumar et al. (2014). The consortium was found to reduce COD and BOD of the effluent by 72.3% and 91.1%, respectively. Substantial reduction in adsorbable organic halides (45%), color (55%), total dissolved solids (22%), and total suspended solids (86%) was observed during the period of 14 h. Some bacterial strains have been reported to degrade lignin derivatives produced during pulping operations (Chandra et al. 2011; Chandra and Bharagava 2013). The color of the kraft mill effluent has been successfully reduced by 26–54% by *Pseudomonas aeruginosa* under anaerobic conditions (Ramsay and Nguyen 2002). Bacterial strains *Streptomycetes badius* and *S. viridosporous* have been found to utilize commercial kraft lignin as a sole carbon source as characterized by amino acid analysis, FTIR, and EDX spectroscopy (Chandra et al. 2011). Two bacterial strains *Pseudomonas putida* and *Acinetobacter calcoaceticus* have been found to reduce 70–80% lignin and COD during the incubation period of 8 days (Murugesan 2003). Bacterium, *Ancylobacter*, and *Methylobacterium* have been found to reduce AOX in the effluent from kraft mill (Keharia and Madamwar 2003). Mixed consortia of aerobic and anaerobic bacteria including *E. coli*, *Bacillus* spp., *Pseudomonas*, *Arthrobacter*, *Sphingomonas*, *Alcaligenes eutrophus*, *Flavobacterium* spp., *Mortierella isabella*, *Zooglea*, and *Chaetomium cochliolidae* have been reported to degrade resin acids (Tiku et al. 2010; Raj et al. 2014). Mishra and Thakur (2010) isolated alkalotolerant *Bacillus* sp. from rom sludge of pulp and paper mill wastewater and assessed the effluent detoxification efficiency of the bacterium by Taguchi approach. The results showed that the color and lignin content of the effluent was reduced by 25–69% and 28–53%, respectively (Mishra and Thakur 2010). Karn et al. (2010) isolated a bacterium *Pseudomonas stutzeri* CL7A from the secondary sludge of the pulp and paper mill effluent. The bacterium was found to use pentachlorophenol as a sole carbon source. The bacterium mineralized high concentration (600 mg/L) of pentachlorophenol and resulted in 66.8% removal of pentachlorophenol from the sludge of pulp and paper mill wastewater during the incubation period of 2 weeks. The results obtained from the study suggested that the bacterium *Pseudomonas stutzeri* CL7A can be effectively used for the removal of pentachlorophenol from pulp and paper mill wastewater (Karn et al. 2010). Chandra et al. (2009) employed two pentachlorophenol-degrading bacterial strains *Bacillus cereus* (ITRC-S6) and *Serratia marcescens* (ITRC-S7) for the detoxification of pulp and paper mill effluent. The bacterial strains were found to reduce PCP (85–90%), lignin (30–42%), BOD (40–70%), COD (50–60%), total phenol (32–40%), and color (45–52%), during the incubation period of 168 h. Increase cell biomass in response to the addition of co-substrate resulted in the liberation of a substantial amount of chloride due to bacterial dechlorination of chlorolignins and chlorophenols leading to a reduction in color, lignin, and toxicity in the effluent.

9.7.1.1 Factors Affecting Treatment of Pulp and Paper Mill Wastewater

The detoxification and reduction of chemical load of the pulp and paper mill wastewater by bacteria depend on several factors like pH, size of inoculum, carbon source, the speed of agitation, and operating mechanism of removal of color. It has been reported that the decolorization of pulp and paper mill wastewater is generally carried out at acidic pH range. At acidic pH negatively charge lignin derivatives and bacterial cells precipitate due to increased protonation thereby leading to a reduction in color. At elevated pH the intensity of color increases (Tiku et al. 2010). Carbon source plays a very crucial role in bioremediation of pulp and paper mill effluent by bacteria. Glucose and sucrose have been reported as optimum sources of carbon which result in 20–25% more removal of color by bacteria (Tiku et al. 2010). The optimum decolorization of pulp and paper mill wastewater has been reported at 150 rpm. The optimum size of inoculum for treatment of pulp and paper mill wastewater has been reported to 7.5% (Singhal and Thakur 2009).

9.7.2 Treatment with Microalgae

Microalgae are a heterogeneous group of heterotrophic and photosynthetic unicellular microorganisms containing chlorophyll and other photosynthetic pigments. Microalgae consist of a group of single-celled eukaryotic organisms (red, brown, and green algae). Microalgae are present in all aquatic ecosystems and perform fixation and cycling of nutrients. The bioremediation potential of microalgae is based on their natural existence and degradation of pollutants present in water ecosystem. Microalgae have been successfully employed for the treatment of domestic wastewater (Olguin 2012) (Table 9.3). Since then they have been used for the removal of different pollutants from the wastewater. Algal species *Microcystis* has been reported to decolorize bleach kraft mill wastewater (Iyovo et al. 2010; Sharma et al. 2014). During the period of 2 months of incubation, the pure and mixed culture of algae was found to reduce the color of the effluent up to 70%. The extent was decolorization was intense during initial 15–20 days. Algal metabolism resulted in the transformation of color containing molecules to colorless compounds with limited degradation and assimilation (Chandra and Singh 2012). Under continuous lighting conditions, a mixed culture of algae containing *Chlamydomonas*, Microcystis, and *Chlorella* was employed to remove adsorbable organic halides and color. The algae were found to remove 70% AOX and 80% color of the wastewater during the incubation period of 30 days. It has been reported by several authors on the basis of analysis of alkaline extraction of the biomass of algae that the algal metabolism is responsible for the removal of the color of the wastewater (Sharma et al. 2014; Chandra and Singh 2012).

9.7.3 Remediation with Fungi

For the treatment of pulp and paper mill effluent, fungal treatment has drawn the significant attention of the researchers. Fungi are promising biological agents for detoxification of pulp and paper mill wastewater as they secrete different ligninolytic enzymes (Table 9.3). Fungi secrete extracellular enzymes and can survive in effluent and sludge with high chemical load (Kamali and Khodaparast 2015). White-rot fungi are the best lignin degraders and are only microbes which completely mineralize lignin into water and carbon dioxide. White-rot fungi have been reported to degrade and decolorize bleach plant effluent at commercial scale. The application of these fungi has been proven to be technically viable in airlift reactors, trickling filters, and fluidized bed reactors used for the treatment of wastewater (Prouty 1990).

In order to find out promising fungi for treatment of pulp and paper mill effluent, several researchers have isolated and screened fungi from diverse environmental sources. Fukuzumi et al. (1977) reported the application of white-rot fungi in the treatment of pulp and paper mill effluent for the first time. The fungi were cultivated in under liquid medium under submerged culture condition containing spent liquor from alkali extraction, vitamins, and nutrients.

Among the different fungi explored, *Tinctoporia* sp. was found to produce best results. Ligninolytic fungus *Phanerochaete chrysosporium*, *Tinctoria borbonica*, *and Trametes versicolor* have been widely reported to degrade lignin effectively along with carbohydrates. Fungus *Aspergillus niger* and *Trichoderma* spp. have been reported to degrade lignin and decolorize wastewater originating from hardwood pulp bleaching (Dashtban et al. 2010; Kamali and Khodaparast 2015). The white-rot fungus *Schizophyllum* commune effectively decolorizes bagasse-based pulp and paper mill wastewater. It also results in the degradation of lignin and significant reduction of BOD and COD in presence of suitable carbon source like sucrose (Dashtban et al. 2010). Abd El-Rahim and Zaki (2005) reported that the fungus *Tinctoria borbonica* results in 90–99% decolorization of kraft waste liquor to a light yellow color during the incubation period of 4 days. The white-rot fungus *P. chrysosporium* was used for the decolorization of pulp and paper mill effluent by Singhal et al. (2005). The authors reported that the color, pH, lignin, COD, and total phenols of the wastewater were significantly reduced. The optimum decolorization was found at pH 5.5.

The lignin-degrading efficiency of some white-rot fungi like *P. chrysosporium*, *T. versicolor*, *P. ostreatus*, and *L. edodes* was investigated by Wu et al. (2005). The researchers cultivated fungi individually on porous plastic media to study the detoxification of black liquor. The fungi were found to degrade lignin by 71% and reduce COD by 48%. It was observed that all the nutritional parameters, viz., carbon source, nitrogen source, pH, and trace elements, significantly affected the decolorization of the effluent and the reduction of the COD. The immobilization of fungal consortium was carried out for the first time by Malaviya and Rathore (2007) for the treatment of pulp and paper mill wastewater. The two basidiomycetes fungi *Merulius aureus* syn and *Phlebia* sp. and a deuteromycetous fungus *Fusarium sambucinum* Fuckel MTCC 3788 were immobilized on nylon mesh. The consortium was used for the treatment of pulp and paper mill effluent in a continuously aerated benchtop bioreactor. The results of the study revealed that the consortium reduced the color, lignin, and COD of the effluent in the order of 78.6%, 79.0%, and 89.4%, respectively, during the incubation period of 4 days. During initial 24 h of treatment, a major reduction in electrical conductivity, total dissolved solids, and salinity of the effluent was observed. Singhal and Thakur (2009) studied bioremediation potential of *Aspergillus nidulans* and *Emericella nidulans* by employing Taguchi method. The fungus was found to decolorize effluent by 31% and reduce lignin by 37%. The optimum pH and temperature for improved decolorization were found between 6–5 and 30–35 °C, respectively. White-rot fungi *P. ostreatus* and *Pleurotus sajor-caju* were tested for degradation of organic matter and removal of color from kraft pulp mill effluent by an activated sludge process by Belém et al. (2008). During the incubation period of 14 days, 57 and 76% reduction in absorbance of the effluent was resulted by both the fungus, respectively, at 460 nm. Both the fungi were found to reduce COD by 67% during the incubation period of 14 days. Wheat bran broth and 100 polyurethane foam were used to study the decolorization of pulp and paper mill wastewater at lab scale using fungus *T. versicolor* under submerged culture conditions by Pendroza et al. (2007).

The culture of the fungus was incubated for 9 days at 25 °C. The fungus resulted in substantial reductions in chlorophenols and 82% reduction in COD and color of the effluent. Three white-rot fungi (*Pleurotus sajor-caju*, *Trametes versicolor*, and *Phanerochaete chrysosporium*) and one soft-rot fungus (*Rhizopus oryzae*) species confirmed their potential for future applications in the biological treatment of effluents derived from the secondary treatment of a bleached kraft pulp mill processing

Eucalyptus globulus. During the incubation period of 10 days, 25–46 and 72–74% reduction in absorbance of the effluent was obtained. The fungi resulted in 74–81% reduction in COD (Freitas 2009). Sharari et al. (2011) used *Phanerochaete chrysosporium* for the treatment of bagasse mill wastewater under submerged culture conditions. During the incubation period of 9 days, 98.5% reduction in BOD and 98.7% reduction in COD were observed. The fungus reduced re in COD and total dissolved solids from 3950 to 575 mg L^{-1} and color from 560 to 111 mg L^{-1} PtCo. White-rot fungus *Gliocladium virens* grows naturally as a soil saprophyte and has been used for bioremediation of pulp and paper mill effluents. The fungus grows very well in the effluent and has been found to result in 42% decolorization followed by a reduction in lignin (52%), cellulose (75%), and BOD (65%) (Kamali and Khodaparast 2015). White-rot fungus *Coriolus versicolor* has also been tested by several researchers as a promising candidate for bioremediation of pulp and paper mill wastewater. The fungus *Coriolus versicolor* secretes oxidoreductase enzyme which results in degradation of lignin via one electron oxidation. The studies have shown that the fungus removes 60% of the color of the pulp and paper mill effluent within 6 days of incubation (Kamali and Khodaparast 2015). Another white-rot fungus *Trametes versicolor* has been studied for its capability to decolorize pulp and paper mill wastewater and has been found to decolorize the effluent by 34% during the incubation period of 3 days in presence of glucose as a co-substrate and urea as source of nitrogen (Kamali and Khodaparast 2015). Kreetachat et al. (2016) studied the decolorization of pulp and paper mill effluents by employing white-rot fungus *Fibrodontia* sp. RCK783S. The experiments were carried out minimal broth medium. The fungus was found to remove color and total organic content by 61.58% and 48.32%, respectively, during the incubation period of 5 days. Lignin-degrading enzyme laccase was responsible for the reduction of color and chemical load of the effluent.

9.7.3.1 Factors Affecting Decolorization of Pulp and Paper Mill Wastewater by Fungi

Different factors pH, temperature, dilution of the effluent, size of the inoculum, concentration of nutrients, and dissolved oxygen affect the process of decolorization and detoxification of pulp and paper mill effluent. Decolorization of pulp and paper mill wastewater by fungi involves a series of complex reactions catalyzed by fungal enzymes. The incorporation of mineral solution into media facilitates the growth and metabolism of fungi as well as the decolorization of the effluent. Fungi generally carry out decolorization of the wastewater under different pH (4–9) and incubation temperature ranging from 27 to 32 °C. However, the optimum pH and temperature for the maximum decolorization are 5 and 27 °C, respectively (Bajpai 1999). Fungi prefer aerobic conditions for decolorization of the effluent. Among different carbon sources, glucose has been documented as an optimum source of carbon required during decolorization of wastewater by fungi. High concentration of nitrogen levels has an inhibitory effect on the process of the biodegradation of lignin and derivatives. Nitrogen-scarce conditions or an abundant supply of nitrogen has no significant effect on the rate of decolorization (Bajpai 1999).

The high concentration of the effluent affects the growth of the fungi. Maximum decolorization of the effluent has been reported when the effluent was diluted by 20% (Zhang et al. 1999). Color removal of pulp and paper mill effluent is proportional to the size of the inoculum. Optimum decolorization of pulp and paper mill effluent has been found with the size of the inoculum ranging from 7.5% to 15% (Eaton et al. 1980; Singhal and Thakur 2009).

9.8 Enzymatic Treatment of Pulp and Paper Mill Wastewater

In recent years, several microorganisms have been investigated for the treatment of colored wastewater from pulp and paper industries. The feasibility of treatment processes using microorganism as a whole cell depends on adaptability and activity of the microorganism selected since the pollutants present in pulp and paper mill effluent are toxic in nature and may have a detrimental effect on growth and survival of the microorganism during the treatment process. A microbial culture-based treatment methods require long incubation time for the treatment of wastewater (Murugesan et al. 2009; Shankar and Shikha 2011). Under such circumstances, enzymatic remediation of wastewater using isolated enzyme has drawn the significant attention of the researchers and suggested as potential alternatives to conventional methods of the treatment of the wastewater. In these methods cell-free enzyme is used to treat wastewater. Enzymatic treatment involves chemical processes based on the activity of biological catalysts. Thus at present, treatment of pulp and paper mill wastewater using microbial enzymes due to its availability and potential decolorization ability against a wide variety of pollutants gaining popularity day by day. Different inligninolytic and ligninolytic enzyme systems have been reported to degrade lignin. Ligninolytic enzyme system of white-rot fungi effectively degrades lignin. This system comprised of three principal enzymes laccase, manganese peroxidase (MnP), and lignin peroxidase (LiP) (Table 9.4).

Forss et al. (1979) investigated the role of fungal laccase in the treatment of pulp and paper mill effluent. Under aerobic conditions, the reaction mixture containing laccase and pulp bleaching wastewater was treated at pH 4.8 for 1 h followed by flocculation with aluminum sulfate. The enzyme was found to substantially reduce the catechols, phenols, guaiacols, and vanillins present in wastewater. Direct removal of chlorophenolic compounds present in pulp and paper mill wastewater using laccases from *Trametes versicolor* was studied by Roy-Arcand and Archibald (1991). It was observed that laccases, secreted by *T. versicolor*, resulted partial elimination of chlorophenols. Horseradish peroxidases (HRP) from *Phanerochaete chrysosporium* have been found to degrade chlorophenolics like tetrachloro guaiacol, pentachlorophenol, 4,5-dichloroguaicol, 4,5,6-trichloroguaicol, and 2,4,6-trichlorophenol. Dezotti et al. (1995) immobilized lignin peroxidase from the

Table 9.4 Ligninolytic enzymes of white-rot fungi

Enzyme	Cofactor	Substrate and mediator	Reaction
Aryl-alcohol oxidase (AAO)	–	Aromatic alcohols (anisyl, veratryl alcohol)	Aromatic alcohols oxidized to aldehydes; H_2O_2 production
Glyoxal oxidase (GLOX)	–	Glyoxal, methyl glyoxal	Glyoxal oxidized to glyoxal acid; H_2O_2 production
Laccase	O_2	Phenols, mediators, e.g., hydroxybenzotriazole or ABTS	Phenols are oxidized to phenoxyl radicals; other reactions in the presence of mediators
Lignin peroxidase (LiP)	H_2O_2	Veratryl alcohol	Aromatic ring oxidized to cation radical
Manganese peroxidase (MnP)	H_2O_2	Mn, organic acids as chelators, thiols, unsaturated fatty acids	Mn(II) oxidized to Mn(III); chelated Mn(III) oxidizes phenolic compounds to/phenoxyl radicals; other reactions in the presence of additional compounds
Other H_2O_2-producing enzymes	–	Many organic compounds	O_2 reduced to H_2O_2
Versatile peroxidase (VP)	H_2O_2	Mn, veratryl alcohol, compounds similar to LiP and MnP	Mn(II) oxidized to Mn(III), oxidation of phenolic and nonphenolic compounds, and dyes

fungus *Chrysonilia sitophila* on activated silica for the removal of color from extraction stage effluent.

Immobilized HRP enzyme brought down COD of the effluent by 73% and color by 12%. Ferrer et al. (1991) reported that immobilized lignin peroxidase decolorizes kraft effluent. Ullah et al. (2000) employed purified laccase from the fungus *Coriolus versicolor* for the removal of pentachlorophenol (PCP) from water at pH 5. Laccase resulted in 60% removal of pentachlorophenol during the treatment time of 72 h. The study demonstrated potential application in remediation of wastewater containing pentachlorophenol. A carbon-based mesoporous magnetic composite was fabricated to immobilize laccase for the treatment of effluent containing phenols and *p*-chlorophenol. Immobilized laccase was found to remove 78% and 8% phenol *p*-chlorophenol, respectively, during the contact time of 1 h (Liu et al. 2011).

9.9 Biotechnological Processes Reducing Chemical Load in Wastewater

Chemical load in pulp and paper mill wastewater can be significantly minimized by using green processes like bio-depitching, biopulping, biobleaching, bio-deinking, etc.

9.9.1 Bio-depitching

During the production of papers, lipophilic extractives like fatty alcohols, resin acids, sterols, and terpenoids (conjugated sterols, triglycerides, and waxes) are generated. Effluent generated during the process contains resin acids like abietic, dehydroabietic and isopimaric, and pimaric acids extractives which are toxic to human health and ecosystem (Bajpai 1999). Use of microorganism is suggested as a greener approach to control pitch problem and can be a potential alternative to conventional treatment.

9.9.2 Biopulping

Pulping of wood can be performed by treating plant wood with white-rot fungi. White-rot fungi are known to degrade lignin constituent of wood with the help of ligninolytic enzyme system. Fungal pretreatment of wood chips prior to mechanical pulping reduces energy consumption, pitch content, and a chemical load of the effluent (Singh et al. 2010).

9.9.3 Biobleaching

Bleaching operations in paper industry generation of large amounts of toxic chlorinated organic compounds in the effluent. Chlorine compounds are harmful to the ozone layer. Biobleaching has emerged as one of the promising technologies which have the capacity to reduce the use of chlorine compounds. In addition, biobleaching is energy and cost incentive technology (Shankar et al. 2015). Fungal enzymes like laccases and xylanases have bleaching properties and can be utilized as biobleaching agents for making brighter paper.

9.9.4 Bio-deinking

Deinking is the industrial process of removing printing ink from paper fibers of recycled paper to make deinked pulp. The key in the deinking process is the ability to detach ink from the fibers. This is achieved by a combination of mechanical action and chemical means. At present, deinking is carried out using large amount deinking chemicals like NaOH, Na_2SiO_3, Na_2CO_3, H_2O_2, chelating agent, and surfactants (Bajpai 2012). Use of a large amount of chemicals makes the process costly and environmentally damaging. Treatment of toxic effluent characterized by high COD values needs costly wastewater treatment (Zhang et al. 2008). Keeping in view disadvantages of chemical deinking, the development of an alternative method for deinking is necessitated. Enzymatic deinking as a potentially efficient solution is proposed as an alternative to the chemical deinking (Ibarra et al. 2012). Over the past few years, enzymatic deinking, as a cost-effective and eco-friendly technology, has attracted a great deal of attention of the researchers (Pathak et al. 2014).

9.10 Conclusions and Future Perspectives

Different physicochemical methods used for the treatment of pulp and paper mill effluent are costly and un-ecofriendly. Biotechnological methods offer better alternatives to physicochemical methods. Among different biological methods, fungi and their enzyme system-based methods are promising technologies to treat pulp and paper mill effluent. However, there is a need to explore novel hyper secretary fungal strains and optimize physicochemical conditions favoring maximum decolorization and detoxification of pulp and paper mill effluent. Utilization of spent fungal biomass in preparation of active fungal biomass is expected to reduce the cost of the treatment process. Fungal decolorization and detoxification of pulp and paper mill wastewater is anaerobic process. For improved detoxification, the use of oxygen should be promoted as ideal gas for fluidization instead of air. Use of oxygen during fungal treatment increases the concentration of ligninolytic enzymes in the medium. A detailed study pertaining to the possible mode of action of ligninolytic enzymes during structural changes at the molecular level in lignin and its derivative compounds present in the effluent is of utmost importance. Pretreatment of pulp and paper mill wastewater with fungi can be carried out before bacterial treatment to increase organic chlorine and chlorolignins with high molecular weight. Pretreatment of wood can be carried out as a substitute for processes of internal modification (oxygen bleaching, high-level chlorine dioxide substitution and modified cooking, etc.) and traditional biological treatment.

Acknowledgments The authors are thankful to Vice Chancellor, Babasaheb Bhimrao Ambedkar University, and Head, Department of Environmental Microbiology, Babasaheb Bhimrao Ambedkar University for rendering necessary facilities. Young Scientist Fellowship (Award No: CST/242/, dated 12/05/2015, Council of Science and Technology Uttar Pradesh Lucknow India) to first author Dr. Shiv Shankar is also gratefully acknowledged.

References

Abd El-Rahim WM, Zaki EA (2005) Functional and molecular characterization of native Egyptian fungi capable of removing textile dyes. Arab J Biotech 8:189–200

Abedin RMA, Taha HM (2008) Antibacterial and antifungal activity of cyanobacteria and Green microalgae. Evaluation of medium components by Plackett-Burman design for antimicrobial activity of Spirulina platensis. Global J Biotechnol Biochem 3(1):22–31

Abhishek A, Dwivedi A, Tandan N, Kumar U (2017) Comparative bacterial degradation and detoxification of model and kraft lignin from pulp paper wastewater and its metabolites. Appl Water Sci 7:757–767

Ali M, Sreekrishnan TR (2001) Aquatic toxicity from pulp and paper mill effluents: a review. Adv Environ Res 5:175–196

Arbib Z, Ruiz J, Álvarez-Díaz P, Garrido-Pérez C, Perales JA (2014) Capability of different microalgae species for phytoremediation processes: Wastewater tertiary treatment, CO2 bio-fixation and low cost biofuels production. Water Res 49:465–474

Arora PK, Srivastava A, Singh VP (2014) Bacterial degradation of nitrophenols and their derivatives. J Hazard Mater 266:42–59

Arora PK, Srivastava A, Garg SK, Singh VP (2018) Recent advances in degradation of chloronitrophenols. Bioresour Technol 250C:902–909

Avsar E, Demirer GN (2008) Cleaner production opportunity assessment study in SEKA Balikesir pulp and paper mill. J Clean Prod 16:422–431

Ayebo A, Breuer GM, Cain T, Wichman MD, Subramanian P, Reynolds SJ (2006) Sterols as biomarkers for waste impact and source characterization in stream sediment. J Environ Health 68:46–50

Bajpai P (1999) Application of Enzymes in the Pulp and Paper Industry. Biotechnol Prog 15(2):147–157

Bajpai P (2012) Biotechnology for pulp and paper processing. Springer US, Boston

Belém A, Panteleitchouk AV, Duarte AC, Rocha-Santos TAP, Freitas AC (2008) Treatment of the effluent from a kraft bleach plant with white rot fungi *Pleurotus sajorcaju* and *Pleurotus ostreatus*. Global NEST J 10(3):426–431

Berryman D, Houde F, De Blios C, O'Shea M (2004) Nonylphenolic compounds in drinking and surface water downstream of treated textile and pulp and paper effluents: a survey and preliminary assessment of their potential effects on public health and aquatic life. Chemosphere 56(3):247–255

Betancur M, Bonelli PR, Velasquez JA, Cukierman AL (2009) Potentiality of lignin from the kraft pulping process for removal of trace nickel from wastewater: effect of demineralisation. Bioresour Technol 100:1130–1137

Bharagava RN, Saxena G, Mulla SI, Patel DK (2017a) Characterization and identification of recalcitrant organic pollutants (ROPs) in tannery wastewater and its phytotoxicity evaluation for environmental safety. Arch Environ Contam Toxicol. https://doi.org/10.1007/s00244-017-0490-x

Bharagava RN, Chowdhary P, Saxena G (2017b) Bioremediation: An ecosustainable green technology: its applications and limitations. In: Bharagava RN (ed) Environmental pollutants and their bioremediation approaches, 1st edn. CRC Press, Taylor & Francis Group, Boca Raton, pp 1–22. https://doi.org/10.1201/9781315173351-2

Bharagava RN, Saxena G, Chowdhary P (2017c) Constructed wetlands: an emerging phytotechnology for degradation and detoxification of industrial wastewaters. In: Bharagava RN (ed) Environmental pollutants and their bioremediation approaches, 1st edn. CRC Press, Taylor & Francis Group, Boca Raton, pp 397–426. https://doi.org/10.1201/9781315173351-15

Brownlee BG, MacInnis GA, Noton LR (1993) Chlorinated anisoles and veratroles in a Canadian river receiving bleached kraft pulp mill effluent. Identification, distribution, and olfactory evaluation. Environ Sci Technol 1093:2450–2455

Bugg TDH, Ahmad M, Hardiman EM, Rahmanpour R (2011) Pathways for degradation of lignin in bacteria and fungi. Nat Prod Rep 28:1883–1896

Chandra R, Bharagava RN (2013) Bacterial degradation of synthetic and kraft lignin by axenic and mixed culture and their metabolic products. J Environ Biol 34:991–999

Chandra R, Singh R (2012) Decolorization and detoxification of rayon grade pulp and paper mill effluent by mixed bacterial culture isolated from pulp-paper mill effluent polluted site. Biochem Eng J 61:49–58

Chandra R, Raj A, Purohit HJ, Kapley A (2007) Characterisation and optimisation of three potential aerobic bacterial strains for kraft lignin degradation from pulp paper waste. Chemosphere 67:839–846

Chandra R, Raj A, Yadav S, Patel DK (2009) Reduction of pollutants in pulp paper mill effluent treated by PCP-degrading bacterial strains. Environ Monit Assess 155(1):1–11

Chandra R, Abhishek A, Sankhwar M (2011) Bacterial decolorization and detoxification of black liquor from rayon grade pulp manufacturing paper industry and detection of their metabolic products. Bioresour Technol 102:6429–6436

Chandra R, Saxena G, Kumar V (2015) Phytoremediation of environmental pollutants: An eco-sustainable green technology to environmental management. In: Chandra R (ed) Advances in biodegradation and bioremediation of industrial waste, 1st edn. CRC Press, Taylor & Francis Group, Boca Raton, pp 1–30. https://doi.org/10.1201/b18218-2

Christianson-Heiska I, Haavisto T, Paranko J, Bergelin E, Isomaa B (2008) Effects of the wood extractives dehydroabietic acid and betulinol on reproductive physiology of zebrafish (Danio rerio) – a two generation study. Aquat Toxicol 86:388–339

Chuphal Y, Kumar V, Thakur IS (2005) Biodegradation and Decolorization of Pulp and Paper Mill Effluent by Anaerobic and Aerobic Microorganisms in a Sequential Bioreactor. World J Microbiol Biotechnol 21(8-9):1439–1445

Criado MR, Pombo da Torre S, Rodríguez Pereiro I, Cela Torrijos R (2004) Optimization of a microwave-assisted derivatization-extraction procedure for the determination of chlorophenols in ash samples. J Chromatogr A 1024:155–163

D'Souza DT, Tiwari R, Sah AK, Raghukumar C (2006) Enhanced production of laccase by a marine fungus during treatment of colored effluents and synthetic dyes. Enzym Microb Technol 38:504–511

Daniel G, Asiegbu F, Johansson M (1998) The saprotrophic wood-degrading abilities of Heterobasidium annosum intersterility groups P and S. Mycol Res 102(8):991–997

Dashtban M, Schraft H, Syed TA, Qin W (2010) Fungal biodegradation and enzymatic modification of lignin. Int J Biochem Mol Biol 1:36–50

de Morais P, Stoichev T, Basto MCP, Vasconcelos MTSD (2012) Extraction and preconcentration techniques for chromatographic determination of chlorophenols in environmental and food samples. Talanta 89:1–11

Dey S, Choudhury MD, Das S (2013) A review on toxicity of paper mill effluent on fish. Bull Environ Pharmacol Life Sci 2:17–23

Dezotti M, Innocentini-Mei LH, Duran N (1995) Silica immobilised enzyme catalysed removal of chlorolignins from kraft effluent. J Biotechnol 43:161–167

Eaton DC, Chang HM, Kirk TK (1980) Decolorization of kraft mill bleach plant effluent. TAPPI J 63:103–106

Farstad L, Naess B (1977) Effects of alkali lignins on some biological systems. Acta Agric Scand 27:123–128

Ferrer I, Dezotti M, Duran N (1991) Decolorization of Kraft effluent by free and immobilized lignin peroxidases and horseradish peroxidase. Biotechnol Lett 13(8):577–582

Forss K, Kokkonen R, Sirelius H, Sagfors PE (1979) How to improve spent sulphite liquor use. Pulp Paper Can 80(12):411–415

Freitas AC, Ferreira F, Costa AM, Pereira R, Antunes SC (2009) Biological treatment of the effluent from a bleached kraft pulp mill using basidiomycete and zygomycete fungi. Sci Total Environ 407:3282–3289

Fukuzumi T, Nishida A, Aoshima K, Minami K (1977) Decolourization of kraft waste liquor with white-rot fungi–I: screening of the fungi and culturing condition for decolourization of kraft waste liquor. Mokuzai Gakkaishi 23(6):290–298

Gautam S, Kaithwas G, Bharagava RN, Saxena G (2017) Pollutants in tannery wastewater, pharmacological effects and bioremediation approaches for human health protection and environmental safety. In: Bharagava RN (ed) Environmental pollutants and their bioremediation approaches, 1st edn. CRC Press, Taylor & Francis Group, Boca Raton, pp 369–396. https://doi.org/10.1201/9781315173351-14

Goutam SP, Saxena G, Singh V, Yadav AK, Bharagava RN (2018) Green synthesis of TiO_2 nanoparticles using leaf extract of *Jatropha curcas* L. for photocatalytic degradation of tannery wastewater. Chem Eng J 336:386–396. https://doi.org/10.1016/j.cej.2017.12.029

Gupta M, Shankar S, Shikha (2015) Role of fungal laccases in lignin degradation and decolorization of pulp and paper mill effluent. In: Govil JN (ed) Recent developments in biotechnology. Studium Press, Houston ISBN: 1-62699-026-3, pp 241–270

Haq I, Jumar S, Raj A, Lohani M, Satyanarayana GNV (2017) Genotoxicity assessment of pulp and paper mill effluent before and after bacterial degradation using *Allium cepa* test. Chemosphere 169:642–650

Hooda R, Bhardwaj NK, Singh P (2015) Screening and identification of lignolytic bacteria for the treatment of pulp and paper mill effluent. Water Air Soil Pollut 226:305

Hossain K, Ismail N (2015) Bioremediation and detoxification of pulp and paper mill effluent: a review. Res J Environ Toxicol 9(3):113–134

Huuha TS, Kurniwan TA, Sillaneaa MET (2010) Removal of silicon from pulping white water using integrated treatment of chemical precipitation and evaporation. Chem Eng J 158:584–592

Ibarra D, Concepción M, Blanco A, Martínez AT, Martínez MJ (2012) Enzymatic deinking of secondary fibers: cellulases/hemicellulases versus laccase-mediator system. J Ind Microbiol Biotechnol 39:1–9

Iyovo GD, Du G, Chen J (2010) Sustainable bioenergy bioprocessing: biomethane production, digestate as biofertilizer and as supplemental feed in algae cultivation to promote algae biofuel commercialization. J Microb Biochem Technol 2:100–106

Kannan K, Oblisami G, Loganathan BG (1990) Enzymology of ligno-cellulose degradation by Pleurotus sajor-caju during growth on paper-mill sludge. Biological Wastes 33(1):1–8

Kamali M, Khodaparast Z (2015) Review on recent developments on pulp and paper mill wastewater treatment. Ecotoxicol Environ Saf 114:326–342

Kansal SK, Singh M, Sud D (2008) Effluent quality at kraft/soda agr o-based paper mill and its treatment using a heterogenous photocatalytic system. Desalination 228:183–190

Karn SK, Chakrabarty SK, Reddy MS (2010) Pentachlorophenol degradation by *Pseudomonas stutzeri* CL7 in the secondary sludge of pulp and paper mill. J Environ Sci 22(10):1608–1612

Karrasch B, Parra O, Cid H, Mehrens H, Pacheco P (2006) Effects of pulp and paper mill effluents on the microplankton and microbial self-purification capabilities of the Bibyo River. Chile Sci Total Environ 359:194–208

Keharia H, Madamwar D (2003) Bioremediation concepts for treatment of dye containing wastewater: a review. Indian J Exp Biol 41:1068–1075

Kostamo A, Holmbom B, Kukkonen JVK (2004) Fate of wood extractives in wastewater treatment plants at kraft pulp mills and mechanical pulp mills. Water Res 38:972–982

Kreetachat T, Chaisan O, Vaithanomsat P (2016) Decolorization of pulp and paper mill effluents using wood rotting fungus *Fibrodontia* sp. RCK783S. Int J Environ Sci Dev 7(5):321–324

Kumar A, Dhall P, Kumar R (2010) Redefining BOD: COD ratio of pulp mill industrial wastewaters in BOD analysis by formulating a specific microbial seed. Int Biodeterior Biodegrad 64:197–202

Kumar V, Dhall P, Naithani S, Kumar A, Kumar R (2014) Biological approach for the treatment of pulp and paper industry effluent in sequence batch reactor. J Bioremed Biodegr 5:218

Kumar S, Saha T, Sharma S (2015a) Treatment of pulp and paper mill effluents using novel biodegradable polymeric flocculants based on anionic polysaccharides: a new way to treat the waste water. Int Res J Eng Technol 2(4):1417–1428

Kumar M, Singh J, Singh MK, Singhal A, Thakur IS (2015b) Investigating the degradation process of kraft lignin by b- proteobacterium, *Pandoraea* sp. ISTKB. Environ Sci Pollut Res 22:15690–157902

Liu T, Hu H, He Z, Ni Y (2011) Treatment of poplar alkaline peroxide mechanical pulping (APMP) effluent with *Aspergillus niger*. Bioresour Technol 102(15):7361–7365

Malaviya P, Rathore VS (2007) Bioremediation of pulp and paper mill effluent by a novel fungal consortium isolated from polluted soil. Bioresour Technol 98(18):3647–3651

Michałowicz J, Duda W (2007) Phenols – sources and toxicity. Pol J Environ Stud 16:347–362

Mishra, Thakur (2010) Isolation and characterization of alkalotolerant bacteria and optimization of process parameters for decolorization and detoxification of pulp and paper mill effluent by Taguchi approach. Biodegradation 21(6):967–978

Min M, Wang L, Li Y, Mohr MJ, Hu B, Zhou W, Chen P, Ruan R (2011) Cultivating Chlorella sp. in a Pilot-Scale Photobioreactor Using Centrate Wastewater for Microalgae Biomass Production and Wastewater Nutrient Removal. Appl Biochem Biotechnol 165(1):123–137

Murugesan K (2003) Bioremediation of paper and pulp effluent. Indian Jexp Biol 41:1239–1248

Murugesan K, Kim YM, Jeon JR, Chang YS (2009) Effect of metal ions on reactive dye decolorization by laccase from *Ganoderma lucidum*. J Hazard Mater 168(1):523–539

Oikari A, Lindström-Seppä P, Kukkonen J (1988) Subchronic metabolic effects and toxicity of a simulated pulp mill effluent on juvenile lake trout, *Salmo trutta mlacustris*. Ecotoxicol Environ Saf 16:202–218

Olguín EJ (2012) Dual purpose microalgae-bacteria-based systems that treat wastewater and produce biodiesel and chemical products within a biorefinery. Biotechnol Adv 30(5):1031–1046

Pathak P, Bhardwaj NK, Singh AK (2014) Production of crude cellulase and xylanase from *Trichoderma harzianum* PPDDN10 NFCCI-2925 and its application in photocopier waste paper recycling. Appl Biochem Biotechnol 172:3776–3797

Pendroza AM, Mosqueda R, Alonso-Vante N, Rodriguez-Vazquez R (2007) Sequential treatment via Trametes versicolor and UV/titanium dioxide/Rux Sey to reduce contaminants in waste water resulting from the bleaching process during paper production. Chemosphere 67(4):793–801

Pokhrel D, Viraghavan T (2004) Treatment of pulp and paper mill wastewater- a review. Sci Total Environ 333:37–58

Prasongsuk S, Lotrakul P, Imai T, Punnapayak H (2009) Decolourization of pulp mill wastewater using thermotolerant white rot fungi. Sci Asia 35:37–41

Priyadarshinee R, Kumar A, Mandal T, Dasguptamandal D (2017) Exploring the diverse potentials of Planococcus sp. TRC1 for the deconstruction of recalcitrant kraft lignin. Int J Environ Sci Te 14(8):1713–1728

Prouty AL (1990) Bench-scale development and evaluation of a fungal bioreactor for color removal from bleach effluents. Appl Microbiol Biotechnol 32(4):490–493

Qu X, Gao WJ, Han MN, Chen A, Liao BQ (2012) Integrated thermophilic submerged aerobic membrane bioreactor and electrochemical oxidation of pulp and paper effluent treatment-toward system closure. Bioresour Technol 116:1–8

Raj A, Kumar S, Haq I, Singh SK (2014) Bioremediation and toxicity reduction in pulp and paper mill effluent by newly isolated ligninolytic *Paenibacillus* sp. Ecol Eng 71:355–362

Rämänen H, Lassila H, Lensu A, Lahti M, Oikari A (2010) Dissolution and spatial distribution of resin acids and retene in sediments contaminated by pulp and paper industry. J Soils Sediments 10:349–358

Ramsay JA, Nguyen T (2002) Decolorization of textile dyes by Trametes versicolor and its effect on dye toxicity. Biotechnol Lett 24:1757–1761

Ratia H, Rämänen H, Lensu A, Oikari A (2013) Betulinol and wood sterols in sediments contaminated by pulp and paper mill effluents: dissolution and spatial distribution. Environ Sci Pollut Res 20:4562–4573

Renberg L, Svanberg O, Bengsson B-E, Sundström G (1980) Chlorinated guajacols and cathecols. Bioaccumulation potential in bleaks (*Alburnus alburnus*, Pisces) and reproductive and toxic effects on the harpacticoid *Nitora snipes* (Crustacea). Chemosphere 9:143–150

Roy-Arcand L, Archibald FS (1991) Direct dechlorination of chlorophenolic compound by laccase from Trametes (Coriolus) versicolor. Enzym Microb Technol 13:194–203

Santos TR, Ferreira F, Silva L, Freitas AC, Pereira R, Diniz M, Castro L, Peres I, Duarte AC (2010) Effects of tertiary treatment by fungi on organic compounds in a kraft pulp mill effluent. Environ Sci Pollut Res 17(4):866–874

Sethunathan N, Megharaj M, Chen ZL, Williams BD, Gareth LR (2004) Algal Degradation of a Known Endocrine Disrupting Insecticide, Î±-Endosulfan, and Its Metabolite, Endosulfan Sulfate, in Liquid Medium and Soil. J Agric Food Chem 52(10):3030–3035

Saxena G, Bharagava RN (2015) Persistent organic pollutants and bacterial communities present during the treatment of tannery wastewater. In: Chandra R (ed) Environmental waste management, 1st edn. CRC Press, Taylor & Francis Group, Boca Raton, pp 217–247. https://doi.org/10.1201/b19243-10

Saxena G, Bharagava RN (2016) Ram Chandra: advances in biodegradation and bioremediation of industrial waste. Clean Techn Environ Policy 18:979–980. https://doi.org/10.1007/s10098-015-1084-9

Saxena G, Bharagava RN (2017) Organic and inorganic pollutants in industrial wastes, their ecotoxicological effects, health hazards and bioremediation approaches. In: Bharagava RN (ed) Environmental pollutants and their bioremediation approaches, 1st edn. CRC Press, Taylor & Francis Group, Boca Raton, pp 23–56. https://doi.org/10.1201/9781315173351-3

Saxena G, Chandra R, Bharagava RN (2016) Environmental pollution, toxicity profile and treatment approaches for tannery wastewater and its chemical pollutants. Rev Environ Contam Toxicol 240:31–69. https://doi.org/10.1007/398_2015_5009

Saxena G, Purchase D, Mulla SI, Saratale GD, Bharagava RN (2018) Phytoremediation of heavy metal-contaminated sites: environmental considerations, field studies, sustainability and future prospects. J Environ Manag

Shankar S, Shikha (2011) Fungal laccases: production and biotechnological applications in pulp and paper industry. In: Jain PK, Gupta VK, Bajpai V (eds) Recent advances in environmental biotechnology. Lap Lambert Academic Publishing Ag & Co. Kg, Dudweller Landstr, Germany., ISBN: 978-3-8443-0687-3, pp 73–111

Shankar S, Shikha (2012) Laccase production and enzymatic modification of lignin by a novel *Peniophora* sp. Appl Biochem Biotechnol 166:1082–1094

Shankar S, Shikha (2014) Effects of metal ions and redox mediator system on decolorization of synthetic dyes by crude laccase from a novel white rot fungus *Peniophora* sp. (NFCCI-2131). Appl Biochem Biotechnol 175(1):635–647

Shankar S, Shikha, Gupta M (2015) Microbial xylanases: production, applications and challenges. In: Gupta VK, Sharma GD, Tuohy MG, Gaur R (eds) The handbook of microbial bioresources. CAB International, Wallingford, pp 313–329 2016, ISBN-13:9781780645216

Sharari M, Latibari AJ, Guillet A, Aurousseau M, Mouhamadou B, Rafeiee G, Mirshokraei A, Parsapaghouh D (2011) Application of the white rot fungus *Phanerochaete chrysosporium* in biotreatment of bagasse effluent. Biodegradation 22(2):421–430

Sharma P, Goel R, Capalash N (2007) Bacterial laccases. World J Microbiol Biotechnol 23:823–832

Sharma R, Chandra S, Singh A, Singh K (2014) Degradation of pulp and paper mill effluents. IIOAB J 5:6–12

Singh J, Kaur A (2015) Vermicompost as a strong buffer and natural adsorbent for reducing transition metals, BOD, COD from industrial effluent. Ecol Eng 74:13–19

Singh S, Singh BB, Chandra R (2009) Biodegradation phenol in batch culture by pure and mixed strains of *Paenibacillus* and *Bacillus cereus*. Polish J Microbiol 58:319–325

Singh P, Sulaiman P, Hashim O, Rupani R, Peng RF (2010) Biopulping of lignocellulosic material using different fungal species: a review. Rev Environ Sci Biotechnol 9:2141–2151

Singh JS, Abhilash PC, Singh HB, Singh RP, Singh DP (2011a) Genetically engineered bacteria: an emerging tool for environmental remediation and future research perspectives. Gene 480:1–9

Singh YP, Dhall P, Mathur RM, Jain RK, Thakur VV (2011b) Bioremediation of pulp and paper mill effluent by tannic acid degrading Enterobacter sp. Water Air Soil Pollut 218:693–701

Singhal A, Thakur IS (2009) Decolorization and detoxification of pulp and paper mill effluent by *Cryptococcus* sp. Biochem Eng J 46:21–27

Singhal V, Kumar A, Rai JPN (2005) Bioremediation of pulp and paper mill effluent with *Phanerochaete chrysosporium*. J Environ Res 26(3):525–529

Sponza DT (2003) Application of toxicity tests into discharges of the pulp-paper industry in Turkey. Ecotoxicol Environ Saf 54:74–86

Tewari PK, Batra VS, Balakrishnan M (2009) Efficient water use in industries: cases from the Indian agro-based pulp and paper mills. J Environ Manag 90:265–273

Tiku DK, Kumar A, Sawhney S, Singh VP, Kumar R (2007) Effectiveness of treatment technologies for wastewater pollution generated by Indian pulp mill. Environ Monit Assess 132:453–466

Tiku DK, Kumar A, Chaturvedi R, Makhijani SD, Manoharan A, kumar R (2010) Holistic bioremediation of pulp mill effluent using autochthonous bacteria. Int Biodeter Biodegad 64(3):173–183

Tremblay L, van der Kraak G (1999) Comparison between the effects of the phytosterol β-sitosterol and pulp and paper mill effluents on sexually immature rainbow trout. Environ Toxicol Chem 18:329–336

Tyagi S, Kumar V, Singh J, Teotia P, Bisht S, Sharma S (2014) Bioremediation of pulp and paper mill effluent by dominant aboriginal microbes and their consortium. Int J Environ Res 8:561–568

Ullah MA, Bedford CT, Evans CS (2000) Reactions of pentachlorophenol with laccase from *Coriolus versicolor*. Appl Microbiol Biotechnol 53:230–234

Veningerová M, Prachar V, Uhnák J, Kovacicová J (1994) Polychlorinated phenols in total diet. Z Lebensm Unters Forsch 199:317–321

Vepsalainen M, Kivisaari H, Pulliainen M, Oikari A, Sillanpaa M (2011) Removal of toxic pollutants from pulp mill effluents by electrocoagulation. Sep Purif Technol 81(2):141–150

Voss RH, Rapsomatiotis A (1985) An improved solvent-extraction based procedure for the gas chromatographic analysis of resin and fatty acids in pulp mill effluents. J Chromatogr 346:205–214

Wang BGL, Ma HZ (2007) Electrochemical oxidation of pulp and paper making wastewater assisted by transition metal modified Kaolin. J Hazard Mater 143:198–205

Wu J, Xiao YZ, Yu H-Q (2005) Degradation of lignin in pulp mill wastewaters by white rot fungi on biofilm. Bioresour Technol 96:1357–1363

Zhang X, Renaud S, Paice M (2008) Cellulase deinking of fresh and aged recycled newsprint/magazines (ONP/OMG). Enzym Microb Technol 43:103–108

Zhang T, Wada S, Vamagishi T, Hiroyasu I, Tatsumi K, Zhao QX (1999) Treatment of bleaching wastewater from pulp-paper plants in China using enzymes and coagulants. J Environ Sci 11(4):480–484

Zwain HM, Hassan SR, Zaman NQ, Aziz HA, Dahlan I (2013) The start-up performance of Modified Anaerobic Baffled Reactor (MABR) for the treatment of recycled paper mill wastewater. J Environ Chem Eng 1:61–64

Chapter 10
Recent Advances in Phytoremediation of Soil Contaminated by Industrial Waste: A Road Map to a Safer Environment

Cassiano A. R. Bernardino, Claudio F. Mahler, Paula Alvarenga, Paula M. L. Castro, Eduardo Ferreira da Silva, and Luís A. B. Novo

Contents

C. A. R. Bernardino · C. F. Mahler
Department of Civil Engineering, COPPE, Federal University of Rio de Janeiro, Rio de Janeiro, RJ, Brazil

P. Alvarenga
LEAF, Department of Sciences and Engineering of Biosystems, School of Agriculture, University of Lisbon, Lisbon, Portugal

P. M. L. Castro
Centre of Biotechnology and Fine Chemistry – Associated Laboratory, Faculty of Biotechnology, Catholic University of Portugal, Porto, Portugal

E. F. da Silva
GeoBioTec Research Center, Department of Geosciences, University of Aveiro, Aveiro, Portugal

L. A. B. Novo (✉)
Centre of Biotechnology and Fine Chemistry – Associated Laboratory, Faculty of Biotechnology, Catholic University of Portugal, Porto, Portugal

GeoBioTec Research Center, Department of Geosciences, University of Aveiro, Aveiro, Portugal
e-mail: novo@ua.pt

G. Saxena, R. N. Bharagava (eds.), *Bioremediation of Industrial Waste for Environmental Safety*, https://doi.org/10.1007/978-981-13-1891-7_10

Abstract The advent of the industrial revolution has boosted human and technological development, but at a price. Contaminants have been increasingly accumulating across wider areas scattered around the world, making pollution a major planetary issue. Among these pollutants, those related to industrial waste are perhaps the most threatening due to their broad variety and capacity to migrate through the air, water, and soil, enter the food chain, and cause a deleterious impact on life and the environment. Finding solutions to deter industrial pollution has thus become the need of the hour. However, conventional techniques have often proven unworkable due to their cost and harmful collateral effects to the environment. In this light, a plant-based technique with the capacity to stabilize, extract, and/or degrade pollutants known as phytoremediation, has emerged as a promising alternative due to its inexpensiveness and environment-friendly character. Here, we review the latest advances on phytoremediation of sites afflicted by industrial pollution and outline the future scope of this green technique.

Keywords Contaminated soil · Phytostabilization · Phytoextraction · Metals · E-waste · Sludge · Tailings

10.1 Introduction

Soil contamination is a global environmental concern due to its negative impact on the environment, human health, soil fertility, and socioeconomic balance. Around the world, about 52 million hectares, representing more than 16% of Earth's total area, are affected by soil contamination to some extent. In Europe alone, there are approximately 2.5 million contaminated sites, whose annual management costs are equivalent to 6.5 billion euros, as estimated by the European Environment Agency (2015). In the United States, more than 530,000 contaminated sites and about 9.3 million hectares are being tracked by the Environmental Protection Agency (2013). Yet, conventional techniques for the remediation of contaminated soils (such as vitrification, soil incineration, excavation and landfilling, soil washing, among others) are often impractical, financially infeasible, and environmentally damaging (Prasad and Freitas 2003; Kushwaha et al. 2016).

In this backdrop, a plant-based technology known as phytoremediation has gained great visibility as an inexpensive and environment-friendly alternative to mitigate soil pollution (Bharagava et al. 2017; Saxena and Bharagava 2017; Chandra et al. 2015). Phytoremediation can be divided into different categories according to the type of pollutant or mitigation process (Table 10.1). The hyperaccumulation of metals was first reported in the 1970s (Jaffré et al. 1976) and refers to the phenomenon through which some plants (usually metallophytes) uptake and accumulate elements to concentrations that are one to three orders of magnitude greater than those measured in the shoots of other plants sharing the same environment (Padmavathiamma and Li 2007). Accordingly, lower limits have been proposed for the levels of the most common metals in the aboveground tissue. Furthermore, in order to classify as a hyperaccumulator, a plant must not only present shoot

Table 10.1 List of soil-related phytoremediation processes

Phytoremediation process	Description
Phytoextraction	Uptake and accumulation of pollutants in the plant aerial parts
Phytostabilization	Reduction of the pollutants mobility in the rhizospheric region by plant roots, bacteria, and amendments
Phytodegradation	Enzymatic degradation of organic pollutants in the plant tissue
Phytostimulation or rhizodegradation	Enhancement of the rhizospheric bacteria activity to promote the degradation of organic pollutants
Phytovolatilization	Uptake of contaminants from soil, followed by their transformation and release to the atmosphere
Phytodesalination	Removal of salts from saline soils via halophytic plants

Adapted from Prasad and Freitas (2003), Padmavathiamma and Li (2007), and Ali et al. (2013)

concentrations of a given metal above the stipulated threshold but also have grown in natural field conditions (van der Ent et al. 2013).

To assess the plant's ability to uptake and accumulate contaminants, two parameters are often used: the bioconcentration factor (BF) and the translocation factor (TF). The former addresses the plants aptitude to uptake and accumulate the contaminant, while the latter hints at the plant's ability to transfer the contaminant from the root to the shoot. Both parameters can be computed through the following formulae:

$$F = \frac{C_p}{C_s} \tag{10.1}$$

$$\mathrm{TF} = \frac{C_a}{C_r} \tag{10.2}$$

where C_p is the concentration of the whole plant, C_s is the concentration in soil (preferably bioavailable), C_a is the concentration in the aboveground plant tissue, and C_r is the concentration in the root (Novo et al. 2018).

In this chapter, we review some of the most recent literature about phytoremediation of soil polluted by the main types of industrial waste (Fig. 10.1).

10.2 Phytoremediation of Mining Waste

Is it estimated that approximately 0.4×10^6 km^2 of land are globally disturbed by mining activities (Hooke and Martín-Duque 2012). Mining wastes can be typically divided into two groups: (1) waste rock generated during the extraction process that does not contain ore minerals, metals, coal, or mineral fuels, or their concentration is subeconomic (sterile tailings), and (2) processing wastes produced during the beneficiation or mineral treatment, which may include tailings, sludge, and

Fig. 10.1 Main types of soil-polluting industrial waste

wastewater from the processing of minerals, coal, and mineral fuel (Lottermoser 2010). Moreover, there is great concern about the release of acid mine drainage (AMD) to surrounding soils. AMD results from the oxidation, and weathering of sulfide ore sand poses a significant threat due to its capacity to pronouncedly decrease pH and mobilize metals through the soil and into groundwater (Alcolea et al. 2012).

Anawar et al. (2011) investigated the tailings of São Domingos Mine in Portugal to evaluate the environmental risks and assess the potential of its native plant species for phytoremediation. Even if low pH and high levels of As, Ag, Cr, Hg, Sn, Sb, Fe, and Zn were found in the tailings, the authors identified several indigenous plant species that managed to normally grow and develop. Hence, due to their tolerance to the hostile mining environment of São Domingos and their ability to revegetate its tailings, the species *Erica australis*, *Cistus ladanifer*, *Lavandula luisierra*, *Daphne gnidium*, *Rumex induratus*, *Juncus* sp., and *Genista hirsutus* were selected as promising candidates for phytostabilization.

Testiat et al. (2013) studied an area in Marseille (France), where a lead smelting unit operated between 1851 and 1921. The authors evaluated soil metal contamination at the site and chose two native plant species (*Globularia alypum* and *Rosmarinus officinalis*) to determine their ability for phytoremediation. The results showed that the soil was heavily contaminated, with concentrations of As, Pb, and Sb ranging between 0.01 and 7.0 g kg^{-1}, 0.58 and 130 g kg^{-1}, and 0.03 and 9.03 g

kg^{-1}, respectively. Furthermore, unusual levels of Cu and Zn were also found, with the uppermost values reaching 7.9 and 65 g kg^{-1}, respectively. Metal levels in plant tissue of *G. alypum* varied between 0.001 and 0.38 g kg^{-1} in root and 0.001–0.18 g kg^{-1} in shoot for As, 0.007–0.088 g kg^{-1} in root and 0.006–0.011 g kg^{-1} in shoot for Cu, 0.020–2.64 g kg^{-1} in root and 0.014–1.35 g kg^{-1} in shoot for Pb, and 0.024–0.77 g kg^{-1} in root and 0.030–1.023 g kg^{-1} in shoot for Zn. Concerning *R. officinalis*, metal concentrations oscillated between 0.002 and 0.029 g kg^{-1} in root and 0.0008–0.033 g kg^{-1} in shoot for As, 0.0001–0.041 g kg^{-1} in root and 0.0001–0.017 g kg^{-1} in shoot for Cu, 0.038–1.15 g kg^{-1} in root and 0.0002–0.153 g kg^{-1} in shoot for Pb, and 0.015–0.98 g kg^{-1} in root and 0.027–0.115 g kg^{-1} in shoot for Zn. Due to low translocation and bioconcentration factor values (<1), *G. alypum* and *R. officinalis* are not suitable for phytoextraction. Still, their remarkable adaptation to the severe conditions of the site suggests that both species could be instrumental for its reclamation through phytostabilization.

A study was conducted in Chenzhou City (China), to evaluate the accumulation of metals and polycyclic aromatic hydrocarbons (PAHs) in surface soil and 18 wild plant species in the vicinity of mining industry facilities (Sun et al. 2014). Samples were collected in the proximity of a coke power station, smelting factories, and a coal mining site. The main products of the ore smelter are Pb, Zn, Cu, Au, and Ag, while the surface coal mining is focused on bituminous coal. The highest soil metal levels were found near the smelting units, with mean values of 251, 34.9, 851, 2315, and 4610 mg kg^{-1} for As, Cd, Cu, Zn, and Pb, respectively. The authors pointed out that the elevated concentrations of metals were likely related to the transport and deposition of volatile particles and aerosols from stack emission or the fugitive emissions. At the coal mining site, the high soil concentrations of As (127 mg kg^{-1}) were probably linked with the emission of unburnt As-holding coal particles during the different coal processing steps. In relation to the PAHs, the measurements of the 16 different compounds showed that the sum of their concentrations ranged from 0.215 mg kg^{-1} (smelting site) to 32.47 mg kg^{-1} (coke power station). Comparisons with background values showed that metal concentrations in plants sampled at the coal mining site were normal for Cd, Cu, Pb, and Zn and relatively high for As (reaching a maximum of 381 mgkg^{-1}). Plants from the smelting area presented elevated concentrations of As, Cd, Cu, Pb, and Zn, with topmost results of 1276, 34.8, 1187, 4430, and 6565 mg kg^{-1}, respectively. These results were aligned with the metal levels found at the corresponding soils, denoting the bioindication potential of the plants. Maximum PAHs concentrations in plants reached 1.49 mg kg^{-1} on *Pteris vittata* at the coke power station, 0.980 mg kg^{-1} on *Stipa bungeana* at the smelting site, and 1.22 mg kg^{-1} on *P. vittata* at the coal mining point. The analysis of the plants' ability to tolerate and accumulate metals and PAHs, including their bioconcentration factors and correlations between contaminant types, suggests the suitability of *P. vittata* and *Pteris cretica* for the joint remediation of As and PAHs; *Boehmeria nivea* for the simultaneous mitigation of Pb, As, and PAHs; and *Miscanthus floridulus* for the concurrent alleviation of Cu and PAHs pollution.

Ruiz Olivares et al. (2013) carried out a study in mine tailings in the arid region of Zimapan (Hidalgo State, Mexico), to appraise the potential of *Ricinus communis* for phytoremediation of metal-polluted sites. Samples from tailings and plants were collected at three different locations presenting high concentrations of Cu, Zn, Mn, Pb, and Cd, due to mining activities that started in 1632, including nearly 40 smelters that have been operating in the region since 1940. The selected sampling points were three sterile mine tailings heaps and an abandoned cultivated area, naturally colonized by *R. communis* plants. While Ni levels were relatively normal (35–89 mg kg^{-1}), the range of concentrations for Cu (80–1248 mg kg^{-1}), Zn (727–15,227 mg kg^{-1}), Mn (136–1655 mg kg^{-1}), Pb (426–3883 mg kg^{-1}), and Cd (17–92 mg kg^{-1}) were very high. Overall, low concentrations of these metals were recorded at the shoots of the plant samples. Still, the high root metal levels, generally low translocation factor values, and the noteworthy mycorrhizal activity that usually promotes metal sequestration in the rhizospheric region highlighted the aptitude of *R. communis* for phytostabilization of metal-contaminated soils.

Yildirim and Sasmaz (2017) collected plant samples from soils contaminated with Ag, As, and Pb at the Gümüsköy mining district (Kütahya, Turkey). The authors assessed metal levels in soil and in 11 native species including *Anchusa arvensis*, *Alyssum saxatile*, *Onosma* sp., *Cynoglossum officinale*, *Phlomis* sp., *Carduus nutans*, *Glaucium flavum*, *Centaurea cyanus*, *Isatis* sp., *Verbascum thapsus*, and *Silene compacta*. Average metal levels in soils, roots, and aboveground tissue of the selected plants were, respectively, 35.93, 10.19, and 11.51 mg kg^{-1} (Ag); 4771, 2320, and 1340 mg kg^{-1} (As); and 4180, 1424, and 1050 mg kg^{-1} (Pb). The combination of the metal accumulation results with the bioconcentration and translocation factor values indicates that the selected plant species could play a relevant role in phytostabilization processes.

10.3 Phytoremediation of Industrial Sludge

Sludge is a semisolid residue generated during wastewater treatment processes. Current biological and physicochemical wastewater treatments produce a large amount of sludge that, after processing, is often disposed in land due to its rich content in nitrogen, phosphorus, and organic matter (Zhang et al. 2017). Yet, industrial sludge resulting from the wastewater of metal plating, wood processing, inorganic pigments, petroleum refining, and tanneries, to name but a few, raises serious concerns because of its elevated metal concentrations (Barakat 2011; Zhang et al. 2017). Sludge related to these units frequently exhibits high levels of cadmium, chromium, copper, lead, and mercury, which are particularly detrimental to human life and the environment. Unfortunately, in developing countries the disposal of industrial sludge on uncultivated lands is relatively common, posing a major risk to surrounding communities (Wu et al. 2015).

Luhach and Chaudhry (2012) carried out an experiment to test the potential of *Jatropha curcas* to attenuate metal pollution in soils polluted with sludge from a petroleum refinery in Haryana, India. The pot experiment employed six different levels of refinery sludge contamination (10, 20, 30 40, 50, and 60%), by mixing the latter with unpolluted agricultural soil. Soil metal levels in the 10% and 60% treatments varied between 1.2 and 5.8 mg kg^{-1} (Cr), 4 and 20 mg kg^{-1} (Cu), 0.08 and 0.48 mg kg^{-1} (Cd), and 9.5 and 45 mg kg^{-1} (Ni). Although these soil metal ranges are not excessively high in comparison with other literature reports involving industrial sludge, promising results were obtained. *J. curcas* endured the toxic effects of all treatments and displayed removal rates above 50% for all elements in the 10, 20, and 30% treatments.

A greenhouse trial investigated the capacity of *Justicia gendarussa* to reclaim soils polluted with metals from textile sludge (Majid et al. 2012). The study used six combinations of soil and sludge: 100% soil (control), 80% sludge +20% soil, 60% sludge +40% soil, 40% sludge +60% soil, 20% sludge +80% soil, and 100% sludge. The concentration of metals in soil varied from 0.46 (control) to 24.16 mg kg^{-1} (100% sludge) for Cu, 7020 (control) to 13693 mg kg^{-1} (100% sludge) for Al, and 9.83 (control) to 53.40 mg kg^{-1} (100% sludge) for Zn. Regardless of the soil sludge content, elemental analysis of the plant tissue revealed low accumulation for all metals, both in root and shoot. Consequently, the bioconcentration factor was also low for all elements (<1). Nevertheless, the plant's biomass yield did not show signs of major decline, even when grown in 100% sludge. These results favor the use of *J. gendarussa* for phytostabilization of sludge-contaminated soils.

Hassani et al. (2015) examined metal levels in soils and plants from an area contaminated by the acidic sludge of the Eshtehard industrial zone in Iran. The concentrations of metals in soil collected at the selected sampling points fluctuated between 169.2 and 556.7 mg kg^{-1} for Cr, 644.3 and 2341.3 mg kg^{-1} for Zn, 11.8 and 88.2 mg kg^{-1} for Cd, 119 and 1425.4 mg kg^{-1} for Pb, and 89.2 and 1522.4 mg kg^{-1} for Ni. Striking accumulation results were observed on *Stipa hohenackeriana*, which exhibited concentrations in shoots of 150 mg kg^{-1} for Cd, 596.2 mg kg^{-1} for Cr, and 1967 mg kg^{-1} for Pb. Accordingly, this species showed bioconcentration and translocation factor values of 3.41 and 1.13 (Cd), 2.51 and 1.36 (Cr), and 1.25 and 1.26 (Pb). In addition, other species also presented remarkable metal levels in their shoots, such as *Papaver piptostigma* (351.5 mg kg^{-1} for Cr and 3648 mg kg^{-1} for Zn), *Achillea tenuifolia* (2585 mg kg^{-1} for Pb and 199.8 mg kg^{-1} for Cd), *Bromus tectorum* (195.9 mg kg^{-1} for Cd and 4760 mg kg^{-1} for Zn), *Pteropyrum aucheri* (8634 mg kg^{-1} for Zn and 3393 mg kg^{-1} for Pb), *Nonnea persica* (3228 mg kg^{-1} for Zn), and *Peganum harmala* (7510 mg kg^{-1} for Ni and 2830 mg kg^{-1} for Pb). Taking into consideration the nature and results of this study, as well as the most recent literature (van der Ent et al. 2013), the following species display hyperaccumulation behavior: *S. hohenackeriana* (Cd, Cr, and Pb), *P. piptostigma* (Cr and Zn), *A. tenuifolia* (Pb and Cd), *Bromus tectorum* (Cd and Zn), *P. aucheri* (Zn and Pb), *N. persica* (Zn), and *P. harmala* (Ni and Pb). Thus, these species may constitute outstanding candidates for the phytoextraction of metal-polluted soils.

Chandra and Kumar (2017) collected sludge and plant samples at a site subjected to disposal of post-methanated distillery sludge in Uttar Pradesh, India. The analysis showed that in addition to organic pollutants, the sludge contained different metals, like Fe (2403 mg kg^{-1}), Zn (210 mg kg^{-1}), Mn (126 mg kg^{-1}), Cu (73.62 mg kg^{-1}), Cr (21.825 mg kg^{-1}), Pb (16.33 mg kg^{-1}), and Ni (13.425 mg kg^{-1}). The authors screened 15 plant species growing on the site, including *Datura stramonium*, *Achyranthes* sp., *Kalanchoe pinnata*, *Trichosanthes dioica*, *Parthenium hysterophorus*, *Cannabis sativa*, *Amaranthus spinosus* L., *Croton bonplandianum*, *Solanum nigrum*, *Ricinus communis*, *Saccharum munja*, *Basella alba*, *Setaria viridis*, *Chenopodium álbum*, and *Blumea lacera*. In general, all species presented low metal concentrations in roots and shoots. Nonetheless, the ability of these plants to thrive in the harsh sludge environment, suggests that they may be useful for phytostabilization.

10.4 Phytoremediation of Radioactive Waste

Radioactivity can be defined as the process through which an unstable atomic nucleus with a surplus of energy and/or mass spontaneously decays by emitting radiation in the form of electromagnetic waves (gamma rays) or streams of alpha, beta, or neutron particles (Lottermoser 2010; Smičiklas and Šljivić-Ivanović 2016). Radioactive pollutants are among the most dangerous hazards due to their long half-lives and deleterious effects on life and the environment. Even if in concentrations that do not pose a direct risk to human health, radioactive pollutants in soil can still reduce agricultural productivity and/or enter the food chain (UNSCEAR 2016; Smičiklas and Šljivić-Ivanović 2016; Urushadze and Manakhov 2017). Main sources of soil radioactive pollution include nuclear weapons testing, nuclear power plants accidents, and the nuclear fuel cycle – from the extraction stage to disposal (Hooda 2010; UNSCEAR 2016; Smičiklas and Šljivić-Ivanović 2016).

Fuhrmann et al. (2002) carried out a field experiment to examine the capability of three plant species to uptake ^{137}Cs and ^{90}Sr from the soil of the hazardous waste management facility at the Brookhaven National Laboratory (USA). The plants, *Amaranthus retroflexus*, *Brassica juncea*, and *Phaseolus acutifolius*, were planted in a series of spatially randomized plots, in soil featuring an average activity of 0.06 kBq kg^{-1} for ^{90}Sr and 11.8 kBq kg^{-1} for ^{137}Cs. The bioconcentration factors for ^{137}Cs in *A. retroflexus*, *B. juncea*, and *P. acutifolius* were 2.58, 0.46, and 0.17, respectively, while for ^{90}Sr, significant increments were found: 6.5, 8.2, and 15.2, respectively. The highest levels of both radionuclides were measured on *A. retroflexus*, which also presented a linear relationship between the concentrations of ^{137}Cs in plant tissue and soil – denoting its suitability for the bioindication of this isotope. In light of these results, the authors have estimated that *A. retroflexus* would take 7 and 18 years for the removal of 50% of ^{90}Sr and ^{137}Cs, respectively, from the soil.

Wang et al. (2012) reported the uptake of ^{133}Cs and ^{88}Sr by *Raphanus sativus* following an outdoor pot study. Soil from the garden center of the Southwest University of Science and Technology at Mianyang (China) was spiked with ^{133}CsCl and ^{88}Sr$(NO_3)_2$, to obtain six different concentrations of both radionuclides in soil: 0, 2.5, 5, 10, 20, and 40 mg kg^{-1}. The results showed that the plant's capability to accumulate ^{88}Sr was higher than that observed for ^{133}Cs. Shoot concentrations of ^{133}Cs and ^{88}Sr ranged from 4.64 to 43.62 mg kg^{-1} and from 75.72 to 133.54 mg kg^{-1}, respectively, between the 2.5 and 40 mg kg^{-1} treatments. Moreover, the translocation factor for ^{88}Sr was greater than 1 on all treatments. The authors concluded that *R. sativus* presents potential for phytoextraction of ^{88}Sr.

A field study was conducted at a uranium mill tailings impoundment in South China to assess the uptake of ^{226}Ra by the native vegetation (Hu et al. 2014). A total of 80 species, from which the majority belonged to the Poaceae and Asteraceae families, were screened. The authors noted that the stability of the vegetation communities was negatively influenced by increasing levels of ^{226}Ra activity in the soil, which varied between 7.32 and 29.52 kBq kg^{-1} in the selected sampling sites. The highest activity of the radionuclide was found in *Pteris multifida* with 150.6 kBq kg^{-1} (ash weight basis) and bioconcentration factor of 9.13, *Pteridium aquilinum* with 122.2 kBq kg^{-1} (ash weight basis) and bioconcentration factor of 7.41, and *Dryopteris scottii* with 105.7 kBq kg^{-1} (ash weight basis) and bioconcentration factor of 6.41. Moreover, these species were abundant in the sampling sites with the highest ^{226}Ra activity (29.52, 26.62, and 24.68 kBq kg^{-1}). Hence, *P. multifida*, *P. aquilinum*, and *D. scottii* are promising prospects for the phytoextraction of ^{226}Ra.

Stojanović et al. (2012) investigated the potential of tobacco plants (*Nicotiana tabacum*) to accumulate uranium. The experiment was conducted in situ at the Kalna-Gabrovnica uranium mine in southeast Serbia. Two cultivars of tobacco (Virginian and Burley) were planted on tailings containing an average of 15.3 mg U kg^{-1} and allowed to grow between late May and September. The results showed that maximum values of U in leaves were 4.18 and 3.50 mg kg^{-1}, for the Burley and Virginian varieties, respectively. Even though further research is necessary, these findings suggest that tobacco plants may be an advantageous tool to remove U from the soil.

10.5 Phytoremediation of Fly Ash

Coal fly ash, an industrial by-product, is mainly derived from coal combustion in thermal power plants. Fly ash contains a considerable amount of elements such as As, Cr, Pb, Ni, Ba, Sr, V, and Zn, and therefore constitutes an important threat for the environment and human health (Nordberg et al. 2009; Kabata-Pendias 2011). Moreover, coal fly ash may also carry radionuclides like ^{222}Ru, ^{238}U, ^{226}Ra, ^{232}Th, and ^{40}K. Soil pollution with fly ash occurs through the atmospheric deposition of particles with an average size of <20 μm and low to medium bulk density (0.54–0.86 g cm^{-3}), high surface area (300–500 m^2 kg^{-1}), and light texture. The majority of ashes are alkaline, although pH values may vary between 1.2 and 12.5 (Yao et al. 2015).

Pandey (2013) examined the suitability of *Ricinus communis* to phytoremediate fly ash disposal sites. The author collected plant samples from a nonpolluted reference site and fly ash polluted site under the influence of the Feroze Gandhi Unchahar Thermal Power Plant in Unchahar, India. The levels of Ni, Cu, Zn, Cd, and Pb of the reference area were 32.29, 25.63, 122.54, 15.18, and 10.77 mg kg^{-1}, while those of the fly ash polluted area were 132.65, 41.13, 91.68, 36.29, and 34.33 mg kg^{-1}, respectively. Analysis of plant tissue showed that in spite of the low shoot metal concentrations (which implied translocation factor values smaller than 1), the bioconcentration factor was greater than 1 for all elements. Furthermore, the capacity of *R. communis* to naturally grow on this fly ash-contaminated environment makes it a potential contender for phytostabilization of soils presenting the same characteristics.

Krgović et al. (2015) determined the levels of a wide range of metals in *Erigeron canadensis* growing on the fly ash landfill of the Kolubara power plant in Serbia. The total metal contents of the fly ash samples were 5100, 40.6, 200.8, 0.30, 7.17, 44.39, 29.48, 8679, 35.07, 12.96, and 19.04 mg kg^{-1} for Al, As, Ba, Cd, Co, Cr, Cu, Fe, Ni, Pb, and Zn, respectively. Concentrations of these metals were mostly low in the plant's roots and shoots. Cd and Zn were the only elements for which the values of the bioconcentration and translocation factors were higher than 1. Considering these results and its dominance in vegetation cover and abundance, *E. canadensis* could be a good solution for phytostabilization of fly ash-contaminated sites.

Ghosh et al. (2015) carried out a pot experiment to test the ability of vetiver grass (*Vetiveria zizanioides*) to phytoremediate fly ash-contaminated soil. Vetiver is well-known for its capacity to endure harsh environments and control soil erosion due to a large root system (Chiu et al. 2005; Zhang et al. 2014). Fly ash was collected from the dumpsite of the Kolaghat Thermal Power Station in West Bengal, India, and used to prepare four different treatments according to the fly ash content: control (garden soil), 25%, 50%, and 100% fly ash. Metal concentrations in fly ash samples were 28.90 mg kg^{-1} (Zn), 23.39 mg kg^{-1} (Pb), 16.29 mg kg^{-1} (Cu), 14.06 mg kg^{-1} (Ni), 0.91 mg kg^{-1} (Cd), and 0.51 mg kg^{-1} (As). The analysis of metal content in plant tissue has shown low accumulation of all elements, particularly in the shoots where values were often below the detection limits. Additional assays showed that the genotoxicity of fly ash declined during the 18-month trial. Therefore, the authors highlighted the usefulness of vetiver not only to photostabilize fly ash by restricting contaminants in the large root system but also to reduce its genotoxic potential. Moreover, the plant can also be safely grazed by animals due to its low metal content in the aerial parts.

10.6 Phytoremediation of E-waste

The last few decades have seen an exponential growth in the production of electronic equipment and gadgets like televisions, refrigerators, washing machines, air conditioners, microcomputers, printers, video game consoles, and mobile phones, to list a few. Altogether, they constitute the main sources of the so-called E-waste and waste electrical and electronic equipment (Robinson 2009; Zeng et al. 2017). These products often encompass components such as circuit boards, fluorescent tubes, cathode ray tube, light-sensitive resistors, corrosion alloys, batteries, monitor glass, switches, and flat-screen monitors, which contain a large number of organic and inorganic contaminants. The former include polybrominated diphenyl ethers (PBDEs), polychlorinated biphenyls (PCBs), dioxins, furans, polycyclic aromatic hydrocarbons (PAHs), polyhalogenated aromatic hydrocarbons (PHAHs), and hydrogen chloride. Inorganic pollutants may comprise up to 60 metals including Pb, Sb, Hg, Cd, Ni, Cr, Cu, Zn, Hg, and platinum group elements, among others, that often leach to soils due to inadequate disposal of raw or processed E-waste in landfills (Robinson 2009; Bhattacharya and Khare 2016; Kumar et al. 2017).

Ye et al. (2014) investigated the combination of vetiver grass and soil washing with tea saponin in a peanut oil-water solvent system to reclaim an E-waste polluted site. The soil, collected at an abandoned E-waste disposal plant in the old district of Jiangnan County (Nanjing City, China), presented both organic and inorganic contamination. The analyses found 198.3 mg kg^{-1} (total organic pollutants), 12.9 mg kg^{-1} (PBDEs), 33.3 mg kg^{-1} (PCBs), 152.1 mg kg^{-1} (PAHs), 350.6 mg kg^{-1} (Pb), and 220.3 mg kg^{-1} (Ni). While two initial washings with the 5.0 mL L^{-1} peanut oil and 5.0 g L^{-1} tea saponin extracted 94.6%, 97.1%, 95.1%, 83.5%, and 87.1% of PBDEs, PCBs, PAHs, Pb, and Ni, respectively, the cultivation of vetiver during 4 months helped degrading the residual PBDEs, PCBs, and PAHs, by 45.1%, 36.2%, and 40.2%, accordingly. Moreover, the authors have also reported that vetiver was able to diminish Pb and Ni in soil by 4.1% and 2.0%, respectively. These results suggest that under the adopted strategy, vetiver may have reduced soil E-waste contamination through phytoextraction, phytodegradation, rhizodegradation, and phytostabilization.

Luo et al. (2015) carried out a field study to determine the potential of *Eucalyptus globulus* to phytoremediate Cd-polluted soil in site subjected to nearly 30 years of E-waste disposal (Southeast Guangdong Province, China). The concentrations of Cd in soil varied from 0.42 to 1.17 mg kg^{-1} (average of 0.67 mg kg^{-1}), exceeding the environmental quality parameters established in Chinese legislation. Seedlings of *E. globulus* were planted in 2010 and allowed to grow for 3 years. Samples collected in the summer of 2013 presented Cd concentrations of 3.85 and 0.41 mg kg^{-1} in the roots and shoots, respectively. The corresponding values of the bioconcentration and translocation factors were 5.75 and 0.11. These findings hint that this species may be suitable for phytostabilization.

A pot study employing soil from the abovementioned site in Guangdong Province (China) has provided new insights about the potential of *E. globulus* for phytoremediation (Luo et al. 2017). The authors chose a multi-technique approach to evaluate the individual and combined effects of a chelating agent (ethylenediaminetetraacetic acid, EDTA), direct current electric fields (2, 4, and 10 V), and foliar application of a phytohormone (cytokinin), on phytoremediation of the E-waste-polluted soil with *E. globulus*. In total, eight treatments comprising different combinations of the previously mentioned variables were examined. At the end of the experiment, the results allowed to infer that the combination of *E. globulus* + EDTA + 4 V electric field + foliar application of cytokinin could decrease the time needed to reduce the excess of Cd, Pb, and Cu in soil from 10.4 (control) to 1.1 years. Moreover, this approach could also significantly lessen the volume of generated leachate.

10.7 Future Directions

Nearly 30 years after its formal inception, phytoremediation has benefited from many breakthroughs that have improved its efficiency and widened the range of potential applications. Yet, while further developments are in the horizon (Basharat et al., 2018), many critics rightfully assert that phytoremediation "advances in the lab but lags in the field" (Beans 2017). Besides greater interaction among academia, industry, investors, society, and governmental bodies, field-scale phytoremediation also needs a conceptual shift that clearly incorporates an economic upside (Conesa et al. 2012). In this view, the use of energy crops for phytoremediation may play a critical role enhancing the process's sustainability, cost-effectiveness, and commercial appeal (Tripathi et al. 2016). Moreover, the ever-increasing number of polluted sites worldwide may also present an additional opportunity for a phytoremediation-related technique: phytomining. The latter uses the ability of plants to retrieve valuable metals from tailings and contaminated and/or mineralized soils, with the purpose of obtaining an economic profit (Novo et al. 2015, 2017). In the context of industrial waste contamination, phytomining could be particularly interesting to recover elements like gold or platinum group metals from mine tailings or E-waste-polluted soils.

Acknowledgments The authors gratefully acknowledge the financial support from the National Council for Scientific and Technological Development of Brazil (CNPq, 141368/2014-4), the ERDF Interreg Sudoe Program (PhytoSUDOE-SOE1/P5/E0189), and the Portuguese Foundation for Science and Technology (FCT, SFRH/BPD/103476/2014).

References

Alcolea A, Vázquez M, Caparrós A (2012) Heavy metal removal of intermittent acid mine drainage with an open limestone channel. Miner Eng 26:86–98. https://doi.org/10.1016/j.mineng.2011.11.006

Ali H, Khan E, Sajad MA (2013) Phytoremediation of heavy metals-Concepts and applications. Chemosphere 91:869–881. https://doi.org/10.1016/j.chemosphere.2013.01.075

Anawar HM, Freitas MC, Canha N, Regina IS (2011) Arsenic, antimony, and other trace element contamination in a mine tailings affected area and uptake by tolerant plant species. Environ Geochem Health 33:353–362. https://doi.org/10.1007/s10653-011-9378-2

Barakat MA (2011) New trends in removing heavy metals from industrial wastewater. Arab J Chem 4:361–377. https://doi.org/10.1016/j.arabjc.2010.07.019

Basharat Z, Novo L, Yasmin A (2018) Genome Editing Weds CRISPR: What Is in It for Phytoremediation? Plants 7: 51. https://doi.org/10.3390/plants7030051

Beans C (2017) Core concept: phytoremediation advances in the lab but lags in the field. Proc Natl Acad Sci 114:7475–7477. https://doi.org/10.1073/pnas.1707883114

Bharagava RN, Chowdhary P, Saxena G (2017) Bioremediation: an ecosustainable green technology: its applications and limitations. In: Bharagava RN (ed) Environmental pollutants and their bioremediation approaches, 1st edn. CRC Press/Taylor & Francis Group, Boca Raton, pp 1–22. https://doi.org/10.1201/9781315173351-2

Bhattacharya A, Khare SK (2016) Sustainable options for mitigation of major toxicants originating from electronic waste. Curr Sci 111:1946–1954. https://doi.org/10.18520/cs/v111/i12/1946-1954
Chandra R, Kumar V (2017) Phytoextraction of heavy metals by potential native plants and their microscopic observation of root growing on stabilised distillery sludge as a prospective tool for in situ phytoremediation of industrial waste. Environ Sci Pollut Res 24:2605–2619. https://doi.org/10.1007/s11356-016-8022-1
Chandra R, Saxena G, Kumar V (2015) Phytoremediation of environmental pollutants: an eco-sustainable green technology to environmental management. In: Chandra R (ed) Advances in biodegradation and bioremediation of industrial waste, 1st edn. CRC Press/Taylor & Francis Group, Boca Raton, pp 1–30. https://doi.org/10.1201/b18218-2
Chiu KK, Ye ZH, Wong MH (2005) Enhanced uptake of As, Zn, and Cu by Vetiveria zizanioides and *Zea mays* using chelating agents. Chemosphere 60:1365–1375. https://doi.org/10.1016/j.chemosphere.2005.02.035
Conesa HM, Evangelou MWH, Robinson BH, Schulin R (2012) A critical view of current state of phytotechnologies to remediate soils: still a promising tool? Sci World J 173829:10. https://doi.org/10.1100/2012/173829
EPA (2013) Protecting and restoring land: making a visible difference in communities: OSWER FY13 end of year accomplishments report
European Environment Agency (2015) Progress in management of contaminated sites—European Environment Agency
Fuhrmann M, Lasat MM, Ebbs SD (2002) Uptake of cesium-137 and strontium-90 from contaminated soil by three plant species; application to phytoremediation. J Environ Qual 31:904–909. https://doi.org/10.2134/jeq2002.9040
Ghosh M, Paul J, Jana A (2015) Use of the grass, *Vetiveria zizanioides* (L.) Nash for detoxification and phytoremediation of soils contaminated with fly ash from thermal power plants. Ecol Eng 74:258–265. https://doi.org/10.1016/j.ecoleng.2014.10.011
Hassani AH, Nouri J, Mehregan I (2015) Phytoremediation of soils contaminated with heavy metals resulting from acidic sludge of Eshtehard industrial town using native pasture plants. J Environ Earth Sci 5:87–94
Hooda PS (2010) Trace elements in soils. Wiley, New York
Hooke RL, Martín-Duque JF (2012) Land transformation by humans: a review. GSA Today 12:4–10. https://doi.org/10.1130/GSAT151A.1
Hu N, Ding D, Li G et al (2014) Vegetation composition and 226Ra uptake by native plant species at a uranium mill tailings impoundment in South China. J Environ Radioact 129:100–106. https://doi.org/10.1016/j.jenvrad.2013.12.012
Jaffré T, Brooks RR, Lee J, Reeves RD (1976) Sebertia acuminata: a hyperaccumulator of nickel from New Caledonia. Science 193:579–580. https://doi.org/10.1126/science.193.4253.579
Kabata-Pendias A (2011) Trace elements in soils and plants, 4th edn. CRC Press, Boca Raton
Krgović R, Trifković J, Milojković-Opsenica D (2015) Phytoextraction of metals by *Erigeron canadensis* L. from fly ash landfill of power plant "Kolubara". Environ Sci Pollut Res 22:10506–10515. https://doi.org/10.1007/s11356-015-4192-5
Kumar A, Holuszko M, Espinosa DCR (2017) E-waste: an overview on generation, collection, legislation and recycling practices. Resour Conserv Recycl 122:32–42. https://doi.org/10.1016/j.resconrec.2017.01.018
Kushwaha A, Rani R, Kumar S, Gautam A (2016) Heavy metal detoxification and tolerance mechanisms in plants: implications for phytoremediation. Environ Rev 24:39–51. https://doi.org/10.1139/er-2015-0010
Lottermoser B (2010) Mine wastes, third. Springer Berlin/Heidelberg, Berlin
Luhach J, Chaudhry S (2012) Phytoremediation potential of *Jatropha curcas* for removal of heavy metals from refinery sludge. Int J Sci Eng Res 3:1–5
Luo J, Qi S, Peng L, Xie X (2015) Phytoremediation potential of cadmium-contaminated soil by *Eucalyptus globulus* under different coppice systems. Bull Environ Contam Toxicol 94:321–325. https://doi.org/10.1007/s00128-014-1450-z

Luo J, Cai L, Qi S (2017) A multi-technique phytoremediation approach to purify metals contaminated soil from e-waste recycling site. J Environ Manag 204:17–22. https://doi.org/10.1016/j.jenvman.2017.08.029

Majid NM, Islam MM, Nap ME (2012) Heavy metal uptake and translocation by Justicia gendarussa Burm F. from textile sludge contaminated soil. Acta Agric Scand Sect B Soil Plant Sci 62:101–108. https://doi.org/10.1080/09064710.2011.579994

Nordberg G, Fowler BA, Nordberg M, Friberg L (2009) Handbook on the toxicology of metals, 3rd edn. Academic, London

Novo LAB, Mahler CF, González L (2015) Plants to harvest rhenium: scientific and economic viability. Environ Chem Lett 13:439–445. https://doi.org/10.1007/s10311-015-0517-3

Novo LAB, Castro PML, Alvarenga P, da Silva EF (2017) Phytomining of Rare and Valuable Metals. In: Ansari AA, Gill SS, Gill R, et al. (eds) Phytoremediation - Management of Environmental Contaminants, vol 5. Springer International Publishing, Cham, pp 469–486. https://doi.org/10.1007/978-3-319-52381-1_18

Novo LAB, Castro PML, Alvarenga P, da Silva EF (2018) Plant growth–promoting rhizobacteria-assisted phytoremediation of mine soils. In: Bio-geotechnologies for mine site rehabilitation. Elsevier, Amsterdam, pp 281–295

Padmavathiamma PK, Li LY (2007) Phytoremediation technology: hyper-accumulation metals in plants. Water Air Soil Pollut 184:105–126. https://doi.org/10.1007/s11270-007-9401-5

Pandey VC (2013) Suitability of *Ricinus communis* L. cultivation for phytoremediation of fly ash disposal sites. Ecol Eng 57:336–341. https://doi.org/10.1016/j.ecoleng.2013.04.054

Prasad MNV, Freitas HM de O (2003) Metal hyperaccumulation in plants – biodiversity prospecting for phytoremediation technology. Electron J Biotechnol 6:285–321

Robinson BH (2009) E-waste: an assessment of global production and environmental impacts. Sci Total Environ 408:183–191. https://doi.org/10.1016/j.scitotenv.2009.09.044

Ruiz Olivares A, Carrillo-González R, González-Chávez MDCA, Soto Hernández RM (2013) Potential of castor bean (*Ricinus communis* L.) for phytoremediation of mine tailings and oil production. J Environ Manag 114:316–323. https://doi.org/10.1016/j.jenvman.2012.10.023

Saxena G, Bharagava RN (2017) Organic and inorganic pollutants in industrial wastes, their ecotoxicological effects, health hazards and bioremediation approaches. In: Bharagava RN (ed) Environmental pollutants and their bioremediation approaches, 1st edn. CRC Press/Taylor & Francis Group, Boca Raton, pp 23–56. https://doi.org/10.1201/9781315173351-3

Smičiklas I, Šljivić-Ivanović M (2016) Radioactive contamination of the soil: assessments of pollutants mobility with implication to remediation strategies. In: Soil contamination – current consequences and further solutions. InTech, pp 253–276

Stojanović MD, Mihajlović ML, Milojković JV (2012) Efficient phytoremediation of uranium mine tailings by tobacco. Environ Chem Lett 10:377–381. https://doi.org/10.1007/s10311-012-0362-6

Sun L, Liao X, Yan X (2014) Evaluation of heavy metal and polycyclic aromatic hydrocarbons accumulation in plants from typical industrial sites: potential candidate in phytoremediation for co-contamination. Environ Sci Pollut Res 21:12494–12504. https://doi.org/10.1007/s11356-014-3171-6

Testiati E, Parinet J, Massiani C (2013) Trace metal and metalloid contamination levels in soils and in two native plant species of a former industrial site: evaluation of the phytostabilization potential. J Hazard Mater 248–249:131–141. https://doi.org/10.1016/j.jhazmat.2012.12.039

Tripathi V, Edrisi SA, Abhilash PC (2016) Towards the coupling of phytoremediation with bioenergy production. Renew Sust Energ Rev 57:1386–1389. https://doi.org/10.1016/j.rser.2015.12.116

UNSCEAR (2016) Effects of atomic radiation. New York

Urushadze TF, Manakhov DV (2017) Radioactive contamination of the soils of Georgia. Ann Agrar Sci 15:375–379. https://doi.org/10.1016/j.aasci.2017.07.009

van der Ent A, Baker AJM, Reeves RD (2013) Hyperaccumulators of metal and metalloid trace elements: facts and fiction. Plant Soil 362:319–334. https://doi.org/10.1007/s11104-012-1287-3

Wang D, Wen F, Xu C et al (2012) The uptake of Cs and Sr from soil to radish (*Raphanus sativus* L.)- potential for phytoextraction and remediation of contaminated soils. J Environ Radioact 110:78–83. https://doi.org/10.1016/j.jenvrad.2012.01.028

Wu Q, Cui Y, Li Q, Sun J (2015) Effective removal of heavy metals from industrial sludge with the aid of a biodegradable chelating ligand GLDA. J Hazard Mater 283:748–754. https://doi.org/10.1016/j.jhazmat.2014.10.027

Yao ZT, Ji XS, Sarker PK (2015) A comprehensive review on the applications of coal fly ash. Earth Sci Rev 141:105–121. https://doi.org/10.1016/j.earscirev.2014.11.016

Ye M, Sun M, Wan J (2014) Evaluation of enhanced soil washing process with tea saponin in a peanut oil-water solvent system for the extraction of PBDEs/PCBs/PAHs and heavy metals from an electronic waste site followed by vetiver grass phytoremediation. J Chem Technol Biotechnol 90:2027–2035. https://doi.org/10.1002/jctb.4512

Yildirim D, Sasmaz A (2017) Phytoremediation of As, Ag, and Pb in contaminated soils using terrestrial plants grown on Gumuskoy mining area (Kutahya Turkey). J Geochem Explor 182:228–234. https://doi.org/10.1016/j.gexplo.2016.11.005

Zeng X, Duan H, Wang F, Li J (2017) Examining environmental management of e-waste: China's experience and lessons. Renew Sust Energ Rev 72:1076–1082. https://doi.org/10.1016/j.rser.2016.10.015

Zhang X, Gao B, Xia H (2014) Effect of cadmium on growth, photosynthesis, mineral nutrition and metal accumulation of bana grass and vetiver grass. Ecotoxicol Environ Saf 106:102–108. https://doi.org/10.1016/j.ecoenv.2014.04.025

Zhang Q, Hu J, Lee DJ (2017) Sludge treatment: current research trends. Bioresour Technol 243:1159–1172. https://doi.org/10.1016/j.biortech.2017.07.070

Chapter 11
Toxicity of Hexavalent Chromium in Environment, Health Threats, and Its Bioremediation and Detoxification from Tannery Wastewater for Environmental Safety

Vidya Laxmi and Garima Kaushik

Contents

Abstract Heavy metal pollution in the environment has become one of the major concerns worldwide due to its long persistency in the environment and highly toxic nature in living beings. Heavy metals are often non-biodegradable in nature and hence persist for a long time in the environment, cause serious soil and water

V. Laxmi · G. Kaushik (✉)
Department of Environmental Science, Central University of Rajasthan, Kishangarh, Ajmer, Rajasthan, India
e-mail: garimakaushik@curaj.ac.in

G. Saxena, R. N. Bharagava (eds.), *Bioremediation of Industrial Waste for Environmental Safety*, https://doi.org/10.1007/978-981-13-1891-7_11

pollution, and create severe health hazards in humans and animals. Among heavy metals, chromium has been considered as the highly toxic inorganic pollutant, which is disposed off in the environment through various natural and anthropogenic sources. Chromium has been also listed as the priority pollutant by many environment and health protection agencies such as the US Environmental Protection Agency and Agency for Toxic Substances and Disease Registry. Hexavalent form of chromium has been reported to cause genotoxic, carcinogenic, and mutagenic effects in living beings. Hence, the removal of chromium from tannery wastewater is urgently required. The physicochemical approaches are currently applied for the removal of chromium from tannery wastewater, which are not eco-friendly in nature and use a huge amount of chemicals to treat tannery wastewater. However, biological approaches can be an eco-friendly alternative to treat and detoxify the chromium-containing tannery wastewater prior to its disposal in the environment. This chapter provides a detailed information about the sources of chromium contamination in the environment and their toxic effects in human, animal, plants, as well as in microbes and bioremediation approaches for its detoxification from tannery wastewater for environmental and public health safety.

Keywords Bioremediation · Detoxification · Health hazards · Hexavalent chromium · Tannery wastewater

11.1 Introduction

Rapid industrialization and globalization have become one of the serious pollution problems worldwide. Industries discharge huge amount of partially treated, mostly not treated, wastewater into the environment, which causes serious soil and water pollution (Shakoori et al. 2000; Arunakumara and Zhang 2008; Kavita and Keharia 2012). Heavy metals such as lead, chromium, cadmium, mercury, nickel, zinc, aluminum, arsenic, copper, and iron are causing severe poisoning conditions in the environment and creating major concerns worldwide because these elements are non-biodegradable and toxic to living beings (Ballantyne et al. 1999; Derek 1999; Dias et al. 2002). Wastewater discharges lead to high oxygen demand and nutrient loading, promoting toxic aquatic ecosystem (Morrison et al. 2001). When the concentration of heavy metal ions in the surroundings rises above a specific threshold level, it inhibits a variety of metabolic activities and proves toxic to most organisms (Hörcsik et al. 2006), and it can create adverse effects on the environment and human health due to their bioavailability and solubility (Smith 2001). Toxic heavy metals are normally found in wastewaters released by many industrial processes, such as those employed in the electroplating, metal finishing, metallurgical, tannery, chemical manufacturing, mining, and battery manufacturing industries (Kang et al. 2007).

One of the most important industries causing water pollution is tannery. There are about 2500–3000 tanneries in India, which processes 600 ktons of hides and skins annually (Noorjahan 2014). Total annual discharge of wastewater from these tanneries is 50–60 MLD liquid wastes and 300 million kg sludge into the environment (Ahamed and Kashif 2014), and these industries spread mostly across Tamil Nadu, West Bengal, Uttar Pradesh, Andhra Pradesh, Karnataka, Rajasthan, and Punjab. In Tamil Nadu alone, there are about 1120 tanneries generating around 100 tons of sludge per day (dry basis). The effluent generated in these tanneries has high amounts of organic pollutants as well as high concentration of inorganic pollutants including heavy metals such as Zn, Pb, Cu, Fe, and Mn along with hexavalent chromium, used during the process (Goutam et al. 2018; Bharagava et al. 2017a; Gautam et al. 2017; Saxena et al. 2016; Saxena and Bharagava 2015; Noorjahan 2014).

The chemicals used in the tanning process, only 20% of chemicals are absorbed by the leather, and the rest (80% of chemicals) are discharged through wastewater into the surroundings (Shashwat 2011) with 2000–5000 ppm Cr(VI) concentration compared to recommended permissible limits of 2 ppm in the effluent (Ahamed and Kashif 2014). Chromium(VI) is very toxic and carcinogenic to the environment, plants, animals, and human beings as well (Belay 2010; Shashwat 2011). Chromium has several oxidation states (di, tri, penta, and hexa); trivalent chromium (Cr(III)), together with the hexavalent state (Cr(VI), can be the main form present in the surroundings and aquatic environments. Although Cr(III) is less toxic than Cr(VI), long-term exposure to Cr(III) is known to cause cancer and allergic skin reactions (Kang et al. 2007) (Fig. 11.1).

Fig. 11.1 Tanning processing with resultant high TDS and chromium rich open disposal effluent systems (Source: http://www.thehindubusinessline.com/multimedia/dynamic/01374/effluents1_jpg_1374965g.jpg, http://m.cdn.blog.hu/zo/zoldella/image/india_bor.jpg)

11.2 Chromium

11.2.1 General Introduction

Chromium is an inorganic transition metal consisting several oxidation states with trivalent and hexavalent species being the most common forms (Mwinyihija 2010). Chromium is a steel-gray, lustrous, hard, brittle metal of Group VI-B of the transition series with anatomic number of 24 and atomic weight of 51.996, and it is found in about 0.0122% of the Earth's crust. It exists in a number of valence states ranging from −2 to +6. However, Cr(III) and Cr(VI) are of significance owing to their stability in natural environments. The commended permissible limit in drinking water is 50 μg/L for Cr(VI) (Kanmani et al. 2012). Cr(VI) is extremely toxic to living organisms causing many diseases like allergies, eczema, irritations, and respiratory tract disorders and sometimes shows mutagenic and carcinogenic effects on biological systems due to its strong oxidizing nature. This toxic action is due to the fact that Cr(VI) complexes can easily penetrate biological membranes and undergo immediate reduction reactions leading to the formation of various reactive intermediates (causes oxidative stress in the organisms) (Focardi et al. 2012) (Table 11.1).

11.2.2 Chemical Properties and Its Forms

Chromium is blue-gray, lustrous, and brittle but relatively hard metal that can be polished to get a better shine. It is oxidized to form a thin protective coating to prevent further corrosion. Chromium does not react with water but it reacts with most acids. It is stable in air. Cr^{6+} reduces into Cr^{3+} in the presence of oxygen and forms a thin layer on the metal surface to protect from further corrosion. Chromium(III) is less toxic and an essential element for human and plant National Institute of Environmental Health Sciences NIEHS (2015) (Table 11.2).

Table 11.1 Chromium general properties

Name	Chromium (Cr)
Element group	Group 6 of the transition metals
Atomic number	24
Molecular weight	51.9961 g mol^{-1}
Electron configuration	[Ar] $3d^5$ $4s^1$
Common oxidation states	+2, +3, +6
Other oxidation states	+1, +4, +5
Common ions	Cr^{2+}, Cr^{3+}, Cr^{6+}

Source: http://chemwiki.ucdavis.edu/chemistry_of_chromium

Table 11.2 Summary of various Cr oxidation numbers, type and environmental behaviour

Valency	Environmental behavior	Remarks
Cr	Unstable	
Cr^{1+}	Unstable	
Cr^{2+}	Readily oxidize to Cr_3 but stable only in the absence of any oxidant	Active under anaerobic condition
Cr^{3+}	Most stable	Considerable energy required to convert to lower or higher state
Cr^{4+}	Forms unstable intermediate reactions to trivalent and oxidation state	Exhibit this phase during oxidation and reduction
Cr^{5+}	Unstable intermediate	Observed during oxidation and reduction
Cr^{6+}	In acidic conditions demonstrates very high positive redox potential and unstable in the presence of electron donors	Strongly oxidizing

Adapted from Mwinyihija (2010)

11.2.3 Use of Chromium As Leather Tanning Agent in Tannery Industry

Partially treated/untreated tanning industry effluents released at nearby water body like a river and lake create contamination load which is a serious concern all over the world than after other issues such as the use of pesticides, solvents, dyes, and processing chemicals which cause toxicity and persistence. The leather tanning industry has been identified as one of the main causes of environmental and water pollution. The dark brown or black color of tannery wastewater has indicated the high content of organic matter, and it may vary according to the used chemicals (Jahan et al. 2015). Tanning industries are especially large contributors of chromium pollution in India. Leather processing requires large amount of chemicals like sodium chloride, chromium sulfate, calcium salts, ammonium salts, sodium sulfide, acids, alkalis, fat liquor, and organic dyes. However, one of the major emerging environmental problems in the tanning industry is the disposal of chromium-contaminated sludge produced as a by-product of wastewater treatment. There are more than 2500 tanneries in India, of which about 80% of them are engaged in chrome tanning with a total processing capacity of 600 ktons of hides/skins per year resulting in the discharge of up to 1500–3000 ppm of chromium compounds in the effluents (Pandi et al. 2009). About 60,000 tons of raw hides and skins are processed every year using ammonium salt sulfate and chloride during deliming and dehairing process. Solvent vapors release a huge quantity of untreated effluent into the open air and water bodies causing pollution (Jahan et al. 2015). Tanning is the reaction process of the collagen fibers in the hides with tannins, chromium, alum, or other chemical agents. Based on the tanning agents, tanning operations are further divided into vegetable tanning and chrome tanning (Midha and Dey 2008). The tanning industry involves the following processes linked closely with pollution load (Fig. 11.2).

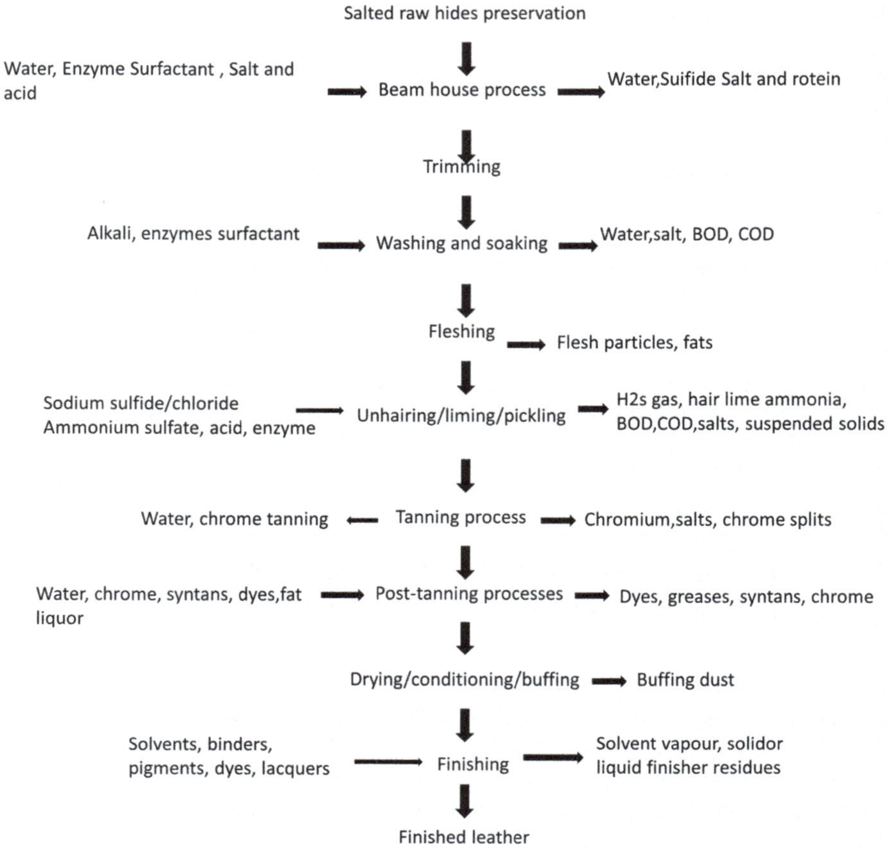

Fig. 11.2 Chromium used as leather tanning agent in tanning process (Source: Midha and Dey 2008 and Mwinyihija 2010)

11.2.4 Sources of Chromium Contamination in Environment

Chromium is a naturally occurring element present in lithosphere environment such as rocks, soils, and volcanic dust. Anthropogenic activities have increased the amount of chromium into the environment, and it causes harmful effects on the environment as well as human beings (Cheng et al. 2014; Joutey et al. 2015). Chromium has widespread applications in various industries such as in aerospace, chemicals, wood preservation, tanning industries, electroplating, textile dying, alloy manufacturing, pulp and paper mills, and ferrochromium production (Kumari et al. 2006; Zhitkovich 2011; Cheng et al. 2014; Joutey et al. 2015). Chromium is an essential element in trace amount for living organisms and plants but toxic and mutagenic at a higher level (Kumari et al. 2006).

11.2.4.1 Natural Source

Chromium is a naturally occurring element present in the whole environment such as rocks, soils, plants, and volcanic dust (Joutey et al. 2015). The primary most important natural source of chromium is the Earth's crust; it is highly concentrated (200 mg kg^{-1}) in igneous rocks and Earth's oceanic crust in serpentinite and ophiolite complexes (Oze et al. 2007; Mishra and Bharagava 2015). Chromium is found in the soil and water by the process of weathering of chromium-containing rocks and diagenetic reactions and readily adsorbs on clay soil minerals with hydroxides [Al(III), Fe(III)] (Kimbrough et al. 1999; Oze et al. 2007). Chromium is also found in less amount in fresh vegetables, and many food supplements, for example, broccoli, cheese, brewer's yeast, molasses, mushrooms, barley, oatmeal, organ meats, egg yolks, and asparagus, also contain small amounts of the trace mineral (Corleone 2016).

11.2.4.2 Anthropogenic Sources

Naturally occurring chromium has optimum concentration in nature present in many food supplements and rocks, but the interference of human activities has increased the chromium concentration in the environment worldwide and causes a serious threat to the environment and human beings (Cheng et al. 2014; Joutey et al. 2015). Major anthropogenic chromium flows through release of various industrial applications such as in refractories and foundry sands for its heat resistance, fossil fuel combustion, waste incineration, electroplating, tanning, textile dyeing, corrosion inhibition, and wood preservation (Johnson et al. 2006; Joutey et al. 2015). Additional areas of application of chromium include wood preservatives and production of chrome pigments (e.g., lead chromate) which are used in paints, printing inks, and anticorrosive materials (Cheng et al. 2014). Total annual global consumption of chromium is 12.5 million tons, of which 85% chromium are consumed by metallurgical industries and the remaining 15% chromium are consumed by other industries. Ferrochromium is used to produce chromium-containing steel and contains 40–80% chromium with iron. It is used in the production of full alloy steel, bearing and high-speed steels, tool steel, superalloy, and welding materials (Mishra and Bharagava 2015).

11.2.4.2.1 Tannery Wastewater As a Major Source of Chromium Contamination

Tannery industries are the main source of chromium pollution and release a huge amount of chromium in the environment. There are more than 2500 tanneries in India, and most of them (nearly 80%) are engaged in chrome tanning process (Srivastava and Thakur 2006; Mishra and Bharagava 2015). Chromium-based processes in tannery industries cause greater concerns about environmental load worldwide. Tannery wastewater contains high concentrations of organic, inorganic, and nitrogenous compounds, chromium, sulfides, and suspended and dissolved solids including toxic compound. In the tanning process, Cr(III) is used as chromium sulfate as a tanning agent, which leads to conversion into hexavalent chromium and is finally released into tannery wastewater (Srivastava and Thakur 2006; Srivastava et al. 2007). The hexavalent chromium is the most toxic and mutagenic element with effluent that is discharged into a nearby water body (rivers, lake, ponds) which causes harmful effect on the environment and human beings. Chromium is used in the tannery industry as an anticorrosive agent. Up to 20% of chemicals including chromium are absorbed by the leathers, and the rest is released as wastes.

11.3 Toxicity of Hexavalent Chromium in Environment

The rapid growth of industries and urban areas has very serious pollution problems worldwide. Industries discharge a huge amount of partially treated, mostly not treated, wastewater into the environment, which causes serious soil and water pollution (Kavita and Keharia 2012; Shakoori et al. 2000; Arunakumara and Zhang 2008). Chromium which is considered as carcinogenic and toxic to animals is released in the environment through both natural processes and human activities. Industries like leather, textile, electroplating, and chemical manufacturing release hexavalent chromium and other heavy metals into the environment leading to water and soil pollution. Chromium causes air pollution through coal manufacturing (Sneddon 2012) (Table 11.3).

Table 11.3 Source of hexavalent chromium and its toxicity in environment

Sources	Application	Toxic effects
Tannery industry	Used in tanning as anticorrosive agent to prevent the leather such as ammonium dichromate, lead chromate, etc.	It is a mutagen and carcinogen that enters into environment to contaminate the air, water, and soil resulting harmful effect on plants and aquatic animals through bioaccumulation in the cell. It also causes alterations in the germination process as well as in the growth of roots, stems, and leaves and inhibition on enzymatic activities and mutagenesis in the plants
Electroplating	Chromic acid and hard chrome salt used in the metal finishing as corrosion-resistant deposit	
Textile	Chromium acetate, chromium nitrate, potassium chromate, and dichromate used in dye manufacturing process	

11.3.1 Toxicity of Hexavalent Chromium in Human Beings and Animals

Hexavalent chromium is considered as carcinogenetic and mutagenic to all living things. Chromium(VI) is the most toxic form and may cause health impact by entering the human or animal body through ingestion, inhalation, and dermal contact which causes allergy, rashes on skin surface (Cr(VI) act as an oxidant), nose irritations and nosebleed, weakened immune system, genetic material alteration (genotoxicity and mutagenicity), and kidney and liver damage and may even go as far as death of the individual (Das and Mishra 2008; Sneddon 2012; Mishra and Bharagava 2015) (Table 11.4).

11.3.2 Toxicity of Hexavalent Chromium in Plants

Chromium is released through various industrial processes including tanning, electroplating, textile, etc. and dumped in nearby water bodies. Soil and water from dumping site contain a high amount of chromium content which is very toxic to plants due to high solubility and permeability to cell membranes. Toxicity of chromium is observed at multiple levels, in plants with a reduction in yield to inhibit the enzymatic activities until mutagenesis (Shanker et al. 2005; Mishra and Bharagava 2015). Chromium toxicity is dependent on metal speciation, which is determinant for its uptake, translocation, and accumulation. The uptake of hexavalent chromium is thought to be an active mechanism performed by carriers for the uptake of essential elements such as sulfate. Chromium also competes with Fe, S, and P for carrier binding (Oliveira 2012). The presence of Cr in the external environment leads to changes in the growth and development pattern of the plant or

Table 11.4 Major health effects related with chromium toxicity

Chromium toxicity	Related effects in human beings and animals
Carcinogenicity	Lung, nasal, and sinus cancer is associated with the exposure through inhalation and ingestion
Genotoxicity and mutagenicity	DNA damage, gene mutation, sister chromatid exchange, chromosomal aberration
Respiratory	Ulceration and perforation of nasal septum and irritation of upper airways
Cardiovascular	Direct effect on blood and changes in bioelectric and mechanical activity of myocardium
Reproductive and developmental	Birth defects and decreased spermatogenesis and histopathological alterations
Dermatotoxicity	Skin lesions and rashes and allergic reactions
Gastrointestinal	Gastrointestinal gastric tract irritation

Adapted from Mishra and Bharagava (2015)

reduces the growth, induces chlorosis in young leaves, reduces seed germination, and causes ultrastructural modification of cell membrane (Panda and Chaudhury 2005; Shanker et al. 2005).

11.3.3 Toxicity of Hexavalent Chromium in Microbes

In trace amount, chromium is an essential element for living organisms because it is involved in electron transport reactions in biological systems. If chromium concentration is high, then it is very toxic to microorganisms. The toxicity is due to the reduction of hexavalent chromium to pentavalent chromium in which free radicals are formed and it causes deoxyribonucleic acid damage. Microorganisms are very sensitive to hexavalent chromium; it is toxic and mutagenic to most of microorganisms and causes cell enlargement and inhibits cell growth. Chromium toxicity causes a reduction in the population of soil microbes like bacteria, actinomycetes, and fungi. Chromium toxicity is caused by entering through a biological membrane by means of sulfate transport system demonstrated in many bacterial cells (Cervantes et al. 2001; Mishra and Bharagava 2015). Around 400–800 μg of hexavalent chromium inhibits the bacterial growth and interacts directly with bacterial DNA causing frameshift mutation and base pair replacement (Fernando et al. 1977). Chromium toxicity on algae and fungi had also been demonstrated. The growth of algae has no effect at 40–100 ppm of hexavalent chromium, whereas some algae are sensitive up to 15–20 ppm of Cr(VI) such as *Chlorella* and *Scenedesmus* (Cervantes et al. 2001; Mishra and Bharagava 2015).

11.4 Bioremediation Approaches for Chromium Detoxification from Tannery Wastewater

Bioremediation technique has been very promising nowadays to treat pollution loads. It includes living organisms like plants and microorganisms for the removal of organic compounds and toxic heavy metals from the environment to save the environment and human health (Saxena et al. 2018; Saxena and Bharagava 2017; Bharagava et al. 2017b, c; Chandra et al. 2015; Singh et al. 2011). It is an eco-friendly approach and includes the following methods.

11.4.1 Bioremediation

Bioremediation or biological treatment is also known as secondary treatment involving microorganisms in the process. The microbes utilize the organic matter as a food nutrient present in the waste and produce carbon dioxides, water, and energy as its metabolites. Bioremediation is a very popular, pollution control technology as compared to conventional physicochemical methods as it is cost-effective and efficient and uses natural systems (such as bacteria, fungi, algae, and plants) to catalyze the degradation or transformation of various toxic chemicals to nontoxic or less harmful forms (Moore 1998). The effluents from residential and industrial discharge constitute a major source of water pollution. The industrial effluents are discharged into open drains which finally joins the rivers (Kumari et al. 2006). Heavy metal/chromium(VI) from wastewater has also been treated conventionally by physicochemical processes like ion exchange, reverse osmosis, microfiltration, ultrafiltration, etc. These physicochemical processes were very costly and unreliable. The biotechnological method that includes microorganisms is a more reliable alternative to remove toxic metal/chromium or transform it into nontoxic or less toxic form (Sharma and Adholeya 2011). At present, the emphasis is given to the utilization of biological adsorbents for the removal and recovery of heavy metal contaminants. Biomass involving pure microbial strains has shown high capacities for the selective uptake of metals from dilute metal-bearing solutions (Kang et al. 2007). Microbial management of industrial or other wastes for the removal of hazardous compounds can be achieved through degradation, biosorption, or bioconversion to less toxic forms (Pillai and Archana 2012). Bacteria, fungi, and marine microalgae have been identified as potential metal biosorbents (Sharma and Adholeya 2011).

11.4.2 Bioremediation of Hexavalent Chromium Using Bacteria

Bacteria, fungi, microalgae, and other microorganisms are playing the most important role in the degradation and detoxification of industrial pollutants (Chandra et al. 2011). A wide variety of bacteria have been reported for reducing Cr(VI) to Cr(III) in various waste channels. Application of Cr(VI)-resistant bacteria for detoxification of Cr(VI) has been considered as an economical, effective, and safe procedure over conventional physical and chemical methods. Enzymatic reduction of chromate by Cr(VI)-resistant bacteria proceeds aerobically, anaerobically, or both. In *Bacillus* sp. ES29, chromate-reducing activity was localized in the soluble fraction of cell-free extract (CFE), which utilizes NADH as the sole electron donors. Focardi et al. (2012) characterized a Cr(VI)-resistant bacteria strain having enormous potential to reduce Cr(VI). *Bacillus* species are the major producers of dehairing proteases, and these enzymes are largely keratinases (Pillai and Archana 2012). A bacterial consortium was developed in the chemostat in the presence of PCP as a sole carbon source and applied with fungus for the degradation of pentachlorophenol and removal of chromium in the sequential bioreactor for effective and efficient removal of major contaminants in tannery effluents (Srivastava et al. 2007). Some chromate-resistant bacteria including *Achromobacter* sp., *Enterobacter cloacae HO1*, *Providencia* sp., *Brucella* sp., and *Bacillus* sp. were reported to be of Cr(VI) reduction ability under both aerobic and anaerobic conditions (Amoozegar et al. 2007).

Table 11.5 Metal removal through bacteria

Bacteria	Metal resistant	Removal method	References
Pseudomonas aeruginosa	Cr(VI), Cr(III)	Biosorption	Kang et al. (2007)
Pseudomonas sp., *Bacillus* sp., *E. coli*	Cr(VI)	Transformation	Cheung and Gu (2007)
Halomonas sp. *TA-04*	Cr(VI)	Bioremediation	Focardi et al. (2012)
Acinetobacter sp.	Cr(VI), PCB	Biosorption	Srivastava et al. (2007)
Nesterenkonia sp., *Achromobacter* sp., *Enterobacter cloacae HO1*, *Providencia* sp., *Brucella* sp., *Bacillus* sp.	Cr(VI), Cr(III)	Bioremediation	Amoozegar et al. (2007)
Bacillus methylotrophicus	Cr(VI)	Bioremediation	Mala et al. (2015)
Arthrobacter sp.	Cr(VI), phenol	Bioremediation	Bhattacharya et al. (2015)
Sporosarcina saromensis M52	Cr(VI)	Bioremediation	Zhao et al. (2016)

Some of the resistant mechanisms are being emphasized in an effective way to remove metal toxicity using bioremediation approaches. A variety of Cr(VI)-resistant bacteria have been screened to clean up the Cr(VI) toxicity (Rahman and Singh 2014). *Providencia* species is a gram-negative chromate-reducing bacteria and could grow and reduce up to 100% chromate at a concentration ranging from 100 to 300 mg/L and 99.31% at a concentration of 400 mg/L. *Providencia* sp. has a great potential for bioremediation of hexavalent chromium and also shows other heavy metals (such as Zn, Cu, Ni, Pb, Hg, Co) tolerance (Thacker et al. 2006). Several investigations have reported that *Pseudomonas aeruginosa* displays efficiency for metal uptake. It was found that the amount of mercury adsorbed by a *P. aeruginosa* biomass sample (180 mg Hg/g dry cells) and also the highest affinity and maximal capacity for uranium (100 mg U/g dry weight) were higher than that bound to a cation exchange resin (100 mg Hg/g dry resin, <100 mg U/g dry weight) (Kang et al. 2007) (Table 11.5).

11.4.3 Bioremediation of Hexavalent Chromium Using Fungi

Fungi have been identified as tremendously most important organisms which have ecological and economic value. It provides valuable products in various sectors such as food, pharma, cosmetics, detergents, etc. Fungi play a very important role in bioremediation or mycoremediation for heavy metal removal from wastes. Fungi remove organic matter from wastes by utilizing it as their nutrient and enzymatic activity. Tannery wastewater is composed of many heavy metals including a high concentration of chromium. Various fungal species have been used for removal of Cr(VI) or to transform Cr(VI) to Cr(III) which is a less toxic form, e.g., *Paecilomyces lilacinus*, *Aspergillus*, *Penicillium*, *Trichoderma*, *Phanerochaete*, *Hirsutella*, etc.

Table 11.6 Metal removal through fungi

Fungi	Metal resistant	Removal method	References
Aspergillus niger	Cr(VI)	Bioremediation	Srivastava and Thakur (2006)
Aspergillus sp. and *Hirsutella* sp.	Cr(VI)	Bioremediation	Srivastava and Thakur (2006)
Paecilomyces lilacinus	Cr(VI)	Bioremediation	Sharma and Adholeya (2011)
Trichoderma sp., *Penicillium spinulosum*, and *Aspergillus niger*	Cu, Zn, Mn, Cd	Bioaccumulation, uptake	Siddiquee et al. (2015)
Penicillium, *Aspergillus*, *Saccharomyces*, *Rhizopus*, and *Mucor*	Cr, Cu, Zn, Cd, Pb, and U	Biosorption	Bishnoi (2005)
Fusarium sp.	Co, Cu, Mn	Biosorption	Verma et al. (2016)
Metarhizium anisopliae, *Phomopsis* sp., *Fusarium oxysporum*, *Saccharomyces cerevisiae*	Cr^{6+}, Pb^{2+}, Hg^{2+}, Cd^{2+}, Cu^{2+}	Biosorption	Zili et al. (2016)

These are some examples of filamentous chromium-resistant fungi, which have the ability to remove or bioaccumulate Cr(VI) or transform (detoxify) Cr(VI) to Cr(III) form through direct enzymatic reactions or indirect with metabolites (Sharma and Adholeya 2011). *Fusarium* sp., *Metarhizium anisopliae*, *Phomopsis* sp., *Aspergillus* sp., *Trichoderma* sp., *Saccharomyces cerevisiae*, and *Penicillium* sp. show resistance to heavy metals (e.g., Cr, Cd, Pb, Hg, Cu, Co, Mn, and Fe) with up to 1000 ppm concentration and have biosorption capabilities (Zili et al. 2016) (Table 11.6).

11.4.4 Bioremediation of Hexavalent Chromium Using Algae

New technologies are being proposed to access the treatment of wastewater. Algae have the potential for bioremediation of wastewater treatment; it has been well-studied over the last 40 years. However, wastewater treatment through algal remediation has much wider scope than its present applications (Ryther et al. 1972; Kuyucak and Volesky 1988; Volesky 1990; Wase et al. 1997). Algae play a crucial role in wastewater treatment. The problems of conventional wastewater treatment can be overcome by using algae. Some of the advantages of algal remediation include cost-effective waste treatment, low-energy requirement, decreased sludge formation, and production of algal biomass for the industrial or commercial level. Algae have the ability to bioaccumulate the heavy metals and remove toxic metals from the wastewater. Sometimes at tertiary treatment stage, algae also have been found to be much efficient in the removal of pathogens. But the heavy metal may get accumulated in the algal cell and disrupt the metabolism by damaging the photosynthetic activities and enzymatic pathways or by reorienting the nutrient transport and availability (Arun et al. 2014). Microalgae can be used to produce high range valuable products such as polysaccharides, lipids, proteins, carotenoids, vitamins, food

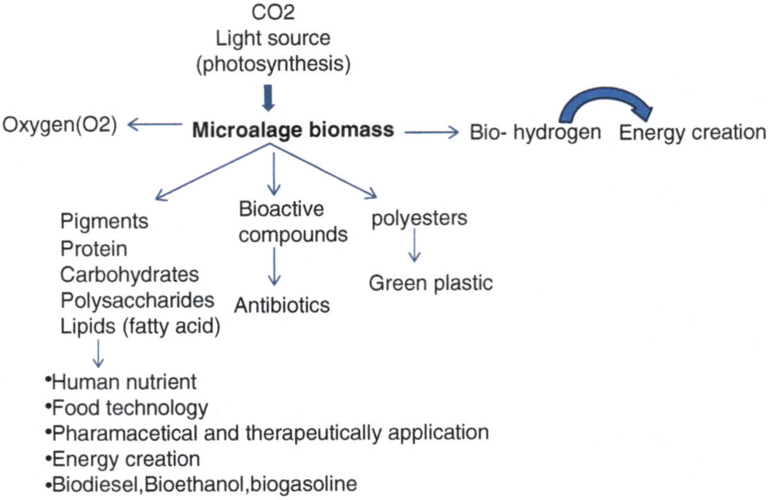

Fig. 11.3 Overview of products synthesized by microalgal strains and areas of application (Source: Koller et al. 2014)

supplements, cosmetics, and pharmaceuticals and for energy production. It also plays an important role as bioelimination of nutrients from wastewater (Vilchez et al. 1997) (Fig. 11.3).

In the global food chain, phytoplankton has a major role, but microalgae are pivotal for the global food chain. Their exceptional capacity for CO_2 fixation illustrates their indispensable significance to sustain Earth's ecosystems. In the field of "white biotechnology," microalgae have been used as promising bio-catalysts and can be implemented for food production, fine chemicals, diverse "green energy carriers," and fuels which can mitigate the present and future world crises for fuel and food and environmental concerns (Koller et al. 2014; Sing et al. 2014). We can use algae instead of using bacteria to clean wastewater because many of the pollutant sources in wastewater are also food sources for algae. Nitrates and phosphates are common components of plant fertilizers. Cell cycles of algae require large amount of nitrates and phosphates like plants. Heavy metals also form an important role in the regular functioning of algae like iron and chromium for photosynthesis and metabolic activity, respectively. The use of algae instead or bacteria has several advantages in bioremediation processes. One major advantage in the removal of pollutants is that this is a process that under light conditions does not need oxygen. Pollutants are removed, are digested by using carbon dioxide, and release oxygen into the environment. Hence, phycoremediation could potentially be coupled with carbon sequestration. Odor issues are not faced in phycoremediation because they do not depend on fouling processes. Phycoremediation has some disadvantages too. The basic problem faced in phycoremediation is competition with bacteria. Since the genome of eukaryotic algae is much larger and has more complex cell machinery, the algae take longer time period to divide (a few hours) as compared to bacteria in optimal conditions. In order to maintain the bacterial populations which could outcompete the algae and push them out in the wastewater treatment plant, all the conditions have to be carefully monitored. Furthermore, algae although sturdy are more sensitive compared to bacteria. So water conditions should be kept in control, and the algal population should not be overexposed to the pollution sources that will kill them (Dwivedi 2012). Microalgae have high limits of salt-tolerant levels because of its ability to accumulate ions intracellularly. The species *Dunaliella tertiolecta* has the salt tolerance level in the range 0.17–1.5 M of NaCl. However, other investigations stated that *D. tertiolecta* can go beyond these limits of NaCl to a range of 0.05–3 M (Fazeli et al. 2006).

Microalgae are not only good source of antioxidant but also play a crucial role in wastewater treatment including food preservation. Since microalgae are a potential new source of antioxidant, two species of microalgae are exploited at present for the production of carotenoid antioxidant commercially, i.e., *Dunaliella* for beta-carotene production and *Haematococcus* for astaxanthin production (Goiris et al. 2015). Microalgae can be used in diversified fields; it can produce phytochemicals which can be employed for human food, aquaculture, animal feed, cosmetics, pharmaceuticals, and possibly for biofuel and its associated by-product (Sing et al. 2014). Because of their higher photosynthetic efficiency, increased biomass production, and rapid growth rate, microalgae are considered to be a good source of biofuel production as compared to other energy crops like rapeseeds and soybeans. Since microalgae biomass can accumulate lipids within their cells similar to vegetable oil, biofuel from microalgae can be considered as a most suitable alternative which has a potential to produce 100 times more oil per acre than any other energy plants (Mubarak et al. 2015) (Table 11.7).

Present reports say that when growing cultures of fungi, bacteria, and marine microalgae are employed, the biosorbed metal levels are often less as compared to intracellular levels of metals. Though it is well-known fact that the cost of chemicals and energy can be substantially reduced with biological treatment as compared to conventional physical and chemical treatment, there are still some problems that arise during their real application in wastewater treatment processes (Sharma and Adholeya 2011).

Table 11.7 Metal removal through algae

Algae/microalgae	Metal resistant	Removal method	References
Spirulina fusiformis	Chromium	Bioaccumulation	Pandi et al. (2009)
Dunaliella sp.	Cr(VI)	Bioaccumulation	Arun et al. (2014)
Eichhornia crassipes	Cr(VI)	Bioremediation	Jahan et al. (2014)
Chlorella salina, C. sorokiniana	Co, Zn, Mn, Pb	Bioaccumulation	Kaparapu and Geddada (2016)
Phormidium ambiguum, Pseudochlorococcum typicum, Scenedesmus quadricauda	Hg, Pb, Cd	Bioremoval	Shanab et al. (2012)
Nannochloropsis, Pavlova lutheri, Tetraselmis chuii, Chaetoceros muelleri	Fe, Mn, Al, La, Ce, Ba	Bioremoval	Richards and Mullins (2013)

11.4.5 Phytoremediation

Hexavalent chromium is the most toxic metal found naturally and anthropogenically in the environment; Cr(VI) is more toxic to the plants than trivalent chromium (Nayak et al. 2015). Phytoremediation is a recently growing eco-friendly process in which plants are used to decontaminate the soil from organic nutrients and heavy metals like chromium. This method is cheap or cost-effective and removed almost completely metals from the contaminated site. Phytoremediation involves five processes like phytotransformation, rhizoremediation, phytostabilization, phytoextraction, and rhizofiltration (Revathi et al. 2013; Ahemad 2015; Gupta and Balomajumder 2015; Mishra and Bharagava 2015). Phytoremediation through common plants like wheat, *Jatropha*, mustard, sunflower, etc. has been accumulating high concentration of chromium. The higher plant can accumulate high amount of chromium concentration from the soil through the root of the plant (Mangkoedihardjo et al. 2008; Mishra 2015). Hexavalent chromium has been removed by entering in the tissues crossing the biological membrane in all living organisms, such as plants and microorganisms, which results in free radicals and reactive oxygen species (ROS) formation (Ahemad 2015). Phytoremediation of chromium using wheat seedling crops shows significant reduction of wheat crop as the chromium concentration increases. Chromium toxicity causes growth reduction, damaged cell wall, and changes in metabolic activity of the plants. Recently few researchers worked for enhancing the phytoremediation process by using some kind of chelators like EDTA, CA, and Zn^{2+} to enhance the bioavailability and uptake of metals or chromium in the plants at higher level treated with high chromium concentration. Chromium is found at a high level after phytoremediation in roots and less at aerial parts (Nayak et al. 2015). On the other hand, plant growth-promoting bacteria (PGPB) also help in phytoremediation with the process of phytostabilization and phytoextraction. In this, the bacteria help to accumulate toxic metal in plant cells and enhance the plant growth (Ahemad 2015).

11.5 Conclusion

This study depicts that the heavy metals such as Zn, Pb, Hg, Cd, Cr, Ni, Co, Fe, etc. cause severe problem to the environment. These heavy metals mainly released from industrial waste into rivers cause water pollution and affect the aquatic ecosystem. Among all heavy metals, chromium is extremely toxic element to living organisms, causing allergies; eczema; lung, nasal, and sinus cancer; etc. and shows genotoxicity and mutagenicity effects on living organisms (biological system). Various treatment technologies have been proposed including physical, chemical, and biological processes. Physical processes are very expensive, and chemical process causes other chemical loads. So the biological processes have been identified most significant in the removal of BOD and COD and transformation of toxic heavy metals into nontoxic or less toxic form. The algae or microalgae are suitable for wastewater treatment than other microorganisms (such as bacteria, fungi, yeast, etc.) because algae and microalgae not only treated the wastewater, but it also works as CO_2-fixing agent; supplies food or nutrients and vitamins naturally; has commercial importance, that is, in cosmetics and pharmaceutical industry; and is suitable alternative biofuel for the future.

Acknowledgment The author is extremely thankful to the Rajiv Gandhi National Fellowship approved by the University Grants Commission (UGC).

References

Ahamed MN, Kashif PM (2014) Safety disposal of tannery effluent sludge: challenges to researchers-a review. Int J Pharm Sci Res 5:733–736

Ahemad M (2015) Enhancing phytoremediation of chromium-stressed soils through plant-growth-promoting bacteria. J Genet Eng Biotechnol 13(1):51–58

Amoozegar MA, Ghasemi A, Razavi MR, Naddaf S (2007) Evaluation of hexavalent chromium reduction by chromate-resistant moderately halophile, *Nesterenkonia sp.* strain MF2. Process Biochem 42:1475–1479

Arun N, Vidyalaxmi, Singh DP (2014) Chromium(VI) induced oxidative stress in halotolerant alga *Dunaliella salina* and *D. tertiolecta* isolated from sambhar salt lake of Rajasthan (India). Cell Mol Biol (Noisy-le-Grand France) 60:90–96

Arunakumara KKIU, Zhang X (2008) Heavy metal bioaccumulation and toxicity with special reference to microalgae. J Ocean Univ China 7:60–64

Ballantyne B, Timothy CM, Tore S (1999) General and applied toxicology, 2nd edn, vol 3. Macmillan Publishers. 2052–2062:2145–2155

Belay AA (2010) Impacts of chromium from tannery effluent and evaluation of alternative treatment options. J Environ Prot 1:53–58

Bharagava RN, Saxena G, Mulla SI, Patel DK (2017a) Characterization and identification of recalcitrant organic pollutants (ROPs) in tannery wastewater and its phytotoxicity evaluation for environmental safety. Arch Environ Contam Toxicol. https://doi.org/10.1007/s00244-017-0490-x

Bharagava RN, Saxena G, Chowdhary P (2017b) Constructed wetlands: an emerging phytotechnology for degradation and detoxification of industrial wastewaters. In: Bharagava RN (ed) Environmental pollutants and their bioremediation approaches, 1st edn. CRC Press, Taylor & Francis Group, Boca Raton, pp 397–426. https://doi.org/10.1201/9781315173351-15

Bharagava RN, Chowdhary P, Saxena G (2017c) Bioremediation: an ecosustainable green technology: its applications and limitations. In: Bharagava RN (ed) Environmental pollutants and their bioremediation approaches, 1st edn. CRC Press, Taylor & Francis Group, Boca Raton, pp 1–22. https://doi.org/10.1201/9781315173351-2

Bhattacharya A, Gupta A, Kaur A, Malik D (2015) Simultaneous bioremediation of phenol and Cr (VI) from tannery wastewater using bacterial consortium. Int J Appl Sci Biotechnol 3(1):50–55

Bishnoi NR (2005) Fungus: an alternative for bioremediation of heavy metal containing wastewater: a review. J Sci Ind Res 64(2):93–100

Cervantes C, Campos-García J, Devars S, Gutiérrez-Corona F, Loza-Tavera H, Torres-Guzmán JC, Moreno- Sánchez R (2001) Interactions of chromium with microorganisms and plants. FEMS Microbiol Rev 25(3):335–347

Chandra R, Bharagava RN, Kapley A, Purohit HJ (2011) Bacterial diversity, organic pollutants and their metabolites in two aeration lagoons of common effluent treatment plant (CETP) during the degradation and detoxification of tannery wastewater. Bioresour Technol 102:2333–2341

Chandra R, Saxena G, Kumar V (2015) Phytoremediation of environmental pollutants: an eco-sustainable green technology to environmental management. In: Chandra R (ed) Advances in biodegradation and bioremediation of industrial waste, 1st edn. CRC Press, Taylor & Francis Group, Boca Raton, pp 1–30. https://doi.org/10.1201/b18218-2
Cheng H, Zhou T, Li Q, Lu L, Lin C (2014) Anthropogenic chromium emissions in China from 1990 to 2009. PLoS One 9(2):87753
Cheung KH, Gu JD (2007) Mechanism of hexavalent chromium detoxification by microorganisms and bioremediation application potential: a review. Int J Biodeter Biodegr 59:8–15
Corleone J (2016) What is the best source of chromium in food? http://www.livestrong.com/article/199306-what-is-the-best-source-of-chromium-in-food/
Das AP, Mishra S (2008) Hexavalent chromium environment pollutant and health hazard. J Environ Res Dev 2:3
Derek WJ (1999) Exposure or absorption and the crucial question of limit for mercury. J Can Dent Assoc 65:42–46
Dias MA, Lacerda ICA, Pimentel PF, De Castro HF, Rosa CA (2002) Removal of heavy metals by an *Aspergillus terreus* strain immobilized in a polyurethane matrix. Lett Appl Microb 34:46–50
Dwivedi S (2012) Bioremediation of heavy metal by algae: current and future perspective. J Adv Lab Res Biol 3:229–233
Fazeli MR, Tofighi H, Samadi N, Jamalifar H (2006) Effects of salinity on β-carotene production by *Dunaliella tertiolecta* DCCBC26 isolated from the Urmia salt lake, north of Iran. Bioresour Technol 97:2453–2456
Fernando L et al (1977) Toxicity and mutagenicity of hexavalent chromium on salmonella typhimurium. Appl Environ Microbiol 33:805–809
Focardi S, Pepi M, Landi G, Gasperini S, Ruta M, Di Biasio P, Focardi SE (2012) Hexavalent chromium reduction by whole cells and cell free extract of the moderate halophilic bacterial strain *Halomonas sp.* TA-04. Int Biodeter Biodegr 66:63–70
Gautam S, Kaithwas G, Bharagava RN, Saxena G (2017) Pollutants in tannery wastewater, pharmacological effects and bioremediation approaches for human health protection and environmental safety. In: Bharagava RN (ed) Environmental pollutants and their bioremediation approaches, 1st edn. CRC Press, Taylor & Francis Group, Boca Raton, pp 369–396. https://doi.org/10.1201/9781315173351-14
Goiris K, Colen VW, Wilches I, León-Tamariz F, De Cooman L, Muylaert K (2015) Impact of nutrient stress on antioxidant production in three species of microalgae. Algal Res 7:51–57
Goutam SP, Saxena G, Singh V, Yadav AK, Bharagava RN (2018) Green synthesis of TiO_2 nanoparticles using leaf extract of *Jatropha curcas* L. for photocatalytic degradation of tannery wastewater. Chem Eng J 336:386–396. https://doi.org/10.1016/j.cej.2017.12.029
Gupta A, Balomajumder C (2015) Phytoremediation of heavy metals and its mechanism: a brief review. J Integr Sci Technol 3(2):51–59
Hörcsik Z, Oláh V, Balogh A, Mészáros I, Simon L, Lakatos G (2006) Effect of chromium (VI) on growth, element and photosynthetic pigment composition of *Chlorella pyrenoidosa*. Acta Biol Szeged 50:19–23
Jahan MAA, Akhtar N, Khan NMS, Roy CK, Islam R, Nurunnabi M (2014) Characterization of tannery wastewater and its treatment by aquatic macrophytes and algae. Bangladesh J Sci Ind Res 49(4):233–242
Jahan MAA, Akhtar N, Khan NMS, Roy CK, Islam R, Nurunnabi M (2015) Characterization of tannery wastewater and its treatment by aquatic macrophytes and algae. Bangladesh J Sci Ind Res 49:233–242
Johnson J, Schewel L, Graedel TE (2006) The contemporary anthropogenic chromium 7 cycle. Environ Sci Technol 40:7060–7069
Joutey NT, Sayel H, Bahafid W, Ghachtouli NE (2015) Mechanisms of hexavalent chromium resistance and removal by microorganisms. Rev Environ Contam Toxicol 233:45–69
Kang SY, Lee JU, Kim KW (2007) Biosorption of Cr (III) and Cr (VI) onto the cell surface of *Pseudomonas aeruginosa*. Biochem Eng J36:54–58

Kanmani P, Aravind J, Preston D (2012) Remediation of chromium contaminants using bacteria. Int J Environ Sci Technol 9:183–193

Kaparapu J, Geddada MNR (2016) Applications of immobilized algae. Algal Biomass Util 7(2):122–128

Kavita B, Keharia H (2012) Reduction of hexavalent chromium by *Ochrobactrum intermedium* BCR400 isolated from a chromium-contaminated soil. 3 Biotech 2:79–87

Kimbrough DE, Cohen Y, Winer AM, Creelman L, Mabuni C (1999) A critical assessment of chromium in the environment. Crit Rev Environ Sci Technol 29(1):1–46

Koller M, Muhr A, Braunegg G (2014) Microalgae as versatile cellular factories for valued products. Algal Res 6:52–63

Kumari SB, Kirubavathy AK, Thirumalnesan R (2006) Stability and water quality criteria of open drainage municipal sewage water at Coimbatore used for irrigation. J Environ Biol 27:709–712

Kuyucak N, Volesky B (1988) Biosorbents for recovery of metals from industrial solutions. Biotechnol Lett 10:137–142

Mala JGS, Sujatha D, Rose C (2015) Inducible chromate reductase exhibiting extracellular activity in Bacillus methylotrophicus for chromium bioremediation. Microbiol Res 170:235–241

Mangkoedihardjo S, Ratnawati R, Alfianti N (2008) Phytoremediation of hexavalent chromium polluted soil using *Pterocarpus indicus* and *Jatropha curcas* L. World Appl Sci J 4(3):338–342

Midha V, Dey A (2008) Biological treatment of tannery wastewater for sulfide removal. Int J Chem Sci 6(2):472–486

Mishra S, Bharagava RN (2015) Toxic and genotoxic effects of hexavalent chromium in environment and its bioremediation strategies. J Environ Sci Health C 34(1):1–32

Moore PD (1998) Essential elements from waste. Nature 333:706

Morrison GO, Fatoki OS, Ekberg A (2001) Assessment of the impact of point river. Water SA 27:475–480

Mubarak M, Shaija A, Suchithra TV (2015) A review on the extraction of lipid from microalgae for biodiesel production. Algal Res 7:117–123

Mwinyihija M (2010) Main pollutants and environmental impacts of the tanning industry. In: Ecotoxicological diagnosis in the tanning industry. Springer, New York, pp 17–35

National Institute of Environmental Health Sciences NIEHS (2015) Hexavalent chromium. https://www.niehs.nih.gov/health/materials/hexavalent_chromium_508.pdf

Nayak AK, Jena RC, Jena S, Bhol R, Patra HK (2015) Phytoremediation of hexavalent chromium by *Triticum aestivum* L. Sci Agric 9(1):16–22

Noorjahan CM (2014) Physicochemical characteristics, identification of bacteria and biodegradation of industrial effluent. J Bioremed Biodegr 5:3

Oliveira H (2012) Chromium as an environmental pollutant: insights on induced plant toxicity. Hindawi Publishing Corporation. J Bot 2012:1–8

Oze C, Bird DK, Fendorf S (2007) Genesis of hexavalent chromium from natural sources in soil and groundwater. Natl Acad Sci USA 104(16):6544–6549

Panda SK, Chaudhury S (2005) Chromium stress in plant. Braz J Plant Physiol 17(1):95–102

Pandi M, Shashirekha V, Swamy M (2009) Bioabsorption of chromium from retan chrome liquor by *cyanobacteria*. Microbiol Res 164:420–428

Pillai P, Archana G (2012) A novel process for biodegradation and effective utilization of chrome shavings, a solid waste generated in tanneries, using chromium resistant *Bacillus subtilis* P 13. Process Biochem 47:2116–2122

Rahman Z, Singh VP (2014) Cr (VI) reduction by *Enterobacter sp.* DU17 isolated from the tannery waste dump site and characterization of the bacterium and the Cr (VI) reductase. Int Biodeter Biodegr 91:97–103

Revathi K, Haribabu TE, Sudha PN (2013) Phytoremediation of Chromium contaminated soil using Sorghum plant. Int J Environ Sci 2:417–428

Richards RG, Mullins BJ (2013) Using microalgae for combined lipid production and heavy metal removal from leachate. Ecol Model 249:59–67

Ryther JH, Dunstan WM, Tenore KR, Huguenin JE (1972) Controlled eutrophication increasing food production from the sea by recycling human wastes. Bioscience 22:144–152

Saxena G, Bharagava RN (2015) Persistent organic pollutants and bacterial communities present during the treatment of tannery wastewater. In: Chandra R (ed) Environmental waste management, 1st edn. CRC Press, Taylor & Francis Group, Boca Raton, pp 217–247. https://doi.org/10.1201/b19243-10

Saxena G, Bharagava RN (2017) Organic and inorganic pollutants in industrial wastes, their ecotoxicological effects, health hazards and bioremediation approaches. In: Bharagava RN (ed) Environmental pollutants and their bioremediation approaches, 1st edn. CRC Press, Taylor & Francis Group, Boca Raton, pp 23–56. https://doi.org/10.1201/9781315173351-3

Saxena G, Chandra R, Bharagava RN (2016) Environmental pollution, toxicity profile and treatment approaches for tannery wastewater and its chemical pollutants. Rev Environ Contam Toxicol 240:31–69. https://doi.org/10.1007/398_2015_5009

Saxena G, Purchase D, Mulla SI, Saratale GD, Bharagava RN (2018) Phytoremediation of heavy metal-contaminated sites: environmental considerations, field studies, sustainability and future prospects. J Environ Manag

Shakoori AR, Makhdoom M, Haq RU (2000) Hexavalent chromium reduction by a dichromate-resistant gram-positive bacterium isolated from effluents of tanneries. Appl Microb Biotechnol 53:348–351

Shanab S, Essa A, Shalaby E (2012) Bioremoval capacity of three heavy metals by some microalgae species (Egyptian Isolates). Plant Signal Behav 7(3):392–399

Shanker AK, Cervantes C, Loza-Taverac H, Avudainayagamd S (2005) Chromium toxicity in plants. Environ Int 31:739–753

Sharma S, Adholeya A (2011) Detoxification and accumulation of chromium from tannery effluent and spent chrome effluent by *Paecilomyces lilacinus* fungi. Int Biodeter Biodegr 65:309–317

Sharma V, Sachdeva MV, Sakhuja N, Arora D (2011) Impact of heavy metals (Chromium and Nickel) on the health of residents of Jagadhri city due to intake of contaminated underground water. Arch Appl Sci Res 3:207–212

Shashwat K (2011) Impact of tannery effluent with special reference to seasonal variation on physico-chemical characteristics of river water at Kanpur (U.P), India. J Environ Anal Toxicol 1:4

Siddiquee S, Rovina K, Azad SA, Naher L, Suryani S, Pasicha C (2015) Heavy metal contaminants removal from wastewater using the potential filamentous fungi biomass: a review. J Microb Biochem Technol 7:384–393

Sing SF, Isdepsky A, Borowitzka MA, Lewis DM (2014) Pilot-scale continuous recycling of growth medium for the mass culture of a halotolerant *Tetraselmis sp.* in raceway ponds under increasing salinity: a novel protocol for commercial microalgal biomass production. Bioresour Technol 161:47–54

Singh JS, Abhilash PC, Singh HB, Singh RP, Singh DP (2011) Genetically engineered bacteria: an emerging tool for environmental remediation and future research perspectives. Gene 480:1–9

Smith WL (2001) Hexavalent chromium reduction and precipitation by sulphate-reducing bacterial biofilms. Environ Geochem Health 23:297–300

Sneddon C (2012) Chromium and its negative effects on the environment. http://serc.carleton.edu/NAGTWorkshops/health/case_studies/chromium.html

Srivastava S, Thakur IS (2006) Isolation and process parameter optimization of *Aspergillus* sp. for removal of chromium from tannery effluent. Bioresour Technol 97:1167–1117

Srivastava S, Ahmad AH, Thakur IS (2007) Removal of chromium and pentachlorophenol from tannery effluents. Bioresour Technol 98:1128–1132

Thacker U, Parikh R, Shouche Y, Madamwar D (2006) Hexavalent chromium reduction by *Providencia sp*. Process Biochem 41:1332–1337

Verma P, Singh S, Verma RK (2016) Heavy metal biosorption by Fusarium strains isolated from iron ore mines overburden soil. Int J EnvironSci Toxicol Res 4(4):61–69

Vilchez C, Garbayo I, Lobato MV, Vega J (1997) Microalgae-mediated chemicals production and wastes removal. Enzyme Microb Technol 20:562–572

Volesky B (1990) Biosorption of heavy metals. CRC press, Boca Raton

Wase DAJ, Forster CF, Ho YS (1997) Low-cost biosorbents: batch processes. In: Biosorbents for metal ions, pp 141–163

Zhao R, Wang B, Cai QT, Li XX, Liu M, Hu D, Fan C (2016) Bioremediation of hexavalent chromium pollution by Sporosarcina saromensis M52 isolated from offshore sediments in Xiamen, China. Biomed Environ Sci 29(2):127–136

Zhitkovich A (2011) Chromium in drinking water: sources, metabolism, and cancer risks. Chem Res Toxicol 24:1617–1629

Zili D, Jinping W, Chunhai J, Cougui C (2016) Isolation of heavy metal-resistant fungi from contaminated soil and co-culturing with rice seedlings. Afr J Microbiol Res 10(28):1080–1085

Chapter 12
Arsenic Contamination in Environment, Ecotoxicological and Health Effects, and Bioremediation Strategies for Its Detoxification

Manoj Kumar, Anoop Yadav, and A. L. Ramanathan

Contents

Abstract The present day global environmental pollution is resultant of modernization, industrialization, urbanization, and several other anthropogenic activities, which involve the huge application of trace metals. Among the trace metals, Arsenic

M. Kumar (✉)
Department of Environmental Science, School of Earth, Environment and Space Studies, Central University of Haryana, Mahendergarh, Haryana, India

School of Environmental Sciences, Jawaharlal Nehru University, New Delhi, India

A. Yadav
Department of Environmental Science, School of Earth, Environment and Space Studies, Central University of Haryana, Mahendergarh, Haryana, India

A. L. Ramanathan
School of Environmental Sciences, Jawaharlal Nehru University, New Delhi, India

G. Saxena, R. N. Bharagava (eds.), *Bioremediation of Industrial Waste for Environmental Safety*, https://doi.org/10.1007/978-981-13-1891-7_12

(As) is known as the leading toxicant to the environment worldwide and having the various toxic effects on human and animal health. Exposure of As causes various types of health effects like dermal and neurological problems, reproductive and pregnancy effects, cardiovascular effects, diabetes mellitus, diseases of the respiratory system, multiorgan cancers, etc. The persistence of As in the environment may pollute or contaminate soils and aqueous system as both natural components or as the result of human activity. In recent years, the development of efficient green chemistry methods for detoxification of trace metal poisoning has become a major focus of researchers. It has been investigated in order to find an eco-friendly and recyclable technique for the removal of trace elements contamination from the natural resources. Bioremediation process in this regards is an option that offers the possibility to reduce or render trace and toxic elements such as As using plants and microbes. Among the various bioremediation processes, phytoremediation and bioremediation using microbes are quite effective. Phytoremediation includes the removal of contaminants with the help of green plants, while the microbial bioremediation includes the removal of trace and toxic elements by microorganisms (bacteria, fungi, yeast and algae) as sorbents. The aim of this chapter is to give an overview of the As contamination in the environment and also the mechanism of removal of the As from the contaminated resources by the potent application of plants and microbes.

Keywords Arsenic contamination · Groundwater · Ecotoxicological · Phytoremediation · Bioremediation · Fluvial regions

12.1 Introduction

Industrial wastewaters is mainly responsible for the heavy metal contamination in the environment (Goutam et al. 2018; Bharagava et al. 2017a; Gautam et al. 2017; Saxena et al. 2016; Saxena and Bharagava 2015). However, Arsenic (As) is both a geogenic and anthropogenic contaminant, which poses a significant threat to human life, health, and social well-being in the current scenario. It is widely spread in a number of areas worldwide, including countries in Asia, North and Latin America, parts of Europe and Africa. Exposure to As in concentrations exceeded from the World Health Organization provisional guide value (WHO 2011) through the consumption of As-enriched groundwater is having adverse impacts on human health (Bhattacharya et al. 2014). The As calamity is severe among Asian countries, especially the Ganga-Meghna-Brahmaputra plain of India and Bangladesh (Chakraborti et al. 2013, 2016).

Identification of the safe aquifers in naturally occurring As-contaminated areas is the main challenge for the scientific community worldwide. There are some chemical and geological processes which are responsible for the mobilization of As; hence the proper knowledge of these processes will be very useful to overcome the situation. Studies have been documented that more than 6 million people are at the risk from drinking As-contaminated water >50 μg/L in West Bengal, while this number is 30–35 million in Bangladesh (Chakraborti et al. 2002; Mandal and Suzuki 2002; Srivastava and Sharma 2013). It has been reported that 0.9 million people from 15 districts of Bihar are at health risk, a state of India which is located in the eastern part of the Gangetic plain (Saha 2009). The situation is also the same in other states of India like Uttar Pradesh, Jharkhand, Chhattisgarh, Assam, West Bengal, and Manipur; these data show that As is posing a serious risk to the population in mass.

Arsenic has been classified as a class (I) human carcinogen due to its sensitivity and mobilization at normal pH range of 6.5–8.5 (range of natural groundwater). The toxicity of the As varies with the species, inorganic and organic. The inorganic species (arsenite and arsenate) are more toxic than the organic (monomethylarsonic acid, dimethylarsinic acid, arsenobetaine etc.). Furthermore, As(III) is much time toxic than the As(V) (Hughes et al. 2011). Inorganic forms of As are dominant in natural water, while the organic species dominate in surface water bodies where the bacterial activities are more prominent like industrial and agricultural runoff (Smedley and Kinniburgh 2002). Thus it is suggested that estimation of the total As would not be very useful to evaluate its human health hazard.

The persistence of As in the environment may pollute or contaminate soils and aqueous stream as both natural components or as the result of human activity. In recent years, the development of efficient green chemistry methods for detoxification or cleaning of trace metal poisoning has become a major focus of researchers. Several investigations have been performed globally in order to find an eco-friendly and recyclable technique for the removal of traces elements contamination from the natural resources. Plants and microbe have the tendency to uptake trace elements from the media termed as process bioremediation. Bioremediation process offers the possibility to reduce or render trace and toxic elements. Among the various bioremediation processes, phytoremediation and bioremediation using microbes are quite effective. Phytoremediation includes the removal of contaminants with the help of green plants, while the microbial bioremediation includes the removal of trace and toxic elements by microorganisms (bacteria, fungi, yeast and algae) as sorbents (Saxena et al. 2018; Bharagava et al. 2017b, c; Saxena and Bharagava 2017; Chandra et al. 2015). The aim of this chapter is to give an overview of the As contamination in the environment and also the mechanism of removal of the As from the contaminated resources by the potent application of plants and microbes.

12.2 Arsenic in the Environment: Occurrence and Sources

Arsenic is a class I carcinogen, also termed as "king of poison." It is found in the earth crust in a trace amount including in air, water, and soils. The wide extent of the As has caused a menace to the lives of several hundred million people in different regions worldwide and has resulted into the world's largest environmental calamity (Smedley and Kinniburgh 2002; Ravenscroft et al. 2009). Arsenic is not newly known but also has been used in ancient time 300 BC in hardening the bronze and use of As compounds as an ulcer treatment.

An average concentration of As (1800 μg/kg) has been reported in the earth crust (Mason 1966). The concentration of As may occur up to five times in shales and alluvium. Arsenic has been reported as a major component in more than 200 minerals, such as arsenides, sulfides, oxides, arsenates, and arsenites including elemental As. Arsenopyrite (FeAsS) is known as the most abundant As ore mineral. High As concentration also has been observed in many oxide minerals and hydrous metal oxide either as part of the mineral structure or as sorbed species which includes a coating on the edges of clay minerals and on the surface of the calcite (Goldberg and Glaubig 1988; Smedley and Kinniburgh 2002). Muds and clay tend to have higher As concentration rather than sand and sandstones (Ravenscroft et al. 2009). An average order of 5000–10,000 μg/kgAs concentration in soils has been reported. Boyle and Jonasson (1973) and Shacklette et al. (1974) quoted an average value of 7200 and 7400 μg/kg, respectively, for American soils. Average As concentration in the stream sediment was found in the range of 5000–8000 μg/kg in England and Wales (AGRG 1978). A considerable concentration (range 1200–2600 μg/kg) has also been reported in sediments of the river Ganges, from Brahmaputra river with a range 1400–5900 μg/kg and from Meghna river with a range 1300–5600 μg/kg (Datta and Subramanian 1997). Arsenic enters into groundwater through weathering processes of As mineral bearing rocks followed by runoff, deposition, and leaching (NIH 2010).

Along with the natural source of As occurrence, there are several anthropogenic ways by which it can enter into the environment which include combustion of coal, application in agricultural practices using As-based pesticides, chromated copper arsenate (CCA) for wood preservation, activities like smelting of base metal ores, and mining activities (Smedley and Kinniburgh 2002).

The concentration of As in freshwater strongly depends on the source of As, local geochemical environment, and the amount available. It can vary by more than 10^3 orders of magnitude. The main factors for mobilization and accumulation of As are rock-water interactions and availability of favorable physical and geochemical conditions in the aquifers (Smedley and Kinniburgh 2002). The presence of As in the aquifers is controlled by many factors that include redox potential, pH, adsorption/desorption, precipitation/dissolution, As speciation, and biological transformation.

12.3 Aqueous Speciation and Toxicity of Arsenic

Arsenic behaves differently among the heavy metalloids and oxyanion-forming elements (e.g., As, Sb, Mo, Cr, Re, Se, and V) in its sensitivity and mobilization at the typical values of pH (6.5–8.5) found in the natural aquifers. Arsenic can be found in several oxidation states (−3, +3, 0 and +5) and forms organic and inorganic species in the environment (Smedley and Kinniburgh 2002). Inorganic forms are dominant in natural water, while organic forms are found in surface water where biological activities take place or where waters are significantly affected by industrial waters containing organic impurities. Inorganic forms are oxyanions of trivalent arsenite As(III) and or pentavalent arsenate As(V), and major organic species are dimethylarsinic acid (DMA) and monomethylarsonic acid (MMA). Inorganic species are more toxic than the organic (NIH 2010).

12.4 Arsenic in Groundwater and Surface Water

Concentration of As higher than WHO guideline (10 μg/L) has been reported in several areas worldwide. Arsenic is widely spread in groundwaters of many states of India including Uttar Pradesh, Bihar, Jharkhand, Chhattisgarh, Assam, West Bengal, and Manipur. More than 170,000 water samples from tube wells were analyzed from all surveyed states of India, and half of the samples had As >10 μg/L. The maximum As concentration was detected in a tube well from West Bengal as 3700 μg/L, which was 370 times higher than the WHO guideline value (10 μg/L) and 74 times higher than the Indian standard of As (50 μg/L) in drinking water. Elevated concentrations of As in groundwater were also detected in other states such as Bihar and Uttar Pradesh. Altogether 13.85 and 6.96 million people from all surveyed states were exposed to As greater than 10 and 50 μg/L, respectively (Chakraborti et al. 2017).

Higher concentration of As has been also observed in the river water samples, mainly in the seasonal river channels or tributaries of the Ganges. The concentration of As varied between the Ganges and its tributaries as follows: Ganges (2–4 μg/L), Bay (23 μg/L), Burhi Gandak (17–24 μg/L), Kamala (10 μg/L), Bagmati (13 μg/L), Santi (5 μg/L), Blan (12 μg/L), and Punpun (6 μg/L) (unpublished). Higher As concentration has not been observed in most studies on the middle and lower Gangetic plain. However, an elevated concentration of As (20–22 μg/L) has been reported in a small stream (river Gobra) in Murshidabad (Stuben et al. 2003) and river Jalangi and upper Ichamati (range 37–101 μg/L) of West Bengal (Mukherjee et al. 2009). It has been suggested that the reduction of Mn in the Gobra while the groundwater discharge served as a source of As in river Jalangi and upper Ichamati is due to elevated As concentrations. In a study done in middle Gangetic plain, it was observed that the river or channels which are small and locally recharged (Burhi Gandak, Bay, Bagmati, and Blan) contain high levels of As, while the river Ganges and its channels contain much fewer concentrations (unpublished). The probable reason for higher concentrations of As may be lean flow into the channels and recharge from the adjacent aquifers or may be upstream anthropogenic inputs. The river Kamala originates in the Himalayas and contains much less As concentration in its upper reaches near Jainagar city of Madhubani district in Bihar state.

12.5 Arsenic in Sediments and Its Relation with the Geological Settings in Fluvial Regions

The Ganges basin, also known for world's largest alluvium sedimentation and deposition, is an end result of the India-Asia plate collision that had started in Palaeogene geological period. In Gangetic plain, the sediment deposition is divided into Pleistocene and Holocene deposition (Revenscroft 2001). A flat topography exists with a north to south slope in the Holocene alluvium Gangetic plain. Acharyya and

Shah (2007) have discussed geomorphologic and quaternary morphostratigraphy of Gangetic plain on 1:50,000 scale, along with field observations to identify fluvial landforms and soil characteristics. The terrain is divided into two types of deposition: the older alluvium (Pleistocene) is characterized by the presence of yellow-brown clay with profuse calcareous and ferruginous concentrations. The newer alluvium (Holocene) is characterized by unoxidized, organic-rich sand, silt, and clay and restricted to low-lying fluvial and fluviolacustrine settings. The major part of Ganga plain consists of interfluves upland terrace surface (Singh 1996). The plain has been incised by dendritic drainage and channels containing a good amount of organic muds of Holocene age (Ravenscroft et al. 2005). Shah (2008) revealed a mineral assemblage (quartz, muscovite, chlorite, kaolinite feldspar, amphibole, and goethite) with the help of XRD studies on soil samples of As-safe older alluvium and As-contaminated newer alluvium from Gangetic plain. In Gangetic plain tube wells are tapped mostly in shallow aquifers, which hold 30% of total replenishable groundwater. Shallow aquifers are the main source of drinking water to fulfill the daily requirements of the local population and remain as the major input for societal development.

Primarily mineral assemblage of quartz, calcite ($CaCO_3$), muscovite, and chlorite with minor amounts of smectite, feldspar, hematite, siderite, goethite, and magnetite was reported in Gangetic plain (Kumar et al. 2016). The significant presence of altered feldspar, chert, and chlorite indicated the sedimentary and metasedimentary origin of sands (Ahmed et al. 2004). The presence of calcite and muscovite in the upper oxidized sediments suggested a close relationship with chelation of the metals in the zone of oxidation (Hasan et al. 2009). Peaks of hematite (Fe_2O_3) and goethite ($FeO(OH)$) were observed throughout the profile, but magnetite (Fe_3O_4) was reported only in the upper yellowish oxidized silty clay. The presence of this secondary mineral was noted in early studies of As in Southeast Asia by Islam et al. (2004) and Anawar et al. (2006). Jönsson and Sherman (2008) investigated on sorption of As to siderite and find As(V) sorbs strongly, but As(III) sorbs weakly. Mumford et al. (2012) reported 184,000 μg/kg of As in a siderite, which included quartz, so the actual concentration in the siderite is probably higher. Hence, siderite was reported as decidedly a sink for As in Gangetic plain.

12.6 Arsenic in Agricultural Soils: A Close Relation with Groundwater Extraction

The total concentration of As observed in agricultural soils ranged from 3527 to 14,690 μg/kg in Gangetic plain (Kumar et al. 2016), and this finding is approximately two to seven times higher than the world's typical average value of 2000 μg/kg for igneous and sedimentary rocks (Mandal and Suzuki 2002). A high variability was observed in the concentration of As in agricultural soils in the Gangetic plain (Kumar et al. 2016). It is known that presence of higher concentration of As in soils may be attributed to geogenic contribution (Kumar et al. 2016; Meharg and Rahman

2003), although few studies indicate higher As concentration in soils than sediments (Chowdhury et al. 1999). This suggests that other sources may contribute to the anomalous high concentration of As in soils. The situation of an elevated As in agricultural soils has been documented in previous studies conducted from As prone areas of Bangladesh, a country in which it has been confirmed that As levels were elevated in zones where As in groundwater used for irrigation was high (Meharg and Rahman 2003). Mean As concentration in agricultural soils has been reported as 101,000 μg/kg in West Bengal, India (Norra et al. 2005). Another study conducted in West Bengal in an As-contaminated area revealed a range of 10,000–35,000 μg/kg in agricultural soils (Sanz et al. 2007). Arsenic concentration of surface soils having a range of 1090–2480 μg/kg and any specific trend was not observed in a concentration with the depth of 2 mbgl (Vicky-Singh et al. 2010). The As concentration in agricultural soils varied from 880 to 4960 μg/kg in Feni district of Bangladesh, while it ranged from 3110 to 8900 for Dhamrai, Bangladesh, and 17,600–65,000 μg/kg for Faridpur, Bangladesh (Ahsan et al. 2009). A range of 8500–10,300 μg/kg in Japanese paddy soils, 2000–4600 μg/kg in South Korean soils, and 6700–9100 μg/kg in soils from Thailand (Mandal and Suzuki 2002) has been observed.

12.7 Arsenic in Dietary Components

In study from central Gangetic basin it was reported that the vegetables with highest mean As concentration were luffa, brinjal, cucumber, ladyfinger, gourd, and green gram with mean values of 800, 492, 399, 375, 268, and 174 μg/kg, respectively (Kumar et al. 2016). In other dietary components, rice, wheat, and maize have the highest mean As concentrations (Kumar et al. 2016). Variation in As and other elements' accumulation in plants depends on many factors, viz., availability of As in water and soils, accumulation capacity by plant, lifespan of plants grown, etc. (Roychowdhury et al. 2003). Among all food categories such as vegetables, rice, wheat, maize, and green gram, vegetables contained elevated concentrations of As. In another study, it was reported that As concentrations (dry wt.) in wheat and maize were 24 μg/kg and 11 μg/kg, respectively, from Maner district of Bihar, India (Singh and Ghosh 2011).

12.8 Ecotoxicological and Health Effects of Arsenic

Arsenic is a toxic element, known as class (I) human carcinogen and widely distributed in the environment as both inorganic and organic forms (Hughes et al. 2011). In general, the inorganic forms (arsenite and arsenate) of As are much more toxic than the organic forms (monomethylarsonic acid, dimethylarsinic acid, arsenobetaine, etc.) of As. Arsenite is generally more toxic than arsenate, and humans are exposed to both forms of inorganic As from water and food. There are many pathways by which As can enter the human body via food chain (ingestion by water and food sources), and occupational exposure is the most common (Rahman et al. 2009). Various inorganic species (arsenite and arsenate) and organic species (methylated anionic species, volatile As hydride, and organo As) in food materials have been reported as the main pathways to human exposure (Momplaisir et al. 2001).

12.9 Bioremediation Strategies for Arsenic Detoxification

Several conventional remediation methods such as solidification and stabilization, soil flushing/washing, electrokinetics, chemical reduction/oxidation, in situ oxidation, thermal desorption, surfactant-enhanced aquifer remediation, vitrification, pneumatic fracturing, excavation/retrieval, landfill, etc. are available (Saxena et al. 1999; Wenzel et al. 1999). Most of these traditional remediation methods used for in situ and ex situ remediation of contaminated sites are uneconomical, unsustainable, and destructive. Bioremediation is the use of plants or microorganisms to decontaminate an environment by transforming or degrading pollutants. In situ bioremediation is a well-established technology for the treatment of contaminated sites, especially when the pollutants are diffused in large areas. Such managed areas are available for safe uses at an economically acceptable price.

Generally, metal contaminants are found in various matrix of the environment such as soils, sediments, and water. Metal contaminants may be present naturally or produced anthropogenically through various industrial processes. The main reason for the remediation of the metal-contaminated site is that metals are nonbiodegradable in nature; however, they can be transmuted through various processes such as sorption, methylation, complexation, and changes in valence state. These conversions affect the movement and bioavailability of metal pollutants in a different matrix.

Arsenic bioremediation is an economical and environment friendly method. Arsenic is a chemical analog to phosphorus in several plants, and it is easily taken up by plants (Tu and Ma 2003). There are several ways and methods of As bioremediation that are discussed in the next sections of this chapter.

12.9.1 Arsenic Remediation Using Plants (Phytoremediation)

The generic term "phytoremediation" contains two words: the Greek prefix phyto means "plant," attached to the Latin root medium which means "to correct or remove an evil." In general, phytoremediation is a set of mechanisms that use different plants to absorb, extract, contain, or immobilize contaminants (organic and inorganic) from different components of the environment (such as air, water, and soil) (Salt et al. 1998). Arsenic is an unnecessary constituent for plants, and inorganic As species are usually extremely toxic for plants. Usually, the As concentrations found in most of the terrestrial plants are below 10 mg kg^{-1} under normal conditions (Matschullat 2000).

Arsenate present in the soil act as phosphate analog in plants by replacing phosphate in several biochemical and also interrupt phosphate metabolism. For example, in some plants arsenate can interrupt mitochondrial oxidative phosphorylation and thus, the production of the nucleotide adenosine triphosphate (ATP), which is the main energy source for cells (Fayiga and Saha 2016). After absorption As can find its way into the fruit/grains of plants and accumulate there. There are several factors which can affect the phytoremediation of As-polluted soil and water, such as bioavailability of As in the soil, redox potential, speciation of As in soil, phosphate concentrations, the presence of co-contaminants, plant age, plant nutrition, rhizosphere characteristics, and biological associations with microbes (Fayiga and Saha 2016). During phytoremediation, plant removes or uptakes metals by using one or more of these mechanisms, i.e., phytoextraction, phytostabilisation, rhizofiltration, phytodegradation, and phytovolatilization.

Arsenic inhibits plant growth, delays seed germination, and causes foliage chlorosis and necrosis (Odutayo et al. 2015). Presence of As in plants may lead to a decrease in crop yield, as As can disturb the uptake and transport of nutrients in plants (Paivoke and Simola 2001). Phytoremediation has a benefit over traditional remediation technology of As-contaminated soils such as burial and chemical stabilization, which may cause long-term health intimidations due to leakage or chemical uncertainty (Allen 2001; Förstner and Haase 1998; Bhattacharya et al. 2007). Thus phytoremediation technology has the possibility to become an environment friendly and economical unconventional remediation technique for As-contaminated sites (Bhattacharya et al. 2007).

Higher plants can bear high levels of As by two basic approaches, which are (a) exclusion, whereby carriage of As is limited, and low, moderately constant As concentrations are retained in the shoot or grain over a wide range of soil concentrations, and (b) accumulation, whereby As is accumulated in less/nontoxic form(s) in upper parts of plant at both high and low soil concentrations (Baker et al. 2000). Some studies conducted on As remediation using various plant species are given in Table 12.1.

Most of the plants do not accumulate As, and its concentrations in leaves or seeds are often below 1 mg/kg. Several authors reported that some tropical and subtropical plant species can tolerate and uptake various inorganic and organic forms of As (Meharg and Hartley-Whitaker 2002). Ma et al. (2001a, b) conducted a screening study on several plant species growing at an As-contaminated site in Florida and reported that *Pteris vittata*, as an As hyperaccumulator fern. The results showed that its leaves can contain an excess of 1% As (dry weight basis). This was due to the capability of the fern to translocate As from the lower parts of plants (roots) to the upper parts (leaves) and accumulate it, due to its ability to uphold high phosphate in its roots (Tu and Ma 2003; Luongo and Ma 2005). When As enters in plants through the roots of plants and through P transporters (Pit and Pst proteins), As(V) is reduced to As(III) before being ejected to cell vacuoles.

In another study, Wang et al. (2007) examined the discrepancy of As accumulation by ferns collected at various sites in south China (Guangxi Province) and

Table 12.1 Selected phytoremediation studies conducted using various plant species

Location/contaminated site	Plant species used	Arsenic concentration (mg kg^{-1})	Remediation time	Study type	References
Udine, Italy	*Populus* and *Salix*	886	2 years	Pot and field	Vamerali et al. (2009)
Tarabana, Municipio Palavecino, Edo. Lara, Venezuela	Water hyacinth (*Eichhornia crassipes*) and lesser duckweed (*Lemna minor*)	0.15	21 days	Greenhouse	Alvarado et al. (2008)
India	Chickpea (*Cicer arietenum)*	10	One season	Pot	Srivastava and Singh (2014)
Mine tailing spill, Aznalcóllar, Spain	*Quercus ilex*, *Olea europaea*, *Populus alba* L., *Phillyrea angustifolia* L., *Pistacia lentiscus* L., *Rosmarinus officinalis* L., *Retama sphaerocarpa* L., *Tamarix africana* Poir.,	129	10 years	Field	Domínguez et al. (2008)
Mine tailing spills, Sanlúcar la Mayor, Spain	*Lupinus albus*	–	3 weeks, 1–6 months	Pot study and field trials	Vázquez et al. (2006)
Sugar Brook, UK	*Salix*, *Populus*, *Alnus*, *Betula*, and *Larix*	4.9–5266	3 years	Field trials	French et al. (2006)
Rixton clay pits, Kidsgrove Merton Bank, UK	*Miscanthus*	60–78	8 months	Laboratory	Hartley et al. (2009)
Uranium mine sites, Saxony, Germany	Duckweed (*Lemna gibba*)	0.15	21 days	Laboratory	Mkandawire and Dudel (2005)
As-contaminated sites, Bangladesh	Water hyacinth (*Eichhornia crassipes* L.), barnyard grass (*Echinochloa crusgalli* L.), and water taro (*Monochoria hastata* L.)	100	45 days	Pot study	Islam et al. (2013)

observed genotypic variations within *P. vittata* that could be useful in breeding improved cultivars. Similarly, Aldrich et al. (2007) studied the accumulation of As(III) and As(V) by mesquite a desert plant species and its possible use for decontamination of As-contaminated soils. During experiment, mesquite seedlings were grown in agar-based medium containing 5 mg/L of either As(III) or As(V). The results of the study revealed that the As(V) concentration was significantly higher than the As(III) concentration in all parts of the plant. It happened as As(V) was reduced to As(III) inside the mesquite plant revealed by X-ray absorption spectroscopy (XAS). Therefore the mesquite plants could be a potential candidate for the phytoremediation of As-contaminated soils in arid regions.

A greenhouse experiment with *P. vittata* showed that frond As concentration increased from 29.4 to 15,861 mg kg^{-1} in soil spiked with 1500 mg kg^{-1} As in 2 weeks showing the high translocation of As in the fern (Ma et al. 2001a, b). In another study, Tu and Ma (2002) reported that the fern (*P. vittata)* was also able to remove As from soils containing different As species at different As concentrations. A very large percentage of the As (75–93%) accumulated in the fronds showing that the fern was able to translocate As within the plant (Ma et al. 2001a, b; Zhang et al. 2002; Lombi et al. 2002; Tu et al. 2002). Many plants have been reported to tolerate As in As-contaminated soils but are not hyperaccumulators because they accumulate As only in their roots (Bondada and Ma 2003; Srivastava et al. 2006).

Alvarado et al. (2008) observed the elimination of As (0.15 mg/L) from polluted water by two aquatic plants, i.e., water hyacinth (*Eichhornia crassipes*) and lesser duckweed (*Lemna minor*). Two plant densities were used: 1 kg/m^2 and 4 kg/m^2 (on a wet basis) for lesser duckweed and water hyacinth, respectively. The results showed that water hyacinth removed 18% of As and the removal rate was 600 mg As/ha/d. The results showed that *L. minor* removed 140 mg As/ha/d (5%) from the As-polluted water.

Salido et al. (2003) performed field and greenhouse trials to assess the performance of phytoremediation of two metals (As and lead) from polluted soil at an EPA Superfund site (*Barber Orchard*). To remove As Chinese brake ferns (*Pteris vittata*) were used in the study. Results showed that the concentration of As in shoots of ferns was about 20 times higher than the soil As concentration under field conditions. According to the study, to bring the As concentration to safe levels (40 mg/kg), it was estimated that 8 years would be required. Results also indicated that reduction in the acid-extractable quantity of soil As and increase in the pH of soil may improve As removal from the contaminated site.

12.9.2 Arsenic Remediation Using Bacteria

Plants often exist in mutual relationships with other microorganisms, especially bacteria and fungi (Fayiga and Saha 2016). It is important to know the effects and role of As accumulation on these microbes. It is also important to know what role these microbes play in As uptake by the fern. Arsenic-resistant bacteria have been isolated in the fronds and rhizosphere of *P. vittata* (Rathinasabapathi et al. 2006; Huang et al. 2010).

Arsenic bioremediation depends on microbial action to detoxify, mobilize, or immobilize As through oxidation-reduction, biomethylation, sorption, and complexation processes (Wang and Zhao 2009). In prokaryotes and unicellular eukaryotes, arsenate (AsV) and arsenite (AsIII) enter the cell through phosphate (Pi) transporters and aquaglyceroporins, respectively. In bacteria, genes localized in the ars operon regulate As resistance. It typically comprises three genes, arsR, arsB, and arsC, encoding three proteins, which convert arsenate to arsenite and extrudes arsenite from the cells (Rosen 2002; Tripathi et al. 2007).

Takeuchi et al. (2007) conducted a study on nine bacterial strains of marine and nonmarine origins for As resistance and removal. The As-resistant and accumulating bacteria are extensively present in the marine and nonmarine aquatic environments. Further, they reported that As-accumulating bacterial species, such as *Marinomonas communis*, were possible candidates for remediation of As-polluted aquatic mediums. During the experiment, *M. communis* indicated the high As tolerance with median effective concentration (EC50) value of 510 mg Asl^{-1} and was capable of removing As from culture medium amended with arsenate. Salmassi et al. (2002) isolated a heterotrophic bacterial strain (*Agrobacterium albertimagni* strain AOL15) which can oxidize As(III) from the surface of aquatic macrophytes collected in a Hot Creek. The isolated bacterial strain can oxidize 585 μM As(III) within 24 h in mannitol medium in laboratory conditions.

Similarly, Liu et al. (2011) piloted a laboratory study for As removal from contaminated soil by using genetically modified bacteria. The study showed that those bacteria which have expressed arsM gene efficiently removed As through volatilization from the As-contaminated soil. Further, it was reported that use of genetically engineered microorganisms is a cost-effective and capable approach for As bioremediation from contaminated sites.

In past years, many microorganisms were isolated from As-contaminated sites, and these were potentially involved in As removal (Routh et al. 2007). Results of the study showed that *A. bolidensis* (a novel gram-positive, facultatively anaerobic, coccus-shaped actinomycete) actively reduced As(V) to As(III) in aqueous media in laboratory conditions.

Dey et al. (2016) isolated two rod-shaped gram-positive bacteria, from As-contaminated groundwater of West Bengal, India. These isolated bacteria (*Bacillus* sp. and *Aneurinibacillus aneurinilyticus*) can tolerate arsenate and arsenite concentration up to 4500 ppm and 550 ppm, respectively. The results showed that the isolates can remove 51.4–51.9% of arsenite and 50.3–53.37% of arsenate, respectively, from a As amended culture media. The isolated bacterial species were As resistant and can be used asa potential candidate for the bioremediation of As.

12.9.3 Arsenic Remediation Using Algae

Phytoremediation is the use of algae (micro or macro) for the biotransformation of contaminants, comprising metals, nutrients, and xenobiotics from industrial liquid effluents and from terrestrial polluted sites. The alga is commonly present on earth and has adapted to a diversity of habitats, with wide tolerance to environmental conditions. Microalgae have the capability of removing environmental toxicants such as heavy metals, hydrocarbons, and pesticides through various mechanisms, such as biosorption, bioconcentration, biotransformation, and biovolatisation.

Becker (1983) observed that some planktonic algae have a great capability for the remediation of metal pollutants from wastewater. These algae absorb or remove heavy metal residues from wastewaters and the separation of the metal-saturated algae from the medium is also done at low cost, resulting in good-quality reusable effluent water and valuable biomass which could be used for various uses. The efficiency of the phytoremediation depends on several parameters, i.e., growth rate of algae, transfer factor achieved by the algae, the quantity of heavy metal in the medium, preferred extent of metal removal from the medium, and metal recovery in relation to investment and operating expenses (Becker 1983).

Microalgae can potentially be used in remediation of As from contaminated media, due to their capability to bioaccumulate and biotransform As (Bahar et al. 2013). Recently, Wang et al. (2013) examined two freshwater green algae species (*Chlamydomonas reinhardtii* and *Scenedesmus obliquus*) to assess the As bioaccumulation kinetics under different conditions. They found that the As bioaccumulation was significantly greater in phosphate-limited conditions as compared to phosphate-enriched conditions. This may be due to the competition between As(V) and phosphate for the identical transporter to enter the cells due to their structural resemblance.

12.9.4 Arsenic Remediation Using Fungi and Yeast

Mycorrhizal fungi and plant association help plants to reduce metal toxicity, by binding metals present in the contaminated soil to fungal hyphae (Koslowsky and Boerner 1989; Gadd 1993). So the plants grown in metal-contaminated soil and infected by fungi show better resistance than normal plants. As a result, plants infected by these fungi show a higher degree of resistance to metals, because the metals are accumulated in fungal hyphae and are not translocated to plant parts (Bradley et al. 1981; Brown and Wilkins 1985; Dehn and Schüepp 1989).

Several other As accumulator fungal species have been reported by different authors, such as *Scopulariopsis brevicaulis* (Gosio 1892), *Phaeolus schweinitzii* (Pearce et al. 1998), *Fusarium oxysporum* (Granchinho et al. 2002), *Sinorhizobium meliloti* (Carrasco et al. 2005), *Neosartorya fischeri*, *Aspergillus clavatus* (Cernansky et al. 2007, 2009), *Aspergillus candidus* (Vala 2010), *Aspergillus niger* (Mukherjee et al. 2010), *Trichoderma* sp., *Neocosmospora* sp., *and Rhizopus* sp. (Srivastava et al. 2011).

In a study, Chen et al. (2007) studied the effect plant phosphorus on arbuscular mycorrhizal fungi Glomus mosseae and As accumulation by *Medicago sativa*. The phosphorous and As interactions were also studied in plants. The results of the study indicate that due to fungal colonization, plant dry weight increased more than six-folds and also considerably increased in total uptake (phosphorus and As contents). Notwithstanding of phosphorus and As addition levels, phosphorous content was two times higher in shoot and root of both the arbuscular mycorrhizal fungi and plants. The results also showed that an As concentration was significantly lesser than corresponding uninoculated controls. The decrease in As concentration in the shoot was due to the "dilution effects" caused by reduced As partitioning to upper parts of the plant and stimulated the growth of arbuscular mycorrhizal fungi.

In another study, the arsenate tolerance level in a fungus (*Aspergillus niger*) was observed by Mukherjee et al. (2010). The results point out that *A. niger* had a great arsenate uptake capacity and tolerance. Further, they reported that the *A. niger* can tolerate oxidative stress by influencing its antioxidative defense mechanism and may be used for the removal of arsenate from contaminated water.

In yeasts, As tolerance is provided by three connecting genes in the cluster ACR1, ACR2, and ACR3: ACR1 encrypts a putative transcription factor; ACR2 encrypts an arsenate reductase, and ACR3 encrypts a plasma membrane AsIII-efflux transporter. This approach confirms the reduction and removal of As(V) from the cytosol to the exterior medium. The additional mechanism that functions in yeast for the removal of cytosolic As is an ABC-type transporter, yeast cadmium factor, which is positioned at the vacuolar membrane and sequesters glutathione conjugates of AsIII (AsIII–GS3) in the vacuole (Ghosh et al. 1999).

12.10 Conclusion

Among all the metalloids, Arsenic (As) is recognized as the leading toxicant worldwide and having various toxic effects on human and animal health as well as on the environment. Exposure of As may be by various routes including direct inhalation from the atmosphere, ingestion through contaminated food and water and dermal absorption. It causes various types of health effects like dermal and neurological involvement, reproductive and pregnancy effects, cardiovascular effects, diabetes mellitus, diseases of the respiratory system, multiorgan cancers etc. The source and occurrence and accumulation of As in the environment could be natural and anthropogenic causing environmental contamination. Due to its harmful effects, it has been investigated in order to find an eco-friendly technique for the removal of As contamination from the natural resources. Bioremediation process in this regards is an option that offers the possibility to reduce or render trace and toxic elements such as As using plants and microbes. Among the various bioremediation processes, phytoremediation and bioremediation using microbes are quite effective. Phytoremediation includes the removal of contaminants with the help of green plants, while the microbial bioremediation includes the removal of trace and toxic elements by microorganisms (bacteria, fungi, yeast and algae) as sorbents.

References

Acharyya SK, Shah BA (2007) Groundwater arsenic contamination affecting different geologic domains in India–a review: influence of geological setting, fluvial geomorphology and quaternary stratigraphy. J Environ Sci Health A 42:1795–1805

AGRG (1978) The Wolfson geochemical atlas of England and Wales. Clarendon Press, Oxford

Ahmed KM, Bhattacharya P, Hasan MA, Akhter SH, Alam SM, Bhuyian MH (2004) Arsenic enrichment in groundwater of the alluvial aquifers in Bangladesh: an overview. Appl Geochem 19:181–200

Ahsan DA, DelValls TA, Blasco J (2009) Distribution of arsenic and trace metals in the floodplain agricultural soil of Bangladesh. Bull Environ Contam Toxicol 82:11–15

Aldrich MV, Peralta-Videa JR, Parsons JG, Gardea-Torresdey JL (2007) Examination of arsenic(III) and (V) uptake by the desert plant species mesquite (*Prosopis* spp.) using X-ray absorption spectroscopy. Sci Total Environ 379(2–3):249–255

Allen A (2001) Containment landfills: the myth of sustainability. Eng Geol 60:3–19

Alvarado S, Guedez M, Lue-Meru MP, Nelson G, Alvaro A, Jesus AC, Gyula Z (2008) Arsenic removal from waters by bioremediation with the aquatic plants water hyacinth (*Eichhornia crassipes*) and Lesser Duckweed (*Lemna minor*). Bioresour Technol 99:8436–8440

Anawar HM, Akai J, Yoshioka T, Konohira E, Lee JY, Fukuhara H, TariKulAlam M, Garcia-Sanchez A (2006) Mobilization of arsenic in groundwater of Bangladesh: evidence from an incubation study. Environ Geochem Health 28:553–565

Bahar MM, Megharaj M, Naidu R (2013) Bioremediation of arsenic-contaminated water: recent advances and future prospects. Water Air Soil Pollut 224:1722

Baker AJ, McGrath SP, Reeves RD, Smith JAC (2000) Metal hyperaccumulator plants: a review of the ecology and physiology of a biological resource for phytoremediation of metal-polluted soils. In: Phytoremediation of contaminated soil and water. Lewis Publishers, Boca Raton, pp 85–107

Becker EW (1983) Limitations of heavy metal removal from waste water by means of algae. Water Res 17(4):459–466

Bharagava RN, Saxena G, Mulla SI, Patel DK (2017a) Characterization and identification of recalcitrant organic pollutants (ROPs) in tannery wastewater and its phytotoxicity evaluation for environmental safety. Arch Environ Contam Toxicol. https://doi.org/10.1007/s00244-017-0490-x

Bharagava RN, Chowdhary P, Saxena G (2017b) Bioremediation: an ecosustainable green technology: its applications and limitations. In: Bharagava RN (ed) Environmental pollutants and their bioremediation approaches, 1st edn. CRC Press/Taylor & Francis Group, Boca Raton, pp 1–22. https://doi.org/10.1201/9781315173351-2

Bharagava RN, Saxena G, Chowdhary P (2017c) Constructed wetlands: an emerging phytotechnology for degradation and detoxification of industrial wastewaters. In: Bharagava RN (ed) Environmental pollutants and their bioremediation approaches, 1st edn. CRC Press/Taylor & Francis Group, Boca Raton, pp 397–426. https://doi.org/10.1201/9781315173351-15

Bhattacharya P, Mukherjee AB, Bundschuh J, Zevenhoven R, Loeppert RH (2007) Arsenic in soil and groundwater environment: biogeochemical interactions, health effects and remediation, Trace metals and other contaminants in the environment (Series Editor Nriagu JO), vol 9. Elsevier, Amsterdam, p 645

Bhattacharya P, Naidu R, Polya DA, Mukherjee A, Bundschuh J, Charlet L (2014) Arsenic in hydrological processes-sources, speciation, bioavailability and management. J Hydrol 518:279–283

Bondada BR, Ma LQ (2003) Tolerance of heavy metals in vascular plants: arsenic hyperaccumulation by Chinese brake fern (*Pteris vittata* L.). In: Chandra S, Srivastava M (eds) Pteridology in the new millenium. Kluwer Academic Publishers, Dordrecht, pp 397–420

Boyle RW, Jonasson IR (1973) The geochemistry of as and its use as an indicator element in geochemical prospecting. J Geochem Explor 2:251–296

Bradley R, Burt AJ, Read DJ (1981) Mycorrhizal infection and resistance to heavy metal toxicity in *Calluna vulgaris*. Nature 292:335–337

Brown MT, Wilkins DA (1985) Zinc tolerance of mycorrhizal Betula. New Phytol 99:101–106

Carrasco JA, Armario P, Pajuelo E, Burgos A, Caviedes MA, López R et al (2005) Isolation and characterization of symbiotically effective rhizobium resistant to arsenic and heavy metals after the toxic spill at the Aznalcóllar pyrite mine. Soil Biol Biochem 37:1131–1140

Cernansky S, Urik M, Sevc J, Khun K (2007) Biosorption and biovolatilization of arsenic by heat-resistant fungi. Environ Sci Pollut Res 14:31–35

Cernansky S, Kolencik M, Sevc J, Urik M, Hiller E (2009) Fungal volatilization of trivalent and pentavalent arsenic under laboratory conditions. Bioresour Technol 100:1037–1040

Chakraborti D, Rahman MM, Paul K, Chowdhury UK, Sengupta MK, Lodh D, Chanda CR, Saha KC, Mukherjee SC (2002) Arsenic calamity in the Indian sub-continent-what lessons have been learned? Talanta 58:3–22

Chakraborti D, Rahman MM, Murrill M, Das R, Patil SG, Sarkar A, Dadapeer HJ, Yendigeri S, Ahmed R, Das KK (2013) Environmental arsenic contamination and its health effects in a historic gold mining area of the Mangalur greenstone belt of Northeastern Karnataka, India. J Hazard Mater 262:1048–1055

Chakraborti D, Rahman MM, Ahamed S, Dutta RN, Pati S, Mukherjee SC (2016) Arsenic groundwater contamination and its health effects in Patna district (capital of Bihar) in the middle Ganga plain, India. Chemosphere 152:520–529

Chakraborti D, Rahman MM, Das B, Chatterjee A, Das D, Nayak B, Pal A, Chowdhury UK, Ahmed S, Biswas BK, Sengupta MK, Hossain MA, Samanta Roy MM, Dutta RN, Saha KC, Mukherjee SC, Pati S, Kar PB, Mukherjee A, Kumar M (2017) Hydrogeol J 25(4):1165–1181

Chandra R, Saxena G, Kumar V (2015) Phytoremediation of environmental pollutants: an eco-sustainable green technology to environmental management. In: Chandra R (ed) Advances in biodegradation and bioremediation of industrial waste, 1st edn. CRC Press/Taylor & Francis Group, Boca Raton, pp 1–30. https://doi.org/10.1201/b18218-2

Chen B, Xiao X, Zhu Y-G, Smith FA, Xie ZM, Smith SE (2007) The arbuscular mycorrhizal fungus *Glomus mosseae* gives contradictory effects on phosphorus and arsenic acquisition by *Medicago sativa* Linn. Sci Total Environ 379:226–234

Chowdhury TR, Basu GK, Mandal BK, Biswas BK, Samanta G, Chowdhury UK, Chanda CR, Lodh D, Roy SL, Saha KC, Roy S (1999) Arsenic poisoning in the Ganges delta. Nature 401:545–546

Datta DK, Subramanian V (1997) Texture and mineralogy of sediments from the Ganges-Brahmaputra-Meghna river system in the Bengal basin, Bangladesh and their environmental implications. Environ Geol 30:181–188

Dehn B, Schüepp H (1989) Influence of VA mycorrhizae on the uptake and distribution of heavy metals in plants. Agricult Ecosyst Environ 29:79–83

Dey U, Chatterjee S, Mondal NK (2016) Isolation and characterization of arsenic-resistant bacteria and possible application in bioremediation. Biotechnol Rep 10:1–7

Dominguez MT, Maranon T, Murillo JM, Schulin R, Robinson BH (2008) Trace element accumulation in woody plants of the Guadiamar Valley, SW Spain: a large-scale phytomanagement case study. Environ Pollut 152:50–59

Fayiga AO, Saha UK (2016) Arsenic hyperaccumulating fern: implications for remediation of arsenic contaminated soils. Geoderma 284:132–143

Finkelman RB, Belkin HE, Zheng B (1999) Health impacts of domestic coal use in China. Proc Natl Acad Sci India Sect A Phys Sci USA 96:3427–3431

Förstner U, Haase I (1998) Geochemical demobilization of metallic pollutants in solid waste-implications for arsenic in waterworks sludges. J Geochem Explor 62:29–36

French CJ, Dickinson NM, Putwain PD (2006) Woody biomass phytoremediation of contaminated brownfield land. Environ Pollut 141:387–395

Gadd GM (1993) Interactions of fungi with toxic metals. New Phytol 124:25–60

Gautam S, Kaithwas G, Bharagava RN, Saxena G (2017) Pollutants in tannery wastewater, pharmacological effects and bioremediation approaches for human health protection and environmental safety. In: Bharagava RN (ed) Environmental pollutants and their bioremediation approaches, 1st edn. CRC Press/Taylor & Francis Group, Boca Raton, pp 369–396. https://doi.org/10.1201/9781315173351-14

Ghosh M, Shen J, Rosen BP (1999) Pathway of As(III) detoxification in *Sachharomyces cerevisiae*. Proc Natl Acad Sci U S A 96:5001–5006

Goldberg S, Glaubig RA (1988) Anion sorption on a calcareous, montmorillonitic soil-arsenic. Soil Sci Soc Am J 52:1297–1300

Gosio B (1892) Action of microphytes on solid compounds of arsenic: a recapitulation. Science 19:104–106

Goutam SP, Saxena G, Singh V, Yadav AK, Bharagava RN (2018) Green synthesis of TiO_2 nanoparticles using leaf extract of *Jatropha curcas* L. for photocatalytic degradation of tannery wastewater. Chem Eng J 336:386–396. https://doi.org/10.1016/j.cej.2017.12.029

Granchinho SCR, Franz CM, Polishchuk E, Cullen WR, Reimer KJ (2002) Transformation of arsenic (V) by the fungus *Fusarium oxysporum* melonis isolated from the alga Fucus gardneri. Appl Organomet Chem 16:721–726

Hartley W, Dickinson NM, Riby P, Lepp NW (2009) Arsenic mobility in brownfield soils amended with green waste compost or biochar and planted with Miscanthus. Environ Pollut 157:2654–2662

Hasan MA, von Brömssen M, Bhattacharya P, Ahmed KM, Sikder AM, Jacks G, Sracek O (2009) Geochemistry and mineralogy of shallow alluvial aquifers in Daudkandi upazila in the Meghna flood plain, Bangladesh. Environ Geol 57:499–511

Huang A, Teplitski M, Rathinasabapathi B, Ma LQ (2010) Characterization of arsenic-resistant bacteria from the rhizosphere of arsenic hyperaccumulator *Pteris vittata* L. Can J Microbiol 56:236–246

Hughes MF, Beck BD, Chen Y, Lewis AS, Thomas DJ (2011) Arsenic exposure and toxicology: a historical perspective. Toxicol Sci 123:305–332

Islam FS, Gault AG, Boothman C, Polya DA, Charnock JM, Chatterjee D, Lloyd JR (2004) Role of metal-reducing bacteria in arsenic release from Bengal delta sediments. Nature 430:68–71

Islam MS, Zaman MWU, Rahman MM (2013) Phytoaccumulation of arsenic from arsenic contaminated soils by *Eichhornia crassipes* L., *Echinochloa crusgalli* L. and *Monochoria hastata* L. in Bangladesh. Int J Environ Prot 3(4):17–27

Jönsson J, Sherman DM (2008) Sorption of As (III) and As (V) to siderite, green rust (fougerite) and magnetite: implications for arsenic release in anoxic groundwaters. Chem Geol 255:173–181

Koslowsky SD, Boerner REJ (1989) Interactive effects of aluminium, phosphorus and mycorrhizae on growth and nutrient uptake of *Panicum virgatum* L (Poaceae). Environ Poll 61:107–125

Kumar M, Rahman MM, Ramanathan AL, Naidua R (2016) Arsenic and other elements in drinking water and dietary components from the middle Gangetic plain of Bihar, India: health risk index. Sci Total Environ 539:125–134

Liu S, Zhang F, Chen J, Sun GX (2011) Arsenic removal from contaminated soil via biovolatilization by genetically engineered bacteria under laboratory conditions. J Environ Sci 23(9):1544–1550

Lombi E, Zhao FJ, Fuhrmann M, Ma LQ, McGrath SP (2002) Arsenic distribution and speciation in the fronds of the hyperaccumulator *Pteris vittata*. New Phytol 156:195–203

Luongo T, Ma L (2005) Characteristics of arsenic accumulation by Pteris and non-Pteris ferns. Plant Soil 277:117–126

Ma L, Komar KM, Tu C (2001a) A fern that hyperaccumulates arsenic-A hardy, versatile, fast-growing plant helps to remove arsenic from contaminated soils. Nature 409:579

Ma LQ, Komar KM, Tu C, Zhang W, Cai Y, Kennelley ED (2001b) A fern that hyper-accumulates arsenic. Nature 409:579

Mandal BK, Suzuki KT (2002) Arsenic round the world: a review. Talanta 58:201–235

Mason B (1966) Principles of geochemistry, 2nd edn. McGraw-Hill, New York

Matschullat J (2000) Arsenic in the geosphere-a review. Sci Total Environ 249:97–312

Meharg AA, Hartley-Whitaker J (2002) Arsenic uptake and metabolism in arsenic resistant and nonresistant plant species. New Phytol 154:29–43

Meharg AA, Rahman MM (2003) Arsenic contamination of Bangladesh paddy field soils: implications for rice contribution to arsenic consumption. Environ Sci Technol 37:229–234

Mkandawire M, Dudel G (2005) Accumulation of arsenic in *Lemna gibba* L. (duckweed) in tailing waters of two abandoned uranium mining sites in Saxony, Germany. Sci Total Environ 336(1–3):81–89

Momplaisir GM, Rosal CG, Heithmar EM (2001) Arsenic speciationmethods for studying the environmental fate of organoarsenic animal-feed additives. U.S. EPA, NERL, Las Vegas, pp 01–11

Mukherjee A, Bhattacharya P, Shi F, Fryar AE, Mukherjee AB, Xie ZM, Jacks G, Bundschuh J (2009) Chemical evolution in the high arsenic groundwater of the Huhhot basin (Inner Mongolia, PR China) and its difference from the western Bengal basin (India). Appl Geochem 24(10):1835–1851

Mukherjee A, Das D, Mondal SK, Biswas R, Das TK, Boujedaini N, Khuda-Bukhsh R (2010) Tolerance of arsenate-induced stress in *Aspergillus niger*, a possible candidate for bioremediation. Ecotoxicol Environ Saf 73:172–182

Mumford AC, Barringer JL, Benzel WM, Reilly PA, Young LO (2012) Microbial transformations of arsenic: mobilization from glauconitic sediments to water. Water Res 46:2859–2868

National Institute of Hydrology (NIH), Roorkee, Central Groundwater Board (CGWB) and Ministry of Water Resources, Government of India (GoI) (2010) Mitigation and remedy of groundwater arsenic menace in India: a vision document by New Delhi. pp 1–7

Norra S, Berner Z, Agarwala P, Wagner F, Chandrasekharam D, Stüben D (2005) Impact of irrigation with As rich groundwater on soil and crops: a geochemical case study in West Bengal Delta Plain, India. Appl Geochem 20:1890–1906

Odutayo OI, Feyisola RT, Godonu KG, Makinde OA (2015) Phytotoxicity level and effects of arsenic phytoextraction using *Helianthus annuus* L. (sunflower). J Nat Sci Res 5(1):37–41

Paivoke AE, Simola LK (2001) Arsenate toxicity to *Pisum sativum*: mineral nutrients, chlorophyll content, and phytase activity. Ecotoxicol Environ Saf 49(2):11–21

Pearce RB, Callow ME, Macaskie LE (1998) Fungal volatilization of arsenic and antimony and the sudden infant death syndrome. FEMS Microbiol Lett 158:261–265

Rahman F, Chen Z, Naidu R (2009) A comparative study of the extractability of arsenic species from silverbeet and amaranth vegetables. Environ Geochem Health 31(1):103–113

Rathinasabapathi B, Raman SB, Kertulis G, Ma LQ (2006) Arsenic-resistant proteobacterium from the phyllosphere of arsenic-hyperaccumulating fern (*Pteris vittata* L.) reduces arsenate to arsenite. Can J Microbiol 52:695–700

Ravenscroft P (2001) Distribution of groundwater arsenic in Bangladesh related to geology. In: Jack G, Bhattacharya P, Khan AA (eds) Groundwater arsenic contamination in the Bengal Delta plain of Bangladesh. KTH Special Publication. TRITA-AMI report 3084. pp 41–56

Ravenscroft P, Burgess WG, Ahmed KM, Burren M, Perrin J (2005) Arsenic in groundwater of the Bengal basin, Bangladesh: distribution, filed relations, and hydrological setting. Hydrogeol J 13:727–751

Ravenscroft P, Brammer H, Richards K (2009) Arsenic pollution: a global synthesis. Wiley-Blackwell, Chichester

Rosen BP (2002) Biochemistry of arsenic detoxification. FEBS Lett 529:86–92

Routh J, Saraswathy A, Collins D (2007) *Arsenicicoccus bolidensis* a novel arsenic reducing actinomycete in contaminated sediments near the Adak mine (northern Sweden): impact on water chemistry. Sci Total Environ 379:216–225

Roychowdhury T, Tokunaga H, Ando M (2003) Survey of arsenic and other heavy metals in food composites and drinking water and estimation of dietary intake by the villagers from an arsenic-affected area of West Bengal, India. Sci Total Environ 308:15–35

Saha D (2009) Arsenic groundwater contamination in parts of middle Ganga plain, Bihar. Curr Sci 97:753–755

Salido AL, Hasty KL, Lim JM, Butcher DJ (2003) Phytoremediation of arsenic and lead in contaminated soil using Chinese brake ferns (*Pteris vittata*) and Indian mustard (*Brassica juncea*). Int J Phytoremediation 5(2):89–103

Salmassi TM, Venkateswaren K, Satomi M, Newman DK, Hering JG (2002) Oxidation of arsenite by *Agrobacterium albertimagni* AOL15, sp. nov., isolated from hot creek, California. Geomicrobiol J 19(1):53–66

Salt DE, Smith RD, Raskin I (1998) Phytoremediation. Ann Rev Plant Physiol Plant Mol Biol 49:643–668

Sanz E, Munoz-Olivas R, Camara C, Sengupta MK, Ahamed S (2007) Arsenic speciation in rice, straw, soil, hair and nails samples from the arsenic-affected areas of Middle and Lower Ganga plain. J Environ Sci Health A Tox Hazard Subst Environ Eng 42:1695–1705

Saxena G, Bharagava RN (2015) Persistent organic pollutants and bacterial communities present during the treatment of tannery wastewater. In: Chandra R (ed) Environmental waste management, 1st edn. CRC Press/Taylor & Francis Group, Boca Raton, pp 217–247. https://doi.org/10.1201/b19243-10

Saxena G, Bharagava RN (2017) Organic and inorganic pollutants in industrial wastes, their ecotoxicological effects, health hazards and bioremediation approaches. In: Bharagava RN (ed) Environmental pollutants and their bioremediation approaches, 1st edn. CRC Press/Taylor & Francis Group, Boca Raton, pp 23–56. https://doi.org/10.1201/9781315173351-3

Saxena PK, Krishnaraj S, Dan T, Perras MR, Vettakkorumakankav NN (1999) Phytoremediation of metal contaminated and polluted soils. In: Prasad MNV, Hagemeyer J (eds) Heavy metal stress in plants-from molecules to ecosystems. Springer, Berlin, pp 305–329

Saxena G, Chandra R, Bharagava RN (2016) Environmental pollution, toxicity profile and treatment approaches for tannery wastewater and its chemical pollutants. Rev Environ Contam Toxicol 240:31–69. https://doi.org/10.1007/398_2015_5009

Saxena G, Purchase D, Mulla SI, Saratale GD, Bharagava RN (2018) Phytoremediation of heavy metal-contaminated sites: environmental considerations, field studies, sustainability and future prospects. J Environ Manag

Shacklette HT, Boerngen JG, Keith JR (1974) Selenium, fluorine, and arsenic in superficial materials of the conterminous United States. US Geol. Surv., Circ, vol 692. US Government Printing Office, Washington, DC

Shah BA (2008) Role of quaternary stratigraphy on arsenic-contaminated groundwater from parts of Middle Ganga Plain, UP–Bihar, India. Environ Geol 35:1553–1561

Singh IB (1996) Late Quaternary sedimentation of Ganga plain foreland basin. Geol Surve India Spec Publ 21:161–172

Singh SK, Ghosh AK (2011) Entry of arsenic into food material-a case study. World Appl Sci J 13:385–390

Smedley PL, Kinniburgh DG (2002) A review of the source, behavior and distribution of arsenic in natural waters. Appl Geochem 17:517–568

Srivastava S, Sharma YK (2013) Arsenic occurrence and accumulation in soil and water of eastern districts of Uttar Pradesh, India. Environ Monit Assess 18:4995–5002

Srivastava S, Singh N (2014) Mitigation approach of arsenic toxicity in chickpea grown in arsenic amended soil with arsenic tolerant plant growth promoting Acinetobacter sp. Ecol Eng 70:146–153

Srivastava M, Ma LQ, Santos J (2006) Three new arsenic hyperaccumulating ferns. Sci Total Environ 364:24–31

Srivastava PK, Vaish A, Dwivedi S, Chakrabarty D, Singh N, Tripathi RD (2011) Biological removal of arsenic pollution by soil fungi. Sci Total Environ 409:2430–2442

Stuben D, Berner Z, Chandrasekharam D, Karmakar J (2003) Arsenic enrichment in groundwater of West Bengal, India: geochemical evidence for mobilization of As under reducing conditions. Appl Geochem 18(9):1417–1434

Takeuchi M, Kawahata H, Gupta LP, Kita N, Morishita Y, Ono Y, Komai T (2007) Arsenic resistance and removal by marine and non-marine bacteria. J Biotechnol 127(3):434–442

Tripathi RD, Srivastava S, Mishra S, Singh N, Tuli R, Gupta DK, Maathuis FJM (2007) Arsenic hazards: strategies for tolerance and remediation by plants. Trends Biotechnol 25(4):158–165

Tu C, Ma LQ (2002) Effects of arsenic concentrations and forms on arsenic uptake by the hyperaccumulator ladder brake. J Environ Qual 31:641–647

Tu C, Ma L (2003) Effects of arsenate and phosphate on their accumulation by an arsenic hyperaccumulator *Pteris vittata* L. Plant Soil 249:373–382

Tu C, Ma LQ, Bondada B (2002) Arsenic accumulation in the hyper-accumulator Chinese brake fern and its utilization potential for phytoremediation. J Environ Qual 31:1671–1675

Vala AK (2010) Tolerance and removal of arsenic by a facultative marine fungus *Aspergillus candidus*. Bioresour Technol 101:2565–2567

Vamerali T, Bandiera M, Coletto L, Zanetti F, Dickinson NM, Mosca G (2009) Phytoremediation trials on metal-and arsenic-contaminated pyrite wastes (Torviscosa, Italy). Environ Pollut 157:887–894

Vazquez S, Agha R, Granado A, Sarro MJ, Esteban E, Penalosa JM, Carpena RO (2006) Use of white lupin plant for phytostabilization of Cd and As polluted acid soil. Water Air Soil Poll 177:349–365

Vicky-Singh, Brar MS, Preeti-Sharma, Malhi SS (2010) Arsenic in water, soil, and rice plants in the Indo-Gangetic plains of northwestern India. Commun Soil Sci Plant Anal 41:1350–1360

Wang S, Zhao X (2009) On the potential of biological treatment for arsenic contaminated soils and groundwater. J Environ Manag 90(8):2367–2376

Wang HB, Wong MH, Lan CY et al (2007) Uptake and accumulation of arsenic by 11 Pteris taxa from Southern China. Environ Pollut 145:225–233

Wang Z, Luo Z, Yan C (2013) Accumulation, transformation, and release of inorganic arsenic by the freshwater cyanobacterium *Microcystis aeruginosa*. Environ Sci Pollut Res 20(10):7286–7295

Wenzel WW, Lombi E, Adriano DC (1999) Biogeochemical processes in the rhizosphere: role in phytoremediation of metal-polluted sites. In: Prasad MNV, Hagemeyer J (eds) Heavy metal stress in plants – from molecules to ecosystems. Springer, Heidelberg/Berlin, pp 273–303

WHO (2011) Guideline for Drinking-water Quality, fourth edn. World Health Organization, Geneva

Zhang W, Cai Y, Tu C, Ma LQ (2002) Arsenic speciation and distribution in an arsenic hyperaccumulating plant. Sci Total Environ 300:167–177

Chapter 13
Organophosphate Pesticides: Impact on Environment, Toxicity, and Their Degradation

Sikandar I. Mulla, Fuad Ameen, Manjunatha P. Talwar, Syed Ali Musstjab Akber Shah Eqani, Ram Naresh Bharagava, Gaurav Saxena, Preeti N. Tallur, and Harichandra Z. Ninnekar

Contents

S. I. Mulla (✉)
Department of Biochemistry, Karnatak University, Dharwad, Karnataka, India

CAS Key Laboratory of Urban Environment and Health, Institute of Urban Environment, Chinese Academy of Sciences, Xiamen, People's Republic of China

F. Ameen
Department of Botany and Microbiology, Faculty of Science, King Saud University, Riyadh, Kingdom of Saudi Arabia

M. P. Talwar · H. Z. Ninnekar
Department of Biochemistry, Karnatak University, Dharwad, Karnataka, India

S. A. M. A. S. Eqani
CAS Key Laboratory of Urban Environment and Health, Institute of Urban Environment, Chinese Academy of Sciences, Xiamen, People's Republic of China

R. N. Bharagava · G. Saxena
Laboratory of Bioremediation and Metagenomics Research (LBMR), Department of Microbiology (DM), Babasaheb Bhimrao Ambedkar University (A Central University), Lucknow, Uttar Pradesh, India

P. N. Tallur
Government Arts and Science College, Karwar, Uttara Kannada, Karnataka, India

G. Saxena, R. N. Bharagava (eds.), *Bioremediation of Industrial Waste for Environmental Safety*, https://doi.org/10.1007/978-981-13-1891-7_13

Abstract Organophosphate pesticides are extensively used for the control of weeds, diseases, and pests of crops. Hence, these insecticides persist in the environs and thereby cause severe pollution problems. Synthetic pesticides including organophosphates insecticides are found to be toxic and/or hazardous to a variety of organisms like living soil biota along with valuable arthropods, fish, birds, human beings, animals, and plants. Organophosphate pesticides might be decontaminated quickly through hydrolysis on exposure to biosphere, which are responsible to be significantly influenced by abiotic and/or biotic factors. The bacterial cultures isolated from various places are the major entities in the environment with a unique capability to break down different organophosphate pesticides for their growth. Additionally, a potential engineered strain(s) application for the bioremediation of organophosphate(s) is of great interest. In the current chapter, the published information on organophosphates impact on environment, toxic effects, and the available results of their degradation are discussed.

Keywords Toxicity · Chlorpyrifos · Methyl parathion · Quinalphos · Profenofos · Degradation

13.1 General Introduction

The green revolution has directed to an upsurge in the food production and, however, triggered many environmental problems with the increased use of agrochemicals (including pesticides). The pesticides are classified into four major groups (Table 13.1). First and foremost are the groups of persistent organochlorine pesticides such as dichlorodiphenyltrichloroethane, heptachlor, hexachlorobenzene, etc. Organochlorine insecticides introduced in the1940s are used in various crop protections from the pests. The extensive use of these insecticides, during the 1950s–1970s, interfere with food and nonfood crops such as corn, wheat, and tobacco. Organochlorine pesticides fluctuate in their mechanisms of toxicity due to their differences in chemical structures. These are also known as lipophilic chemicals, and their accumulation in the higher trophic levels leads to biomagnifications with the food chain (Poon et al. 2005). For example, increased concentrations of dichlorodiphenyltrichloroethane and its metabolites have been found in soil, water, and sediment samples (Bould 1995; Miersma et al. 2003; Shen et al. 2005; Yanez et al. 2002).

Table 13.1 Major classes of pesticides

Pesticides classes	Examples	Chemical name	Structure
Organochlorine pesticides	DTT	1,1′-(2,2,2-trichloroethane-1,1-diyl)bis(4-chlorobenzene)	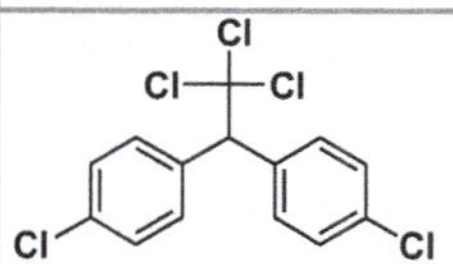
	Dicofol	2,2,2-trichloro-1,1-bis(4-chlorophenyl)ethanol	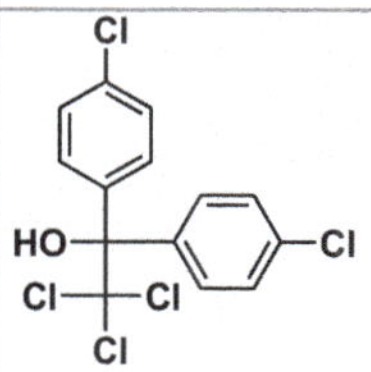
Organophosphate pesticides	Parathion	O,O-diethyl O-(4-nitrophenyl) phosphorothioate	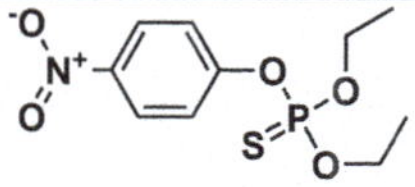
	Diazinon	O,O-diethyl O-[4-methyl-6-(propan-2-yl)pyrimidin-2-yl] phosphorothioate	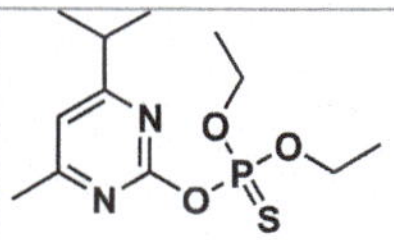
Diethyl 2-dimethoxy	Malathion	2-[(dimethoxyphosphorothioyl) sulfanyl]butanedioate, diethyl	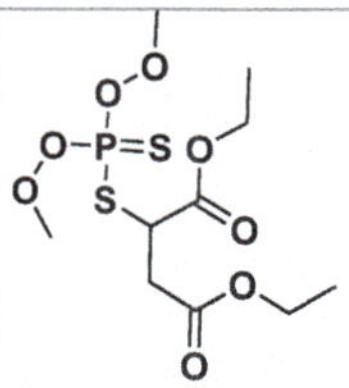
	Profenofos	O-(4-bromo-2-chlorophenyl) O-ethyl S-propyl phosphorothioate	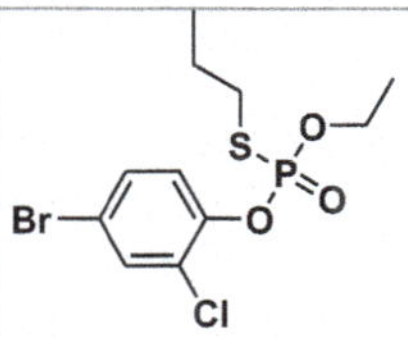
	Quinalphos	*O,O*-diethyl *O*-quinoxalin-2-yl phosphorothioate	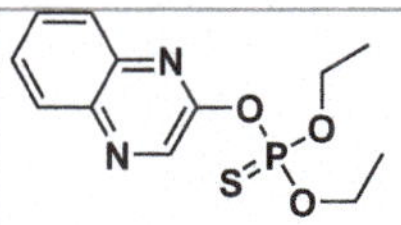
	Methyl parathion	O,O-dimethyl O-4-nitrophenylphosphorothioate	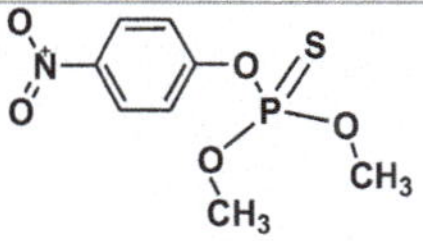
	Chlorpyrifos	O,O-di ethyl-O-(3,5,6-trichloro-2-pyridinyl)- phosphorothioate	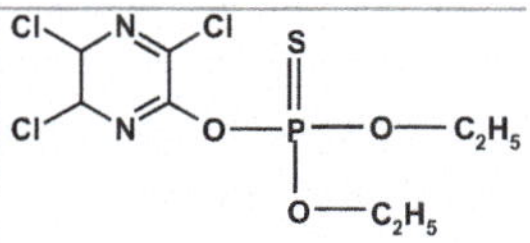

(continued)

Table 13.1 (continued)

Pesticides classes	Examples	Chemical name	Structure
Carbamate pesticides	Carbendazim	Methyl 1H-benzimidazol-2-ylcarbamate	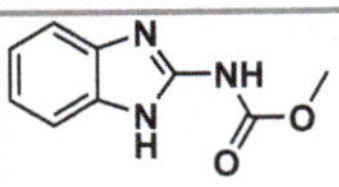
	Carbofuran	2,2-dimethyl-2,3-dihydro-1-benzofuran-7-yl methylcarbamate	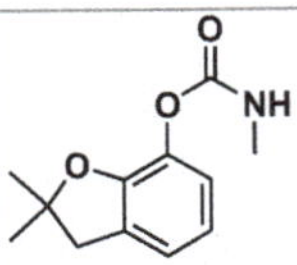
	Carbosulfan	2,2-dimethyl-2,3-dihydro-1-benzofuran-7-yl [(dibutylamino)sulfanyl] methylcarbamate	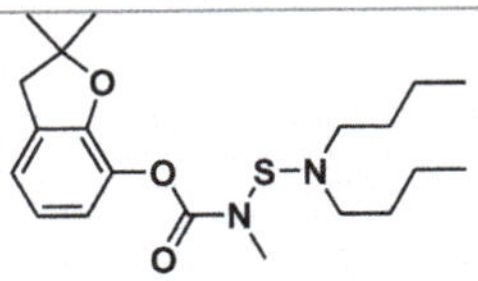
Pyrethroid pesticides	Cyphenothrin	Cyano(3-phenoxyphenyl) methyl 2,2-dimethyl-3-(2-methylprop-1-en-1-yl) cyclopropanecarboxylate	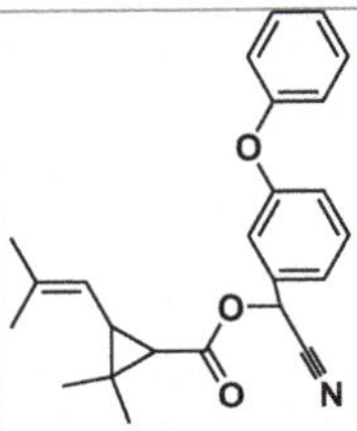
	Cypermethrin	[Cyano-(3-phenoxyphenyl) methyl]3-(2,2-dichloroethenyl)-2,2-dimethylcyclopropane-1-carboxylate	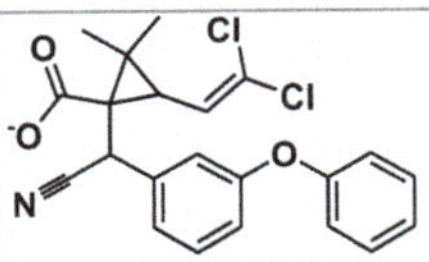

Organophosphates are the second major group of pesticides. The important organophosphate pesticides are malathion, methyl parathion, diazinon, endosulfan, dimethoate, chlorpyrifos, quinalphos, profenofos, and monocrotophos. The third group is carbamate insecticides, based on the carbonic acid. The most recently developed and least persistent of these insecticides belong to pyrethroids, which are derived from the chrysanthemum. In addition to the natural group of insecticides collectively called pyrethrins, some synthetic pyrethroids like cypermethrin, deltamethrin, and fenvalerate insecticides are available under various brand names in the marketplace. These insecticides have rapid knockdown effects and are most frequently used against flying insects (e.g., as aerosols for the control of household insects like flies, mosquitos, etc.). Pesticides with varied chemical nature have been used around the world in the agricultural sector for crop protection from pests, resulting in increased agricultural productivity (Kuo and Regan 1999). On the other hand, their extensive usage leads to the contamination of environmental surroundings (Barcelo 1991).

13.2 Organophosphate Pesticides as Environmental Pollutants

Constantly growing human population significantly depends on agriculture (which represents the world's largest terrestrial biome) for food and nourishment (Mugni et al. 2016). Hence, for food safety, agrochemicals (pesticides, herbicides, and fungicides) are often used in crop production. These agrochemicals, especially pesticides, help to enhance the production of crops by protecting from pests in the course of pre- and post-harvest (Abhilash and Singh 2009). Among the four groups of pesticides, organophosphates are widely used. Some of these pesticides history, half-life period and uses are provided in Table 13.2.

The organophosphate pesticides are used to save crops from pests; however, most of their unused portion as well as their by-products is driven to waste and remains contaminant in the soil, thereby causing loss of fertility, acidification of soil, nitrate leaching, increased resistance of weed species, and loss of biodiversity (Mohapatra 2008; Tilman et al. 2002; Verma et al. 2013).

13.2.1 Chlorpyrifos as an Environmental Pollutant

Chlorpyrifos is introduced in the year 1965 by Dow Chemical Company, USA, and is known by many trade names (including Dursban and Lorsban). The World Health Organization classified chlorpyrifos as class II moderately toxic chemical. It is a

Table 13.2 History, half-life period, and uses of organophosphate pesticides

Pesticide name	Introduction (year)	Half-life period in soil (Days)	Uses
Chlorpyrifos	1965	10–120	Controls Coleoptera, Diptera, Homoptera, and Lepidoptera in soil and on foliage in over large number of crops including rice, cotton, oilseeds, pulses, vegetables, and plantation
Methyl parathion	1949	25–130	Methyl parathion controls boll weevils and many biting or sucking insect pests of agricultural crops, primarily on cotton. It kills insects by contact or stomach and respiratory action
Quinalphos	1969	29–60	Quinalphos applied for controls caterpillars on fruit trees, cotton, vegetables, and peanuts; scale insect on fruit trees and pest complex on rice and also controls aphids, bollworms, borers, leafhoppers, mites, thrips, etc.
Profenofos	1982	7–15	It controls the tobacco budworm, cotton bollworm, armyworm, whiteflies, spider mites, plant bugs, and fleahoppers. Profenofos also control lepidopteron species (the worm complex) at varying rates

broad-spectrum chlorinated organophosphate insecticide (Yadav et al. 2016). It is used in agriculture as a nematicide and acaricide for pest control on various crops. The chlorpyrifos persists for long period in soil and water, because of its nonpolar nature and readily soluble in organic solvents. In addition to the unused chlorpyrifos applied directly in the surroundings, pollution of soil can also be generated in the progress of handling the insecticide in the farmyard as well as in the containers (Yadav et al. 2016). Moreover, due to its slow degradation rate, chlorpyrifos can persist for long periods in soil and thereby affect a substantial risk to the ecosystem (Kulshrestha and Kumari 2011; Singh and Walker 2006; Yadav et al. 2016).

13.2.2 Methyl Parathion as an Environmental Pollutant

Methyl parathion (an insecticide) is extensively used in agriculture crops, primarily cotton, emulsion concentrate, granular food packing, and pest control management, because of its effectiveness toward insect pests (Abhijith et al. 2016). Nevertheless, the uncontrolled usage of methyl parathion may cause potential risk to the aquatic organisms and interfere with the general health, reproductive, and developmental process (Rico et al. 2010). Methyl parathion was detected in many water samples (Diagne et al. 2007). In addition, the accumulation of methyl parathion and its residues in various components of aquatic surroundings has been reported (Diagne et al. 2007; Huang et al. 2011). It is also polluted dairy products (Patnaik and Padhy 2016; Srivastava et al. 2011). On the basis of methyl parathion toxic effect and residue concentration, it has been classified as extremely hazardous and is listed in the HazDat database of chemicals detected in surface and/or groundwater at National Priorities List (NPL) sites (WHO 2004), as a result, encouraging numerous nations to ban or control its usage. Though, methyl parathion is still misused in several developed nations (Ghosh et al. 2010).

13.2.3 Quinalphos as an Environmental Pollutant

Quinalphos is a synthetic, non-systemic, and broad-spectrum organophospate pesticide and used extensively to control pests of a variety of crops such as cotton, paddy, peanuts, coffee, cocoa, soya beans, tea plantation, vegetables, and fruit trees for controls of caterpillars, scale insect, aphids, bollworms, borers, leafhoppers, mites, and thrips (Talwar et al. 2014). However, merely 1% of the used chemical (pesticide) interacted with target insect, whereas the rest of the chemical floats into the environmental surroundings (Gangireddygari et al. 2017). The large-scale usage of quinalphos poses a health hazard to animals and human beings, because of its persistence in the soil and crops (Katti and Verma 1992; Talwar et al. 2014).

13.2.4 Profenofos as an Environmental Pollutant

Profenofos is a non-systemic and broad-spectrum organophosphate insecticide. It is widely used to control lepidopteron insects, whiteflies, aphids, hoppers, and spider mites from a variety of crops including cotton, corn, sugar beet, soybeans, potatoes, vegetables, and tobaccos (EPA 2012; Reddy and Rao 2008; Talwar and Ninnekar 2015). Profenofos is a contaminant in a wide range of aquatic and terrestrial ecosystems (Safiatou et al. 2007; Talwar and Ninnekar 2015). Harnpicharnchai et al. (2013) reported that the average value of profenofos in soil was about 0.041 mg kg^{-1} in summers whereas 0.016 mg kg^{-1} in winters. In addition, profenofos pesticide residue was also detected in water, sediments, as well as in muscle tissues of *Cyprinus carpio* (Mahboob et al. 2013).

13.3 Toxicity of Pesticides

In most instances, various pesticides affect the human beings and animals health due to their capability to interact with living system especially endocrine system (Munoz-de-Toro et al. 2006). Moreover, some of these insecticides were easily transferred from nursing mothers to children through breast milk (Munoz-de-Toro et al. 2006). Carbamate pesticides are related to organophosphates by their mode of action, but the dose required to produce minimum poisoning symptoms and mortality in human beings is higher for carbamate compounds than for organophosphate compounds (Goldberg et al. 1963; Vandekar et al. 1971).

13.3.1 Toxicity of Organophosphate Pesticides

Organophosphates are the one of a major group of pesticides. These chemicals are neurotoxic that act by inhibiting acetylcholine esterase in the central and peripheral nervous system, resulting in choline and acetate formation (Elersek and Filipic 2011). Further, nerves are significantly enhanced and blocked. This suppression leads to convulsion, paralysis, and lastly death for insects and mammals (Singh and Walker 2006). Additionally, organophosphates also bear the potentiality to cause genotoxic and carcinogenic effects (Kaushik and Kaushik 2007).

13.3.1.1 Toxicity of Chlorpyrifos Pesticide

Chlorpyrifos is moderately toxic to human beings, because, it acts on the nervous system by inhibiting acetylcholinesterase activity (Reiss et al. 2012; Schuh et al. 2002). There are reports of genotoxic and mutagenic effects of chlorpyrifos in human beings (Sandal and Yilmaz 2011; Sobti et al. 1992) and rat (Ojha et al. 2013). Nasr et al. (2016) reported that the chlorpyrifos has the tendency to affect significant oxidative damage in brain and kidney of rat. There is an increased risk of various cancers in pesticide applicators, in particular colorectal (Lee et al. 2007), breast (Engel et al. 2005), lymphoma (Karunanayake et al. 2012), prostate (Alavanja et al. 2003), hematopoietic, leukemia, and brain cancers (Lee et al. 2004). Additionally, there is an evidence of immunotoxicity, including the effects on lymphocytes (Blakley et al. 1999) and thymocytes (Prakash et al. 2009). This epidemiological evidence has been linked to neurological effects, persistent developmental disorders, as well as autoimmune disorders. However, many countries have recognized the hazards of chlorpyrifos and have slowly limited or banned their usage. Recently, Jegede et al. (2017) reported that changes in temperature can influence the toxicity of chlorpyrifos toward soil microarthropods.

13.3.1.2 Toxicity of Methyl Parathion Pesticide

Human beings exposed to methyl parathion reported headaches, nausea, sleeplessness, diarrhea, restlessness, breathing problem, dizziness, abdominal cramps, excessive sweating, and mental confusion (Rubin et al. 2002). The toxicity of methyl parathion is associated with hindering acetylcholinesterase (the enzyme responsible for the hydrolysis of the acetylcholine) in mammals especially human beings and pests leading to severe health complications (Liu et al. 2016b). In previous studies, researchers reported that when fish are exposed to methyl parathion, changes were observed in acetylcholinesterase activity, hematological and biochemical parameters (Duquesne and Kuester 2010; Uzunhisarcikli et al. 2007). Moreover, Abhijith et al. (2016) reported that an acute and sublethal dose of methyl parathion induces substantial variations in the enzymatic profiles (in *Catla catla*).

13.3.1.3 Toxicity of Quinalphos Pesticide

Quinalphos is an insecticide affecting acetylcholinesterase inhibition with interaction and also on stomach and respiratory system (Yashwantha et al. 2016). The toxicological effects of quinalphos in rats and other animals have been well documented (Dwivedi et al. 1998). For example, quinalphos (at doses of 1.5 mg kg^{-1} body weight) administered to pregnant rats produced inhibition of acetylcholinesterase activity in fetal brain and placenta, indicating a possible transfer of pesticide from dams to fetuses (Srivastava et al. 1992). In addition, it is also adversely affects the activity of testicular steroidogenic enzymes and thereby causes degeneration of germ cell and reduction in sperm count (Ray et al. 1992). However, quinalphos is primarily metabolized by desertification to quinoxalin-2-ol and phosphorothioate, of that approximately 87% of quinoxalin-2-ol is excreted through urine and the remaining exists in the bile duct. Debnath and Mandal (2000) reported that quinalphos is an environmental xenoestrogenic insecticide, which interferes with the expression of the sex hormones leading to abnormalities in mammals. Moreover, quinalphos is also showed at certain concentration; it becomes toxic in female reproduction (Khera et al. 2016). In another study, a research group reported that quinalphos will be hazardous to silver barb, *Barbonymus gonionotus* (Sadiqul et al. 2016).

13.3.1.4 Toxicity of Profenofos Pesticide

The presence of profenofos residue in the soil poses high environmental risk due to its adverse impact on biosphere (Fosu-Mensah et al. 2016; He et al. 2010). Thus, human populations are certainly exposed to profenofos residue and its by-products. For example, a study reported the presence of profenofos and its intermediate (4-bromo-2-chlorophenol) in human plasma and urine (Gotoh et al. 2001). In another study, a research group demonstrated *in vitro* toxic profile of profenofos by using lymphocytes from peripheral blood samples of healthy human donors (Prabhavathy Das et al. 2006). In addition, profenofos is also highly toxic to fish and invertebrates (Talwar and Ninnekar 2015). The high-level exposure to profenofos causes hepatocellular injury (Gomes et al. 1999). Moreover, high doses of the profenofos induced tissue vacuolization, hemorrhage, and hyperplasia of kupffer cells in the liver. In adddion, swelling of Bowman's capsules and tubular degeneration in the kidney were also documented (Fawzy et al. 2007). It is also able to induce oxidative stress; this may be an earlier diagnostic index in profenofos poisoning (Lin et al. 2003). Likewise, Ruparrelia et al. (1986) reported that semi-static exposure of profenofos was used to understand the toxic effect in aquatic environment, with the special importance on behavioral, morphological, and target enzyme interaction and bioaccumulation of the toxicant in various areas of the body of *Oreochromis mossambicus (Tilapia)*. Furthermore, in chromosomal experimental investigation, samples of the metaphase plates were treated with sublethal doses of profenofos shown in satellite links and chromatid disruptions and gaps, demonstrating the effect of profenofos on chromosomes (Kushwaha et al. 2016).

13.4 Bacterial Degradation of Organophosphate Pesticides

Bioremediation is a process in which microorganisms and plants are used as biological mediators to detoxify toxic/hazardous organic and inorganic chemicals into less risky smaller compounds (Bharagava et al. 2017a, b; Saxena and Bharagava 2017; Chandra et al. 2015; Liu et al. 2007). It is an environmental-friendly and greatly effectual method that can be used as a substitute to chemical and physical methods (Gilani et al. 2016). Pesticide pollutants can be degraded either by biotic and/or abiotic pathways. However, biodegradation of such chemicals by organisms is the primary mechanism in different soils. Hence, it is an advantageous process in the developmental strategies for bioremediation of pesticides contaminated soil, sediment, and water (Qiu et al. 2006). Numerous reports are available on degradation of different class of pesticides (Mulla et al. 2016; Tallur et al. 2015; Talwar and Ninnekar 2015). The successful removal of pesticides (including chlorpyrifos, endosulfan, methyl parathion, coumaphos, ethoprop, parathion, diazinon, and dimethoate) by bacteria has been reported (Singh and Walker 2006; Zheng et al. 2013). Isolation of pure bacterial cultures capable of degrading organophosphate pesticides has gained significant attention, because, these bacteria are easily accessible and offer an environmental-friendly method of in situ reclamation (Ortiz-Hernández and Sánchez-Salinas 2010).

The hydrolysis is the most significant step in organophosphate pesticides catabolism, which causes compounds more exposed to further biodegradation, and the mechanism of hydrolysis along with its kinetic characteristics is well presented in literature (Ortiz-Hernández and Sánchez-Salinas 2010). Bacterial isolates having the ability to degrade organophosphate pesticides by metabolically and/or co-metabolically are listed in Table 13.3.

13.4.1 Bacterial Degradation of Chlorpyrifos

Previous results revealed that in *Flavobacterium* sp. and *Pseudomonas diminuta*, chlorpyrifos degraded co-metabolically in culture medium (Serdar et al. 1982; Sethunathan and Yoshida 1973). In contrast, these strains do not have the ability to utilize chlorpyrifos as a carbon source. The degradation of chlorpyrifos was mediated by soil microorganisms and greatly influenced by abiotic factors (Price et al. 2001). Furthermore, the isolated *Enterobacter* sp. strain B-14 from Australian soil could transform chlorpyrifos to diethylthiophosphoric acid and 3,5,6-trichloro-2-pyridinol (Fig. 13.1) (Singh and Walker 2006).

The isolated *Alcaligenes faecalis* DSP3 (Yang et al. 2005) and *Stenotrophomonas*YC1 (Yang et al. 2006) were shown to be capable of degrading chlorpyrifos and 3,5,6-trichloro-2-pyridinol. In another study, a bacterial strain, *Serratia* sp. (isolated from an activated sludge), can transform chlorpyrifos to 3,5,6-trichloro-2-pyridinol (Xu et al. 2007). Additionally, enhanced degradation of chlorpyrifos by bacterial strain *Arthrobacter*spxz-3 has been reported (Qian et al. 2007). Moreover, the bacterial strains, *Stenotrophomonas* sp. YC-1 and *Sphingomonas* sp. Dsp-2 (isolated from a wastewater effluent of a pesticide-producing division), are correspondingly capable of chlorpyrifos degradation (100%) within a day (Li et al. 2007; Yang et al. 2006). But, *Paracoccus* sp. TRP (isolated from activated sludge sample) mineralizes completely at a given concentration of chlorpyrifos within 4 days. In contrast, a bacterium, *Serratia* sp., is capable to mineralize the same concentration of chlorpyrifos within 18 h only which indicates bacterial strain *Serratia* sp. is highly efficient than *Paracoccus* sp. (Xu et al. 2007, 2008). Additionally, Li and research group isolated various pure bacterial cultures (*Stenotrophomonas* sp., *Bacillus* sp., and *Brevundimonas* sp.) having the ability to degrade chlorpyrifos (Li et al. 2008). Later, Anwar et al. (2009) isolated a bacterium *Bacillus pumilus* strain C2A1 from soil and was found greatly effective

Table 13.3 Bacterial cultures having the capability to degrade organophosphate pesticides either by metabolically and/or co-metabolically

Pesticide	Organisms	References
Chlorpyrifos	*Achromobacter xylosoxidans* (JCp4)	Akbar and Sultan (2016)
	Acinetobacter sp. strain MemCl4	Pailan et al. (2016)
	Acinetobacter calcoaceticus	Akbar et al. (2014)
	Alcaligenes faecalis	Yang et al. (2005)
	Bacillus cereus	Liu et al. (2012)
	Bacillus cereus strain ATCC14579	Ishag et al. (2016)
	Bacillus licheniformis	Zhu et al. (2010)
	Bacillus pumilus	Anwar et al. (2009)
	Bacillus safensis strain FO-36b	Ishag et al. (2016)
	Bacillus sp.	Li et al. (2008)
	Bacillus subtilis	Lakshmi et al. (2008)
	Bacillus subtilis subsp. *inaquosorum* strain KCTC13429	Ishag et al. (2016)
	Brevundimonas sp.	Li et al. (2008)
	Brucella melitensis	Lakshmi et al. (2008)
	Cupriavidus sp.	Lu et al. (2013)
	Enterobacter sp.	Singh et al. (2003)
	Flavobacterium sp. ATCC27551	Mallick et al. (1999)
	Klebsiella sp.	Ghanem et al. (2007)
	Lactobacillus brevis WCP902	Cho et al. (2009)
	Lactobacillus plantarum WCP931	Cho et al. (2009)
	Lactobacillus sakei WCP904	Cho et al. (2009)
	Leuconostoc mesenteroides WCP907	Cho et al. (2009)
	Micrococcus sp.	Guha et al. (1997)
	Ochrobactrum sp. FCp1	Akbar and Sultan (2016)
	Ochrobactrum sp. JAS2	Abraham and Silambarasan (2016)
	Pseudomonas sp.	Yadav et al. (2014)
	Pseudomonas kilonensis SRK1	Khalid et al. (2016)
	Pseudomonas mendocina	Akbar et al. (2014)
	Pseudomonas putida	John et al. (2016)
	Pseudomonas putida KT2440	Gong et al. (2016a)
	Ralstonia sp.	Li et al. (2010)
	Rhizobium sp.	Rayu et al. (2017)
	Serratia	Xu et al. (2007)
	Serratia marcescens	Cycon et al. (2013)
	Sphingomonas sp.	Li et al. (2008)
	Sphingomonas strain HJY	Feng et al. (2017)
	Staphylococcus warneri	John et al. (2016)
	Stenotrophomonas sp. G1	Deng et al. (2015)
	Stenotrophomonas maltophilia	John et al. (2016)

(continued)

Table 13.3 (continued)

Pesticide	Organisms	References
	Stenotrophomonas maltophilia MHFENV20	Dubey and Fulekar (2012)
	Xanthomonas sp.	Rayu et al. (2017)
Methyl parathion	*Acinetobacter radioresistens* USTB-04	Liu et al. (2007)
	Bacillus sp.	Sharmila et al. (1989)
	Burkholderia jiangsuensis	Liu et al. (2016b)
	Citrobacter freundii	Pino and Peñuela (2011)
	Flavobacterium sp.	Pino and Peñuela (2011)
	Flavobacterium balustinum	Somara and Siddavattam (1995)
	Klebsiella sp.	Pino and Peñuela (2011)
	Proteus sp.	Pino and Peñuela (2011)
	Proteus vulgaris	Pino and Peñuela (2011)
	Pseudomonas sp.	Chaudhry et al. (1988)
	Plesiomonas sp. M6	Zhongli et al. (2001)
	Pseudomonas putida	Rani and Lalithakumari (1994)
	Pseudomonas putida X3	Zhang et al. (2016)
	Pseudomonas putida KT2440	Gong et al. (2016b)
	Pseudomonas sp. R1	Sharmila Begum and Arundhati (2016)
	Pseudomonas sp. R2	Sharmila Begum and Arundhati (2016)
	Pseudomonas sp. R3	Sharmila Begum and Arundhati (2016)
	Pseudomonas sp. WBC	Yali et al. (2002)
	Serratia sp. strain DS001	Pakala et al. (2007)
	Stenotrophomonas sp. G1	Deng et al. (2015)
Quinalphos	*Bacillus*	Dhanjal et al. (2014)
	Bacillus thuringiensis	Gangireddygari et al. (2017)
	Ochrobactrum sp.	Talwar et al. (2014)
	Pseudomonas	Pawar and Mali (2014)
	Pseudomonas spp.	Dhanjal et al. (2014)
	Pseudomonas sp.	Nair et al. (2015)
	Pseudomonas aeruginosa Q10	Nair et al. (2015)
	Serratia sp.	Nair et al. (2015)
Profenofos	*Bacillus subtilis*	Salunkhe et al. (2013)
	Burkholderia gladioli	Malghani et al. (2009b)
	Pseudomonas sp.	Salunkhe et al. (2013)
	Pseudomonas aeruginosa strain PF2	Siripattanakul-Ratpukdi et al. (2015)
	Pseudomonas aeruginosa strain PF3	Siripattanakul-Ratpukdi et al. (2015)

(continued)

Table 13.3 (continued)

Pesticide	Organisms	References
	Pseudomonas plecoglossicida strain PF1	Siripattanakul-Ratpukdi et al. (2015)
	Pseudomonas putida	Malghani et al. (2009b)
	Pseudomonas putida (DB17) isolate	
	Pseudoxanthomonas suwonensis strain HNM	Talwar and Ninnekar (2015)
	Stenotrophomonas sp. G1	Deng et al. (2015)

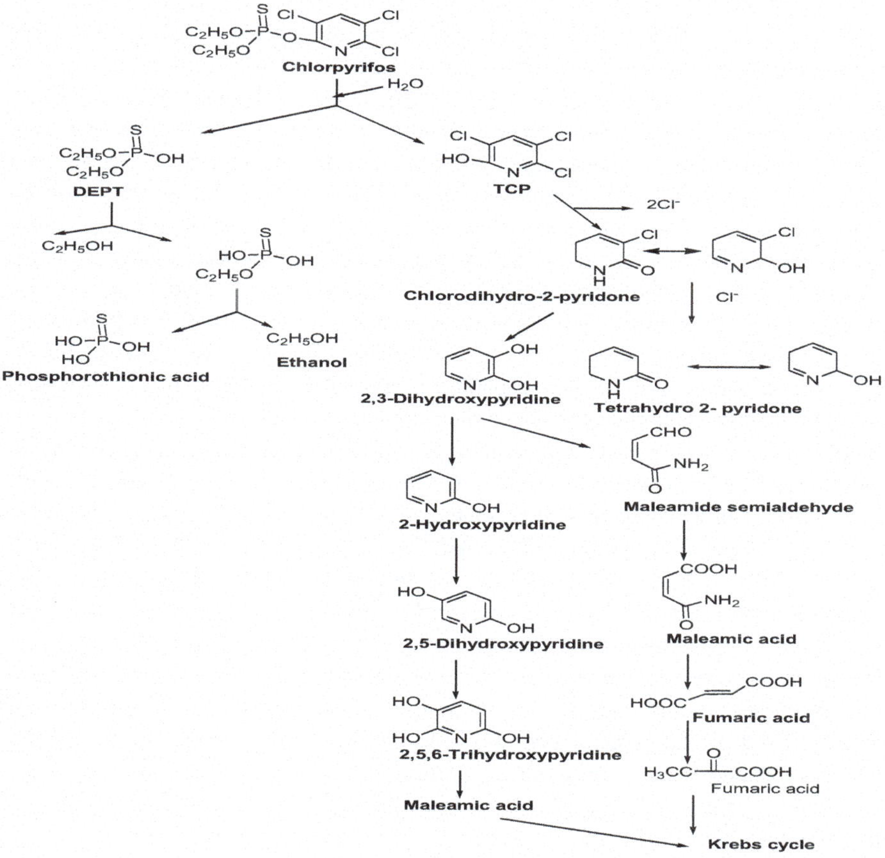

Fig. 13.1 Bacterial degradation of chlorpyrifos (Adapted from Xu et al. 2007; Yadav et al. 2016)

in degrading chlorpyrifos and its hydrolysis by-product 3,5,6-trichloro-2-pyridinol. Dubey and Fulekar (2012) studied *Stenotrophomonas maltophilia* MHF ENV20 (isolated from the *Pennisetum* rhizosphere) potentiality for chlorpyrifos degradation. They reported that the presence of *mpd* gene makes *Stenotrophomonas maltophilia* MHF ENV20 to survive at higher concentration of chlorpyrifos. Cycon et al. (2013) demonstrated that *Serratia marcescens* was competent of degrading chlorpyrifos (at rate constant between 0.017 and 0.052 d^{-1} with $T_{1/2}$ of 13.6–37 days) in various types of soils. In another study, a research group isolated two bacterial strains, namely, *Achromobacter xylosoxidans* JCp4 and *Ochrobactrum* sp. FCp1, demonstrating chlorpyrifos-degradation potential. The authors reported that these organisms were capable to degrade 84.4% and 78.6% of the initial concentration of chlorpyrifos (100 mg L^{-1}) within 10 days (Akbar and Sultan 2016). Abraham and Silambarasan (2016) studied biodegradation of chlorpyrifos and its by-product 3,5,6-trichloro-2-pyridinol by a novel bacterium, *Ochrobactrum* sp. JAS2 (isolated from paddy rhizosphere soil). They reported *mpd* gene responsible for organophosphorus hydrolase production was identified in the bacterium, *Ochrobactrum* sp. JAS2 (Abraham and Silambarasan 2016). On the other hand, Ishag et al. (2016) experimental results revealed that α and β half-lives (days) of chlorpyrifos in *Bacillus safensis* culture were 2.13 and 4.76, respectively. On the other hand, *Bacillus subtilis* as well as *Bacillus cereus* cultures values were 4.09, 9.45, and 4.33, 9.99 for chlorpyrifos, respectively. They also reported that during degradation of chlorpyrifos, no metabolites were detected in *Bacillus subtilis* subsp. *inaquosorum* strain KCTC 13429 as well as *Bacillus cereus* strain ATCC14579 culture medium (Ishag et al. 2016). Conversely, a key intermediate (hydroxy O-ethyl O-3,5,6-trichloropyridin-2-ylphosphorothioate) was detected after biodegradation by *Bacillus safensis* strain FO-36b culture medium (Ishag et al. 2016). Furthermore, a research group reported that the engineered MB285 strain (a solvent-tolerant bacterium, *Pseudomonas putida*) was capable of completely mineralizing chlorpyrifos through direct biodegradation and two intermediates, namely, 3,5,6-trichloro-2-pyridinol and diethyl phosphate, appeared in the culture medium (Liu et al. 2016a). In another study, a bacterial strain (*Acinetobacter* sp. strain MemCl4) having the ability to utilize chlorpyrifos as a sole source of carbon was isolated by enrichment culture technique from an agricultural soil sample, and 3,5,6 trichloro-2-pyridinol was identified as a major intermediate of chlorpyrifos catabolism (Pailan et al. 2016). Rayu et al. (2017) isolated *Xanthomonas* sp., *Pseudomonas* sp., and *Rhizobium* sp. from sugarcane farm soils by enrichment method and reported all three isolates completely mineralize chlorpyrifos (10 mg L^{-1}) in mineral salt media as a sole source of carbon and nitrogen. Recently, Feng et al. (2017) demonstrated chlorpyrifos degradation using endophytic bacterium, *Sphingomonas* sp. strain HJY that was isolated from Chinese chives (*Allium tuberosum* Rottl. ex Spreng). They reported that strain HJY-*gfp* inoculated in Chinese chives showed higher degradation of chlorpyrifos inside the plants than in noninoculated plants.

13.4.2 Bacterial Degradation of Methyl Parathion

Studies on the degradation of methyl parathion by different microorganisms have been reported in the literature (Singh and Walker 2006). Previously, Chaudhry et al. (1988) isolated a bacterium *Pseudomonas* sp. that can co-metabolically degrade methyl parathion. Thereafter, Rani and Lalithakumari (1994) isolated a bacterium

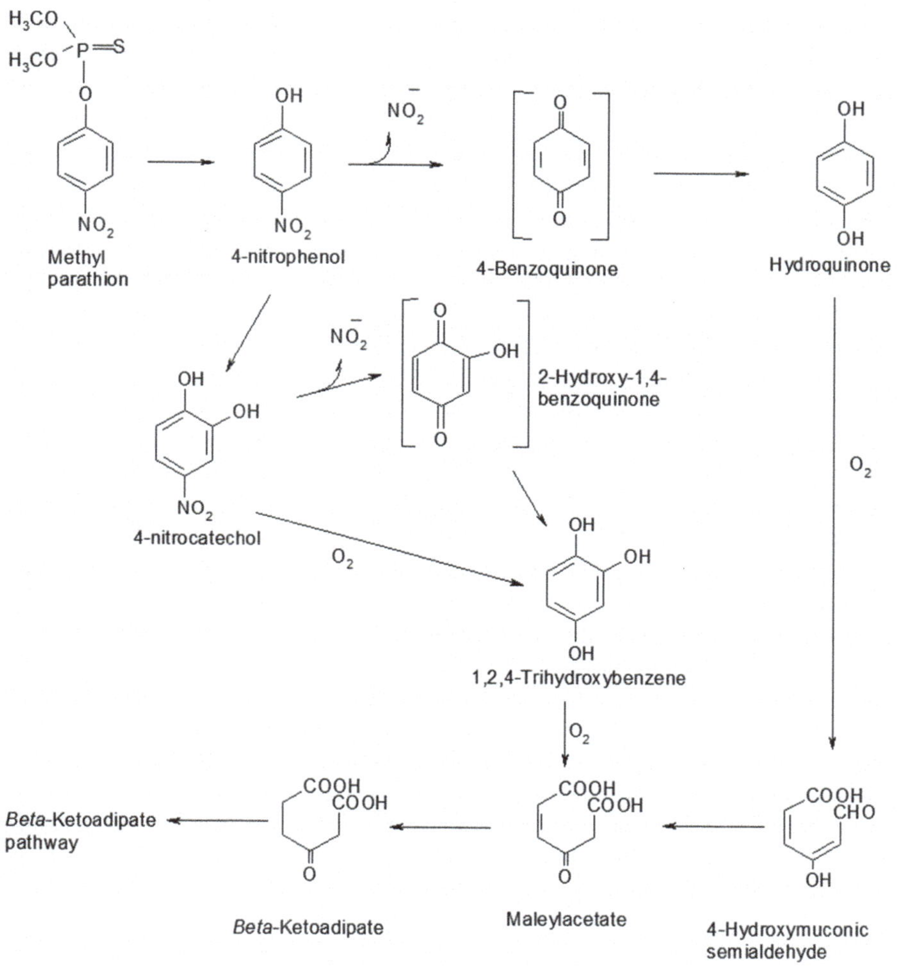

Fig. 13.2 Bacterial degradation of methyl parathion (Adapted from Singh and Walker 2006)

(*Pseudomonas putida*) that can hydrolyze methyl parathion as well as utilize *p*-nitrophenol as a source of carbon and energy (Fig. 13.2).

Later, Somara and Siddavattam (1995) reported that *Flavobacterium balustinum* can also utilize methyl parathion as a sole source of carbon. Additionally, methyl parathion degradation by free- and immobilized-cells of the bacterium (*Pseudomonas* sp.) on sodium alginate beads was studied and reported (Ramanathan and Lalithakumari 1996). On the other hand, Charoensri et al. (2001) studied methyl parathion degradation rates at different conditions including inoculum sizes of bacteria, with and without glucose, pH, salinity, concentrations of methyl parathion, and the metabolism of *p*-nitrophenol. In *Plesiomonas* sp. strain M6 isolate, methyl parathion was transformed to dimethyl phosphorothioate and *p*-nitrophenol by hydrolysis; however, further degradation of *p*-nitrophenol was not observed (Zhongli et al. 2001). Yali et al. (2002) reported *Pseudomonas* sp. WBC (isolated from polluted soils around a Chinese pesticide factory) was capable to mineralize methyl parathion completely and can utilize it as a sole source of carbon and nitrogen. In addition, a soil bacterium, *Serratia* sp. strain DS001, capable of utilizing methyl parathion as the sole source of carbon was isolated by selective enrichment technique. In *Serratia* sp. strain DS001, *p*-nitrophenol and dimethylthiophosphoric acid were observed as main by-products of methyl parathion catabolism (Pakala et al. 2007). In another study, a newly isolated bacterium, *Acinetobacter radioresistens* USTB-04 was used for the degradation of methyl parathion. In a bacterium, methyl parathion (1200 mg L^{-1}) was completely degraded; however, no intermediate was observed during the degradation (Liu et al. 2007). Pino and Peñuela (2011) demonstrated the degradation of the pesticide methyl parathion (150 mg L^{-1}) by bacterial consortium achieved by selective enrichment from highly polluted soils in Moravia (Medellin, Colombia). They reported in the presence of glucose 98% of methyl parathion degradation achieved within 120 h. Additionally, Zhao et al. (2014) investigated an influence of kaolinite and goethite on microbial degradation of methyl parathion. They observed during methyl parathion degradation catabolic activities of *Pseudomonas putida* cells were increased by the presence of kaolinite and decreased by the presence of goethite. On the other hand, Gong et al. (2016b) reported metabolic engineering of *Pseudomonas putida* KT2440 for complete mineralization of methyl parathion. They observed that the strain was genetically stable and its growth was not inhibited. Furthermore, the engineered strain showed higher degradation of spiked methyl parathion (50 mg kg^{-1} soil) in soil samples. In another study, a research group reported that the genetically engineered *Pseudomonas putida* X3 strain can utilize methyl parathion as a sole source of carbon for growth. In an engineered X3 strain, methyl parathion was hydrolyzed to *p*-nitrophenol. However, no further degradation was observed, this might be due to the lack of *p*-nitrophenol degrading genes in X3 strain (Zhang et al. 2016).

13.4.3 Bacterial Degradation of Quinalphos

The hydrolysis of the ester bond connecting the aromatic moiety to dimethyl phosphorothioate in quinalphos leads to 2-hydroxyquinoxaline, which has also been identified as the key metabolite (Fig. 13.3).

Pawar and Mali (2014) experimental results revealed that *Pseudomonas* strain can degrade quinalphos up to 90.4% in the presence of co-substrate (glucose) whereas up to 38.2% observed in the absence of glucose. Moreover, Dhanjal et al. (2014) were isolated *Bacillus* and *Pseudomonas* sp. from different contaminated soils having the ability to degrade quinalphos. They reported that more than 80% of quinalphos was degraded within 17 days in the presence of isolated bacteria; however, no intermediates were observed in the course of the biodegradation process. An organism having the ability to degrade quinalphos was isolated and identified as

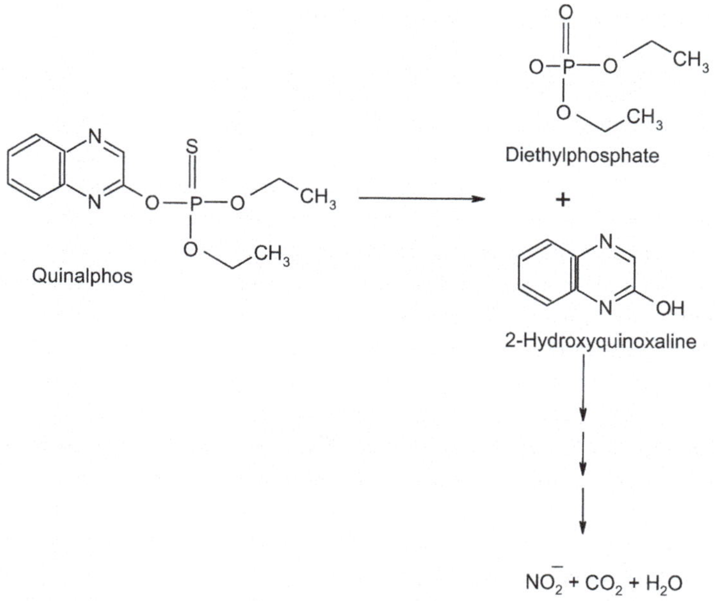

Fig. 13.3 Bacterial degradation of quinalphos (Adapted from Talwar et al. 2014)

Ochrobactrum sp. strain HZM from the pesticide-contaminated soil samples by enrichment on quinalphos as a sole source carbon (Talwar et al. 2014). They reported isolated *Ochrobactrum* sp. strain HZM can utilize various organophosphate pesticides like quinalphos, profenofos, methyl parathion, and chlorpyrifos as carbon sources. Furthermore, they also reported 84.61% of quinalphos degradation (in *Ochrobactrum* sp. strain HZM) can be achieved under the optimum pH 7 and 27 °C by response surface methodology. The degradation of quinalphos in *Ochrobactrum* sp. strain HZM proceeds via hydrolysis to yield 2-hydroxyquinoxaline and diethyl phosphate. Additionally, the gene responsible for organophosphate hydrolase was detected in *Ochrobactrum* sp. strain HZM by PCR technique. Nair et al. (2015) isolated 12 different bacterial strains (having the ability to grow on quinalphos) of which 3 competent isolates such as *Pseudomonas* sp., *Serratia* sp., and *Pseudomonas aeruginosa* degraded quinalphos (at a given concentration) up to 86%, 82%, and 94%, respectively. In *Pseudomonas aeruginosa*, 2-hydroxyquinoxaline and phosphorothioic acid were accumulated during quinalphos degradation (Nair et al. 2015). Recently, Gangireddygari et al. (2017) studied the effect of environmental factors on quinalphos degradation in *Bacillus thuringiensis*. They reported that highest quinalphos degradation was achieved by using an inoculum of 1.0 O.D with optimum pH (6.5–7.5) and 35–37 °C. Furthermore, there results also revealed that addition of yeast extract slightly improves quinalphos degradation rate (Gangireddygari et al. 2017).

13.4.4 Bacterial Degradation of Profenofos

Profenofos has been reported to be degraded by few bacterial strains, *Pseudomonas aeruginosa* (Malghani et al. 2009a), *Pseudomonas putida*, *Burkholderia gladioli* (Malghani et al. 2009b), *Bacillus subtilis* (Salunkhe et al. 2013), and *Stenotrophomonas* sp. G1 (Deng et al. 2015). 4-Bromo-2-chlorophenol was identified as the major intermediate during profenofos catabolism (Fig. 13.4).

On the other hand, this intermediate (4-bromo-2-chlorophenol) offers a sensitive and precise biomarker of profenofos contact (Dadson et al. 2013). The profenofos degradation by *Bacillus subtilis* has been studied in the vineyard soil, but environmental pH of vineyard soil impacts on degradation of profenofos. In addition, degradation is faster in alkaline than the acidic environments; not only soil pH, physicochemical properties of soil, and the microbial diversity may also affect the degradation of profenofos (Salunkhe et al. 2013). In another study, Siripattanakul-Ratpukdi et al. (2015) isolated three bacterial strains, *Pseudomonas plecoglossicida* strain PF1, *Pseudomonas aeruginosa* strain PF2, and *Pseudomonas aeruginosa* strain PF3 having the ability to degrade profenofos. These bacterial strains individually degrade profenofos (20 mg L^{-1}) up to 95.0%, 93.1%, and 95.3% within 96 h, respectively. On the other hand, Talwar and Ninnekar (2015) studied profenofos degradation by free- and immobilized-cells of *Pseudoxanthomonas suwonensis* strain HNM (isolated from pesticide-contaminated soil samples by enrichment technique) in sodium alginate, sodium alginate-polyvinyl alcohol, and sodium alginate-bentonite clay matrices, and they reported that the sodium alginate-bentonite clay immobilized cells showed enhanced degradation rate of profenofos than freely suspended cells and other matrices (Talwar and Ninnekar 2015). Furthermore, Abdullah et al. (2016) studied degradation of profenofos by endogenous bacterial isolates. Their results revealed that isolate DB17 (*Pseudomonas putida)* showed the maximum efficacy to degrade profenofos. Furthermore, in DB 17 isolate, a gene responsible for organophosphate pesticide was detected.

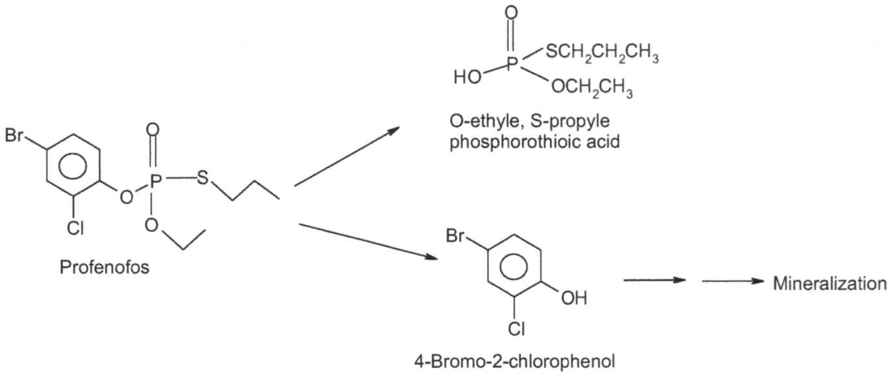

Fig. 13.4 Bacterial degradation of profenofos (Adapted from Talwar and Ninnekar 2015; Kushwaha et al. 2016)

13.5 Conclusion

In view of the extensive pollution of environmental surroundings caused by organophosphate compounds usage along with their toxicity toward biological living systems, considerable attention has been paid to understanding organophosphate pesticides degradation. Biotic mediators (especially bacteria) have a possibility to degrade pesticides into their less toxic by-products. Several bacterial strains that can decompose organophosphate insecticides via metabolism and/or co-metabolism have been isolated and demonstrated. The usage of microbes (biological mediators) is highly efficient as they are environmentally friendly and inexpensive. Certain biological mediators (bacteria) could degrade numerous organophosphate compounds, and some could degrade either single or a small number of such compounds. The organophosphate pesticides hydrolysis decreases the toxicity toward human beings and animals. However, the impact of the subsequent decomposition intermediates on environmental surroundings has not been completely investigated. The mechanisms of different organophosphate pesticides degradation pathways are not yet fully investigated. Hence, this part of investigation issues needs concentrated efforts, as intermediates of several organophosphates catabolism are contaminants and might have a harmful impact on the environmental surroundings as well as nontarget living organisms. Additionally, bioremediation of organophosphates can be further enhanced by the use of engineered microorganisms.

References

Abdullah RR, Ghani SBA, Sukar NA (2016) Degradation of profenofos and λ-cyhalothrin using endogenous bacterial isolates and detection of the responsible genes. J Bioremed Biodegr 7:360

Abhijith BD, Ramesh M, Poopal RK (2016) Responses of metabolic and antioxidant enzymatic activities in gill, liver and plasma of *Catla catla* during methyl parathion exposure. J Basic Appl Zool 77:31–40

Abhilash PC, Singh N (2009) Pesticide use and application: an Indian scenario. J Hazard Mater 165:1–12

Abraham J, Silambarasan S (2016) Biodegradation of chlorpyrifos and its hydrolysis product 3,5,6-trichloro-2-pyridinol using a novel bacterium *Ochrobactrum* sp. JAS2: a proposal of its metabolic pathway. Pestic Biochem Physiol 126:13–21

Akbar S, Sultan S (2016) Soil bacteria showing a potential of chlorpyrifos degradation and plant growth enhancement. Braz J Microbiol 47(3):563–570

Akbar S, Sultan S, Kertesz M (2014) Bacterial community analysis in chlorpyrifos enrichment cultures via DGGE and use of bacterial consortium for CP biodegradation. World J Microbiol Biotechnol 30:2755–2766

Alavanja MC, Samanic C, Dosemeci M, Lubin J, Tarone R, Lynch CF, Knott C, Thomas K, Hoppin JA, Barker J, Coble J, Sandler DP, Blair A (2003) Use of agricultural pesticides and prostate cancer risk in the agricultural health study cohort. Am J Epidemiol 157:800–814

Anwar S, Liaquat F, Khan QM, Khalid ZM, Iqbal S (2009) Biodegradation of chlorpyrifos and its hydrolysis product 3,5,6-trichloro-2-pyridinol by *Bacillus pumilus* strain C2A1. J Hazard Mater 168(1):400–405

Barcelo D (1991) Occurrence, handling and chromatographic determination of pesticides in the aquatic environment. A review. Analyst 116(7):681–689

Bharagava RN, Chowdhary P, Saxena G (2017a) Bioremediation: an ecosustainable green technology: its applications and limitations. In: Bharagava RN (ed) Environmental pollutants and their bioremediation approaches, 1st edn. CRC Press/Taylor & Francis Group, Boca Raton, pp 1–22. https://doi.org/10.1201/9781315173351-2

Bharagava RN, Saxena G, Chowdhary P (2017b) Constructed wetlands: an emerging phytotechnology for degradation and detoxification of industrial wastewaters. In: Bharagava RN (ed) Environmental pollutants and their bioremediation approaches, 1st edn. CRC Press/Taylor & Francis Group, Boca Raton, pp 397–426. https://doi.org/10.1201/9781315173351-15

Blakley BR, Yole MJ, Brousseau P, Boermans H, Fournier M (1999) Effect of chlorpyrifos on immune function in rats. Vet Hum Toxicol 41(3):140–144

Bould HL (1995) DDT residues in the environment-a review with a New Zealand perspective. N Z J Agric Res 38:257–277

Chandra R, Saxena G, Kumar V (2015) Phytoremediation of environmental pollutants: an eco-sustainable green technology to environmental management. In: Chandra R (ed) Advances in biodegradation and bioremediation of industrial waste, 1st edn. CRC Press/Taylor & Francis Group, Boca Raton, pp 1–30. https://doi.org/10.1201/b18218-2

Charoensri K, Esuchart U, Nouwarath S, Pairote P (2001) Degradation of methyl parathion in an aqueous medium by soil bacteria. Sci Asia 27:261–271

Chaudhry GR, Ali AN, Wheeler WB (1988) Isolation of a methyl parathion-degrading *Pseudomonas* sp. that possesses DNA homologous to the *opd* gene from a *Flavobacterium* sp. Appl Environ Microbiol 54:288–293

Cho KM, Math RK, Islam SM, Lim WJ, Hong SY, Kim JM, Yun MG, Cho JJ, Yun HD (2009) Biodegradation of chlorpyrifos by lactic acid bacteria during kimchi fermentation. J Agric Food Chem 57(5):1882–1889

Cycon M, Zmijowska A, Wojcik M, Piotrowska-Seget Z (2013) Biodegradation and bioremediation potential of diazinon-degrading *Serratia marcescens* to remove other organophosphorus pesticides from soils. J Environ Manag 117:7–16

Dadson OA, Ellison CA, Singleton ST, Chi L-H, McGarrigle BP, Lein PJ, Farahat FM, Farahat T, Olson JR (2013) Metabolism of profenofos to 4-bromo-2-chlorophenol, a specific and sensitive exposure biomarker. Toxicology 306:35–39

Debnath D, Mandal TK (2000) Study of quinalphos (an environmental oestrogenic insecticide) formulation (Ekalux 25 E.C.)-induced damage of the testicular tissues and antioxidant defence systems in Sprague-Dawley albino rats. J Appl Toxicol 20(3):197–204

Deng S, Chen Y, Wang D, Shi T, Wu X, Ma X, Li X, Hua R, Tang X, Li QX (2015) Rapid biodegradation of organophosphorus pesticides by *Stenotrophomonas* sp. G1. J Hazard Mater 297:17–24

Dhanjal NIK, Kaur P, Sud D, Cameotra SS (2014) Persistence and biodegradation of quinalphos using soil microbes. Water Environ Res 86:457–461

Diagne M, Oturan N, Oturan MA (2007) Removal of methyl parathion from water by electrochemically generated Fenton's reagent. Chemosphere 66(5):841–848

Dubey KK, Fulekar MH (2012) Chlorpyrifos bioremediation in *Pennisetum* rhizosphere by a novel potential degrader *Stenotrophomonas maltophilia* MHF ENV20. World J Microbiol Biotechnol 28(4):1715–1725

Duquesne S, Kuester E (2010) Biochemical, metabolic, and behavioural responses and recovery of *Daphnia magna* after exposure to an organophosphate. Ecotoxicol Environ Saf 73:353–359

Dwivedi PD, Das M, Khanna SK (1998) Role of cytochrome *P-450* in quinalphos toxicity: effect on hepatic and brain antioxidant enzymes in rats. Food Chem Toxicol 36(5):437–444

Elersek T, Filipic M (2011) Organophosphorous pesticides – mechanisms of their toxicity. In: Stoytcheva M (ed) Pesticides – the impacts of pesticides exposure. InTech

Engel LS, Hill DA, Hoppin JA, Lubin JH, Lynch CF, Pierce J, Samanic C, Sandler DP, Blair A, Alavanja MC (2005) Pesticide use and breast cancer risk among farmers' wives in the agricultural health study. Am J Epidemiol 161:121–135

EPA (2012) Environmental protection agency reregistration eligibility decision for profenofos. http://www.epa.gov/oppsrrd1/REDs/profenofos_red.pdf

Fawzy I, Iman Z, Hamza A (2007) The effect of an Organophosphorus insecticide on the hepatic, renal and pulmonary tissues of mice fetuses Egypt. J Med Lab Sci 16:99–113

Feng F, Ge J, Li Y, Cheng J, Zhong J, Yu X (2017) Isolation, colonization, and Chlorpyrifos degradation mediation of the endophytic bacterium *Sphingomonas* strain HJY in Chinese chives (*Allium tuberosum*). J Agric Food Chem 65(6):1131–1138

Fosu-Mensah BY, Okoffo ED, Darko G, Gordon C (2016) Organophosphorus pesticide residues in soils and drinking water sources from cocoa producing areas in Ghana. Environ Syst Res 5:10

Gangireddygari VSR, Kalva PK, Ntushelo K, Bangeppagari M, Djami Tchatchou A, Bontha RR (2017) Influence of environmental factors on biodegradation of quinalphos by *Bacillus thuringiensis*. Environ Sci Eur 29(1):11

Ghanem I, Orfi M, Shamma M (2007) Biodegradation of chlorpyrifos by *Klebsiella* sp. isolated from an activated sludge sample of waste water treatment plant in Damascus. Folia Microbiol 52(4):423–427

Ghosh PG, Sawant NA, Patil SN, Aglave BA (2010) Microbial biodegradation of organophosphate pesticides. Int J Biotechnol Biochem 6:871–876

Gilani RA, Rafique M, Rehman A, Munis MFH, ur Rehman S, Chaudhary HJ (2016) Biodegradation of chlorpyrifos by bacterial genus *Pseudomonas*. J Basic Microbiol 56:105–119

Goldberg ME, Johnson HE, Knaak JB, Smyth HFJ (1963) Psychopharmacological effects of reversible cholinesterase inhibition induced by *N*-methyl-3-isopropyl-phenyl carbamate (compound 10854). J Pharm exp Ther 141:244–252

Gomes J, Dawodu AH, Lloyd O, Revitt DM, Anilal SV (1999) Hepatic injury and disturbed amino acid metabolism in mice following prolonged exposure to organophosphorus pesticides. Hum Exp Toxicol 18(1):33–37

Gong T, Liu R, Che Y, Xu X, Zhao F, Yu H, Song C, Liu Y, Yang C (2016a) Engineering *Pseudomonas putida* KT2440 for simultaneous degradation of carbofuran and chlorpyrifos. Microb Biotechnol 9(6):792–800

Gong T, Liu R, Zuo Z, Che Y, Yu H, Song C, Yang C (2016b) Metabolic engineering of *Pseudomonas putida* KT2440 for complete mineralization of methyl parathion and gamma-hexachlorocyclohexane. ACS Synth Biol 5(5):434–442

Gotoh M, Sakata M, Endo T, Hayashi H, Seno H, Suzuki O (2001) Profenofos metabolites in human poisoning. Forensic Sci Int 116(2–3):221–226

Guha A, Kumari B, Bora TC, Roy MK (1997) Possible involvement of plasmids in degradation of malathion and chlorpyrifos by *Micrococcus* sp. Folia Microbiol 42:574–576

Harnpicharnchai K, Chaiear N, Charerntanyarak L (2013) Residues of organophosphate pesticides used in vegetable cultivation in ambient air, surface water and soil in Bueng Niam subdistrict, Khon Kaen, Thailand. Southeast Asian J Trop Med Pub Health 44:1088–1097

He J, Fan M, Liu X (2010) Environmental behavior of profenofos under paddy field conditions. Bull Environ Contam Toxicol 84(6):771–774

Huang QY, Huang L, Huang HQ (2011) Proteomic analysis of methyl parathion-responsive proteins in zebrafish (*Danio rerio*) brain. Comp Biochem Physiol C Toxicol Pharmacol 153(1):67–74

Ishag AESA, Abdelbagi AO, Hammad AMA, Elsheikh EAE, Elsaid OE, Hur J-H, Laing MD (2016) Biodegradation of chlorpyrifos, malathion and dimethoate by three strains of bacteria isolated from pesticide-polluted soils in Sudan. J Agric Food Chem 64:8491–8498

Jegede OO, Owojori OJ, Rombke J (2017) Temperature influences the toxicity of deltamethrin, chlorpyrifos and dimethoate to the predatory mite *Hypoaspis aculeifer* (Acari) and the springtail *Folsomia candida* (Collembola). Ecotoxicol Environ Saf 140:214–221. https://doi.org/10.1016/j.ecoenv.2017.02.046

John EM, Sreekumar J, Jisha MS (2016) Optimization of Chlorpyrifos degradation by assembled bacterial consortium using response surface methodology. Soil Sedimentol Contam 25:668–682

Karunanayake CP, Spinelli JJ, McLaughlin JR, Dosman JA, Pahwa P, McDuffie HH (2012) Hodgkin lymphoma and pesticides exposure in men: a Canadian case-control study. J Agromedicine 17(1):30–39

Katti G, Verma S (1992) Persistence of quinalphos against pests under Indian conditions. Pestic Inf 18:37–40

Kaushik P, Kaushik G (2007) An assessment of structure and toxicity correlation in organochlorine pesticides. J Hazard Mater 143(1–2):102–111

Khalid S, Hashmi I, Khan SJ (2016) Bacterial assisted degradation of chlorpyrifos: the key role of environmental conditions, trace metals and organic solvents. J Environ Manag 168:1–9

Khera KS, Kaur J, Sangha GK (2016) Reproductive toxicity of quinalphos on female albino rats: effects on ovary and uterus. Indian J Anim Res 50:537–543

Kulshrestha G, Kumari A (2011) Fungal degradation of chlorpyrifos by *Acremonium* sp. strain (GFRC-1) isolated from a laboratory-enriched red agricultural soil. Biol Fertil Soils 47:219–225

Kuo W, Regan R (1999) Removal of pesticides from rinsate by adsorption using agricultural residuals as medium. J Environ Sci Health B 34:431–447

Lakshmi CV, Kumar M, Khanna S (2008) Biotransformation of chlorpyrifos and bioremediation of contaminated soil. Int Biodeterior Biodegrad 62:204–209

Lee W, Blair A, Hoppin JA, Lubin JH, Rusiecki JA, Sandler DP, Dosemeci M, Alavanja MCR (2004) Cancer incidence among pesticide applicators exposed to chlorpyrifos in the agricultural health study. J Nat Cancer Inst 96:1781–1789

Lee WJ, Sandler DP, Blair A, Samanic C, Cross AJ, Alavanja MCR (2007) Pesticide use and colorectal cancer risk in the agricultural health study. Int J Cancer 121:339–346

Li X, He J, Li S (2007) Isolation of a chlorpyrifos-degrading bacterium, *Sphingomonas* sp. strain Dsp-2, and cloning of the mpd gene. Res Microbiol 158(2):143–149

Li X, Jiang J, Gu L, Ali SW, He J, Li S (2008) Diversity of chlorpyrifos-degrading bacteria isolated from chlorpyrifos-contaminated samples. Int Biodeterior Biodegrad 62:331–335

Li J, Liu J, Shen W, Zhao X, Hou Y, Cao H, Cui Z (2010) Isolation and characterization of 3,5,6-t richloro-2-pyridinol-degrading *Ralstonia* sp. strain T6. Bioresour Technol 101(19):7479–7483

Lin L, Liu J, Zhang K, Chen Y (2003) An experimental study of the effects of profenofos on antioxidase in rabbits. Wei Sheng Yan Jiu 32(5):434–435

Liu FY, Hong MZ, Liu DM, Li YW, Shou PS, Yan H, Shi GQ (2007) Biodegradation of methyl parathion by *Acinetobacter radioresistens* USTB-04. J Environ Sci (China) 19(10):1257–1260

Liu Z, Chen X, Shi Y, Su Z (2012) Bacterial degradation of Chlorpyrifos by *Bacillus cereus*. Adv Mater Res 356–360:676–680

Liu J, Tan L, Wang J, Wang Z, Ni H, Li L (2016a) Complete biodegradation of chlorpyrifos by engineered Pseudomonas putida cells expressing surface-immobilized laccases. Chemosphere 157:200–207. https://doi.org/10.1016/j.chemosphere.2016.05.031

Liu XY, Chen FF, Li CX, Luo XJ, Chen Q, Bai YP, Xu JH (2016b) Improved efficiency of a novel methyl parathion hydrolase using consensus approach. Enzym Microb Technol 93:11–17

Lu P, Li Q, Liu H, Feng Z, Yan X, Hong Q, Li S (2013) Biodegradation of chlorpyrifos and 3,5,6-trichloro-2-pyridinol by *Cupriavidus* sp. DT-1. Bioresour Technol 127:337–342

Mahboob S, Niazi F, Sultana S, Ahmad Z (2013) Assessment of pesticide residues in water, sediments and muscles of *Cyprinus carpio* from head Balloki in the River Ravi. Life Sci J 10:32–38

Malghani S, Chatterjee N, Hu X, Zejiao L (2009a) Isolation and characterization of a profenofos degrading bacterium. J Environ Sci (China) 21:1591–1597

Malghani S, Chatterjee N, Yu HX, Luo Z (2009b) Isolation and identification of Profenofos degrading bacteria. Braz J Microbiol 40:893–900

Mallick BK, Banerji A, Shakli NA, Sethunathan NN (1999) Bacterial degradation of chlorpyrifos in pure culture and in soil. Bull Environ Contam Toxicol 62:48–55

Miersma NA, Pepper CB, Anderson TA (2003) Organochlorine pesticides in elementary school yards along the Texas-Mexico border. Environ Pollut 126(1):65–71

Mohapatra PK (2008) Textbook of environmental microbiology. I.K. International Publishing House Pvt. Ltd, New Delhi

Mugni H, Paracampo A, Demetrio P, Pardi M, Bulus G, Ronco A, Bonetto C (2016) Toxicity persistence of Chlorpyrifos in runoff from experimental soybean plots to the non-target amphipod *Hyalella curvispina*: effect of crop management. Arch Environ Contam Toxicol 70(2):257–264

Mulla SI, Wang H, Sun Q, Hu A, Yu CP (2016) Characterization of triclosan metabolism in *Sphingomonas* sp. strain YL-JM2C. Sci Rep 6:21965

Munoz-de-Toro M, Beldomenico HR, Garcia SR, Stoker C, De Jesus JJ, Beldomenico PM, Ramos JG, Luque EH (2006) Organochlorine levels in adipose tissue of women from a littoral region of Argentina. Environ Res 102(1):107–112

Nair AM, Rebello S, Rishad KS, Asok AK, Jisha MS (2015) Biosurfactant facilitated biodegradation of quinalphos at high concentrations by *Pseudomonas aeruginosa* Q10. Soil Sediment Contam 24:542–553

Nasr HM, El-Demerdash FM, El-Nagar WA (2016) Neuro and renal toxicity induced by chlorpyrifos and abamectin in rats: toxicity of insecticide mixture. Environ Sci Pollut Res Int 23(2):1852–1859

Ojha A, Yaduvanshi SK, Pant SC, Lomash V, Srivastava N (2013) Evaluation of DNA damage and cytotoxicity induced by three commonly used organophosphate pesticides individually and in mixture, in rat tissues. Environ Toxicol 28:543–552

Ortiz-Hernández ML, Sánchez-Salinas E (2010) Biodegradation of the organophosphate pesticide tetrachlorvinphos by bacteria isolated from agricultural soils in México. Rev Int Contam Ambient 26:27–38

Pailan S, Sengupta K, Ganguly U, Saha P (2016) Evidence of biodegradation of chlorpyrifos by a newly isolated heavy metal-tolerant bacterium *Acinetobacter* sp. strain MemCl4. Environ Earth Sci 75:1019

Pakala SB, Gorla P, Pinjari AB, Krovidi RK, Baru R, Yanamandra M, Merrick M, Siddavattam D (2007) Biodegradation of methyl parathion and *p*-nitrophenol: evidence for the presence of a *p*-nitrophenol 2-hydroxylase in a Gram-negative *Serratia* sp. strain DS001. Appl Microbiol Biotechnol 73(6):1452–1462

Patnaik R, Padhy RN (2016) Evaluation of geno-toxicity of methyl parathion and chlorpyrifos to human liver carcinoma cell line (HepG2). Environ Sci Pollut Res Int 23(9):8492–8499

Pawar KR, Mali GV (2014) Biodegradation of Quinolphos insecticide by *Pseudomonas* strain isolated from grape rhizosphere soils. Int J Curr Microbiol App Sci 3:606–613

Pino N, Peñuela G (2011) Simultaneous degradation of the pesticides methyl parathion and chlorpyrifos by an isolated bacterial consortium from a contaminated site. Int Biodeterior Biodegrad 65:827–831

Poon BH, Leung CK, Wong CK, Wong MH (2005) Polychlorinated biphenyls and organochlorine pesticides in human adipose tissue and breast milk collected in Hong Kong. Arch Environ Contam Toxicol 49(2):274–282

Prabhavathy Das G, Pasha Shaik A, Jamil K (2006) Cytotoxicity and genotoxicity induced by the pesticide profenofos on cultured human peripheral blood lymphocytes. Drug Chem Toxicol 29(3):313–312

Prakash A, Khan S, Aggarwal M, Telang AG, Malik JK (2009) Chlorpyrifos induces apoptosis in murine thymocytes. Toxicol Lett 189:S83

Price OR, Walker A, Wood M, Oliver MA (2001) Using geostatistics to evaluate spatial variation in pesticide/soil interactions. In: Walker A (ed) Pesticide behaviour in soil and water. vol 78. British Crop Protection Council, Farnham, pp 233–238

Qian B, Zhu LS, Xie H, Wang J, Liu W, Xu QF, Song Y, Xu RJ (2007) Isolation and degrading characters of chlorpyrifos degrading bacteria XZ-3. Huan Jing KeXue 28(12):2827–2832

Qiu XH, Bai WQ, Zhong QZ, Li M, He FQ, Li BT (2006) Isolation and characterization of a bacterial strain of the genus *Ochrobactrum* with methyl parathion mineralizing activity. J Appl Microbiol 101(5):986–994

Ramanathan MP, Lalithakumari D (1996) Methylparathion degradation by *Pseudomonas* sp. A3 immobilized in sodium alginate beads. World J Microbiol Biotechnol 12:107–108

Rani NL, Lalitha-kumari D (1994) Degradation of methyl parathion by *Pseudomonas putida*. Can J Microbiol 4:1000–1004

Ray A, Chatterjee S, Ghosh S, Bhattacharya K, Pakrashi A, Deb C (1992) Quinalphos-induced suppression of spermatogenesis, plasma gonadotrophins, testicular testosterone production and secretion in adult rats. Environ Res 57(2):181–189

Rayu S, Nielsen UN, Nazaries L, Singh BK (2017) Isolation and molecular characterization of novel Chlorpyrifos and 3,5,6-trichloro-2-pyridinol-degrading bacteria from sugarcane farm soils. Front Microbiol 8:518

Reddy NC, Rao JV (2008) Biological response of earthworm, *Eisenia foetida* (Savigny) to an organophosphorous pesticide, profenofos. Ecotox Environ Safe 71:574–582

Reiss R, Neal B, Lamb JC, Juberg DR (2012) Acetylcholinesterase inhibition dose-response modeling for chlorpyrifos and chlorpyrifos-oxon. Regul Toxicol Pharmacol 63(1):124–131

Rico EP, de Oliveira DL, Rosemberg DB, Mussulini BH, Bonan CD, Dias RD, Wofchuk S, Souza DO, Bogo MR (2010) Expression and functional analysis of Na^+-dependent glutamate transporters from zebrafish brain. Brain Res Bull 81(4–5):517–523

Rubin C, Esteban E, Kieszak S, Hill RH Jr, Dunlop B, Yacovac R, Trottier J, Boylan K, Tomasewski T, Pearce K (2002) Assessment of human exposure and human health effects after indoor application of methyl parathion in Lorain County, Ohio, 1995–1996. Environ Health Perspect 110:1047–1051

Ruparrelia SG, Verma Y, Kasyap SK, Chatterjee BB (1986) A new approach for the use of standard fish toxicological study. In: Dalela RC, Madhysta MN, Joseph MM (eds) Environmental biology, coastal ecosystem. Academy of Environmental Biology, Muzzafarnagar, pp 89–92

Sadiqul IM, Ferdous Z, Nannu MT, Mostakim GM, Rahman MK (2016) Acute exposure to a quinalphos containing insecticide (convoy) causes genetic damage and nuclear changes in peripheral erythrocytes of silver barb, *Barbonymus gonionotus*. Environ Pollut 219:949–956

Safiatou BD, Jean MC, Donald EM (2007) Pesticide residues in soil and water from four cotton growing area of Mali West Africa. J Agric Food Environ Sci 1(1)

Salunkhe VP, Sawant IS, Banerjee K, Rajguru YR, Wadkar PN, Oulkar DP, Naik DG, Sawant SD (2013) Biodegradation of profenofos by *Bacillus subtilis* isolated from grapevines (*Vitis vinifera*). J Agric Food Chem 61:7195–7202

Sandal S, Yilmaz B (2011) Genotoxic effects of chlorpyrifos, cypermethrin, endosulfan and 2,4-D on human peripheral lymphocytes cultured from smokers and nonsmokers. Environ Toxicol 26(5):433–442

Saxena G, Bharagava RN (2017) Organic and inorganic pollutants in industrial wastes, their ecotoxicological effects, health hazards and bioremediation approaches. In: Bharagava RN (ed) Environmental pollutants and their bioremediation approaches, 1st edn. CRC Press/Taylor & Francis Group, Boca Raton, pp 23–56. https://doi.org/10.1201/9781315173351-3

Schuh RA, Lein PJ, Beckles RA, Jett DA (2002) Noncholinesterase mechanisms of chlorpyrifos neurotoxicity: altered phosphorylation of Ca^{2+}/cAMP response element binding protein in cultured neurons. Toxicol Appl Pharmacol 182(2):176–185

Serdar CM, Gibson DT, Munnecke DM, Lancaster JH (1982) Plasmid involvement in parathion hydrolysis by *Pseudomonas diminuta*. Appl Environ Microbiol 44(1):246–249

Sethunathan N, Yoshida T (1973) A *Flavobacterium* that degrades diazinon and parathion. Can J Microbiol 19:873–875

Sharmila Begum S, Arundhati A (2016) A study of bioremediation of methyl parathion *in vitro* using potential *Pseudomonas* sp. isolated from agricultural soil, Visakhapatnam, India. Int J Curr Microbiol App Sci 5:464–474

Sharmila M, Ramanand K, Sethunathan N (1989) Effect of yeast extract on the degradation of organophosphorus insecticides by soil enrichment and bacterial cultures. Can J Microbiol 35:1105–1110

Shen L, Wania F, Lei YD, Teixeira C, Muir DC, Bidleman TF (2005) Atmospheric distribution and long-range transport behavior of organochlorine pesticides in North America. Environ Sci Technol 39(2):409–420

Singh BK, Walker A (2006) Microbial degradation of organophosphorus compounds. FEMS Microbiol Rev 30(3):428–471

Singh BK, Walker A, Morgan JAW, Wright DJ (2003) Effect of soil pH on the biodegradation of chlorpyrifos and isolation of a chlorpyrifos-degrading bacterium. Appl Environ Microbiol 69:5198–5206

Siripattanakul-Ratpukdi S, Vangnai AS, Sangthean P, Singkibut S (2015) Profenofos insecticide degradation by novel microbial consortium and isolates enriched from contaminated chili farm soil. Environ Sci Pollut Res Int 22:320–328

Sobti RC, Krishan A, Pfaffenberger CD (1992) Cytokinetic and cytogenetic effects of some agricultural chemicals on human lymphoid cells in vitro: organophosphates. Mutat Res 102:89–102

Somara S, Siddavattam D (1995) Plasmid mediated organophosphate pesticide degradation by *Flavobacterium balustinum*. Biochem Mol Biol Int 36:627–631

Srivastava MK, Raizada RB, Dikshith TS (1992) Fetotoxic response of technical quinalphos in rats. Vet Hum Toxicol 34(2):131–133

Srivastava S, Narvi SS, Prasad SC (2011) Levels of select organophosphates in human colostrum and mature milk samples in rural region of Faizabad district, Uttar Pradesh, India. Hum Exp Toxicol 30:1458–1463

Tallur PN, Mulla SI, Megadi VB, Talwar MP, Ninnekar HZ (2015) Biodegradation of cypermethrin by immobilized cells of *Micrococcus* sp. strain CPN 1. Braz J Microbiol 46(3):667–672

Talwar MP, Ninnekar HZ (2015) Biodegradation of pesticide profenofos by the free and immobilized cells of *Pseudoxanthomonas suwonensis* strain HNM. J Basic Microbiol 55(9):1094–1103

Talwar MP, Mulla SI, Ninnekar HZ (2014) Biodegradation of organophosphate pesticide quinalphos by *Ochrobactrum* sp. strain HZM. J Appl Microbiol 117(5):1283–1292

Tilman D, Cassman KG, Matson PA, Naylor R, Polasky S (2002) Agricultural sustainability and intensive production practices. Nature 418:671–677

Uzunhisarcikli M, Kalender Y, Dirican K, Kalender S, Ogutcu A, Buyukkomurcu F (2007) Acute, subacute and subchronic administration of methyl parathion-induced testicular damage in male rats and protective role of vitamins C and E. Pestic Biochem Physiol 87:115–122

Vandekar M, Plestina R, Wilhelm K (1971) Toxicity of carbamates for mammals. Bull World Health Organ 44:241–249

Kushwaha M, Verma S, Chatterjee S (2016) Profenofos, an acetylcholinesterase-inhibiting organophosphorus pesticide: a short review of its usage, toxicity, and biodegradation. J Environ Qual 45(5):1478–1489

Verma P, Verma P, Sagar R (2013) Variations in N mineralization and herbaceous species diversity due to sites, seasons, and N treatments in a seasonally dry tropical environment of India. For Ecol Manag 297:15–26

WHO (2004) Methyl parathion in drinking-water. WHO/SDE/WSH/03.04/106. http://www.who.int/water_sanitation_health/dwq/chemicals/methylparathion.pdf

Xu G, Li Y, Zheng W, Peng X, Li W, Yan Y (2007) Mineralization of chlorpyrifos by co-culture of *Serratia* and *Trichosporon* spp. Biotechnol Lett 29(10):1469–1473

Xu G, Zheng W, Li Y, Wang S, Zhang J, Yan Y (2008) Biodegradation of chlorpyrifos and 3,5,6-trichloro-2-pyridinol by a newly isolated *Paracoccus* sp. strain TRP. Int Biodeterior Biodegrad 62:51–56

Yadav M, Srivastva N, Singh RS, Upadhyay SN, Dubey SK (2014) Biodegradation of chlorpyrifos by *Pseudomonas* sp. in a continuous packed bed bioreactor. Bioresour Technol 165:265–269

Yadav M, Shukla AK, Srivastva N, Upadhyay SN, Dubey SK (2016) Utilization of microbial community potential for removal of chlorpyrifos: a review. Crit Rev Biotechnol 36(4):727–742

Yali C, Xianen Z, Hong L, W Y XX (2002) Study on *Pseudomonas* sp. WBC-3 capable of complete degradation of methyl parathion. Wei Sheng Wu Xue Bao 42:490–497

Yanez L, Ortiz-Perez D, Batres LE, Borja-Aburto VH, Diaz-Barriga F (2002) Levels of dichlorodiphenyltrichloroethane and deltamethrin in humans and environmental samples in malarious areas of Mexico. Environ Res 88(3):174–181

Yang L, Zhao YH, Zhang BX, Yang CH, Zhang X (2005) Isolation and characterization of a chlorpyrifos and 3,5,6-trichloro-2-pyridinol degrading bacterium. FEMS Microbiol Lett 251(1):67–73

Yang C, Liu N, Guo X, Qiao C (2006) Cloning of *mpd* gene from a chlorpyrifos-degrading bacterium and use of this strain in bioremediation of contaminated soil. FEMS Microbiol Lett 265:118–125

Yashwantha B, Pamanji R, Venkateswara Rao J (2016) Toxicomorphomics and toxicokinetics of quinalphos on embryonic development of zebrafish (*Danio rerio*) and its binding affinity towards hatching enzyme. Aquat Toxicol 180:155–163

Zhang R, Xu X, Chen W, Huang Q (2016) Genetically engineered *Pseudomonas putida* X3 strain and its potential ability to bioremediate soil microcosms contaminated with methyl parathion and cadmium. Appl Microbiol Biotechnol 100(4):1987–1997. https://doi.org/10.1007/s00253-015-7099-7

Zhao G, Huang Q, Rong X, Cai P, Liang W, Dai K (2014) Interfacial interaction between methyl parathion-degrading bacteria and minerals is important in biodegradation. Biodegradation 25:1–9

Zheng Y, Long L, Fan Y, Gan J, Fang J, Jin W (2013) A review on the detoxification of organophosphorus compounds by microorganisms. Afr J Microbiol Res 7:2127–2134

Zhongli C, Shunpeng L, Guoping F (2001) Isolation of methyl parathion-degrading strain M6 and cloning of the methyl parathion hydrolase gene. Appl Environ Microbiol 67:4922–4925

Zhu J, Zhao Y, Qiu J (2010) Isolation and application of a chlorpyrifos-degrading *Bacillus licheniformis* ZHU-1. Afr J Microbiol Res 4:2716–2719

Chapter 14
Constructed Wetlands: An Eco-sustainable Phytotechnology for Degradation and Detoxification of Industrial Wastewaters

Mathews Simon Mthembu, Christine Akinyi Odinga, Faizal Bux, and Feroz Mahomed Swalaha

Contents

Abstract Globally, water quality is deteriorating at alarming levels and sanitation; infrastructures are also crumbling at an alarming rate due to technology and management challenges. While infrastructure inadequacy and poor maintenance of the existing structures continue to be a major driving force, industrialization and population increase have played a major role in the crisis of water shortage and wastewater treatment. The inability to recycle industrial wastewater is of particular importance to the socioeconomic development of the country. The water recycling challenges are even more prevalent in poor and developing countries where industrialization, coupled with limited resources and technologies for wastewater recla-

M. S. Mthembu (✉)
Department of Biochemistry and Microbiology, Faculty of Science and Agriculture, University of Zululand, Richards Bay, South Africa
e-mail: MthembuMS@unizulu.ac.za

C. A. Odinga · F. Bux · F. M. Swalaha
Institute for Water and Wastewater Technology, Department of Biotechnology and Food Technology, Durban University of Technology, Durban, South Africa
e-mail: faizalb@dut.ac.za; fswalaha@dut.ac.za

G. Saxena, R. N. Bharagava (eds.), *Bioremediation of Industrial Waste for Environmental Safety*, https://doi.org/10.1007/978-981-13-1891-7_14

mation, is high. There is an urgent need for the development and implementation of innovative industrial wastewater management system that will be both cost-effective and environmentally friendly and be able to reduce industrial contaminants to the levels that will pose no harm to the communities, thus contributing to resolving industrial wastewater treatment constraints in developing countries and, in particular, in the remote poor areas of the developing countries. Phytotechnology has been studied and developed for this purpose and has proved a success in the treatment of both domestic and industrial wastewaters.

Keywords Constructed wetlands · Industrial wastewater · Phytotechnology · Rhizofiltration · Environmental sustainability · Macrophytes

14.1 Introduction

Unsustainable use of freshwater resources, unplanned settlements in urban localities, industrialization, modern agricultural practices, and forest destruction have led to increased water demand and pollution of surface water and groundwater reserves (Sayadi et al. 2012; Mosteo et al. 2013). Among them, industrial wastewaters are considered as the major sources of environmental pollution and toxicity in living beings due to their highly toxic organic and inorganic constituents (Goutam et al. 2018; Bharagava et al. 2017a, b, c; Gautam et al. 2017; Saxena and Bharagava 2015, 2017; Saxena et al. 2016). Therefore, there is a need to conserve the available water resources and recycle wastewater in order to reduce the water stress in areas that are experiencing water scarcity (Kayyali and Jamrah 1999). Economic instability has contributed to the wastewater treatment and recycling challenges in developing countries (Vera et al. 2011; Abdel-Raouf et al. 2012). It has been estimated that developing countries will run out of the water by 2050 (Netherlands Environmental Assessment Agency 2012; OECD 2012). This is a cause for concern not only to communities and industries that depend on water supply for production but also a challenge to the water practitioners in the sector in terms of the application of alternative means of wastewater recycling. Phytoremediation has been used for many decades in many countries with varying degrees of success in the secondary and tertiary level of wastewater treatment, with many experiments being on domestic wastewater. The uses of these systems have also been reported in the primary stages (Chong-Bang et al. 2010; Yongjun et al. 2010). These different outcomes were due to limited information available about the role played by the whole system as well as by each component involved in the treatment technology. Successes in the application of phytotechnology in industrial wastewater have been reported (Vymazal 2009; Chong-Bang et al. 2010; Yongjun et al. 2010; Litter et al. 2012). This technology is engineered to copy natural processes aimed to remove pollutants from municipal and industrial wastewaters as well as from mine drainage (Stefanakis et al. 2011). Natural processes employed include vegetation, soil, and microbial activities to treat contaminated water. The relationship and interactions between plants and microbial assemble highlight the importance of the performance of the phytotechnology in the detoxification of industrial wastewater (Vymazal 2005). However, other characteristics that define the ability and the potential of this technology such as construction and combination of different systems, flow characteristics, loading rate, effect of different operational parameters, and the use of different macrophytes need to be considered in the development, implementation, and practical application of the phytotechnology (Stefanakis et al. 2011).

14.2 Design Characteristics of a Phytoremediation Technology

There are three main types of constructed phytoremediation systems characterized by different configuration design and operation. They are surface flow (SF), subsurface flow (SSF), and vertical flow (VF) systems. These systems are constructed in a closed basin with a substrate and the bottom covered by a rubber lining (geotextile) to ensure that the processes occurring are completely waterproof (Mthembu et al. 2013). This is essential in any environment where leakage of water from the system can have adverse effects, e.g., contamination of groundwater. The substrates of the systems are gravel and sand or lava stones that contribute to nutrient retention in these systems (Farroqi et al. 2008). Advances in engineering and technology have now permitted construction of a multi-designed system functioning as vertical, horizontal, as well as subsurface in one system. These types of hybrid systems represent a new trend and an emerging phytotechnology tool in wastewater treatment relying on macrophytes and sand as a medium. Although the design of these systems may be expensive, their success in contaminant reduction and detoxification in wastewater may offer more advantages than conventional systems.

Hybrid wetlands consist of a series of other systems that are used in combination (Sayadi et al. 2012) in order to achieve higher treatment efficiencies (Vymazal and Kropfelova 2005; Baskar et al. 2009; Vymazal 2010; Nilsson et al. 2012). The optimum combination of the different types of constructed wetlands (CWs) depends largely on the target pollutants (Vymazal 2009). For example, HF CWs are normally characterized by low levels of dissolved oxygen, leading to slow nitrification processes, while VF systems are designed to transport oxygen at a greater capacity, providing a better environment for nitrification (Kalipci 2011). In order to enhance the removal of ammonia, HF CWs may be used in combination with VF CWs in well-engineered systems (Vymazal 2009, 2011). A more recent development is a technology that embodies the combination of free water surface (FWS) CWs and subsurface flow CWs and FWS CWs with subsurface flow CWs (Borkar and Mahatme 2010). This combination is designed with a focus on efficient removal of organic material and suspended matter (Vymazal 2011) and is ideal for the applications in an industrial setting for the purpose of industrial wastewater reclamation.

14.3 Applications of Constructed Wetlands in Wastewater Treatment

Previous studies have shown a high-treatment efficiency of constructed rhizofiltration systems with regard to organic and inorganic nutrient removal as well as removal of other general contaminants such as total suspended solids (TSS) and microbial contaminants from wastewater (Cooper et al. 1996; Yadav et al. 2010; Vymazal 2011). Regular monitoring of systems had shown close to 100% removal of total coliforms and organic pollutants (Shrestha et al. 2003). Although average removal efficiencies of nitrogen and phosphate have been reported, a significant difference in removal efficiency is observed among plant species as well as among different types of rhizofiltration system configurations (Yeh et al. 2009). The main mechanisms leading to contaminant removal in these systems are microbial activities. However, plants also have a huge role in contaminant removal in wastewater. They take up nutrients and incorporate them into plant tissue and thus increase in plant biomass (Zhang et al. 2007). Plants also create suitable conditions for the processes leading to nutrient removal to occur (Dunabin and Browmer 1992; Wu et al. 2011).

Treatment mechanisms involved in a constructed wetland system are the hydraulic retention time (HRT) and volumetric flow rate (VFR) of water. Increased HRT coupled with reduced VFR is believed to provide maximum exposure of water to the root surface of the plant, thus providing sufficient time for the uptake of the nutrient ions and other chemical changes (Srivastava et al. 2008; Steudle 2000).

14.4 Phytoremediation in Constructed Wetlands Systems for Industrial Applications

Phytoremediation is a low-risk and attractive cleanup method for domestic, environmental, and industrial pollution. The process uses plants to clean up contaminated environments. Macrophytes can help clean up many types of contaminants including metals, pesticides, explosives, and oil. However, they work best where contaminant levels are low because high concentrations may limit plant growth and the process may take too long. Phytotechnology takes advantage of natural plant processes and requires less equipment and labor than other methods since macrophytes perform most of the work. Also, the site can be cleaned up without digging up and hauling soil or pumping groundwater, which saves energy. Among other things, macrophytes used during phytoremediation help control soil erosion, make a site more attractive, reduce noise, and improve surrounding air quality.

Phytotechnology is known to achieve better treatment of wastewater including the removal of heavy metals through detoxification, uptake by the macrophytes, and subsequent removal from water (Boutilier et al. 2011). In this process, macrophytes act as biofilters or rhizofilters due to their ability to withstand harsh effluent conditions to transform and remove sediments and pollutants (Saxena et al. 2018; Chandra et al. 2015; Nedunuri et al. 2014). Rhizofiltration removes or reduces microorganisms by means of chemical, biological, and physical processes that include filtration, sedimentation, adsorption, oxidation, direct killing by plant toxins and predation by other microorganisms, natural die-off, temperature, and ultraviolet radiation (Liu 2009; Abdel-Raouf et al. 2012). Employing this technology in industrial settings, microorganisms such as *E. coli* can be greatly reduced to non-detectable levels (McCarthy et al. 2011; Odinga et al. 2013). Macrophytes, through phytoremediation, play a major role in wastewater treatment efficiencies (Ebrahimi et al. 2013). This treatment ability is based on the particular plant species, metabolic activities within the plant, and pollution tolerance rates (Vymazal 2011; Haarstad et al. 2012; Tuttolomondo et al. 2014). Macrophytes' root systems are resilient to various shocks from pollution loading, climatic variations, and pollutants that have very high salinity contents with reported levels of up to 11,500 mg/l as total dissolved solids. Pollutant removal through phytotechnology is also based on water depth, hydraulic loading, hydraulic efficiency, and influent concentration (Haarstad et al. 2012).

A study by Odinga et al. (2011) assessed the effectiveness of an HF CW planted with *Echinochloa pyramidalis* and *Cyperus papyrus* in removing nutrients from sugar factory effluent. The wetland system was designed to accommodate 8 wetland ponds with an HF system and measuring between 1: 20 and 21 ml × 3 mW × 1 mD. The eight CWs were planted with two different macrophytes, *E. pyramidalis* L. and *C. papyrus* L., in an alternate sequence. All wetland cells were waterproofed at the base with high-density polyethylene and lined with fine sand (Cooper et al. 2005). The effluent was introduced into the wetland at different velocities with a retention time of 7 days. Findings from the study indicated that bacteria and fungi formed biofilms on the surfaces of the plants' rhizomes and utilized the contaminants, leading to their reduction in water. Adsorption onto the soil matrix also played a major role in pollution reduction from the wastewater. In a study conducted by Valigore et al. (2012), the effects of hydraulic and solids retention times on productivity and settleability of microbial biomass (microalgae) grown on primary-treated wastewater as a biofuel feedstock concluded that pH levels increased when there was more light that increased the rate of photosynthesis. The dissolved oxygen (DO) levels increased during the same period as the microalgae used more of the CO_2, and DO increase during the light periods when CO_2 was consumed. The study also noted that shorter HRT yielded low DO levels and reduced bacterial growth and that the levels were increased by the microbes and atmospheric oxygen release.

Depending on the underlying processes, applicability, and type of contaminant, phytoremediation industries may be achieved through:

Phytodegradation: It is also known as phytotransformation. It is the breakdown of contaminants taken up by macrophytes through metabolic processes within the plant or the breakdown of contaminants surrounding the plant through the effect of compounds such as enzymes produced by the plants. During this process, complex organic pollutants are degraded into simple molecules and are taken up by the macrophytes in order to form their own cell mass. Macrophytes may contain complex chemical proteins and enzymes that may catalyze and accelerate chemical reactions. In industrial wastewaters, some enzymes may contribute to breaking down and convert ammunition wastes and degrade chlorinated solvents such as trichloroethylene (TCE), while others degrade herbicides.

Phytostimulation: It is a plant-assisted bioremediation or degradation where the breakdown of contaminants in the rhizosphere achieved through microbial activity enhanced by the presence of plant roots. This process is a much slower process than phytodegradation. Microorganisms such as yeast, fungi, and bacteria consume and digest organic substances for nutrition and energy. Certain microorganisms can digest organic substances such as fuels or solvents that are hazardous to humans and break them down into harmless products in a process called biodegradation. Natural substances released by the plant roots like sugars, alcohols, and acids may contain organic carbon that provides food for soil microorganisms, and the additional nutrients enhance their activity. Biodegradation is also aided by the way macrophytes loosen the soil and transport water to the area.

Phytovolatilisation: It is the uptake through transpiration of a contaminant by macrophytes, with the release of the contaminant or a modified form of the contaminant from the plant to the atmosphere. It occurs as growing macrophytes take up water and the organic contaminants. Some of these contaminants can pass through the plants to the leaves and evaporate, or volatilize, into the atmosphere.

Phytoextraction: It is also known as phytoaccumulation and refers to the uptake of metals from soil by macrophytes' roots into aboveground portions of plants. High accumulator macrophytes, hyperaccumulators, absorb unusually large amounts of metals in comparison with other plants. After the plants have been allowed to grow for some time, they are harvested and either incinerated or composted to recycle the metals. This procedure may be repeated as necessary to bring soil contaminant levels down to allowable limits. If plants are incinerated, the ash must be disposed of in a hazardous waste landfill, but the volume of ash will be less than 10% of the volume that would be created if the contaminated soil itself were dug up for treatment. Heavy metals such as nickel, zinc, and copper are best removed through the process of phytoextraction.

Rhizofiltration: It is the adsorption or precipitation onto plant roots of contaminants that are in the solution surrounding the root zone. Rhizofiltration is similar to phytoextraction, but the plants are used to clean up contaminated groundwater rather than soil. The plants to be used for cleanup are raised in greenhouses with their roots in water. Contaminated water is either collected from a waste site or brought to the plants, or the plants are planted in the contaminated area, where the roots then take up water and the contaminants dissolved in it. As the roots become saturated with contaminants, they are harvested.

Phytostabilization: It is the use of certain plant species to immobilize contaminants in the soil and groundwater through absorption and accumulation by roots, adsorption onto roots, or precipitation within the rhizosphere. This process reduces the mobility of the contaminant and prevents migration to the groundwater or air and also reduces bioavailability for entry into the food chain. This technique can be used to reestablish a vegetative cover at sites where natural vegetation is lacking due to high metal concentration in surface soils or physical disturbances to surficial materials. Metal-tolerant species can be used to restore vegetation to the sites, thereby decreasing the potential migration of contamination through wind erosion and transport of exposed surface soils and leaching of soil contamination to groundwater.

14.5 Adaptive Strategies of Macrophytes to Various Growth Processes

Different macrophytes have different survival strategies in water. These survival strategies are essential in macrophytes to absorb and, subsequently, reduce contaminants in wastewater. While there are different types of macrophytes, the two tested macrophytes for application in industrial wastewater are:

Phragmites australis L.: The common weed, *Phragmites australis*, is a tall (1.5–4.0 m) coarse perennial grass found primarily in brackish and freshwater wetlands (Lee and Scholz 2007; Odinga et al. 2013). It is distributed worldwide through endemic in North America (Lee and Scholz 2007). In wetlands, *Phragmites australis* is sometimes considered a nuisance species because once established, it colonizes a region, thus acting like a climax species (Swearingen and Saltonstall 2010; Dolinar and Gaberscik 2010). *Phragmites australis* has many adaptive features that make it suitable as the macrophyte of choice for phytotechnology (Vymazal and Kropfelova 2005). These range from reproduction, apical dominance, mechanical adaptations, and gaseous exchange (Dolinar and Gaberscik 2010). The rapid growth of this macrophyte up to 10 m under optimum conditions may be attained because its horizontal stems exhibit strong apical dominance (Verboven et al. 2011; Sayadi et al. 2012). Rapid growth enhanced the update of nutrients, leading to the treatment efficiency of wastewater (Kropfelova et al. 2009; Howard 2010).

Phragmites species possess abundant aerenchyma cells which are responsible for internal gas exchange. This increases the suitability of this macrophyte in enhancing oxygenation of the rhizosphere in the treatment system (Shimamura et al. 2010; Malik et al. 2010; Shiono et al. 2010). Light attenuation by the elevated *Phragmites australis* leaf canopy imparts increasing stress to shorter understory wetland plant species. Therefore *P. australis* remains the dominant species as water treatment is achieved (Vymazal and Kropfelova 2009; Kalipci 2011; Lum et al. 2014). This macrophyte tends to form a dense thatch elevated above the soil (Howard 2010; Gill et al. 2014). This dense thatch assists in moisture conservation. When this occurs, the macrophytes' population is protected from competition from other plants, thus improving its phytoremediation performance (Howard 2010; Luca et al. 2011; Kalipci 2011).

Kyllinga nemoralis L.: Available literature on the potential of *Kyllinga nemoralis* to remove heavy metals from wastewater is scanty. A sedge of the family Cyperaceae, which forms a mat and grows up to 50 cm high, *Kyllinga nemoralis* is considered a common weed (Majumder 2013; Odinga et al. 2013). The perennial weed spreads by means of creeping rhizomes which contain allopathic oils (Kawabata et al. 1994). This plant mainly propagates through rhizome cuttings and seeds (Majumder 2013). *Kyllinga nemoralis* possess an achene fruit which measures about 1.2–1.5 mm long by 0.5–0.7 mm wide. The leaves contain active chemicals, the essential oils (β-selinene, terpenes α-cyperone, and α-humulene), while the rhizomes possess glycosides, triterpenoids, and flavonoids (Majumder 2013). These compositions render the roots and leaves of *Kyllinga nemoralis* to be highly medicinal and contribute to a microbial reduction in water. Greenhouse studies indicate that this macrophyte cannot withstand drought conditions (Nedunuri et al. 2014), a characteristic that potentially increases its use in phytotechnology. In a study by Sindhu et al. (2014), *Kyllinga nemoralis* root extract was successfully used to inhibit the growth of pathogenic bacteria. Hence, it is a wetland plant that may be used in phytoremediation processes to inhibit the growth of bacteria and fungi in wastewater. Ray and George (2009) studied metal deposition and accumulation on Cyperaceae family along the roadside. They discovered that the *Kyllinga* species accumulated metals which further confirmed that this macrophyte has the potential to remediate and detoxify heavy metals in wastewater.

14.6 Macrophytes and Their Role in Phytotechnology

Macrophytes may only grow in water or in soil that is permanently saturated with water. This characteristic makes macrophytes an important component of the rhizofiltration process during wastewater treatment. The growth of aquatic macrophytes requires the presence of nutrients such as nitrogen and phosphorus for their growth and reproduction. A constructed rhizofilter receives wastewater which is rich in various types of nutrients required by aquatic macrophytes for growth (Keddy 2000). Most of the natural aquatic macrophytes reproduce by flowering and seed settling, while many other types are capable of extensive asexual reproduction by means of rhizomes and fragments (Hutchinson 1975). Aquatic macrophytes require warm weather for growth. They usually grow at a much higher rate during warm seasons. Major factors such as nutrient availability and salinity need to be monitored during their growth since unacceptable levels may retard growth (Keddy 2000).

The growth of many macrophytes requires the presence of a root system growing within an organic-rich soil substrate. Macrophyte roots hanging within a rhizosphere or beneath have been demonstrated to provide an extensive surface area for the attachment of biofilm growth and entrapment of fine suspended particles (Bourn 1932; Hochereutiner 1986). Macrophytes use their root systems to release oxygen into the rhizosphere required by aerobic microorganisms in order to degrade contaminants in the rhizosphere. The amount of oxygen released in the subtropical region is higher and decreases with distance from the root apex (Armstrong 1979). Another important role of roots is the absorption of carbon dioxide from the substrate which serves as a rich source of this dissolved gas. Some macrophytes have shown rapid growth in the absence of any substrate when the carbon dioxide-enriched air is bubbled through a nutrient medium (Bristow 1974).

The mechanisms of nutrient removal from industrial wastewater in constructed rhizofiltration systems by macrophytes have been attributed to phytodegradation, phytostimulation, phytovolatilization, rhizofiltration, phytostabilization, and precipitation. Macrophytes are able to tolerate high concentrations of nutrients and heavy metals and, in some cases, accumulate them in their tissues (Stottmeister et al. 2003). Macrophytes take up these pollutants in order to form their cell mass (Kalbar et al. 2012). The most reactive zones for rhizofiltration are in the rhizosphere, where all physicochemical and biological processes occur. These processes are induced by macrophytes' interactions with microorganisms, soil matrix, and contaminants.

Macrophytes are responsible for approximately 90% of oxygen transport available in the rhizosphere (Vymazal 2011). Oxygen and nitrogen transport stimulates aerobic and anoxic decomposition of organic matter, respectively, as well as promote the growth of nitrifying bacteria and periphyton in the soil medium (Zhang et al. 2007). The added oxygen enhances nutrient transformation and thus contributes to higher treatment efficiency of the phytotechnology.

14.7 Microorganisms and Their Role in Phytotechnology

Microbial biofilms have been reported to play a prominent role in the transformation, mineralization, and subsequent removal of organic and inorganic contaminants in the process of phytotechnology (Moeseneder et al. 1993; Moura et al. 2007). These nutrients are metabolized in various ways. In SSF-constructed rhizofiltration systems, aerobic microbial-driven processes occur predominantly near plant roots as well as on root surfaces. These are mainly responsible for nitrogen removal. In the areas that are largely oxygen-free, denitrification, sulfate reduction, and methanogenesis occur, which removes nitrogen, sulfates, phosphates, and carbon, respectively. The combination of compost degradation and microbial biofilms in heterotrophic microbial growth is responsible for oxygen removal from the rhizofilter units and thereby promotes the formation of hydrogen sulfide in industries. Sulfate-reducing bacteria degrade and reduce nutrients that contain sulfates and produce hydrogen sulfide in the process (Lee and Yang 2010).

Both autotrophs and heterotrophs have been found to incorporate ammonia and convert it to amino acids and proteins (Vymazal 2007). However, in industries, this removal mechanism is less significant compared to microbial transformation. Nitrification-denitrification is the main microbial nitrogen removal mechanism by microorganisms (Stottmeister et al. 2003). In an industrial setting, depending on the type of contaminants, nitrogen compounds are continually transformed from inorganic to organic compounds and back from organic to inorganic through volatilization, ammonification, nitrification, nitrate ammonification, denitrification, and nitrogen fixation. All these transformations are necessary for the rhizofiltration ecosystem to function successfully, and all chemical changes are controlled by enzymes produced by microorganisms. Thus, microorganisms are essential in the phytosystem if contaminants are to be completely degraded and removed from wastewater (Lizotter et al. 2001).

14.8 Mechanisms of Contaminant Removals Using Phytotechnology in Industries

Combinations of biological, chemical, and physical processes have been shown to be responsible for the removal of contaminants in the rhizotechnology. Biologically, both macrophytes and microorganisms play a synergistic role in the removal of contaminants by transforming and/or accumulating them and converting them into their cell mass. Wastewater treatment within a constructed rhizofilter occurs as the water passes through the rhizofiltration media and plant roots. Interactions between water and rhizomes lead to rhizofiltration and sedimentation while that of microorganisms and contaminants cause biodegradation and bioaugmentation. Root hairs and rootlets provide aerobic environment which supports the activities of aerobic microorganisms. Aerobic and anaerobic microorganisms facilitate the decomposition of organic matter and inorganic substances through degradation and nutrient uptake. During these processes, nitrogen is liberated from the system through microbial nitrification and subsequent denitrification processes. Organic nitrogen, nitrate, ammonia, ammonium, and nitrogen gases are the most common forms of nitrogenous compounds liberated during treatment process (Cooper et al. 1996). These compounds are essential for plant growth and development. However, it is important that these are removed whenever they exceed allowable limits since some may be toxic to aquatic life (Brisson and Chazarenc 2009). Suspended solids are removed by settling in the water column in surface flow systems or are physically filtered out by the medium within subsurface flow filters. Pathogens are removed by filtration and adsorption onto biofilms or plant roots. Heavy metals and phosphates are removed by either plant uptake or through sedimentation (Hunter et al. 2001). Macrophytes themselves may also need to be removed from time to time from the system in order to prevent die-off in the system and reintroduction or recycling of the contaminants.

Phytotechnology is based on mechanisms of phytofiltration. The pollutant rhizofiltration works through phytostabilization and phytoextraction, with treatment mechanisms based on absorption and concentration of contaminants by the roots of the plants where microbial activities are involved in the breakdown of organic contaminants (Bouasria et al. 2012). Reports from previous studies indicate that plants used in rhizofilters are effective in the removal of environmental toxic heavy metals such as Cu^{2+}, Cd^{2+}, Cr^{6+}, Ni^{2+}, Pb^{2+}, and Zn^{2+}(Akpor and Muchie 2010). These heavy metals are usually obtained in industrial wastewater. Rhizofilters are designed to remove pollutants (especially metals) using the roots of wetland plants (Elias et al.

2014). The hydraulic retention time (HRT) and volumetric flow rate (VFR) of water are also essential in phytoremediation. Increased HRT coupled with reduced VFR is believed to provide maximum exposure of water to the root surface of the plant, and thus providing sufficient time for the uptake of the nutrient ions and other chemical changes (Steudle 2000; Srivastava et al. 2008).

In order to understand more about the complexities of what happens when industrial wastewater pollutants are degraded in a constructed rhizofiltration system during treatment, it is essential to know more about macrophytes and their interactions with microbial community structure as well as their abundance in the system. The following section discusses the detailed mechanisms of nutrient, pathogen, and heavy metal removal in the wetland phytotechnology:

Nutrient removal: Rhizofilters may be used to remove nitrogen and phosphorus from industrial wastewater. The contaminants are transformed and reduced by microorganisms that are attached to the filter media and root system of the macrophytes. Microorganisms in the system may also utilize nutrients for their growth through several mechanisms while reducing nutrients in the process. Rhizofilter system carries out wastewater treatment as the influent passes through the treatment system. The rhizosphere provides aerobic conditions which support the activities of aerobic microorganisms. Microorganisms in the system facilitate decomposition of organic matter as well as inorganic substances. Filtration, uptake by roots, adsorption, absorption, precipitation, nitrification, and denitrification are all processes responsible for the nutrient reduction in phytotechnology in industries. Rhizofiltration systems have shown reduction efficiencies ranging between 21% for phosphorus and 85% for nitrogen (expressed as plant biomass which accounts for 53% PO_4-P removal and 50% NO_x-N removal) (Greenway and Woolley 2001). The denitrification process may remove between 60% and 70% total nitrogen, with 20–30% of the removal process derived from plant uptake (Lee 2011). The main priority is to reduce the level of excessive concentrations of organic nutrients that may cause oxygen deprivation when consumed (Culp et al. 1986). Ammonia, nitrate, nitrite, phosphates, and orthophosphates are the main organic nutrients present in industrial wastewater that need to be reduced. While rhizofiltration medium plays a vital role in the removal of phosphorus through sorption, its action may be greatly influenced by external factors which may decrease its ability. Therefore, it is essential to understand what and how those factors may affect the performance of the treatment efficiency of the rhizofiltration unit (Yoon et al. 2001).

Parameters such as temperature may either have beneficial or negative effects on macrophytes, microbial growth, and other processes in the system. *Phragmites australis* and *Kyllinga nemoralis* require temperatures between 12 and 25 °C for optimal growth. When temperatures are outside these ranges, the absorption ability of macrophytes to nutrients decrease and thus reduce treatment efficiency.

Pathogens removal: Municipal wastewater is composed of a number of microorganisms, some of which are pathogenic and may adversely affect human life when ingested (Chaidez et al. 2014). The fate of bacteria, yeast, and nematodes, once released from their hosts and into wastewater systems and aquatic or natural environments, has been investigated by a number of scholars (Stevik et al. 2004;

Garcia et al. 2010; Chaidez et al. 2014; Wu et al. 2014). Their presence and quantity in wastewater is a major determinant of the quality of wastewater discharged into various ecosystems (Varela and Manaia 2013). This is because some of the pathogenic microorganism survive the treatment system, whereas others have adaptations that compromise their removal depending on the type of wastewater treatment applied (Al-Jaboobi et al. 2013). Some pathogens are very resilient and change their forms (form cysts, e.g., *Ascaris lumbricoides*) in order to survive harsh environmental conditions (Abbadi et al. 2012; Rani et al. 2014; Odinga et al. 2013).

For successful applications in wastewater treatment, rhizofiltration systems should have the ability to remove pathogens in wastewater. Previous research indicates that rhizofiltration systems have an ability to reduce pathogens with varying but significant degrees of effectiveness (Karim et al. 2004). Though industrial wastewater may not have many pathogens compared to domestic wastewater, microbial water quality improvements using rhizotechnology have been reported, with some studies documenting up to 57% reduction of total coliforms, 62% of fecal coliforms, 98% reduction of most species of *Giardia*, 87% of most *Cryptosporidium* spp., and 38% of coliphage (Stottmeister et al. 2003; Karim et al. 2004). Rhizofiltration systems are also capable of reducing human pathogenic viruses from wastewater. Viruses associated with large particles leave the water column and settle into the bottom sediments, while some that are adsorbed on colloidal particles tend to stay suspended in water before filtered (Karim et al. 2004).

Pathogen removal through rhizofiltration systems was also reported by Greenway (2005) and Boutilier et al. (2011). They documented enteropathogenic *Escherichia coli* removal efficiencies to be between 52% and 99.9%, respectively. To mitigate elimination variability, it may be necessary to better understand what pathogen removal mechanisms dominate within the system in an industrial setting and how these mechanisms may be intensified through the manipulation of the rhizofilter operational parameters at optimum levels. Many mechanisms have been associated with the removal of pathogen employing rhizotechnology. These include physical (filtration, sedimentation, adsorption, and aggregation), biological (consumed by protozoa, lytic bacteria, bacteriophages, natural death), and chemical (oxidative damage, the influence of toxins from other microorganisms and plants) processes. Sedimentation remains a leading mechanism responsible for pathogen reduction in these systems (Karim et al. 2004). This has been demonstrated by many studies which found total coliforms, fecal coliforms, and *Salmonella* had concentrated in sediments of contaminated surface water in wetland systems (Greenway 2005; Boutilier et al. 2011). They also demonstrated that revival of such organisms from sediments was easier than in water column itself. Chauret et al. (1998) observed higher numbers of fecal coliforms in marine sediments than in overlying water. They also found that about 90% of *Salmonella* isolates from sediments showed high recovery in sediments than in water. *Escherichia coli* was also demonstrated to survive longer in sediments than in overlying water.

Accumulation of microorganisms, pathogens in particular, in sediments of constructed rhizofiltration systems designed for wastewater treatment means that these systems could be used for elimination of pathogens from influents. However, removal of pathogens using sedimentation process could also pose some serious threats as the bottom sediments of the system could serve as a potential reservoir of human pathogens. These reservoirs might be released back into the water column by events such as storm and thereby released with effluent to the river during floods or through increased runoffs (Pedescoll et al. 2011).

Plants have also been found to reduce pathogens in rhizofiltration systems. Macrophytes like *Mentha aquatica*, *Phragmites australis*, and *Scorchilacustric* were found to inhibit the growth of *E. coli* (Stottmeister et al. 2003). Other than a bactericidal effect of the macrophytes, other mechanisms and indirect mechanisms of removal such as adsorption, aggregation, and filtration were also shown to be involved in the reduction and subsequent removal of pathogens. It could be concluded from the literature study that the combined action of physical, chemical, and biological processes are required to achieve high-reduction efficiencies of the pathogen in the rhizofiltration technology.

Heavy metal removal: Heavy metals are naturally occurring elements that are potentially hazardous to both terrestrial and aquatic environments when present above permissible limits (Rawat et al. 2012). Their toxicity may be attributed to the fact that they cannot be biologically biodegraded (Thayaparan et al. 2013), and thus when their presence in the environment exceeds acceptable limits, they are considered as pollutants (Lum et al. 2014). The main sources of metal pollution in the environment arise from battery manufacturing factories, mining activities, e-waste, smelting activities, automobile emissions, paints (Mukesh and Thakur 2013), atmospheric deposition and fossil fuel combustion, industries, etc. (Zhang et al. 2011; Song and Li 2014). Traces of heavy metals usually occur in water where some get immobilized in the sediment and form complexes with oxides and hydroxides of iron (Fe) and other particulate matter in the sediment. It is, therefore, necessary to efficiently treat wastewater suspected of having metal concentration beyond acceptable limits. Efficient wastewater treatment in conformity with the set standards will eliminate potential toxicological effects on humans, animals, and aquatic environment (Baysal et al. 2013).

While heavy metals may be found in industrial wastewater and mine drainages, small quantities may be detected in municipal or domestic wastewaters as well. The main heavy metals associated with wastewater and produced by mines and industries are chromium, iron, mercury, copper, lead, cadmium, and zinc. These heavy metals may be removed from wastewater through phytoremediation employing a variety of methods including filtration and sedimentation, adsorption, uptake into plant material, and precipitation by geochemical processes (Stottmeister et al. 2003). Removal efficiencies of heavy metals by rhizofiltration have been reported to be up to 100% (Romero et al. 2011). Other possible removal efficiencies through the system as reported by Sheoran and Sheoran (2006) are 75–99% for cadmium, 26% for lead, 76% for silver, and 67% for zinc, while COD, BOD, and TSS were removed at an efficiency of between 75% and 80%. Metals were demonstrated to accumulate in the leaves, shoots, rhizomes with roots, and lateral roots having the highest content, while the lowest concentrations were found within the shoots.

In SF systems used to treat mine drainage, Fe (II) is oxidized to Fe (III) by abiotic and microbial oxidation. In this system, other inorganic substances such as arsenic may also precipitate. Iron may also be immobilized in the anoxic soil matrix by microbial dissimilatory sulfate reduction, producing hydrogen sulfide. Most heavy metals are taken up and accumulated within the plants. A higher percentage of the metal ions adsorbed and absorbed by the plants are concentrated in the plant roots and less commonly in the stems. Only a few heavy metals like mercury are able to translocate to the leaves (Romero et al. 2011). Different plant species have differing abilities to take up heavy metals (Mitsch and Jorgensen 2004). Some species of plants have high biomasses which enhance their phytoremediation capacity. Plants like *Polygonum punctatum* have been proposed as copper and zinc biomonitors and phytoremediators and are useful as rhizofilters for the treatment of industrial wastewaters and mine drainages.

Previous studies by Stottmeister et al. (2003), Sheoran and Sheoran (2006), and Romero et al. (2011) indicated that from a technological point of view, heavy metal accumulation by macrophytes was insignificant when considering treatment of industrial wastewater and mine drainages. This was because the number of heavy metals that could be accumulated by plants was found to be far too small when compared to the total load in wastewater. In a rhizofiltration system filter, sediments and precipitation are the leading means of heavy metal removal.

14.9 Phytotechnology in Third-World Countries in the Twenty-First Century

Industrial wastewater treatment is a growing concern worldwide, with issues spanning environmental, economic, and fiscal sectors. Of specific concern is the contamination of groundwater with heavy metals as a side effect of societal and industrial pollution. While moderate quantities of certain contaminants are acceptable for consumption, larger quantities may lead to acute poisoning (Dushenkov et al. 1995), especially in countries where water resources are scarce and/or treatment and purification techniques are primitive. New technologies are a necessity to prevent these serious health crises in such cases.

Phytotechnology is a cost-effective technology for the remediation of industrial wastewater to the required effluent discharge standards, with low labor requirements. Large and established facilities of the twenty-first century may reduce costs by adapting to this natural water recycling technique as an alternative to current methods, while small populations devoid of such resources can apply rhizofiltration on a local and manageable scale. In the case of small industries, no big expensive infrastructure may be necessary as phytotechnology may be set on site. In the developed countries, the current state of the art in wastewater treatment systems generally involves significant investments in terms of financial strains and infrastructure where before any treatment occurs, water must be moved from its native location, often requiring an initial investment in mileages of the pipeline. Once at the treatment plant, taking an advantage of the abundance of the technologies of the twenty-first century, a myriad of techniques ranging from aeration, digestion, flocculation, sedimentation, filtration, sludge treatment, and disinfection is applied before reaching a final product with each step requiring additional technology and manpower to operate. While the complete process is effective, the resources needed to carry out this process are too great for certain populations. For example, poor and developing countries may lack the fiscal acuity to invest in the infrastructure required for such extensive treatments. While current estimates for the construction of a full operation filtration system range into the hundreds of millions of dollars depending on the volume of water to be treated and the level of treatment required, this is in stark contrast to few dollars that may be spent per thousands of gallons of wastewater that may be treated using rhizofiltration technology. This cost-effective advantage may be associated with the way in which the system may be implemented.

14.10 Human Health Concerns in Applications of Phytotechnology

Wetlands that receive primary effluent are known to produce noxious odors and should be restricted only to authorized personnel. Human and animal health concerns become significant especially when the water is to be reused or discharged into sensitive receiving rivers, ponds, and lakes. Moreover, whenever there are water shortage and need for irrigation using wastewater, hygiene concerns and infection from potential pollutants become a major concern on the direct users of the treated water (Lilach et al. 2010). Crops grown through irrigation by treated wastewater may contain pathogenic microorganisms based on the treatment method applied. While in industries the main threats may arise from nutrients and heavy metal pollutants (depending on the industry operation), other health concerns may include risks of infection by protozoan parasites like *Cryptosporidium*, *Giardia* spp., and *Ascaris lumbricoides*. However, CWs have demonstrated effective removal of protozoan parasites mainly with horizontal and vertical subsurface flow gravel-based systems (Redder et al. 2010). Other risks may include skin contact and potential transmission of viral and other air-borne diseases normally associated with wastewater. Mosquito populations also pose health effects in the environments around these systems especially in warmer regions where the encephalitis mosquito (*Culex tarsalis*) is dominant and breeds well by extension due to the wet conditions provided by the wetland. The mosquito population may be controlled by a two-tiered pond design which concentrates the prey in smaller areas for consumption by predators such as the mosquitofish (US EPA 1999), and also the introduction of macroinvertebrate predators into the wetland system as suggested in a study by Greenway (2005).

14.11 Conclusion

Alternative and effective ways of managing increasing amounts of wastewaters generated in industries are imperative. Currently experienced drought in most parts of the world, particularly in African countries, accompanied by fast depletion as well as escalating of freshwater sources may be compensated by rhizofiltration technology in wastewater circling to ensure the availability of freshwater that would meet the industrial and population needs. Rhizofiltration systems are a promising, cheap, and effective wastewater treatment tool using local resources in small and mainly rural underdeveloped nations. These systems have been in operation for a while as wastewater treatment methods. However, the lack of knowledge associated with their applications has derailed the progress of their application as alternatives in wastewater treatment. The role played by different components during treatment needs to be understood if these systems are to be applied for industrial purposes. With more research, recent emergent or construction of hybrid systems may contribute to achieving higher contaminant reductions with little investments. While this technology may be of great benefit to everyone who may use it, developing countries may greatly benefit from this technology because of its low capital cost and energy requirements compared to cost of installing the conventional systems. The system may tolerate both great and small volumes of water and varying contaminant levels.

Some wetlands may offer an ideal disposal system for sludge generated from conventional treatment systems to be used for purposes such as nutrient source. Research findings on CWs planted with aquatic macrophytes show that both submerged and free floating designs are able to effectively improve wastewater quality by removing and reducing heavy metals and pathogens. Selection of the most appropriate macrophyte for systems efficiency remains a challenge to many researchers since macrophyte phytoremediation characteristics are specific to the toxins and the antimicrobial properties of each plant are also varied.

References

Abbadi J, Saleh R, Nusseibeh S, Qurie M, Khamis M, Karaman R, Scrano I, Bufo SA (2012) Microbial removal from secondary treated wastewater using a hybrid system of ultrafiltration and reverse osmosis. Environ Sci Eng 1:853–869

Abdel-Raouf N, Al-Homaidan AA, Ibraheem IBM (2012) Microalgae and wastewater treatment. Saudi J Biol Sci 19:257–275

Akpor OB, Muchie M (2010) Bioremediation of polluted wastewater influent: phosphorus and nitrogen removal. Sci Res Essays 5:3222–3230

Al-Jaboobi M, Bouksaim M, Tijane MH, El-Ariqi S (2013) Agricultural quality evaluation of wastewater, used in Yemen vegetables production. Mid-East J Sci Res 16:667–677

Armstrong W (1979) Aeration in higher plants. Adv Bot Res 7:225–331

Baskar G, Deeptha V, Rahman Abdul A (2009) Treatment of wastewater from kitchen in an institution hostel mess using constructed wetland. Intern J Recent Trends Eng 1:54–58

Baysal A, Ozbek N, Akman S (2013) Determination of trace metals in waste water and their removal processes. https://doi.org/10.5772/52025

Bharagava RN, Saxena G, Mulla SI, Patel DK (2017a) Characterization and identification of recalcitrant organic pollutants (ROPs) in tannery wastewater and its phytotoxicity evaluation for environmental safety. Arch Environ Contam Toxicol. https://doi.org/10.1007/s00244-017-0490-x

Bharagava RN, Saxena G, Chowdhary P (2017b) Constructed wetlands: an emerging phytotechnology for degradation and detoxification of industrial wastewaters. In: Bharagava RN (ed) Environmental pollutants and their bioremediation approaches, 1st edn. CRC Press/Taylor & Francis Group, Boca Raton, pp 397–426. https://doi.org/10.1201/9781315173351-15

Bharagava RN, Chowdhary P, Saxena G (2017c) Bioremediation: an ecosustainable green technology: its applications and limitations. In: Bharagava RN (ed) Environmental pollutants and their bioremediation approaches, 1st edn. CRC Press/Taylor & Francis Group, Boca Raton, pp 1–22. https://doi.org/10.1201/9781315173351-2

Borkar RP, Mahatme PS (2010) Wastewater treatment with vertical flow constructed wetland. Int J Environ Sci 2:590–603

Bouasria A, Mustafa T, Bello FD, Zinger L, Lemperire G, Geremia RA, Choler P (2012) Changes in root-associated microbial communities are determined by species-specific plant growth responses to stress and disturbance. Eur J Soil Biol 52:59–66

Bourn WS (1932) Ecological and physiological studies on certain aquatic angiosperms. Contrib Boyce Thompson Inst 4:425–496

Boutilier L, Jamieson R, Gordon R, Lake C (2011) Modeling *E. coli* fate and transport in treatment wetlands using the water quality analysis and simulation program. J Environ Sci Health A Tox Hazard Subst Environ Eng 46:680–691

Brisson J, Chazarenc F (2009) Maximizing pollutant removal in constructed wetlands: should we pay more attention to macrophyte species selection? Sci Total Environ 407:3923–3930

Bristow JM (1974) Nitrogen fixation in the rhizosphere of freshwater angiosperms. Can J Bot 54:217–221

Chaidez C, Soto-Beltran M, Gerba CP, Tamimi AH (2014) Reduction of risk of Salmonella infection from kitchen cleaning clothes by use of sodium hypochlorite disinfectant cleaner. Lett Appl Micro 59:487–492

Chandra R, Saxena G, Kumar V (2015) Phytoremediation of environmental pollutants: an ecosustainable green technology to environmental management. In: Chandra R (ed) Advances in biodegradation and bioremediation of industrial waste, 1st edn. CRC Press/Taylor & Francis Group, Boca Raton, pp 1–30. https://doi.org/10.1201/b18218-2

Chauret C, Bagwell B, Meybodi S, Nolan K, Chen P, Springthorpe S, Sattar S (1998) Effect of microbial antagonism on the survival of Cryptosporidium pavumoocysts in fresh waters. In: Proceedings of IAQW 19th biennial international conference, Canada

Chong-Bang Z, Wang J, Liu LW, Zhu SX, Ge HL, Chang SX, Chang J, Ge Y (2010) Effect of plant diversity on microbial biomass and community metabolic profiles in a full-scale constructed wetland. Ecol Eng 36:62–68

Cooper PF, Job GD, Green MB, Shutes RBE (1996) Reed beds and constructed wetland for wastewater treatment, WRC report, report no. 17/0014, Swindon, pp 47–54

Cooper DJ, Griffin P, Cooper PF (2005) Factors affecting the longevity of subsurface horizontal flow systems operating as tertiary treatment for sewage effluent. Water Sci Tech 51(9):127–135

Culp GL, Wesner GM, Culp RL (1986) Handbook of public water system. Cambridge University Press, Cambridge

Dolinar N, Gaberscik A (2010) Mycorrhizal colonization and growth of Phragmites australis in an intermittent wetland. Aquat Bot 93:93–98

Dunabin JS, Browmer KH (1992) Potential use of constructed wetlands for treatment of industrial wastewater containing metals. Sci Total Environ 3:151–168

Dushenkov V, Motto H, Raskin I, Kumar PN (1995) Rhizofiltration: the use of plants to remove heavy metals from aqueous streams. Environ Sci Tech 30:1239–1245

Ebrahimi A, Taheri E, Ehrampoush MH, Nasiri S, Jalali F, Soltani R, Fatehizadeh A (2013) Efficiency of constructed wetland vegetated with Cyperus alternifolius applied for municipal wastewater treatment. J Environ Public Health 2013:1–5

Elias SH, Mohamed M, Nor-Anuar A, Muda K, Hassan MAHM, Othman MN, Chelliapan S (2014) Ceramic industry wastewater treatment by rhizofiltration system – application of water hyacinth bioremediation. 11OABJ 5:6–14

Farroqi IH, Busheer F, Chaudhari RJ (2008) Constructed wetland systems (CWS) for wastewater treatments. In: Proceedings of Taal 2007: the 12th world lake conference, pp 1004–1009

Garcia J, Rousseau DPL, Morato J, Lesage E, Matamoros V, Bayona JM (2010) Contaminant removal processes in subsurface-flow constructed wetlands: a review. Crit Rev Environ Sci Tech 40:561–661

Gautam S, Kaithwas G, Bharagava RN, Saxena G (2017) Pollutants in tannery wastewater, pharmacological effects and bioremediation approaches for human health protection and environmental safety. In: Bharagava RN (ed) Environmental pollutants and their bioremediation approaches, 1st edn. CRC Press/Taylor & Francis Group, Boca Raton, pp 369–396. https://doi.org/10.1201/9781315173351-14

Gill LW, Ring P, Higgins NMP, Johnston PM (2014) Accumulation of heavy metals in a constructed wetland treating road runoff. Ecol Eng 70:133–139

Goutam SP, Saxena G, Singh V, Yadav AK, Bharagava RN (2018) Green synthesis of TiO_2 nanoparticles using leaf extract of *Jatropha curcas* L. for photocatalytic degradation of tannery wastewater. Chem Eng J 336:386–396. https://doi.org/10.1016/j.cej.2017.12.029

Greenway M (2005) The role of constructed wetlands secondary effluent treatment and water reuse in subtropical and arid Australia. Ecol Eng 25:501–509

Greenway M, Woolley A (2001) Constructed wetlands in Queensland: performance efficiency and nutrient bioaccumulation. Ecol Eng 12:39–55

Haarstad K, Bavor HJ, Maehlum T (2012) Organic and metallic pollutants in water treatment and natural wetlands: a review. Water Sci Technol 65:77–99

Hochereutiner G (1986) Etodessur les phaneragamesaquatique do rhone et du part de geneve. Renue Genet Bacteriol 8:158–268

Howard RJ (2010) Intraspecific variation in growth of marsh macrophytes in response to salinity and soil type: implications for wetland restoration. Estuar Coasts 33:127–138

Hunter R, Combs D, George D (2001) Nitrogen, phosphorous and organic carbon removal in simulated wetland treatment systems. Arch Environ Contam Toxicol 41:274–281

Hutchinson GE (1975) Treatise on limnology, limnological botany. John Wesley, New York

Kalbar PP, Karmakar S, Asolekar SR (2012) Selection of appropriate wastewater treatment technology: scenario based multiple-attribute decision-making approach. Environ Monit 113:158–169

Kalipci E (2011) Investigation of decontamination effect of Phragmites australis for Konya domestic wastewater treatment. Med Plants Res 5:6571–6577

Karim MR, Manshadi FD, Karpiscak MM, Gerba CP (2004) The persistence and removal of enteric pathogens in constructed wetlands. Water Res 38:1831–1837

Kawabata O, Nishimoto RK, Tang CS (1994) Interference of two Kyllinga species (Kyllinga nemoralis and Kyllinga brevifolia) on Bermudagrass (Cynodon dactylon) growth. Weed Technol 8:83–86

Kayyali M, Jamrah IA (1999) A study of selected pathogens in municipal wastewater treatment plant effluents. Int J Environ Health Res 9:321–328

Keddy PA (2000) Wetland ecology: principles and conservation. Cambridge University Press, Cambridge

Kropfelova L, Vymazal J, Svehla J, Stichova J (2009) Removal of trace elements in three horizontal sub-surface flow constructed wetlands in the Czech Republic. Environ Pollut 157:1186–1194

Lee KH (2011) Intergrating carbon footprint into supply chain management: the case of Hyundai Motor Company in the automobile industry. Clean Prod 119:1216–1237

Lee B, Scholz M (2007) What is the role of Phragmites australis in experimental constructed wetland filters treating urban runoff? Ecol Eng 29:87–95

Lee M, Yang M (2010) Rhizofiltration using sunflower (Helianthus annuss L.) and bean (*Phaseolus vulgaris* L. var vulgaris) to remediate uranium contaminated groundwater. Hazard Mater 173:589–596

Lilach IK, Yitzhak H, Dana M, Avital G, Dror M (2010) Microbial population and activity in wetland microcosms constructed for improving treated municipal wastewater. Environ Biol 59:700–709

Litter MI, Alarcon-Herrera MT, Arenas MJ, Aviles M, Dias LE, Cornejo L, Caceres RE, Cipriani HN (2012) Small-scale household methods to remove arsenic from water for drinking purposes in Latin America. Sci Environ 429:107–122

Liu ZH (2009) Removal mechanisms for endocrine disrupting compounds (EDCs) in wastewater treatment -physical means, biodegradation, and chemical advanced oxidation: a review. Sci Total Environ 407:731–748

LizotterJr RE, Moore MT, Locke MA, Kroger R (2001) Role of vegetation in a constructed wetland on nutrients-pesticide mixture toxicity to Hyalellaazteca. Arch Environ Contam Toxicol 60:261–271

Luca GAD, Maine MA, Mufarrege MM, Hadad HR, Sánchez GC, Bonetto C (2011) Metal retention and distribution in the sediment of a constructed wetland for industrial wastewater treatment. Ecol Eng 37:1267–1275

Lum AF, Ngwa ESA, Chikoye D, Suh CE (2014) Phytoremediation potential of weeds in heavy metal contaminated soils of the Bassa Industrial Zone of Douala, Cameroon. Int J Phytoremediation 16:302–319

Majumder P (2013) Investigation of anthelmintic activity of an ignored plant 'KyllingaNemoralis' tuber – a potential hope. Intern J Pharma Biol Sci 4:45–52

Malik AI, Islam AKMR, Colmer TD (2010) Transfer of the barrier to radial oxygen loss in roots of Hordeum marinum to wheat (Triticumaestivum): evaluation of four H. marinum–wheat amphiploids. New Phytol 190:499–508

Mccarthy G, Lawlor PG, Gutierrez M, Gardiner GE (2011) Removal of Salmonella and indicator micro-organisms in integrated constructed wetlands treating agricultural wastewater. J Environ Sci Health A Toxic Hazard Subst Environ Eng 46:764–770

Mitsch WJ, Jorgensen SE (2004) Ecological engineering and ecosystem restoration, treatment wetlands. Wiley, Hoboken

Moeseneder MM, Winter C, Arrieta JM, Herndl GJ (1993) Terminal-restriction length polymorphism (T-RFLP) screening of a marine archaeal clone library to determine the different phylotypes. Micro Methods 44:159–172

Mosteo R, Ormad MP, Goñi P, Rodríguez-Chueca J, García A, Clavel A (2013) Identification of pathogen bacteria and protozoa in treated urban wastewaters discharged in the Ebro River (Spain): water reuse possibilities. Water Sci Tech 68:575–583

Moura A, Taca OM, Henriques I, Dias J, Ferreira P, Correia A (2007) Characterization of bacterial diversity in two aerated lagoons of a wastewater treatment plant using PCR-DGGE analysis. Microbiol Res 164:550–569

Mthembu MS, Odinga CA, Swalaha FM, Bux F (2013) Constructed wetlands: a future alternative wastewater treatment technology. Afr J Biotech 12:4542–4553

Mukesh P, Thakur LS (2013) Heavy metal Cu, Ni and Zn: toxicity, health hazards and their removal techniques by low cost adsorbents: a short overview. Intern J Plant Anim Environ Sci 3:143–157

Nedunuri KV, Lowell C, Meade W, Vonderheide AP, Shann JR (2014) Management practices and phytoremediation by native grasses. Int J Phytoremediation 12:200–214

Netherlands Environmental Assessment Agency (2012) The environment and the planet by 2050. NEAA Publishing, Netherlands, pp 1–55

Nilsson E, Sha L, Qian W, Leedo M (2012) Constructed wetlands for wastewater treatment. VVANOI decentralized water and wastewater treatment

Odinga C, Otieno F, Adeyemo J (2011) Investigating the effectiveness of aquatic plants (Echinocloa L. and Cyperus L.) in removing nutrients from wastewater: the case of Chemelil constructed wetland-Kenya. Intern J Phys Sci 6:3960–3970

Odinga CA, Swalaha FM, Otieno FAO, Ranjith KR, Bux F (2013) Investigating the efficiency of constructed wetlands in the removal of heavy metals and enteric pathogens from wastewater. Environ Tech Rev 2:1–16

OECD (2012) OECD environmental outlook to 2050: the consequences of inaction. OECD Publishing, Paris, pp 1–350

Pedescoll A, Corzo A, Alvarez E, Garcia J, Purgagut J (2011) The effect of primary treatment and flow regime on clogging development in horizontal subsurface flow constructed wetlands. Exp Eval Water Res 45:3579–3589

Rani N, Vajpayee P, Bhatti S, Singh S, Shanker R, Gupta K (2014) Quantification of Salmonella Typhi in water and sediments by molecular-beacon based qPCR. Ecotoxi Environ Saf 108:58–64

Rawat K, Fulekar MH, Pathak B (2012) Rhizofiltration a green technology for remediation of heavy metals. Intern J Innov Biol Sci 2:193–199

Ray JG, George J (2009) Phytosociology of roadside communities to identify ecological potentials of tolerant species. J Ecol Nat Environ 1:184–190

Redder A, Matthias D, Daeschlein G, Baeder-Bederski O, Mkoch C, Uller RM, Exner M, Borneff-Lipp M (2010) Constructed wetlands-are they safe in reducing protozoan parasites? J Hyg Environ Health 213:72–77

Romero SE, Nunez LJ, Negrete M, Rios JEA, Hadad HR, Maine MA (2011) Hg, Cu, Cd, and Zn accumulation in macrophytes growing in tropical wetlands. Water Air Soil Pollut 216:361–373

Saxena G, Bharagava RN (2015) Persistent organic pollutants and bacterial communities present during the treatment of tannery wastewater. In: Chandra R (ed) Environmental waste management, 1st edn. CRC Press/Taylor & Francis Group, Boca Raton, pp 217–247. https://doi.org/10.1201/b19243-10

Saxena G, Bharagava RN (2017) Organic and inorganic pollutants in industrial wastes, their ecotoxicological effects, health hazards and bioremediation approaches. In: Bharagava RN (ed) Environmental pollutants and their bioremediation approaches, 1st edn. CRC Press/Taylor & Francis Group, Boca Raton, pp 23–56. https://doi.org/10.1201/9781315173351-3

Saxena G, Chandra R, Bharagava RN (2016) Environmental pollution, toxicity profile and treatment approaches for tannery wastewater and its chemical pollutants. Rev Environ Contam Toxicol 240:31–69. https://doi.org/10.1007/398_2015_5009

Saxena G, Purchase D, Mulla SI, Saratale GD, Bharagava RN (2018) Phytoremediation of heavy metal-contaminated sites: environmental considerations, field studies, sustainability and future prospects. J Environ Manag

Sayadi MH, Kargar R, Doosti MR, Salehi H (2012) Hybrid constructed wetlands for wastewater treatment: a worldwide review. Proceedings of the Interna Academy of Eco and Environ Sc Stanford California. pp 204–222

Sheoran AS, Sheoran V (2006) Heavy metal removal mechanism of acid mine drainage in wetlands: a critical review. Min Eng 19:105–116

Shimamura S, Yamamoto R, Nakamura T, Shimada S, Komatsu S (2010) Stem hypertrophic lenticels and secondary aerenchyma enable oxygen transport to roots of soybean in flooded soil. Ann Bot 106:277–284

Shiono K, Ogawa S, Yamazaki S, Isoda H, Fujimura T, Nakazono M, Colmer TD (2010) Contrasting dynamics of radial O2-loss barrier induction and aerenchyma formation in rice roots of two lengths. Ann Bot 1–11. https://doi.org/10.1093/aob/mcq221

Shrestha RR, Tuladhar B, Shrestha P (2003) Experiences with application of constructed wetlands for wastewater treatment and reuse in Nepal. ENPHO Magazine, Environ and Public Health Organisation, Kathmandu

Sindhu T, Rajamanikandan S, Srinivasan P (2014) In vitro antioxidant and antibacterial activities of methanol extract of Kyllinga nemoralis. Indian J Pharma Sci 76:170–174

Song Q, Li J (2014) Environmental effects of heavy metals derived from the e-waste recycling activities in China: a systematic review. Waste Manag. https://doi.org/10.1016/j.wasman.2014
Srivastava J, Gupta A, Chandra H (2008) Managing water quality with aquatic macrophytes. Environ Sci J. https://doi.org/10.1007/s11157-008-9135-x
Stefanakis AI, Akratos CS, Tsihrintzis VA (2011) Effect of wastewater step-feeding on removal efficiency of pilot-scale horizontal subsurface flow constructed wetlands. Ecol Eng 37:431–443
Steudle E (2000) Water uptake by roots: effects of water deficit. J Exp Bot 51:1531–1542
Stevik TK, Aa K, Ausland G, Hanssen JF (2004) Retention and removal of pathogenic bacteria in wastewater percolating through porous media: a review. Water Res 38:1355–1367
Stottmeister U, Wießner A, Kuschk P, Kappelmeyer U (2003) Effects of plants and microorganisms in constructed wetlands for wastewater treatment. Biotech Adv 22:93–117
Swearingen J, Saltonstall K (2010) Phragmites field guide: distinguishing native and exotic forms of common reed (Phragmites australis) in the United States. Plant conservation alliance, weeds gone wild. http://www.nps.gov/plant/alien/pubs/index.htm
Thayaparan M, Iqbal SS, Chathuranga PKD, Iqbal MCM (2013) Rhizofiltration of Pb by Azollapinnata. Intern J Environ Sci 3. doi:https://doi.org/10.6088/ijes.2013030600002
Tuttolomondo T, Licata M, Leto C, Leone R, Bella SL (2014) Effect of plant species on water balance in a pilot-scale horizontal subsurface flow constructed wetland planted with Arundodonax L. and Cyperus alternifolius L.-Two-year tests in a Mediterranean environment in the West of Sicily (Italy). Ecol Eng 74:79–92
United States Environmental Protection Agency (USEPA) (1999) Methods for collection, storage and manipulation of sediments for chemical and toxicological analyses: technical manual, EPA 823-B-01-002. Office of Water, United States Environmental Protection Agency, Washington, DC
Valigore JM, Gostomski PA, Wareham DG, O'Sullivan AD (2012) Effects of hydraulic and solids retention times on productivity and settle ability of microbial (microalgal-bacterial) biomass grown on primary treated wastewater as a biofuel feedstock. Water Res 46:2957–2964
Varela AR, Manaia CM (2013) Human health implications of clinically relevant bacteria in wastewater habitats. Environ Sci Pollut Res Intern 20:3550–3569
Vera I, García J, Sáez K, Moragas L, Vidal G (2011) Performance evaluation of eight years' experience of constructed wetland systems in Catalonia as alternative treatment for small communities. Ecol Eng 37:364–371
Verboven P, Pedersen O, Herremans E, Ho QT, Nicolaı¨ BM, Colmer TD, Teakle N (2011) Root aeration via aerenchymatous phellem: three-dimensional micro-imaging and radial O_2 profiles in Melilotus siculus. New Phytol 193:420–431
Vymazal J (2005) Horizontal sub-surface flow and hybrid constructed wetlands systems for wastewater treatment. Ecol Eng 25:478–490
Vymazal J (2007) Removal of nutrients in various types of constructed wetlands. Sci Total Environ 380:48–65
Vymazal J (2009) The use constructed wetlands with horizontal sub-surface flow for various types of wastewater. Ecol Eng 35:1–17
Vymazal J (2010) Constructed wetlands for wastewater treatment. Water Res 2:530–549
Vymazal J (2011) Plants used in constructed wetlands with horizontal subsurface flow: a review. Hydrolobiologia 674:133–156
Vymazal J, Kropfelova L (2005) Growth of Phragmites australis and Phalaris arundinacea in constructed wetlands for wastewater treatment in the Czech Republic. Ecol Eng 25:606–621
Vymazal J, Kropfelova L (2009) Removal of organics in constructed wetland with horizontal subsurface flow: a review of the field experience. Sci Total Environ 407:3911–3922
Wu H, Zhang J, Li P, Zhang J, Xie H, Zhang B (2011) Nutrient removal in constructed microcosm wetlands for treating polluted river water in Northern China. Ecol Eng 37:560–568
Wu S, Kuschk P, Brix H, Vymazal J, Dong R (2014) Development of constructed wetlands in performance intensifications for wastewater treatment: a nitrogen and organic matter targeted review. Water Res 57:40–55

Yadav AK, Kumar N, Sreekrishnan TR, Satya S, Bishnoi NR (2010) Removal of chromium and nickel from aqueous solution in constructed wetland: mass balance, adsorption-desorption and FTIR study. Chem Eng 160:122–128

Yeh TY, Chou CC, Pan C (2009) Heavy metal removal within pilot-scale constructed wetlands receiving river water contaminated by confined swine operations. Desalination 249:368–373

Yongjun Z, Liu B, Zhang W, Hu C, An S (2010) Nutrient removal in constructed microcosm wetlands for treating polluted river water. Ecol Eng 36:167–177

Yoon CG, Kwun SK, Ham JM (2001) Feasibility study of a constructed wetland for sewage treatment in a Korean rural community. Environ Sci Health A 36:1101–1112

Zhang Z, Rengel Z, Meney K (2007) Nutrient removal from simulated wastewater using Camma indica and Schoemoplectus validus in mono-and mixed-culture in wetland microcosms. Water Air Soil Pollut 183:95–105

Zhang C, Qiao Q, Piper JD, Huang B (2011) Assessment of heavy metal pollution from a Fe-smelting plant in urban river sediments using environmental magnetic and geochemical methods. Environ Pollut 159:3057–3070

Chapter 15
Nano-bioremediation: A New Age Technology for the Treatment of Dyes in Textile Effluents

Kadapakkam Nandabalan Yogalakshmi, Anamika Das, Gini Rani, Vijay Jaswal, and Jatinder Singh Randhawa

Contents

Abstract Heterogeneity, recalcitrance, and ubiquitous persistence of textile effluent make it the foulest industrial pollutant which poses a serious threat to soil and water bodies. Textile effluent like all other industrial effluents contributes considerably to environmental pollution. Presence of huge amount of water-soluble unfixed dyes, heavy metals, acids, alkalis, inorganic and organic salts has resulted in a highly concentrated, colored, and complex high strength wastewater that resists degradation. The conventional treatment methods including biological and physico-

K. N. Yogalakshmi (✉) · A. Das · G. Rani · V. Jaswal · J. S. Randhawa
Centre for Environmental Science and Technology, School of Environment and Earth Sciences, Central University of Punjab, Bathinda, Punjab, India

G. Saxena, R. N. Bharagava (eds.), *Bioremediation of Industrial Waste for Environmental Safety*, https://doi.org/10.1007/978-981-13-1891-7_15

chemical methods for treatment of textile waste are not convincing enough because of low biodegradability of dyes, fouling of filters, high pressure requirement, and generation of sludge containing iron hydroxide. Hence, it has become imperative to seek alternative advanced technology which must be essentially environment competent. Exquisite properties are shown to be possessed by the nanoparticles making it an efficient technology for cleanup of environmental pollutants. Nanoremediation is an upcoming field of research with huge prospects in the treatment of environmental contaminants. In addition to its high reactivity with the contaminants, they also act as suitable carriers for immobilization of whole cells and enzymes. Effluent treatment aided by enzymes has been demonstrated to be effective for recalcitrant pollutants and requires moderate reaction conditions, making them environmentally sound. The generic enzymes sought for the treatment of textile pollutants include most of the peroxidase, cytochrome reductase (Fe III), and oxidoreductase. This chapter extensively covers current know-how of nanoparticles as a carrier for several enzymes for the degradation of pollutants present in textile wastewater. The role of nanoparticle in the removal of dyes is also highlighted.

Keywords Nanomaterials · Textile wastewater · Bioremediation · Enzymes · Pollutant removal

15.1 Introduction

The textile industry is one of the most flourishing industries across the globe. According to Global Industry Analysts (GIA) report, the worth of textile industry at present is around $480 billion and is expected to reach $700 billion, shortly. It is the second largest employment generator in many nations after agriculture. In modern scenario, clothing became much more than the need of mankind and became the tool to represent personality and impression on others. People are very conscious about their clothing in terms of quality, colors, designs, and cost. Henceforth, in order to fulfill the global demand, textile industry extensively uses resources like high power, water, and raw material which contribute to pollution of air, water, and soil (MoT 2001).

India produces around 700 different types of dyes and its intermediaries. These dyes and pigments contain hazardous metals such as mercury, chromium, nickel, copper, and cobalt which usually escape the treatment systems because of their incapability to degrade recalcitrant dyestuff. Moreover, textile industries deal with varieties of hazardous chemicals in their various processes starting from preliminary preparatory processes followed by spinning, weaving, and finishing, causing a potential threat to the environment (U.S. Environmental Protection Agency 1997). Although we have various traditional and current technologies to cope up with the existing textile pollution, still the expected effluent quality could not be achieved due to certain limitation. The Bureau of Indian Standards (BIS 10500) has laid down drinking water color standard at the desirable limit of 5 color units and 25 color units to be the maximum permissible limit in the absence of the alternate source. Hence, in order to match the drinking and effluent quality standard laid down by the CPCB and BIS, different physicochemical, membrane technologies and other advanced techniques are encouraged to adopt. These techniques however because of associated high cost and maintenance have forced the research and development to seek a cost-effective and environmentally friendly alternative to answering the problem of dye degradation.

Recently nanoparticles have gathered a lot of attention in many fields of research, and textile effluent treatment is one among them. Nanoparticles have shown to play an important role in the treatment of textile effluent up to the satisfactory standards. Lately, numbers of studies have revealed the effectiveness of various nanomaterials in decolorization of a broad range of dyes and colorants. In this chapter, we will go through the commonly used pollution remediation techniques and the application of nanomaterials as a new combating tool in treating textile-generated pollutants. This chapter also discusses the potential applicability of nanomaterials in enzyme immobilization, acting as a substrate.

15.2 Textile Industry and Its Potential Impact on the Environment

In many countries, the textile industry is one of the major contributors in terms of economy and employment. In India, it serves as the second largest employment-generating sector after agriculture providing employment for over 45–60 million individuals directly and indirectly (Wararkar et al. 2016). According to the annual report 2016–2017 of the textile ministry, around 2% of India's total GDP is contributed by the textile industries. The Indian textile industry comprises of large- and small-scale units across the nation that are involved in the production of traditional and fashionable items for national and international market. Briefly, they are involved in the production of raw and synthetic fibers and occupy the top positions such as the largest producer of jute and second largest producer of silk, cellulose fiber, and yarn. Ludhiana, Mumbai, Surat, Ahmadabad, Coimbatore, and Kanpur are the major textile hubs of India in addition to numerous small towns (Garg and Kaushik 2008; Verma et al. 2012). Apart from being an economic contributor, textile industries also play a vital role in environmental pollution. The textile industries have a wide spectrum of manufacturing activities and processing units that consume a huge amount of power, water, and chemicals. Annually around 1 trillion kilowatts per hour of power is consumed by textile industries. Water is the backbone of the textile industry processes. Beginning with the washing of raw fiber to bleaching, dyeing, and washing of final product, excess amount of water is used. According to the Textile Export Report 2010, the weaving of one T-shirt consumes 700 gal of freshwater. Roughly 200 l of water is consumed to produce 1 kg of fabric. Textile industries are categorized under highly polluting energy-intensive industries as they contribute to 20% freshwater pollution. Broadly, textile waste is classified into four major categories, hard to treat, hazardous or toxic, dispersible, and high volume waste. Textile effluent is usually colored containing high BOD, COD, and total dissolved solids (TDS). In cotton pretreatment process, the average COD concentration reached up to 3000 mg/L, and in case of the dyeing process, COD remains at 1000 mg/L. The other major characteristics of textile industry wastewater are mentioned in Table 15.1. Color is one of the major problems of textile wastewater as they are more stable and resist degradation. Removal of dyes is very difficult even at low concentration. Likewise, use of Glauber and common salt increases the TDS of textile effluent to 17,000 mg/L. Textile effluent is highly distinctive in nature because of the large varieties of chemicals used in different processes. The various salts, surfactants, ionic metals, biocides, toxic anions, dyes, detergents, emulsifiers, and dispersants used in textile industries are a source of aquatic toxicity. Metals enter into the textile waste stream from two sources: dyes and through impurities of chemicals. Only the small proportion of these chemicals such as dyes and finishing agents remain attached to the textiles; the rest enter the processed water (Moustafa 2008).

Excessive use of dyes and other chemicals leads to a huge amount of waste generated that when discharged into water bodies causes pollution. The released chemicals adversely affect the receiving water body due to its high organic matter, detergents and soaps, oil and grease, sulfide, TDS, TSS, and alkalinity (Pala and Tokat 2002) which comprehensively lead to depletion of dissolved oxygen in a water body, increased turbidity, and reduced light penetration for photosynthetic processes. Most of the aquatic organisms are highly sensitive toward pH. The pH of textile wastewater inhibits the growth of aquatic organisms. Dyes present in the wastewater act as a reflector on the surface of water body and hence reduce the light penetration and oxygen diffusion (Wang et al. 2011).

The textile processes also contribute to air pollution during dye and polymerization sequence of finishing operations. Gaseous emission is the second largest pollution problem (after effluent quality) of the textile industry. In textile industries air emission is classified on the basis of their nature into two types: emissions from a point source (boiler, storage tank, and ovens) and the second type a diffusive source which includes warehouses, spills, wastewater treatment, and solvent-based processes. Nitrogen oxides, sulfur dioxides, and volatile organic components are the most common air pollutants generated from the boilers in textile industries. Oil

Table 15.1 Characteristics of textile industry wastewater

Sr. no.	Parameters	Value
1	pH	7.0–9.0
2	Color (Pt-Co)	50–2500
3	Chloride (mg/L)	1000–1600
4	BOD (mg/L)	80–6000
5	COD (mg/L)	150–12,000
6	Total suspended solids (TSS (mg/L)	15–8000
7	Total dissolved solids (TDS (mg/L)	2900–3100
8	Total Kjeldahl nitrogen (mg/L)	70–80

Adapted from Al-Kdasi et al. (2004)

mists are produced when textile material contain spinning oil and during thermal degradation of the processing material. Processes like carbonization and volatilization produce acid mists which are highly corrosive in nature. Depending upon the concentration and amount used in printing and dyeing operations, solvent vapors consist of excessive toxic chemicals such as mineral turpentine oil (MTO), chlorofluorohydrocarbons, ethyl acetate, formaldehyde, mono- and dichlorobenzene, styrene, and hexane (Wang et al. 2011). Generally, environmental hazards occur throughout the whole production chain in textile industries, but wet processing causes a serious threat to the environment. In order to achieve the market demands in all aspects, a huge amount of chemical is used in textile wet processes. The solid waste generated from the textile industry accounts for 5% landfill space. Hence in order to remediate textile effluent, a number of technologies are used; the prominent ones are discussed in the following sections.

15.3 Textile Wastewater Treatment Methods and Technologies

15.3.1 Physical Techniques

Physical methods of textile wastewater treatment are based on the principle of natural sedimentation, electrical attraction, or van der Waals forces. They utilize physical barriers that help in restricting the passage of target pollutants from the wastewater. Physical treatment mechanisms do not cause any change in the chemistry of the target pollutant; however in some cases, e.g., vaporization, the physical structure of the target substance is changed, and in case of filtration, the dispersed substances agglomerate. The common physical treatment methods used for textile wastewater treatment include coagulation and flocculation, ion exchange, adsorption, and membrane filtrations.

15.3.1.1 Coagulation/Flocculation

The process of coagulation and flocculation is carried out to counterbalance the charges by trapping or bridging the particles together to form gelatinous agglomerates, large enough to settle down or restricted in a filter. Coagulation/flocculation is widely employed in the textile industry to treat wastewater because of its low detention time, simple operation and maintenance, and less capital cost. In this technique, coagulants like aluminum sulfate ($Al_2(SO_4)_3$ $18H_2O$), lime ($Ca(OH)_2$), ferric sulfate ($Fe_2(SO_4)_3$ $7H_2O$), and ferric chloride ($FeCl_3$ $7H_2O$) bind with the dispersed dyes and other pollutants to remove them through the action of electrostatic force, sorption, and bridging. The process of sorption and bridging which aid in coagulating/flocculating dispersed dyes and colloids from the wastewater occurs due to

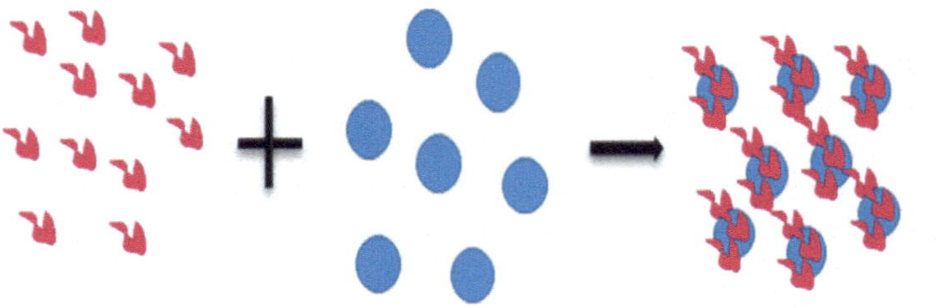

Fig. 15.1 Mechanism of coagulation/flocculation

protonated amine groups and high molecular weight polymer, respectively (Guibal and Roussy 2007). Coagulation and flocculation decrease the suspended matter, dissolved substances, colloidal particles, coloring agents, and non-settable particles from the effluent. The process of coagulation/flocculation is explained in Fig. 15.1. Aluminum-containing coagulants are assumed to cause severe health issues; hence environment-friendly coagulants such as chitosan (cellulosic polymer) were also used. The flocs produced from chitosan coagulation are efficient as they are relatively larger with faster settling velocity. The quality of flocs is enhanced by the presence of inorganic solutes in the mineral suspension (Pan et al. 1999). By this technique, the COD removal of the textile wastewater varied in the range of 50–70% when the concentration of the coagulant/flocculants is optimized (Vera et al. 2005).

Moreover, the chemical coagulants also showed 80–100% removal of dye. For instance, El-Gohary and Tawfik (2009) removed 100% of colorants from the textile wastewater using magnesium chloride-based coagulant with an optimum dosage of 120 mg/L at an optimum pH of 11. Similarly, alum, using commercial cationic flocculent, showed to treat 98% of reactive as well as acid dyes at an optimized dosage of 20 mg/L and at neutral pH (Golob et al. 2005).The pH of the textile wastewater influences the coagulation process. The pH 4 of the wastewater is considered optimum in some cases as it increases the sorption capacity when it interacts electrostatically with the anions in the solution. Optimum pH leads to rapid and larger floc formation resulting in easy settling, decreased turbidity, and high COD removal (72.5%) (Guzman et al. 2003). Unlike chitosan which works best at low pH, lime is shown to be highly effective at higher pH. Lime coagulant resulted in 100% decolorization and 50% COD removal. Magnesium chloride along with lime showed complete decolorization and COD reduction by 40% at pH 11 (Torres et al. 2010). The demerits associated with this process are sludge problems that arise from the usage of coagulants and secondly the requirement of requisite chemical addition to maintaining optimum pH for effective removal of dyes. In some cases, coagulants like gelatinous hydroxide tend to get solubilized in the wastewater. Likewise, electrochemical coagulation requires excessive conductivity in wastewater. Sometimes there is the formation of a non-permeable oxide layer on the cathode, and due to its oxidation, anode gets dissolved in the wastewater, limiting the performance of electrocoagulation cell. The requirement of external power supply makes the process more expensive.

15.3.1.2 Ion Exchange

Ion exchange removes undesirable positively and negatively charged ions from the effluent. In this process, the external ionic solution is exchanged with the ions attached to the functional groups electrostatically. Most of the dyes in textile industries are either cationic or anionic, so theoretically they can be removed through the process of ion exchange. Textile wastewaters which contain different dispersed dyes are passed through ion exchange resins till all the ion exchanging active sites are saturated. Once it gets saturated, it is backwashed to remove all the contaminants, and it is further reconditioned. There are various types of ion exchangers: organic and inorganic ionic exchangers that are naturally and synthetically produced with desirable chemical and physical properties.

Composite ion exchangers are one that has more than one ion exchanger combined with either organic, inorganic, or another ion exchanger (Dyer and Hudson 1993). The acidic dyes (anionic and reactive) and basic dyes (cationic) can be efficiently removed through this method, but the removal of other dispersed (nonionic) dyes is negligible. The total exchange capacity of an ion exchange resin is dependent on the number of the functional groups present on it. The process of ion exchange is affected by the concentration of the functional groups which influence the distribution coefficient and variations in the physical factors. High distribution coefficient is always desired. Ion exchange process can be executed in different ways: batch mode, packed bed column, continuous loop, and combined with membrane technology. This technique is being used since years for decontamination and separation of the ion-containing aqueous solution. However, the ion exchange resins are not efficient enough to accommodate the entire range of dyes, so they are not widely applied. The sorbent used in this method is not lost when regenerated, and it is effective (Robinson et al. 2001). Ion exchange is a reliable technology to remove impurities from water and other aqueous phases. Laszlo in 1996 prepared anion exchanger from sugarcane bagasse by quaternizing and cross-linking it. It was simple, cheap and promptly trapped dyestuff. Ion exchange process removed 25–50% organic content and 25% turbidity (Metcalf and Eddy Inc 2003). However, they are not effective for solutions containing lower pH or when the concentration of iron, manganese, and aluminum is high. Other limitations of the process include high TDS and suspended solids. The waste generated in this process is quite concentrated and requires careful disposal. Moreover, the regeneration of sorbent requires organic solvents such as methanol which adds to the operational cost (Joshi et al. 2004).

15.3.1.3 Adsorption

Adsorption is an equilibrium separation technology where the pollutant in the solution is adsorbed onto the adsorbent, which is depicted in Fig. 15.2. This process is widely used in textile wastewater treatment for the removal of dyes and other toxic chemicals such as pesticides, phenols, and cyanides. Adsorption mechanism is studied using different adsorption isotherm models including Langmuir, Freundlich, and Dubinin-Radushkevich adsorption isotherm (Freundlich 1906; Langmuir 1916; Dubinin and Radushkevich 1947). Adsorption isotherm data is crucial to predict and compare the performance of adsorption. It is essentially important for optimizing the adsorption pathways and to estimate the capacity of adsorbents (Thompson et al. 2009). Modeling of adsorption data expresses the process of adsorption at constant pH and temperature (Foo and Hameed 2010).

Among the different adsorbents, activated carbon is most commonly used due to its porosity and large surface area per unit mass. Activated carbon outfitted the adsorption isotherm models (Langmuir and Freundlich) and showed greater efficiency in decolorization of textile wastewater containing various dyes (Qu et al. 2008). Large pore size facilitates diffusion of organic pollutant to reach the active sites. The surface area and pore size of the activated carbon are 750 m^2/g and 2 μm, respectively (Lee et al. 2000). Adsorbents having high affinity and capability of regeneration are always preferred for the purpose of decolorization. Even though the adsorption capacity of the activated carbon is very high, it is too expensive; henceforth low-cost natural adsorbents and other by-products from industries have been given much attention.

In India and China, adsorbents derived from biomass derivatives such as bamboo and agricultural waste have been shown to treat textile wastewater effectively (Karthikeyan et al. 2008; Chan et al. 2008). Other natural adsorbents include flash, teak bark powder, seeds of avocado and pear, bentonite and natural clay. Moreover the different range of polysaccharides such as chitin, chitosan, locust bean gum, tamarind gum, cornstarch, wheat starch, pectin, and alginic acid have been used as

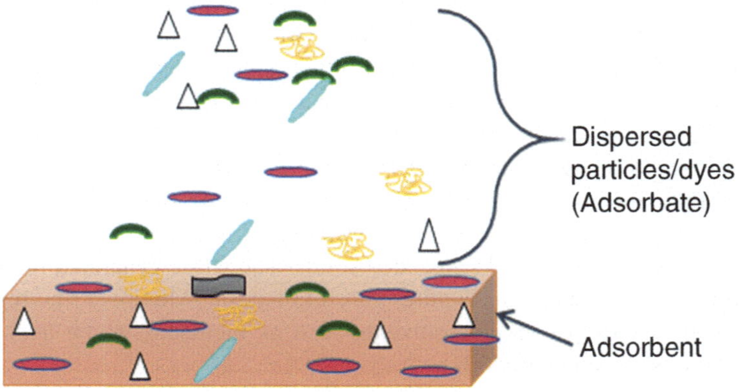

Fig. 15.2 Process of adsorption

an adsorbent. The usage of chitosan showed 94.2%, 92.2% and 99% removal of reactive, direct and acid dyes, respectively (Blackburn 2004). Rahman and Akter (2016) were able to achieve maximum dye removal of 96–97% at the optimum chitin dosage of 2 g and contact time of 60 min. But the absorptivity of chitosan and chitin showed slow kinetics with adsorption capacity of 0.45 mol/kg for acid red dye, and in the case of chitin, the adsorption capacity of 0.45 and 0.13 mol/kg was observed for acid blue dye and acid red dye, respectively (Joshi et al. 2004). Moreover, the major problem associated with this technique is sludge production containing saturated dye materials which pose disposal issue. Regeneration is another major problem of the adsorbents. Thermal or chemical regeneration results in partial recovery of the blocked sites. Further frequent regeneration is required for the adsorbent.

15.3.1.4 Membrane Filtration

Membrane filtration is widely used in process industries to yield high-quality pollutant- and salt-free final product, the permeate. It is an efficient technology to treat textile effluent as it removes dyes, salts such as chlorides, sulfates, silicates, and many more. The product water obtained from membrane processes is of high quality that it can be used for indirect potable water purposes. A general dead-end membrane filtration process is shown in Fig. 15.3. Microfiltration, ultrafiltration, nanofiltration, and reverse osmosis are the popularly used membrane processes. Microfiltration among these is economically viable and has the biggest pore size of around 0.1 μ which can serve the purpose of pretreatment of textile effluent before passing through nanofiltration or reverse osmosis (RO).

Microfiltration membrane, however, removes significant varieties of dispersing and vat dyes. Polypropylene microfiltration membrane is reported to remove direct red 2 and salts from the feed at a feed conductivity of less than 500 mμ Siemens (Porter and Gomes 2000). All other dyes except the reactive dyes are removed by ultrafiltration membranes. Moreover, the use of ultrafiltration membranes is limited due to the problems of rapid fouling, unlike nanofiltration membranes that reject all types of dyes with minimum membrane clogging (Jiraratananon et al. 2000). RO technique which yields high-quality permeate possesses a potential to retain more than 90% of salts and pollutants. The reverse osmosis (0.69–6.9 MPa) and ultrafiltration (0.069–0.69 MPa) are high-pressure processes with the ability to withstand high flux rates, high temperatures, and pH. In order to pass pure water through the RO membrane, a fair amount of pressure is required to surpass the osmotic gradient of the feed and force water to pass from the reject compartment into clear water channel through the membrane. Around 95–100% decolorization of colored effluent is achieved in RO process (Joshi et al. 2004; Hul 1999). Generally, in RO system, the wastewater stream is pumped through the pretreatment compartment which removes the suspended solids and ions such as iron and magnesium to avoid membrane fouling. RO membranes are available in different modules like a spiral wound; tubular, hollow fiber; and disk module. They are sensitive to oxidants like chlorine

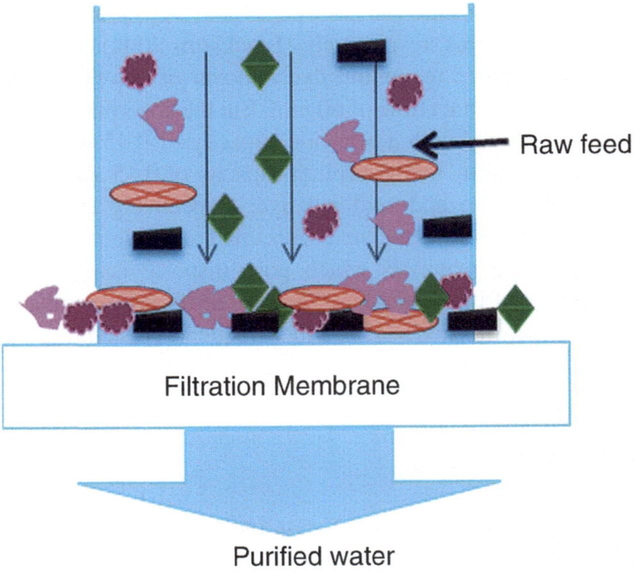

Fig. 15.3 Dead-end membrane filtration

and ozone; hence it should be ensured that they are removed. Moreover, the RO membrane gets fouled due to organic and inorganic substances, colloidal materials, and microorganisms, so their materials and particles have to be removed from the feed prior to passing through the membrane. The demerits of this system include expensive membranes, the elevated working pressure to drive wastewater through the membrane, and shorter lifetime of the membrane as the membrane pores get clogged frequently.

15.3.2 Chemical Treatment Methods

15.3.2.1 Oxidation Process

In this method, dyes are degraded with the help of chemicals. The oxidation process is categorized into an advanced oxidation process (AOP) and chemical oxidation. Irrespective of the type, AOPs use hydroxyl radical, a strong oxidant with an oxidizing potential of 2.33 V to degrade harsh pollutants. The generation of hydroxyl free radical is accelerated by combined ozone, UV radiation, hydrogen peroxide, electron beam irradiation, ultrasound, and titanium dioxide. Among these, ozone/hydrogen peroxide, ozone/UV, and hydrogen peroxide/UV are prominent in oxidizing textile wastewater. In AOP, the organic compounds are mineralized into CO_2, water, and mineral acids with the help of hydroxyl radicals. Hydroxyl radicals are capable of oxidizing majority of complex organic and inorganic compounds present in

Table 15.2 Treatment of textile wastewater using advanced oxidation process

Sr. no.	Type of dyes/textile wastewater	Removal efficiency of dyes (%)	Operating condition	References
1	Reactive yellow 84	80	–	Koch et al. (2002)
2	Biotreated textile wastewater	99	COD_{in} 160 mg/L	Gianluca and Nicola (2001)
		95	COD_{in} 203 mg/L	
3	Simulated reactive dye bath effluent	90	pH 7	Mehmet and Hasan (2002)
4	Disperse dyes	10	pH 8	Lidia et al. (2001)
5	Dye house wastewater	99	Time: 2 h	Percowski and Kos (2003)
6	Reactive dye	80–82	H_2O_2/UV and time: 2 h	Georgiou et al. (2002)
7	Polyester fiber dyeing process	99	O_3/H_2O_2/UV and time: 90 min	Azbar et al. (2004)

textile wastewater with a higher rate of reaction constant. Table 15.2 summarizes decolorization of various dyes using advanced oxidation process. Ultraviolet radiations in the range of $\lambda = 254–578$ nm showed 47% treatment efficiency when textile effluent was exposed to UV for 1–3 h (Ledakowicz et al. 2001). AOP and chemical oxidation can be used individually or synergistically (hybrid AOP) to degrade toxic compounds, dyes, pesticides, and other pollutants. AOP also includes photocatalytic oxidation in which sunlight is used to activate semiconductor catalyst and Fenton reaction. In Fenton oxidation, hydrogen peroxide and ferrous ion produce OH radical that decolorizes a broad range of dyes in short reaction time (Ay et al. 2008). However, Fenton process leads to huge amount of iron sludge production because of flocculation of reagent and dye molecule. In chemical oxidation technique, ozone and hydrogen peroxide form tough nonselective hydroxyl radicals at high pH. These radicals efficiently break down the conjugated double bond and complex aromatic ring structure of dyes. The efficiency of chemical oxidation is lower compared to AOP.

15.3.2.2 Ozonation

Ozone is one of the strongest oxidants which reacts with a large number of organic compounds either directly or by forming secondary hydroxyl free radicals and is widely used to disinfect potable water (Baig and Liechti 2001). It is popularly used in industries for the oxidation of organic and inorganic compounds and for the removal of dyes and odors. Ozonation breaks complex organic compounds into simpler compounds. Ozone is provided to the wastewater through either turbine mixer or diffuser tubes. The process diagram of ozonation is shown in Fig. 15.4. An ozone dose of 300 mg/dm^3 showed improved biodegradability index (1.6 times) of textile effluent (Mehmet and Hasan 2002). Ozone showed the removal of all dyes except vat dyes and nonsoluble disperse dyes which are too slow to react (Namboodri

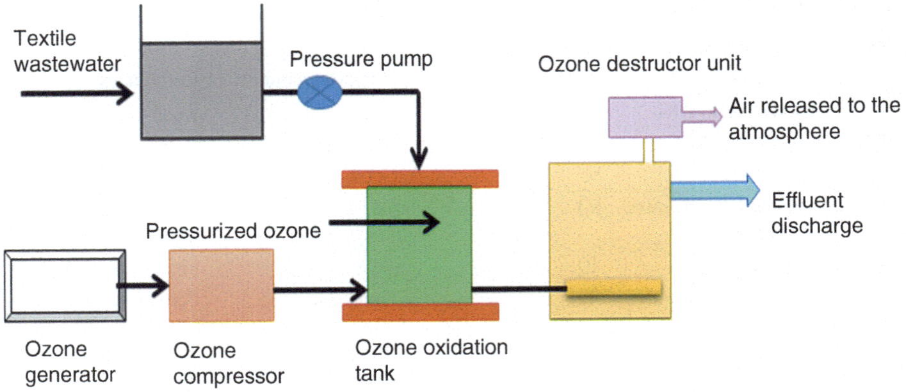

Fig. 15.4 Textile wastewater treatment using ozonation process

et al. 1994). Biotreated textile water showed 99% decolorization efficiency when exposed to ozone for 40–60 min (Ahmet et al. 2003).

Rajeshwari (2000) stated that color removal efficiency was dependent on the temperature. Likewise, increase in pH also showed enhanced oxidation of dyes in textile wastewater. Increased decolorization of simulated reactive dye bath effluent was recorded at pH 7 and ozone concentration of 2970 mg/L (Arslan and Isil 2002). Other examples of textile wastewater treatment using oxidation process are mentioned in Table 15.2. An ozone dosage of 2 mg/L has been demonstrated to remove color and other complex pollutants. Moreover, the biodegradability of textile wastewater is improved in addition to the reduction in COD (Ramasamyaa et al. 2001). As ozone is used in the gaseous state, it results in no sludge production. Water-soluble dyes are rapidly decolorized, while insoluble dyes take a longer time to oxidize. Prior to ozonation, pretreatment is required to avoid excessive ozone consumption by the pollutants present in the wastewater.

15.3.2.3 Electrochemical Methods

In electrochemical anodic oxidation processes, the pollutants are either directly oxidized at the surface of the electrode (direct electrolysis) or the oxidation occurs with the help of oxidizing species generated electrochemically (indirect electrolysis). The energy supplied to an electrochemical reactor plays an important role in an electrochemical process. It transfers electroactive particles from the electrolytic solution (feed) to the electrode and electrons between the feed and the electrode. The electrical energy is converted into chemical energy at the interface of the electrode resulting in oxidation of the pollutants. The operation of the electrochemical process does not require high temperature; hence the use of additional chemicals is avoided (Kim et al. 2002). As this process is non-specific, it acts on a wider range of dyes and pollutants, and it does not form any by-products. Around

100% decolourization was achieved in 6 min of electrolysis during the dyeing process (Vlyssides et al. 2000). PbO_2-coated titanium subtract insoluble anode (TSIA) showed 100% removal of anthraquinone dyes (disperse blue, 26) through electrochemical oxidation. The electrolysis reaction conducted for 3 h and 66 min removed 90–93% of colors at the current density of 4.5 A/dm^2 at neutral pH (Manisankar et al. 2001). Jethava et al. (2001) recorded 95.2% of color removal at the current density of 29.9 mA/cm^2 and flow rate of 5 ml/min. The utilization of chemicals in this process is nearly zero. Moreover, there is no sludge production, and this process does not yield any hazardous metabolite.

15.3.3 Biological Treatment Methods

In biological treatment, microorganisms degrade organic dyes through the aerobic or anaerobic process (Saxena and Bharagava 2015, 2017; Gautam et al. 2017; Bharagava et al. 2017; Saxena et al. 2016). Microbes degrade and utilize the organics as an energy source. Different dyes contained in textile effluent are harmful to the aerobic microorganisms and cause problems of sludge bulking, sludge rising, and flocculation. The aerobic biological process is reported to be not efficient enough to degrade textile dyes especially azo dyes (Ibrahim et al. 2009). Moreover, it requires larger space and higher hydraulic retention time. Aeration is required which can lead to the generation of unknown oxidation compound which can add the color intensity of the effluent. A general biological aeration process is shown in Fig. 15.5. The problem of flocculation too restricts the biodegradation of dyes. Ibrahim et al. (2009) reported COD and color removal of 55% and 70%, respectively, after 6 days of the time period in anaerobic reactor using microbial flocs.

Anaerobic microbes are slow growers and hence require a longer time to acclimatize. The treatment of textile dyes in anaerobic condition generates more toxic aromatic amines during break down of azo dye by azoreductase. Aromatic amine produced in anaerobic condition mineralizes into non-harmful compound when

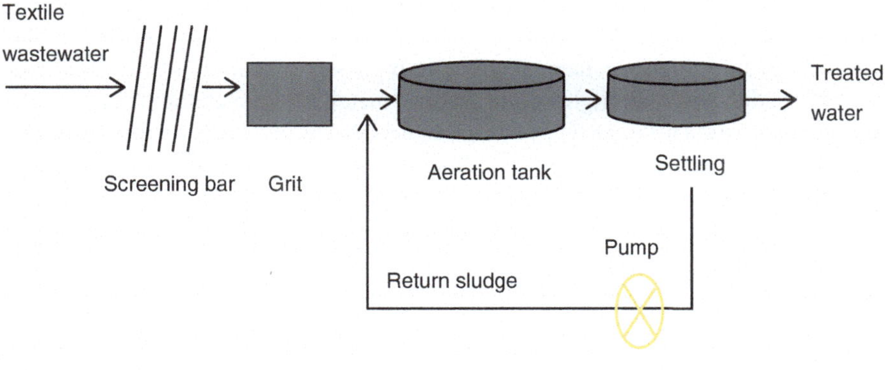

Fig. 15.5 Extended aeration process

exposed to air; hence anaerobic system integrated with the aerobic system can be used to achieve effective treatment. Methanogenic biogas can be produced in the anaerobic process if the COD of the textile effluent is higher than 3000 mg/L.

The anaerobic microbes which maintain the reducing condition for the decolorization of the textile dyes are sensitive to toxic shocks and hence restrict the color removal process. Some reports show the inhibitory effect of the dyes on methanogenesis. The dye itself can affect effective color removal in addition to other inhibitory compounds (heavy metals, sulfide, and salts) that inhibit biological treatment system. Textile effluents are shown to hinder nitrifying bacteria in aerobic treatment systems, thus affecting nitrogen removal process. The COD/N ratio should be considered in the anaerobic treatment of textile effluents. Further, the nutrient additives should also be minimized in wastewater treatment. Thus in order to avoid problems faced in biological treatment systems, some more reliable and promising methods are focused on. The role of nanoparticles in wastewater remediation is the prominent option nowadays; the details of which are described in the following section.

15.4 Role of Nanoparticles in Textile Wastewater Treatment

Due to large-scale production and extensive application, dyes have turned into serious pollutants when improperly handled and disposed off, creating grave public health and environmental problems. One of the problems that textile industry is facing is related with the incomplete exhaustion of dyes onto textile fibers from an aqueous dyeing process and the need to implement innovative and sustainable effluent treatment methods to remove color. Additionally, legislation on the limits of color discharge has increasingly turned rigid. But as textile effluent exhibits low biodegradability, conventional biological treatment methods fail to degrade dyes. Hence advanced methods like chemical treatments, membrane filtration, etc. are employed to treat textile wastewater. But high operation and maintenance cost prove to be the limiting factor in the widespread application. So, more energy and cost-effective techniques are sought after to curb the issue of textile effluent treatment. Nanoparticles have recently been shown to be promising materials in treating textile dyes because nanoparticles come with unique physicochemical properties like ordered structure, high mechanical strength, high aspect ratio, high electrical and thermal conductivity, ultralightweight feature, metallic/semimetallic behavior, and high surface area. Different types of nanomaterials are involved in remediation of textile wastewater. Table 15.3 enlists different types of nanoparticles/nanocomposites utilized in degrading textile effluent. Magnetic iron nanoparticles, carbon nanotubes, zinc nanoparticles, TiO_2, and metal oxide nanoparticles are some of the popularly used nanomaterials.

Table 15.3 Textile effluent degradation using different nanomaterials

Sr. no.	Name of the nanoparticles	Contaminant	Degradation/ adsorption	Adsorbate amount	References
1	γ-Fe_2O_3	Acridine orange dye	98% and 59 mg/g	5 mg	Qadri et al. (2009)
2	Fe_3O_4	Coccine (acidic dye)	65%	0.1 g	Weng et al. (2010)
3	Fe_3O_4	Procion red	30.5 mg/g	0.8 g/L	Hariani et al. (2013)
4	Fe_3O_4	Congo red [CR (azo dye)], Coomassie Brilliant Blue R-250 [CBB (anionic dye)], and acridine orange [AO (cationic dye)]	0.056, 0.082, and 0.078 mg/g	0.02 g	Chaudhary et al. (2013)
5	GO-Fe_3O_4 magnetic nanohybrids	Methylene blue	97.3% and ~72.9 mg/g	10 mg	Huong et al. (2016)
6	Ferrite MNPs	Direct red 31	86%	0.4 g	Mahmoodi et al. (2014)
		Direct red 23	67%		
		Direct green 6	89%		
7	γ-FeOOH	Reactive orange 29	36.3%	500 mg/L	Sheydaei and Khataee (2015)
8	γ-Fe_2O_3	Real textile wastewater	363.6 mmol/g	0.1 g	Nassar et al. (2015)
9	Prussian blue (PB)-modified γ-Fe_2O_3	Methylene blue (MB) dye	100%	0.3 g/L	Wang and Huang (2011)
		COD	53.6%		
		TOC	35%		
10	Fe_3O_4 hollow nanospheres	Neutral red dye	90%	0.05 g	Iram et al. (2010)
11	Multi-walled carbon nanotubes (MWCNT) and single-walled carbon nanotubes (SWCNT)	Reactive blue 4	502.5 and 567.7 mg/g	30 mg	Machado et al. (2014)
12	Carbon materials	Mordant yellow 10 (MY10) and reactive red 120	~85%	0.1 g/L	Pereira et al. (2014)
			70%		
13	Cr-doped ZnS nanoparticles	Methyl orange	74.28%	0.2 mol%	Eyasu et al. (2013)
14	MgO nanoparticles	Acid red 73 total COD	100%	0.8 g/L	Jorfi et al. (2016)
		TOC removal	98.3%		
			86.9%		

(continued)

Table 15.3 (continued)

Sr. no.	Name of the nanoparticles	Contaminant	Degradation/ adsorption	Adsorbate amount	References
15	Nano-zinc oxide	Methyl orange dye	84%	1 g/L	Hemapriyamvadha and Sivasankar (2015)
		Real textile effluent (COD)	41% (1:1 dilution)		
16	Amine-functionalized silica nanoparticles	Acid red 14	434 mg/g,	0.05 g	Mahmoodi (2011)
		Acid black 1	250 mg/g	0.07 g	
		Acid blue 25	167 mg/g	0.07 g	
17	Ammonium-functionalized silica nanoparticles	Methyl orange	105.4 mg/g	0.1 g	Liu et al. (2013)
18	Silver nanoparticles loaded on activated carbon	Methyl orange	98.9%	0.02 g	Karimi et al. (2012)
19	Ag@TiO_2 core-shell nanoparticles	Reactive blue 220	100%	1 g/L	Khanna and Shetty (2014)
20	TiO_2 nanoparticles	Rhodamine and malachite green	72.1%	1.5 g/L	Chanathaworn et al. (2012)
			76.3%		
21	TiO_2 nanoparticles	Methyl orange	92.8%	55 mg	Rashed and El-Amin (2007)
22	CdO-modified ZnO nanocomposites	Methylene blue	98%	–	Saravanan et al. (2015)
		Methyl orange	93%		
		Textile effluent	85%		

15.4.1 Metal Oxide Nanoparticles

Metal particles when downsized to nanoscale often show remarkable optical, electrical, magnetic, and chemical catalytic properties. Metal elements can form a diverse group of oxide compounds and can take up varieties of geometrical configurations with a specific electronic structure which can exhibit either metallic, insulator, or semiconductor nature (Wyckoff 1964). Particle size influences the structural characteristics like lattice symmetry and cell parameters, surface free energy, and stress of any material. The metal oxide nanoparticles show unique chemical and physical properties because of their minute size, high dense corner and edge surface size. The decrease in the particle size causes increase in the number of surfaces, and as a result, the interface atoms create stress/strain and also result in thermodynamic changes which lead to the associated structural changes (Cammarata 2009). The surface properties of nanostructured metal oxides get highly altered producing remarkable sorption capacity. Widely used metal oxide nanoparticles are composed of materials such as titanium, iron, aluminum, zinc, and cerium.

Metal oxide nanoparticles are also used as adsorbents as they are cost-effective and efficient enough in adsorbing pollutants. To enhance this ability (i.e) to adsorb pollutants from wastewater, iron oxide nano-sorbent was designed to treat wastewater

(Zhang et al. 2013). Organic dyes from the wastewater have been removed from the wastewater using nanostructured tungsten oxide (WO_2). The reactive yellow dyes from the textile effluent were successfully removed by zinc and aluminum layered double hydroxide nano-sorbent (El-Safty et al. 2012)

15.4.1.1 Nanostructured Metal Oxide as Photocatalysts

Photocatalysts are known to accelerate chemical reaction on absorbing solar radiation or any other source of light having a certain wavelength (McNaught and Wilkinson 1997). Nanostructured catalysts which are sensitive to light exposure tend to degrade pollutants in wastewater. ZnO, CdS, and TiO_2 are some of the prominent semiconductors which produce electron-hole pair on absorbing light with energy higher than its bandgap energy. These electron-hole pairs have a tendency to react with an aqueous solution and produce highly reactive species such as oxidizing and reducing radicals. These radicals which are mainly hydroxyl or superoxides destruct any type of organic, inorganic, or heavy metal pollutants present in wastewater through some secondary reactions (Kumar et al. 2017). Nanomaterial semiconductors come with a huge surface area and hence possess elevated photocatalytic activity compared to normal semiconductors.

TiO_2 nanoparticles are well-known semiconductor photocatalysts that are widely used for removing pollutants from the wastewater. TiO_2 have been reported to purify water by destroying (photodegradation) microbial pathogens within few hours of exposure to simulated solar light (Chaturvedi et al. 2012). TiO_2 as nano-photocatalyst has shown to remove dyestuff successfully from the waste stream. TiO_2 is generally used in powdered form or as a film for the purpose of photocatalytic degradation of the pollutants. However, TiO_2 in the form of nanotubes has shown improved photoactivity and effective photodegradation of textile dye compared to TiO_2 powder. Increased photoactivity of TiO_2 nanotube is due to efficient separation of charges, which is possible because of the lack of grain boundaries, commonly found in TiO_2 films (Liu et al. 2008). TiO_2 nanotubes prepared by the anodization of titanium foils have shown to catalyze photodegradation of textile dyes (Sohn et al. 2008). The size of TiO_2 nanotubes influences the photocatalytic activity, i.e., shorter nanotubes show better efficiency compared to larger ones (Liu et al. 2008). Moreover, Wang (2000) showed photodegradation of eight different types of commercial dyes including methyl orange in the suspension of TiO_2 under UV light irradiation. A complete decolorization of methyl orange was demonstrated in the TiO_2 suspension activated by fluorescent lamp, halogen lamp, and solar irradiation (Rashed and El-Amin 2007).

ZnO comes with large bandgap compared to all semiconductors, and it is nonhazardous in nature. They find wide applicability because of its high thermal stability and low cost (Lin et al. 2014). ZnO nanoparticles are prepared by thermal evaporation method and chemical deposition method. They have been examined to degrade methyl orange in aqueous solution as photocatalysts for the UV-induced degradation. The photocatalytic activity in ZnO is size dependent, and the particle size of 50 nm showed utmost photocatalytic activity (Wang et al. 2007). As the bandgap in ZnO is larger, it requires UV activation in order to achieve good photocatalytic activity. The photocatalytic activity of ZnO is enhanced by preventing electron-hole recombination process, which is mainly achieved by doping ZnO with metals like Ag, Cd, Cu, Co, Fe, or Mn and nonmetals like N, S, or F. These heterogenous ZnO composites can be ZnO/metal, ZnO/polymer, and ZnO/metal oxide. ZnO modified with CdO showed to be effective enough in degradation of methylene blue in visible light (Saravanan et al. 2011). Shifting of photocatalytic activity from ultraviolet to visible light and enhanced degradation efficiency of the dye by ZnO/CdO composite can be attributed to the suppression of back reaction between ZnO and CdO leading to the generation of a large number of charge carriers (Saravnan et al. 2011). Silver, when coupled with ZnO, expresses improved photocatalytic activity in visible light because of lower bandgap energy and its capability to hinder quick electron-hole recombination. In case of silver coupled with ZnO, the amount of Ag, its size, and its surface area strongly influence the photocatalytic activity of ZnO. The ZnO and Ag in the ratio of 90:10 demonstrated improved degradation of textile dyes under few hours of irradiation by visible light (Saravnan et al. 2013). The activity of binary photocatalyst (ZnO/Ag and ZnO/CdO) is surpassed by the synergy of ZnO/Ag/CdO nanocomposite which was prepared in combination with 80% ZnO, 10% Ag, and 10% CdO using thermal decomposition method for the photocatalytic degradation of textile effluent under the irradiation of visible light (Saravnan et al. 2015).

15.4.1.2 Magnetic Iron Nanoparticles

Magnetic nanoparticles are popular among researchers for their multiple properties. Low toxicity, low cost, magnetic properties such as high magnetic susceptibility, superparamagnetic property, high coercivity, and low Curie temperature are few of its unique properties which makes it more attractive in various applications (Nam and Tratnyek 2000; Bigg and Judd 2002). Magnetic iron nanoparticles are prepared by varied types of methods such as water-in-oil microemulsion, polyol, coprecipitation, sol-gel, sonolysis, and hydrothermal methods. Each method results in magnetic iron nanoparticles of different properties (shape, average size, size distribution, crystallinity, magnetic properties, dispersibility). Among these methods, the coprecipitation method is the commonly used method due to numerous advantages associated with it. Advantages of coprecipitation over other methods include cost-effectiveness, good homogeneity, high purity of the product, and no requirement of heat treatment and organic solvents. Figure 15.6 shows the SEM image of magnetic iron nanoparticles (MNPs) prepared using coprecipitation. Coprecipitation method of nanoparticle synthesis does not have control over the particle size, unlike other methods. The size and morphology control were achieved better by hydrothermal method, and sonochemical method gave unusual magnetic properties to magnetic nanoparticles.

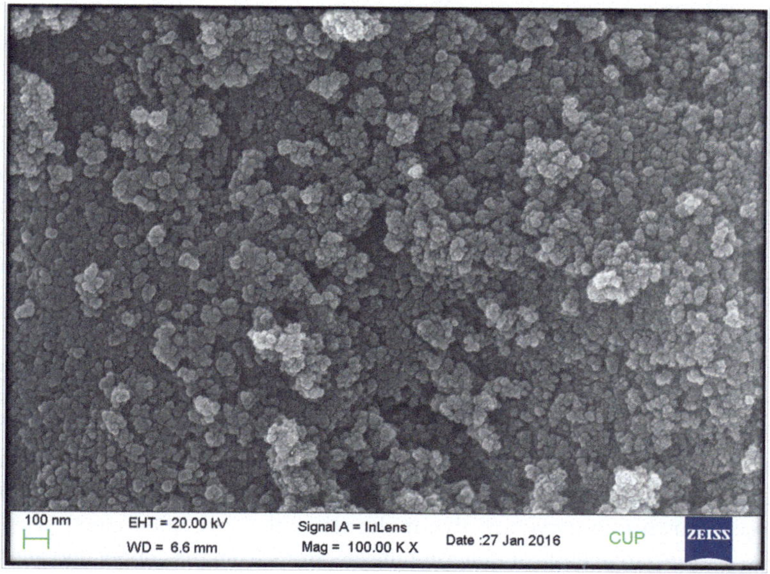

Fig. 15.6 SEM image of MNPs

Magnetic iron nanoparticles are being popularly investigated in various environmental applications. Iron nanoparticles are the most common metallic reducing agent (reductant) used popularly for removal of dyes and phenolic compounds. The dyes are shown to be removed by magnetic nanoparticles via a reduction mechanism. The dye is diffused into the solution and metal surface where adsorption occurs. Iron oxide has shown to have excellent photocatalytic activity because of its narrow bandgap (2.2 eV). It is activated by visible light and captures around 40% of the solar radiation, and it is a very low-cost material available. On absorbing visible light, the excited electrons and subsequent holes activate nearby chemical species and instigate a chemical reaction. Porous hematite nanorods expressed high photocatalytic activity by removing methylene blue, eosin B, and methyl orange from the aqueous phase (Liu et al. 2015).

15.4.2 *Copper Nanoparticles*

Metal nanoparticles like Cu nanoparticles have higher Fermi potential which leads to decrease in their reduction potential values. Hence they can catalyze many electron transfer reactions. Metal nanoparticles can effectively drive those reactions which are thermodynamically favorable but kinetically restricted in the absence of a catalyst (Pradhan et al. 2002). Copper-based catalysts because of their unique catalytic properties have been used to treat a wide range of wastewaters. Out of different compounds of copper, copper sulfide (CuS) has gathered much attention because of its low cost and unique physical, chemical, electronic, and optical properties. CuS, a metal chalcogenide, is an important p-type of semiconductor which occurs in a wide range of stoichiometry. Based on the stoichiometric ratios of copper to sulfide, these CuS can either be direct or indirect bandgap semiconductors. A vast range of morphologies can be prepared from CuS such as nanocubes, nanoplates, nanotubes, nanoflowers, nanospheres, nanorods, and complex hierarchical nanostructures. Good crystalline CuS has shown to successfully degrade dye under visible light irradiation (Meng et al. 2012). CuS and graphene oxide (GO) nanocomposite prepared through hydrothermal method have shown nearly 100% of the photodegradation of methylene blue dye (Saranya et al. 2014). This is because Cus and GO exhibit high photocatalytic activity in dye degradation under the visible light because of the narrow bandgap. Moreover, GO can transform the electronic structure of semiconductor and metal nanoparticles and also stimulate easy transfer of electrons from the excited dye molecule or the conduction band of the semiconductor to the GO because of the equalization of large energy level formed at the interface which results in enhanced photocatalysis. Zerovalent copper nanoparticles have been shown to remove azo dye molecule as they are good electron donors. Moreover, the azo dye molecules accept the electrons from the copper and get transformed into transitional products on combining with H^+ ion. The combination of copper ferrite ($CuFe_2O_4$) nanocomposites also effectively reduced and decolorized textile dyes through the photocatalytic activity (Mahmoodi 2011).

15.4.3 CdS Nanoparticles

CdS is also a well-known semiconductor having a bandgap of 2.42 eV and can be operated at wavelength <495 nm (Khallaf et al. 2008). CdS nanoparticles have attracted intensive interest as a photocatalyst for treatment of industrial dyes in wastewater (Tristao et al. 2006; Zhu et al. 2009).

15.4.4 Gold Nanoparticles (AuNPs)

Gold nanoparticles are known to be an effective catalyst in reducing textile dyes (Brust and Gordillo 2012). Stable Au nanoparticles are prepared from the reduction of Au^{3+} to zerovalent Au with an aqueous starch solution at 70 °C. Here starch served as stabilizing reducing agent and gold nanoparticles in combination with activated carbon have shown remarkable adsorption of Congo red dye (Ghaedi et al. 2011). AuNP has effectively shown to degrade methylene blue and other dyes (Cheval et al. 2012). The catalytic activity of AuNP is highly dependent on particle size, shape, and charge. Positively charged AuNP was shown to degrade around 98% of the methylene blue dye in a microbial fuel cell with a power output of around twofold (30–40 mW/m^2). A complete oxidative mineralization of methylene blue was achieved with approximately 100% recovery of positively charged AuNP (Han et al. 2008).

15.4.5 Carbon Nanotubes

Carbon nanotubes have a modifiable surface chemistry, and its adsorption sights are highly assessable besides possessing high specific surface area. The active sites get reduced because CNT needs to be stabilized in aqueous suspension in order to circumvent clumping due to their hydrophobic surface. Contaminants can be detected, preconcentrated, as well as adsorbed by them (Pan and Xing 2008). Electrostatic attraction and chemical bonding play a vital role in the adsorption of metal ions by CNTs (Rao et al. 2008). They also possess antimicrobial characteristics. They generate oxidative stress on them and destroy the cell membrane (Liu et al. 2013).

The adsorption mechanism for organic compounds varies depending on the properties of the compound of interest (e.g., polar vs. nonpolar). Estimating the adsorption behavior of organics on CNTs is a difficult task (Pan and Xing 2008); therefore different interactions among organics and CNTs have been projected. Electrostatic interactions, π–π stacking interactions, hydrophobic interactions (Ai et al. 2011), van der Waals forces (Yang et al. 2008), and hydrogen bonding interactions are few of them. These interactions might act alone or sometimes together. Usually, there are three types of methods used for the preparation of carbon nanotubes, viz., electric arc, laser ablation, and pyrolysis. Among them, electric arc and laser ablation were the most frequently used methods, but they are associated with only lab-scale production, while pyrolysis is the frequently used commercial method. This method has been employed for the preparation of the longest known threadlike bundles of nanotubes (NT) and macroscopic ribbons ("mats") of interwoven NT (Endo et al. 1995; Niu et al. 1997).

Reactive blue 4 textile dye was removed using multi-walled carbon nanotubes (MWCNT) and single-walled carbon nanotubes (SWCNT) as adsorbents from the aqueous solution. The equilibrium isotherm for contact time was obtained at 4 h for both the adsorbents at 298–323 K. The best fit for the equilibrium data of reactive blue 4 dye was given by the Liu isotherm model. The maximum sorption capacity for adsorption was 502.5 and 567.7 mg g^{-1} for MWCNT and SWCNT, respectively, which occurred at 323 K (Machado et al. 2014). One of the feasible redox mediators is activated carbon (AC) because it shows an advantage in comparison with soluble ones (e.g., anthraquinone disulfonate and anthraquinone-2-sulfonate) (Van Der Zee et al. 2003; Mezohegyi et al. 2010; Pereira et al. 2010). The biodegradation of azo dye and textile wastewater was treated with the samples of microporous thermal-treated AC (ACH2) and mesoporous carbons: xerogels (CXA, CXB) and carbon nanotubes (CNT). The carbon materials used has been reported to show ~85% color removal for Mordant yellow 10 (MY10) and 70% for reactive red 120 (RR120). The degradation of acid orange 10 (AO10) was obtained when CXB and CNT were used together, where 98% of color removal was achieved. The degradation rate achieved was in an order of control < CXA< CXB < CNT for MY10 and RR120. The dyes were reduced to corresponding aromatic amines which was confirmed through HPLC analysis (Pereira et al. 2014).

15.5 Nanoparticles as a Carrier for Enzyme Immobilization

The immobilization of enzymes involves confining of enzyme molecules to a distinct region of space without affecting their catalytic properties by fixing them onto or within some suitable carrier material. Immobilization allows repeated and continued use of enzymes (Brena and Batista-Viera 2006). Nanoparticles were considered as a perfect support material for enzyme immobilization due to the high surface area per unit mass, high enzyme loading capability, less diffusional limitations, and mass transfer resistance (McHenry and Laughlin 2000; Lei and Ju 2012; Hwang and Gu 2013). Nanoparticles can easily diffuse through macromolecular substrates where the diffusion problem is more relevant making it an ideal carrier. When enzyme-immobilized nanoparticles were dispersed in aqueous solutions, it follows a Brownian movement proving the efficiency of immobilized enzyme in comparison to free enzyme. Moreover, immobilization of enzymes onto the nanoparticles reduces the protein unfolding and improves the performance and stability of the immobilized enzyme.

Garcia-Galan et al. (2011) reported that the purity of the enzyme impairs the volumetric activity when nonporous nanoparticle supports were used for immobilization. In addition to all the abovementioned advantages of nanomaterials as a carrier for enzyme immobilization, they are also associated with the benefits of maintaining the regulation of the protein/enzyme orientation onto it. They are devoid of external diffusion problems associated with porous supports. This behavior of nanoparticles makes them an efficient choice for their usage in solid-liquid systems at industry level on a large-scale process (Colombié et al. 2001).

Enzyme immobilization has been carried out for different enzymes, viz., lysozyme, glucose oxidase, aminopeptidase, and alcohol dehydrogenase on different nanoparticle carriers such as polymeric nanoparticles, porous nanoparticles, magnetic nanoparticles, metal nanoparticles, and metal oxide nanoparticles. Generally, the enzymes are immobilized onto the carriers by reversible (adsorption, chelation, disulfide bond, ionic bonding, and affinity bonding) and irreversible methods (covalent coupling, entrapment or microencapsulation, and cross-linking). However, cross-linking and covalent bonding are the most commonly used enzyme immobilization methods for nanoparticle carriers. This method is different from other immobilization techniques as it does not require any support (Shi et al. 1997; Honda et al. 2006). Here, by means of multifunctional reagents, the leading three-dimensional cross-linked aggregates form covalent bonds between the enzyme molecules. This method is also termed as carrier-free immobilization. Glutaraldehyde is the most commonly used reagent for the process of cross-linking. An enzyme is coupled forming covalent bond with other protein molecules using bi- or multifunctional coupling agent (Sheldon 2007; Tran and Balkus 2011). The coupling agent cross-links the enzyme with support material (Tischer and Wedekind 1999). In this mode of immobilization, the biocatalyst formed occupies 100% weight of the protein in it, as the coupling agent has negligible molecular weight as compared to an enzyme (Schoevaart et al. 2004). Structural and functional property of enzymes remains undisturbed during the process of immobilization. Cross-linking agents such as bis-diazobenzidine, glutaraldehyde, and hexamethylene diisocyanate are used as bi-/multifunctional agents (Singh 2009). Glutaraldehyde is the most widely used bifunctional cross-linker, because it facilitates the formation of stable inter- and intra-subunit covalent bonds as it is soluble in aqueous solvents (Datta et al. 2013).

Covalent bonding is a highly stable and most frequently used method of immobilization as the enzyme strongly binds to the support through covalent binding. The enzyme immobilized on matrix involving covalent binding does not release enzyme while being used in the solution due to the stability of the bonds formed (Guisan 2006). It involves highly specific reaction so that no contaminant on the carrier material appears and also no leaking of the enzyme is observed in the reaction mixture. Thermal stability and the half-life of enzyme increase when various support materials like mesoporous silica, chitosan, etc. are used for covalent attachment (Hsieh et al. 2000; Ispas et al. 2009). The covalent attachment of enzymes to carriers is due to amino acids like aspartic acid, arginine, and histidine present in their side chain (D'Souza 1998). Different functional groups like imidazole, indolyl, phenolic hydroxyl, etc. are responsible for their degree of reactivity (Singh 2009). Brena and Batista-Viera (2006) reported that covalent bonding alters the catalytic activity of the enzyme thereby minimizing its activity. Substrate analogs can be used for a coupling reaction to enhance the activity of the enzymes (Mattiasson and Kaul 1991). Fu et al. (2011) found out that if peptide-modified surfaces were used to immobilize enzyme linkage then higher specific activity and stability can be achieved with controlled protein orientation. To achieve enzyme immobilization by covalent bonding, functionalization followed by activation of supports using specific activating agents (such as inorganic and organic halides, carbodiimides, glutaraldehyde, etc.) is vital (Zucca and Sanjust 2014).

15.5.1 Surface-Activating Agents

To covalently bind the enzyme to the nanoparticles, surface activation/modification is very essential. So far several chemicals have been used to modify the surface of the nanoparticles. Some of them include glutaraldehyde, carbodiimides, and cyanuric chloride. Carbodiimides are the chemical activators which facilitate the enzyme in binding with support material via carboxyl group. An intermediate is formed when carbodiimides react with carboxyl which gets stabilized when it reacts with amines, forming a peptide bond, without a spacer. EDAC is the most popular form of carbodiimides, and it serves various purposes, for example, peptide synthesis, cross-linking, and immune conjugate preparation. An advantage associated with

EDAC activation is that no such lengthy linker species are involved which ultimately reduces the hydrodynamic radius of the nanoparticles (Thanha and Greena 2010). Cyanuric chloride is also used to activate the surface of the nanoparticles. It is a triazine which activates the OH group of the support material for covalent bonding of the enzyme (Moreno and Sinisterra 1994; Alcántara et al. 1998). In one of the studies conducted by Moreno and his co-workers (1994), it was observed that cyanuric chloride increased the stability of the enzyme. Glutaraldehyde, a colorless to pale straw-colored linear 5-carbon dialdehyde, is also used as a surface activator. It is soluble in water, alcohol, and organic solvents at all proportions having a pungent smell (Migneault et al. 2004). Glutaraldehyde is an acidic aqueous solution (pH 3.0–4.0) with the ability to generate thermally and chemically stable cross-links than other aldehydes (Nimni et al. 1987).

15.6 Role of Enzyme-Immobilized Nanoparticles in the Removal of Contaminants of Textile Wastewater

Enzyme-immobilized nanoparticles have been popularly reported for the treatment of textile wastewater. Table 15.4 summarizes the list of enzyme-immobilized nanoparticles in textile wastewater treatment. A variety of dyes have been decolorized by the enzyme-immobilized nanoparticles. The mechanism adopted by the

Table 15.4 List of enzyme-immobilized nanoparticles in textile wastewater treatment

Sr. no.	Enzyme	Nanoparticles	Dye	Dye degradation (%)	Reusability	References
1	Laccase	Fe_3O_4/SiO_2	Procion red MX-5B	80	Yes	Wang et al. (2013)
			Azophloxine	79		
2	Horseradish peroxidase	Phospholipid-templated titania	Phenol	93	Yes	Jiang et al. (2014)
			2-Chlorophenol	88		
			Direct Black 38	80		
3	Lignin peroxidase	Magnetic (Fe_3O_4)	Blue azo dye	54	Yes	Darwesh et al. (2015)
4	Glucose oxidase	Magnetic (Fe_3O_4)	Acid yellow	62	Yes	Aber et al. (2016)
5	Laccase	Magnetic poly (p-phenylenediamine) (PpPD/Fe_3O_4)	Blue 19	80	Yes	Liu et al. (2016)
6	Laccase	Magnetic graphene oxide (MGO)	Crystal violet	95	Yes	Chen et al. (2017)
			Malachite green	96		
			Brilliant green	91		
7	Laccase	EDTA-Cu (II)	Indigo carmine	28	Yes	Fernandes et al. (2017)
			Congo red	21		

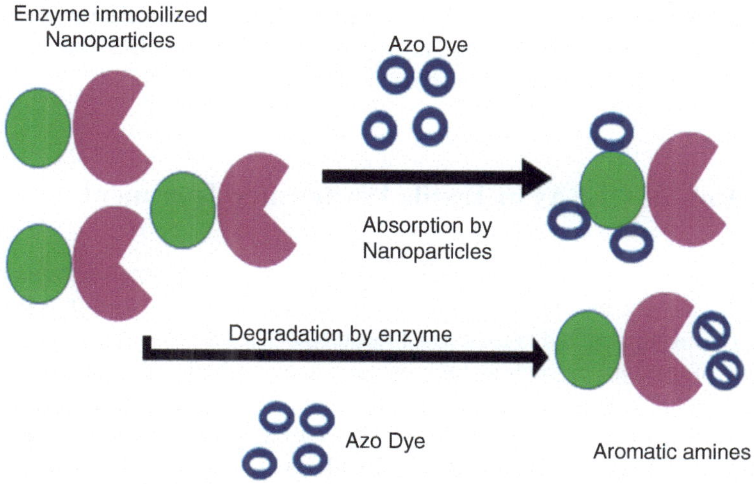

Fig. 15.7 General mechanism involved in degradation of dye using enzyme-immobilized nanoparticles

enzyme-immobilized nanoparticles for dye degradation is the absorption of the dye by the support material followed by degradation (Wang et al. 2013). Liu et al. (2016) reported that dye decolorization occurred by developing a surface microenvironment that allows electrical conductivity. The absorption of dyes occurs on the surface of nanoparticles, and enzyme degrades azo dye into simpler aromatic amines. Figure 15.7 describes the general mechanism involved in the degradation and absorption of azo dyes.

Fe_3O_4/SiO_2 nanoparticles were used as the support for laccase immobilization that showed more than 80% decolorization of the azo dyes. Laccase immobilized on titania nanoparticle, magnetic graphene oxide, and EDTA-Cu (II) chelating magnetic nanoparticles were also used in dye decolorization studies, and the efficiency was in the range of 91–97%. The decolorization was achieved within 20–60 min incubation. Surface functionalization and high surface to volume ratio facilitated the high mass transfer and easy removal of the dye from reaction matrix. pH is one important parameter that influences the laccase enzyme activity. Laccase enzyme showed the best activity at pH 3 (Das et al. 2017; Chen et al. 2017). Similar to laccase, other enzymes, horseradish peroxidase and lignin peroxidase immobilized on phospholipid-templated titania and magnetic (Fe_3O_4) nanoparticles, respectively, were also used for dye and phenolic compound degradation. Unlike laccase, these enzymes showed slightly decreased decolorization in the range of 65–79% at longer incubation period (Jiang et al. 2014; Darwesh et al. 2015). Glucose oxidase enzyme on Fe_3O_4 magnetic nanoparticles particles was also used in dye decolorization. However, they also showed decreased decolorization (63%) compared to laccase enzyme (Aber et al. 2016). In contrast to laccase, glucose oxidase shows better activity at slightly acidic pH 6.5. Immobilization improves the stability of enzymes at a wider pH range. The shift in pH might be due to the variations in the conformation of the enzyme during covalent bond formation leading to changes in the micro-environment of the enzyme (Chiou and Wu 2004). The immobilized nanoparticles showed high stability and activity even up to ten cycles of reusability.

15.7 Role of CLEAs in Textile Wastewater Treatment

Cross-linked enzyme aggregates (CLEAs) are enzyme aggregates prepared by cross-linking of the enzyme aggregates with a difunctional cross-linker. It does not require a solid support and is used as a biocatalyst. The CLEAs are backed up with a number of advantages such as high enzyme activity, high pH, thermal stability, improved functional and operational stability, and reduced mass transfer problems (Sheldon 2007; Bayramoglu et al. 2010). Low cost and environment-friendly characteristic of CLEAs have also favored its widespread application. CLEAs can be prepared from crude enzyme extracts without any need of purification and provide a platform for synergistic action of the multiple enzymes leading to cascading reactions (Dalal et al. 2007; Sheldon 2011). Furthermore, the CLEAs exhibited excellent recovery due to easy separation (Sheldon 2011).

CLEAs are easily separated by filtration or centrifugation and have low production cost, and it is a carrier-free immobilization (Sheldon 2011). However, the number of enzymes immobilized by CLEAs technology is limited. There might be two reasons: (a) CLEAs might not be as effective as expected for enzymes with low Lys residue contents since cross-linking occurs superficially at the surface of enzyme resulting in low mechanical strength (Sheldon 2007). It even releases enzyme into the media during the reaction (Dong et al. 2010). (b) The clumping of CLEA clusters during separation of CLEAs from the reaction mixture by centrifugation or filtration results in internal mass transfer limitations particularly in case of enzymes acting on macromolecular substrates (Wang et al. 2011).

Different methods have been developed to solve the problem of enzymes which are having less content of Lys residue, i.e., enzyme co-aggregation with a polymer containing several free amino groups, e.g., polylysine (Yamaguchi et al. 2011) or polyethyleneimine (Vaidya et al. 2012), can result in the production of mechanically stable CLEAs. Further, Talekar et al. (2012) developed a novel strategy to prepare magnetic cross-linked enzyme aggregates in order to overcome the limitations of CLEAs. The magnetic cross-linked enzyme aggregates (M-CLEAs) were prepared by the addition of amino-functionalized magnetite nanoparticles as an additive into enzyme solution followed by its precipitation to form aggregates. A cross-linking of enzyme aggregates and nanoparticles occurs during this process. Nanoparticles due to their smaller size and high surface area facilitate more coating of amino groups on their surface (Hu et al. 2009). Addition of such amino-functionalized magnetic nanoparticles into the low Lys enzyme solution results in non-leachable and mechanically stable CLEAs. The magnetic properties of the nanoparticles will facilitate better solid-liquid separation. Sinirlioglu et al. (2013) and Kumar et al. (2014) prepared CLEAs of laccase enzyme to study the decolorization of malachite green dye. The magnetic combi-CLEAs showed double the dye removal efficiency (93%) of the free enzymes (47%). The pH and temperature range also increased in comparison to free enzymes. The CLEAs showed a maximum reusability of eight cycles (Zhou et al. 2016). Similar to laccase, CLEAs with horseradish peroxidase were also used for anthraquinone dye, C. I. acid violet 109 decolorization.

15.8 Conclusion

In every aspect of human activity, water plays a fundamental role, and day by day, this resource is getting scarce in various corners of the world. Textile wastewater is one of the major water polluting industry. Many technologies and methods have been designed, explored and used for wastewater treatment. Enzyme assisted wastewater treatment is one of the facet applied for textile effluent treatment. However due to the cost of enzyme production, purification and reusability their application is limited or prohibited. At present, large-scale treatment of effluent using enzymes is not economically viable. From the various reports and studies conducted by researchers, it can be established that nanomaterials can act as efficient material for enzyme immobilization. The separation difficulty of nanoparticles from the treated solution may result in the loss of nanoparticles. To overcome the separation problems and increase their reusability, magnetic nanoparticles and magnetic CLEAs have been explored. These method proved their efficiency in batch studies and their upscaling requires more detailed study. Forthcoming interdisciplinary approach to wastewater treatment is the confluence of nanoscience and enzyme technology. Biocatalysts can be utilized to their maximum potential due to such groundbreaking applications.

15.9 Future Outlooks

Nanoremediation holds a promising role in the future for treatment of any kinds of the waste stream. Many aspects of nanotechnology in bioremediation need to be explored as it is still in its infant stage. Numerous varieties of nanocomposites can be tested to achieve the desired effects. Separate bio-nanomaterial for different waste can also be looked after so that efficient degradation can be obtained in a lesser span of time. The activity of crude enzyme preparations needs to be accentuated for its optimization and also on the advancement of enzyme reusability to thwart the high start-up and running costs.

References

Aber S, Mahmoudikia E, Karimi A, Mahdizadeh F (2016) Immobilization of glucose oxidase on Fe_3O_4 magnetic nanoparticles and its application in the removal of acid yellow 12. Water Air Soil Pollut 227(3):1–11

Ahmet B, Ayfer Y, Doris L, Nese N, Antonius K (2003) Ozonation of high strength segregated effluents from a woollen textile dyeing and finishing plant. Dyes Pigments 58:93–98

Ai L, Zhang C, Liao F, Wang Y, Li M, Meng L, Jiang J (2011) Removal of methylene blue from aqueous solution with magnetite loaded multi-wall carbon nanotube: kinetic, isotherm and mechanism analysis. J Hazard Mater 198:282–290

Alcantara AR, Borreguero I, Lopez-Belmonte MT, Sinisterra JV (1998) Covalent immobilization of crude and partially-purified upases onto inorganic supports: stability and hyperactivation. Biotechnol Prog 15:571–576

Al-Kdasi A, Idris A, Saed K, Guan CT (2004) Treatment of textile wastewater by advanced oxidation processes-a review. Global NEST J 6(3):222–230

Arslan IA, Isil AB (2002) The effect of pre-ozonation on the H_2O_2/UV-C treatment of raw and biologically pre-treated textile industry wastewater. Water Sci Technol 45:297–304

Ay F, Catalkaya EC, Kargi F (2008) Advanced oxidation of Direct Red (DR 28) by Fenton treatment. Environ Eng Sci 25(10):1455–1462

Azbar N, Yonar T, Kestioglu K (2004) Comparison of various advanced oxidation processes and chemical treatment methods for COD and color removal from a polyester and acetate fiber dyeing effluent. Chemosphere 55(1):35–43

Baig S, Liechti (2001) Ozone treatment for biorefractory COD removal. Water Sci Technol 43:197–204

Bayramoglu G, Yilmaz M, Arica MY (2010) Reversible immobilization of laccase to poly(4-vinylpyridine) grafted and Cu(II) chelated magnetic beads: biodegradation of reactive dyes. Bioresour Technol 101:6615–6621

Bharagava RN, Chowdhary P, Saxena G (2017) Bioremediation: an ecosustainable green technology: its applications and limitations. In: Bharagava RN (ed) Environmental pollutants and their bioremediation approaches, 1st edn. CRC Press/Taylor & Francis Group, Boca Raton, pp 1–22. https://doi.org/10.1201/9781315173351-2

Bigg T, Judd SJ (2002) Reductive degradation of azo dyes in aqueous solution by zero-valent iron. IAHS Publ 275:383–390

Blackburn RS (2004) Natural polysaccharides and their interactions with dye molecules: applications in effluent treatment. Environ Sci Technol 38(18):4905–4909

Brena BM, Batista-Viera F (2006) Immobilization of enzymes. In: Methods in biotechnology: immobilization of enzymes and cells. Humana Press, Totowa, pp 15–30

Brust M, Gordillo GJ (2012) Electrocatalytic hydrogen redox chemistry on gold nanoparticles. J Am Chem Soc 134(7):3318–3321

Cammarata RC (2009) Generalized Thermodynamics of Surfaces with Applications to Small Solid Systems. Solid State Phys 61:1–75

Chan LS, Cheung WH, McKay G (2008) Adsorption of acid dyes by bamboo derived activated carbon. Desalination 218(1–3):304–312

Chanathaworn J, Bunyakan C, Wiyaratn W, Chungsiriporn J (2012) Photocatalytic decolorization of basic dye by TiO2 nanoparticle in photoreactor. Songklanakarin J Sci Technol 34(2)

Chaturvedi S, Dave PN, Shah NK (2012) Applications of nano-catalyst in new era. J Saudi Chem Soc 16(3):307–325

Chaudhary GR, Saharan P, Kumar A, Mehta SK, Mor S, Umar A (2013) Adsorption studies of cationic, anionic and azo-dyes via monodispersed Fe_3O_4 nanoparticles. J Nanosci Nanotechnol 13(5):3240–3245

Chen J, Leng J, Yang X, Liao L, Liu L, Xiao A (2017) Enhanced performance of magnetic graphene oxide-immobilized laccase and its application for the decolorization of dyes. Molecules 22(221):1–11

Cheval N, Gindy N, Flowkes C, Fahmi A (2012) Polyamide 66 microspheres metallised with in situ synthesized gold nanoparticles for catalytic application. Nanoscale Res Lett 7(1):182–190

Chiou SH, Wu WT (2004) Immobilization of *Candida rugosa* lipase on chitosan with activation of the hydroxyl groups. Biomaterials 25(2):197–204

Colombié S, Gaunand A, Lindet B (2001) Lysozyme inactivation under mechanical stirring: effect of physical and molecular interfaces. Enzym Microb Technol 28:820–826

D'Souza SF (1998) Immobilized enzymes in bioprocess. Curr Sci 77(1):69–79

Dalal S, Kapoor M, Gupta MN (2007) Preparation and characterization of combi-CLEAs catalyzing multiple non-cascade reactions. J Mol Catal B Enzym 44:128–132

Darwesh OM, Moawad H, Barakat OS, Abd El-Rahim WM (2015) Bioremediation of textile Reactive Blue azo dye residues using nanobiotechnology approaches. Res J Pharm Biol Chem Sci 6(1):1202–1211

Das A, Singh J, Yogalakshmi KN (2017) Laccase immobilized magnetic iron nanoparticles: Fabrication and its performance evaluation in chlorpyrifos degradation. Int Biodeter Biodegr 117:183–189

Datta S, Rene CL, Rajaram YRS (2013) Enzyme immobilization: an overview on techniques and support materials. 3 Biotech 3(1):1–9

Dong T, Zhao L, Huang Y, Tan X (2010) Preparation of cross-linked aggregates of aminoacylase from *Aspergillus melleus* by using bovine serum albumin as an inert additive. Bioresour Technol 101:6569–6571

Dubinin MM, Radushkevich LV (1947) The equation of the characteristic curve of the activated charcoal. Proc Acad Sci USSR Phys Chem Sect 55:331–337

Dyer HJ, Hudson PA (1993) Williams, ion exchange processes: advances and applications. Royal Society of Chemistry, Cambridge

El-Gohary F, Tawfik A (2009) Decolorization and COD reduction of disperse and reactive dyes wastewater using chemical-coagulation followed by sequential batch reactor (SBR) process. Desalination 249(3):1159–1164

El-Safty SA, Shenashen MA, Ismael M, Khairy M, Awual MR (2012) Optical mesosensors for monitoring and removal of ultra-trace concentration of Zn (II) and Cu (II) ions from water. Analyst 137(22):5278–5290

Endo M, Takeuchi K, Kobori K, Takahashi K, Kroto HW, Sarkar A (1995) Pyrolytic carbon nanotubes from vapor-grown carbon fibers. Carbon 33(7):873–881

Eyasu A, Yadav OP, Bachheti RK (2013) Photocatalytic degradation of methyl orange dye using Cr-doped ZnS nanoparticles under visible radiation. Int J ChemTech Res 5(4):1452–1461

Fernandes RA, Daniel-da-Silva AL, Tavares AP, Xavier AM (2017) EDTA-Cu (II) chelating magnetic nanoparticles as a support for laccase immobilization. Chem Eng Sci 158:599–605

Foo KY, Hameed BH (2010) Insights into the modeling of adsorption isotherm systems. Chem Eng J 156:2–10

Freundlich HMF (1906) Over the adsorption in solution. J Phys Chem 57:385–471

Fu J, Reinhold J, Woodbury NW (2011) Peptide-modified surfaces for enzyme immobilization. PLoS One 6(4):1–6

Garcia-Galan C, Berenguer-Murcia A, Fernandez-Lafuente R, Rodrigues RC (2011) Potential of different enzyme immobilization strategies to improve enzyme performance. Adv Synth Catal 353(16):2885–2904

Garg VK, Kaushik P (2008) Influence of textile mill wastewater irrigation on the growth of sorghum cultivars. Appl Ecol Environ Res 6(2):1–2

Gautam S, Kaithwas G, Bharagava RN, Saxena G (2017) Pollutants in tannery wastewater, pharmacological effects and bioremediation approaches for human health protection and environmental safety. In: Bharagava RN (ed) Environmental pollutants and their bioremediation approaches, 1st edn. CRC Press/Taylor & Francis Group, Boca Raton, pp 369–396. https://doi.org/10.1201/9781315173351-14

Georgiou D, Melidis P, Aivasidis A, Gimouhopoulos K (2002) Degradation of azo-reactive dyes by ultraviolet radiation in the presence of hydrogen peroxide. Dyes Pigments 52(2):69–78

Ghaedi M, Ramazani S, Roosta M (2011) Gold nanoparticle loaded activated carbon as novel adsorbent for the removal of Congo red. Indian J Sci Technol 4(10):1208–1217

Gianluca C, Nicola R (2001) Technical note, the treatment and reuse of wastewater in the textile industry means of ozonation and electroflocculation. Water Res 35:567–572

Golob V, Vinder A, Simonic M (2005) Efficiency of the coagulation/flocculation method for the treatment of dyebath effluents. Dyes Pigments 67:93–97

Guibal E, Roussy J (2007) Coagulation and flocculation of dye containing solutions using a biopolymer (chitosan). React Funct Polym 67:33–42

Guisan JM (2006) Immobilization of enzymes as the 21st century begins. In: Immobilization of enzymes and cells, 2nd edn, vol 22. Humana Press, Totowa, pp 1–13

Guzman J, Saucedo I, Revilla J, Navarro R, Guibal E (2003) Copper sorption by chitosan in the presence of citrate ions: influence of metal speciation on sorption mechanism and uptake capacities. Int J Biol Macromol 33:57–65

Han R, Ding D, Xu Y, Zou W, Wang Y, Li Y, Zou L (2008) Use of rice husk for adsorption of Congo red from aqueous solution in column mode. Bioresour Technol 99:2938–2946

Hemapriyamvadha, R. and Sivasankar, T. (2015). Sonophotocatalytic treatment of methyl orange dye and real textile effluent using synthesised nano-zinc oxide. Coloration Technology 131(2):110–119

Hariani PL, Faizal M, Setiabudidaya D (2013) Synthesis and properties of Fe_3O_4 nanoparticles by co-precipitation method to removal procion dye. Int J Environ Sci Dev 4(3):336–340

Honda T, Miyazaki M, Nakamura H, Maeda H (2006) Facile preparation of an enzyme-immobilized microreactor using a cross-linking enzyme membrane on a microchannel surface. Adv Synth Catal 348(15):2163–2171

Hsieh HJ, Liu PC, Liao WJ (2000) Immobilization of invertase via carbohydrate moiety on chitosan to enhance its thermal stability. Biotechnol Lett 22(18):1459–1464

Hul JP (1999) Membrane separation in textile washing processes. Universiteit Twente, Enschede

Hu B, Pan J, Yu H-L, Liu J-W, Xu J-H (2009) Immobilization of *Serratia marcescens* lipase onto amino-functionalized magnetic nanoparticles for repeated use in enzymatic synthesis of Diltiazem intermediate. Process Biochem 44(9):1019–1024

Huong PTL, Huyen NT, Giang CD, Nguyen T, Phan VN, Van Quy N, Huy TQ, Hue DTM, Chinh HD, Le A-T (2016) Facile Synthesis and Excellent Adsorption Property of GO-Fe₃O₄ Magnetic Nanohybrids for Removal of Organic Dyes. J Nanosci Nanotechnol 16(9):9544–9556

Hwang ET, Gu MB (2013) Enzyme stabilization by nano/microsized hybrid materials. Eng Life Sci 13(1):49–61

Ibrahim Z, Amin MFM, Yahya A, Aris A, Umor NA, Muda K, Sofian NS (2009) Characterisation of microbial flocs formed from raw textile wastewater in aerobic biofilm reactor (ABR). Water Sci Technol 60(3):683–688

Iram M, Guo C, Guan YP, Ishfaq A, Liu HZ (2010) Adsorption and magnetic removal of neutral red dye from aqueous solution using Fe_3O_4 hollow nanospheres. J Hazard Mater 181(1–3):1039–1050

Ispas C, Sokolov I, Andreescu S (2009) Enzyme-functionalized mesoporous silica for bioanalytical applications. Anal Bioanal Chem 393(2):543–554

Jethava AD, Gohil MS, Vaghela SS, Gour PM, Susarla VRKS, Ramachandraiah G, Ghosh PK (2001) Electrochemical treatment of dye effluents: decolouration and reduction of COD of reactive azo dyes of two leading textile industries. In: Proceedings of national conference on recent advances in waste management, BHU, Varanasi. pp 377–380

Jiang Y, Tang W, Gao J, Zhou L, He Y (2014) Immobilization of horseradish peroxidase in phospholipid-templated titania and its applications in phenolic compounds and dye removal. Enzym Microb Technol 55:1–6

Jiraratananon R, Sungpet A, Luangsowan P (2000) Performance evaluation of nanofiltration membranes for treatment of effluents containing reactive dye and salt. Desalination 130(2):177–183

Jorfi S, Barzegar G, Ahmadi M, Soltani RDC, Takdastan A, Saeedi R, Abtahi M (2016) Enhanced coagulation-photocatalytic treatment of Acid red 73 dye and real textile wastewater using UVA/synthesized MgO nanoparticles. J Environ Manag 177:111–118
Joshi M, Bansal R, Purwar R (2004) Colour removal from textile effluents. Indian J Fibre Text Res 29:239–259
Karimi H, Mousavi S, Sadeghian B (2012) Silver nanoparticle loaded on activated carbon as efficient adsorbent for removal of methyl orange. Indian J Sci Technol 5(3):2346–2353
Karthikeyan S, Sivakumar P, Palanisamy PN (2008) Novel activated carbons from agricultural wastes and their characterization. EJ Chem 5(2):409–426
Khallaf H, Oladeji IO, Chow L (2008) Optimization of chemical bath deposited CdS thin films using nitrilotriacetic acid as a complexing agent. Thin Solid Films 516(18):5967–5973
Khanna A, Shetty VK (2014) Solar light induced photocatalytic degradation of Reactive Blue 220 (RB-220) dye with highly efficient Ag@TiO_2 core–shell nanoparticles: a comparison with UV photocatalysis. Sol Energy 99:67–76
Kim TH, Park C, Lee J, Shin EB, Kim S (2002) Pilot scale treatment of textile wastewater by combined processes (fluidized biofilm process–chemical coagulation–electrochemical oxidation). J Hazard Mater 112:5–103
Koch M, Yediler A, Lienert D, Insel G, Kettrup A (2002) Ozonation of hydrolyzed azo dye reactive yellow 84 (CI). Chemosphere 46(1):109–113
Kumar VV, Sivanesan S, Cabana H (2014) Magnetic cross-linked laccase aggregates-bioremediation tool for decolorization of distinct classes of recalcitrant dyes. Sci Total Environ 487:830–839
Kumar PS, Narayan AS, Dutta A (2017) Nanochemicals and effluent treatment in textile industries. In: Textiles and clothing sustainability. Springer, Singapore, pp 57–96
Langmuir I (1916) The constitution and fundamental properties of solids and liquids. J Am Chem Soc 38:2221–2295
Ledakowicz S, Solecka M, Zylla R (2001) Biodegradation, decolourisation and detoxification of textile wastewater enhanced by advanced oxidation processes. J Biotechnol 89(2–3):175–184
Lee J, Yoon S, Oh SM, Shin CH, Hyeon T (2000) Development of a new mesoporous carbon using an HMS aluminosilicate template. Adv Mater 12(5):359–362
Lei J, Ju H (2012) Signal amplification using functional nanomaterials for biosensing. Chem Soc Rev 41(6):2122–2134
Lin ST, Thirumavalavan M, Jiang TY, Lee JF (2014) Synthesis of ZnO/Zn nano photocatalyst using modified polysaccharides for photodegradation of dyes. Carbohydr Polym 105:1–9
Liu Z, Zhang X, Nishimoto S, Jin M, Tryk D, Murakami T, Fujishima A (2008) Highly ordered TiO_2 nanotube arrays with controllable length for photoelectrocatalytic degradation of phenol. J Phys Chem C 112:253–259
Liu X, Wang M, Zhang S, Pan B (2013) Application potential of carbon nanotubes in water treatment: a review. J Environ Sci (China) 25:S1263–S1280
Liu Y, Yang W, Liu H (2015) Azobenzene-functionalized cage silsesquioxanes as inorganic–organic hybrid, photoresponsive, nanoscale, building blocks. Chem Eur J 21(12):4731–4738
Liu Y, Yan M, Geng Y, Huang J (2016) Laccase immobilization on poly (p-phenylenediamine)/Fe_3O_4 nanocomposite for reactive blue 19 dye removal. Appl Sci 6(232):1–13
Machado FM, Bergmann CP, Lima EC, Adebayo MA, Fagan SB (2014) Adsorption of a textile dye from aqueous solutions by carbon nanotubes. Mater Res 17:153–160
Mahmoodi NM (2011) Photocatalytic ozonation of dyes using copper ferrite nanoparticle prepared by co-precipitation method. Desalination 279(1):332–337
Mahmoodi NM, Abdi J, Bastani D (2014) Direct dyes removal using modified magnetic ferrite nanoparticle. J Environ Health Sci Eng 12(1):1–10
Manisankar P, Viswanathan S, Rani C (2001) Electrochemical decolourization of disperse-blue-26 dye Εφφλuent. Proceedings of national conference on recent advances in waste management BHU, Varanasi. pp 385–388

Mattiasson B, Kaul R (1991) Determination of coupling yields and handling of labile proteins in immobilization technology. Bioprocess Technol 14:161–179

McHenry ME, Laughlin DE (2000) Nano-scale materials development for future magnetic applications. Acta Mater 48(1):223–238

McNaught AD, Wilkinson A (1997) Compendium of chemical terminology: IUPAC. Blackwell Science, Oxford

Mehmet FS, Hasan ZS (2002) Ozone treatment of textile effluents and dyes: effect of applied ozone dose, pH and dye concentration. J Chem Technol Biotechnol 77:842–850

Meng F, Hong Z, Arndt J, Li M, Zhi M, Yang F, Wu N (2012) Visible light photocatalytic activity of nitrogen-doped $La_2Ti_2O_7$ nanosheets originating from band gap narrowing. Nano Res 5(3):213–221

Metcalf, Eddy Inc (2003) Wastewater engineering treatment and reuse, 4th edn. McGrawHill, New York

Mezohegyi G, Gonçalves F, Órfão JJ, Fabregat A, Fortuny A, Font J, Bengoa C, Stuber F (2010) Tailored activated carbons as catalysts in biodecolourisation of textile azo dyes. Appl Catal B 94(1):179–185

Migneault I, Dartiguenave C, Bertrand MJ, Waldron KC (2004) Glutaraldehyde: behavior in aqueous solution, reaction with proteins, and application to enzyme crosslinking. Biotechniques 37(5):790–806

Moreno JM, Sinisterra JV (1994) Immobilization of lipase from *candida cylindracea* on inorganic supports. J Mol Catal 93:357–369

MoT (2001) Census of textile power processing industry in India, vol VII. Textile Committee, Ministry of Textiles, Government of India, Mumbai

Moustafa S (2008) Environmental impacts of textile industries. Process analysis of textile manufacturing. J Agric Food Chem 52:1331–1337

Nam S, Tratnyek PG (2000) Reduction of azo dyes with zero-valent iron. Water Res 34(6):1837–1845

Namboodri CG, Perkins WS, Walsh WK (1994) Decolorizing dyes with chlorine and ozone – part-I. Am Dyest Rep 83(3):17–26

Nassar NN, Marei NN, Vitale G, Arar LA (2015) Adsorptive removal of dyes from synthetic and real textile wastewater using magnetic iron oxide nanoparticles: thermodynamic and mechanistic insights. Can J Chem Eng 93(11):1965–1974

Nimni ME, Cheung D, Strates B, Kodama M, Sheikh K (1987) Chemically modified collagen: a natural biomaterial for tissue replacement. J Biomed Mater Res 21:741–771

Niu C, Sichel EK, Hoch R, Moy D, Tennent H (1997) High power electrochemical capacitors based on carbon nanotube electrodes. Appl Phys Lett 70(11):1480–1482

Pala A, Tokat E (2002) Color removal from cotton textile industry wastewater in an activated sludge system with various additives. Water Res 36(11):2920–2925

Pan B, Xing BS (2008) Adsorption mechanisms of organic chemicals on carbon nanotubes. Environ Sci Technol 42:9005–9013

Pan JR, Huang CP, Chen SC, Chung YC (1999) Evaluation of a modified chitosan biopolymer for coagulation of colloidal particles. Colloids Surf A 147:359–364

Pereira L, Pereira R, Pereira MFR, Van der Zee FP, Cervantes FJ, Alves MM (2010) Thermal modification of activated carbon surface chemistry improves its capacity as redox mediator for azo dye reduction. J Hazard Mater 183(1):931–939

Pereira RA, Pereira MFR, Alves MM, Pereira L (2014) Carbon based materials as novel redox mediators for dye wastewater biodegradation. Appl Catal B 144:713–720

Perkowski J, Kos L (2003) Decolouration of model dye house wastewater with advanced oxidation processes. Fibres Text East Eur 11:67–71

Porter JJ, Gomes AC (2000) The rejection of anionic dyes and salt from water solutions using a polypropylene microfilter. Desalination 128(1):81–90

Pradhan N, Pal A, Pal T (2002) Silver nanoparticle catalyzed reduction of aromatic nitro compounds. Colloids Surf A Physicochem Eng Asp 196(2):247–257

Qadri S, Ganoe A, Haik Y (2009) Removal and recovery of acridine orange from solutions by use of magnetic nanoparticles. J Hazard Mater 169(1):318–323

Qu S, Huang F, Yu S, Chen G, Kong J (2008) Magnetic removal of dyes from aqueous solution using multi-walled carbon nanotubes filled with Fe2O3 particles. J Hazard Mater 160:643–647

Rahman FBA, Akter M (2016) Removal of dyes form textile wastewater by adsorption using shrimp Shell. Int J Waste Resour 6:244

Rajeswari KR (2000) Ozonation treatment of textile dyes wastewater using plasma ozoniser. PhD thesis, University Malaysia, Malaysia

Ramasamyaa RK, Rahmana NA, San WC (2001) Effect of temperature on the ozonation of textile waste effluent. Color Technol 117(2):95–97

Rao GP, Lu C, Su F (2008) Sorption of divalent metal ions from aqueous solution by carbon nanotubes: a review. Sep Purif Technol 58:224–231

Rashed MN, El-Amin AA (2007) Photocatalytic degradation of methyl orange in aqueous TiO_2 under different solar irradiation sources. Int J Phys Sci 2(3):73–81

Robinson T, McMullan G, Marchant R, Nigam P (2001) Remediation of dyes in textile effluent: a critical review on current treatment technologies with a proposed alternative. Bioresour Technol 77:247–255

Saranya M, Santhosh C, Ramachandran R, Kollu P, Saravanan P, Vinoba M, Jeong SK, Grace AN (2014) Hydrothermal growth of CuS nanostructures and its photocatalytic properties. Powder Technol 252:25–32

Saravanan R, Shankar H, Prakash T, Narayanan V, Stephen A (2011) ZnO/CdO composite nanorods for photocatalytic degradation of methylene blue under visible light. Mater Chem Phys 125(1):277–280

Saravanan R, Thirumal E, Gupta VK, Narayanan V, Stephen A (2013) The photocatalytic activity of ZnO prepared by simple thermal decomposition method at various temperatures. J Mol Liq 177:394–401

Saravanan R, Gupta VK, Mosquera E, Gracia F, Narayanan V, Stephen A (2015) Visible light induced degradation of methyl orange using β-Ag 0.333 V2O5 nanorod catalysts by facile thermal decomposition method. J Saudi Chem Soc 19(5):521–527

Saxena G, Bharagava RN (2015) Persistent organic pollutants and bacterial communities present during the treatment of tannery wastewater. In: Chandra R (ed) Environmental waste management, 1st edn. CRC Press/Taylor & Francis Group, Boca Raton, pp 217–247. https://doi.org/10.1201/b19243-10

Saxena G, Bharagava RN (2017) Organic and inorganic pollutants in industrial wastes, their ecotoxicological effects, health hazards and bioremediation approaches. In: Bharagava RN (ed) Environmental pollutants and their bioremediation approaches, 1st edn. CRC Press/Taylor & Francis Group, Boca Raton, pp 23–56. https://doi.org/10.1201/9781315173351-3

Saxena G, Chandra R, Bharagava RN (2016) Environmental pollution, toxicity profile and treatment approaches for tannery wastewater and its chemical pollutants. Rev Environ Contam Toxicol 240:31–69. https://doi.org/10.1007/398_2015_5009

Schoevaart R, Wolbers MW, Golubovic M, Ottens M, Kieboom APG, van Rantwijk F, van der Wielen LAM, Sheldon RA (2004) Preparation, optimization and structures of cross-linked enzyme aggregates (CLEAs). Biotechnol Bioeng 87:754–762. https://doi.org/10.1002/bit.20184

Sheldon RA (2007) Enzyme immobilization: the quest for optimum performance. Adv Synth Catal 349(8–9):1289–1307

Sheldon RA (2011) Characteristic features and biotechnological applications of cross-linked enzyme aggregates (CLEAs). Appl Microbiol Biotechnol 92(3):467–477

Sheydaei M, Khataee A (2015) Sonocatalytic decolorization of textile wastewater using synthesized γ-FeOOH nanoparticles. Ultrason Sonochem 27:616–622

Shi Y, Jin F, Wu Y, Yan F, Yu X, Quan Y (1997) Improvement of immobilized cells through permealizing and cross-linking. Chin J Biotechnol 13(1):111–113

Singh BD (2009) Biotechnology expanding horizons. Kalyani, Ludhiana

Sinirlioglu ZA, Sinirlioglu D, Akbas F (2013) Preparation and characterization of stable cross-linked enzyme aggregates of novel laccase enzyme from *Shewanella putrefaciens* and using malachite green decolorization. Bioresour Technol 146:807–811

Sohn Y, Smith Y, Misra M, Subramanian V (2008) Electrochemically assisted photocatalytic degradation of methyl orange using anodized titanium dioxide nanotubes. Appl Catal B Environ 84:372–378

Szpyrkowicz L, Juzzolino C, Kaul SN (2001) A Comparative study on oxidation of disperse dyes by electrochemical process, ozone, hypochlorite and fenton reagent. Water Res 35(9):2129–2136

Talekar S, Waingade S, Gaikwad V, Patil S, Nagavekar N (2012) Preparation and characterization of cross linked enzyme aggregates (CLEAs) of *Bacillus amyloliquefaciens* alpha amylase. J Biochem Technol 3(4):349–353

Thanha TKN, Greena LAW (2010) Functionalisation of nanoparticles for biomedical applications. Nano Today 5:213–230

Thompson G, Swain J, Kay M, Forster CF (2009) The treatment of pulp and paper mill effluent: a review. Bioresour Technol 77:275–286

Tischer W, Wedekind F (1999) Immobilized enzymes: methods and applications. In: Biocatalysis-from discovery to application. Springer, Berlin, pp. 95–126

Torres MR, Bouzán CG, Crespi M (2010) Combination of coagulation–flocculation and nanofiltration techniques for dye removal and water reuse in textile effluents. Desalination 252:53–59

Tran DN, Balkus KJ Jr (2011) Perspective of recent progress in immobilization of enzymes. ACS Catal 1(8):956–968

Tristão JC, Magalhães F, Corio P, Sansiviero MTC (2006) Electronic characterization and photocatalytic properties of CdS/TiO_2 semiconductor composite. J Photochem Photobiol A Chem 181(2):152–157

U.S. Environmental Protection Agency (1997) Profile of the Textile Industry. Office of Compliance Sector Notebook Project. U.S. Environmental Protection Agency, Washington, D.C., p 40

Vaidya BK, Kuwar SS, Golegaonkar SB, Nene SN (2012) Preparation of cross-linked enzyme aggregates of l-aminoacylase via co-aggregation with polyethyleneimine. J Mol Catal B Enzym 74(3):184–191

Van Der Zee FP, Bisschops IA, Lettinga G, Field JA (2003) Activated carbon as an electron acceptor and redox mediator during the anaerobic biotransformation of azo dyes. Environ Sci Technol 37(2):402–408

Vera G, Aleksandra V, Marjana S (2005) Efficiency of the coagulation/flocculation method for the treatment of dye bath effluents. Dyes Pigments 67:93–97

Verma AK, Dash RR, Bhunia P (2012) A review on chemical coagulation/flocculation technologies for removal of colour from textile wastewaters. J Environ Manag 93(1):154–168

Vlyssides AG, Papaioannou D, Loizidoy M, Karlis PK, Zorpas AA (2000) Testing an electrochemical method for treatment of textile dye wastewater. Waste Manag 20:569–574

Wang Y (2000) Solar photocatalytic degradation of eight commercial dyes in TiO2 suspension. Water Res 34(3):990–994

Wang H, Huang Y (2011) Prussian-blue-modified iron oxide magnetic nanoparticles as effective peroxidase-like catalysts to degrade methylene blue with H_2O_2. J Hazard Mater 191(1):163–169

Wang H, Xie C, Zhang W, Cai S, Yang Z, Gui Y (2007) Comparison of dye degradation efficiency using ZnO powders with various size scales. J Hazard Mater 141(3):645–652

Wang Z, Xue M, Huang K, Liu Z (2011) Textile dyeing wastewater treatment. In: Advances in treating textile effluent. InTech, pp 1–27. ISBN: 978-953-307-704-8

Wang H, Zhang W, Zhao J, Xu L, Zhou C, Chang L, Wang L (2013) Rapid decolorization of phenolic azo dyes by immobilized laccase with Fe_3O_4/SiO_2 nanoparticles as support. Ind Eng Chem Res 52(12):4401–4407

Wararkar P, Kothari M, Wararkar K (2016) Performance analysis of supply chain management technique at textile industry in India in accordance with Indian textile market. Int J Text Eng Process 2(3):34–39

Weng CH, Lin YT, Yeh CL, Sharma YC (2010) Magnetic Fe_3O_4 nanoparticles for adsorptive removal of acid dye (new coccine) from aqueous solutions. Water Sci Technol 62(4):844–851
Wyckoff RWG (1964) Crystal structures, 2nd edn. Wiley, New York
Yamaguchi M, Namiki Y, Okada H, Mori Y, Furukawa H, Wang J, Ohkusu M, Kawamoto S (2011) Structome of *Saccharomyces cerevisiae* determined by freeze-substitution and serial ultrathin-sectioning electron microscopy. J Electron Microsc 60(5):321–335
Yang K, Wu W, Jing Q, Zhu L (2008) Aqueous adsorption of aniline, phenol, and their substitutes by multi-walled carbon nanotubes. Environ Sci Technol 42(21):7931–7936
Zhang J, Zhai S, Li S, Xiao Z, Song Y, An Q, Tian G (2013) Pb(II) removal of Fe3O4@SiO2-NH2 core-shell nanomaterials prepared via a controllable sol-gel process. Chem Eng J 215-216:461–471
Zhou L, Tang W, Jiang Y, Ma L, He Y, Gao J (2016) Magnetic combined cross-linked enzyme aggregates of horseradish peroxidase and glucose oxidase: an efficient biocatalyst for dye decolourization. RSC Adv 6(93):90061–90068
Zhu H, Jiang R, Xiao L, Chang Y, Guan Y, Li X, Zeng G (2009) Photocatalytic decolorization and degradation of Congo Red on innovative crosslinked chitosan/nano-CdS composite catalyst under visible light irradiation. J Hazard Mater 169(1):933–940
Zucca P, Sanjust E (2014) Inorganic materials as supports for covalent enzyme immobilization: methods and mechanisms. Molecules 19(9):14139–14194

Chapter 16
Green Synthesis of Nanoparticles and Their Applications in Water and Wastewater Treatment

Surya Pratap Goutam, Gaurav Saxena, Diptarka Roy, Anil Kumar Yadav, and Ram Naresh Bharagava

Contents

S. P. Goutam · D. Roy · A. K. Yadav (✉)
Advanced Materials Research Laboratory, Department of Applied Physics (DAP), School for Physical Sciences (SPS), Babasaheb Bhimrao Ambedkar University (A Central University), Lucknow, Uttar Pradesh, India

G. Saxena · R. N. Bharagava
Laboratory of Bioremediation and Metagenomics Research (LBMR), Department of Microbiology (DM), Babasaheb Bhimrao Ambedkar University (A Central University), Lucknow, Uttar Pradesh, India

G. Saxena, R. N. Bharagava (eds.), *Bioremediation of Industrial Waste for Environmental Safety*, https://doi.org/10.1007/978-981-13-1891-7_16

Abstract Green synthesis of nanoparticles (NPs) is an emerging research trend in green nanotechnology as this method is nontoxic or less toxic, eco-friendly, efficient, and cost-effective as compared to other conventional physical and chemical methods. Green synthesis of NPs employs various biological agents such as plants, bacteria, algae, and fungi, but nowadays plant-based green synthesis of NPs is gaining more attention among researchers from around the world. A variety of green synthesized NPs are currently being used in water and wastewater treatment due to their high efficiency and biocompatible nature. Green synthesized NPs are highly proficient for recycling and removal of heavy metal from wastewaters without loss of their stability and degradation of a variety of organic pollutants from wastewaters and, thus, purify the wastewaters for reuse and recycling and could solve various water quality issues worldwide. However, regeneration and reusability are the main challenges to researchers and scientist yet in the green synthesis approach as a technology transfer from laboratory scale to commercial applications. In this chapter, we discussed the green synthesis approach for NPs and their applications in water and wastewater treatment and dye degradation from wastewaters. Further, challenges and issues related to the use of green synthesized NPs in water and wastewater treatment are also discussed.

Keywords Green nanotechnology · Green synthesis · Nanoparticles · Wastewater treatment · Environmental protection

16.1 Introduction

Water is the necessary element for all life forms and covers two-thirds of earth's surface. It has variable physical, chemical, or biological properties after the addition of different organic/inorganic materials, pathogens, heavy metals, or other pollutants which make it harmful to the whole ecosystem. (Bora and Dutta 2014). Moreover, the water on earth is a precious and limited resource, which is boundlessly recycled in the water cycle. Hence, the sustainable management of water resources is immediately required due to their increasing contamination with various organic and inorganic pollutants discharged along with urban wastewaters including industrial wastewaters.

The global challenge of the twenty-first century is to overcome the lack of clean water and to provide safe and clean water for all living beings to survive. The scarcity of treated and clean water influenced the living standard of our society and deducted the equal opportunities of development in the society (Qadir et al. 2007). According to a report of the World Health Organization (WHO) in 2012, 780 million people from all around the world still lack access to safe and clean drinking water supply (Qu et al. 2013). In most of the developing countries, 90% of all diseases arise due to the use of contaminated water for drinking purpose (WHO Report 2012). According to the United Nations World Water Development Report (2017), over the past 5 years, clean water scarcity has become a major risk to the surviving human population (UN Water Development Report 2017). In a report of the World Economic Forum (WEF) in 2016, it is reported that water scarcity was evaluated as the universal risk of premier concern for individuals and economies for the next 10 years (http://unesco.org/images/0024/002471/247153e.pdf). In the countries including India, China, Australia, Western South America, Central America, and Western North America, around 500 million people live in areas where water consumption exceeds by a double factor than treated water resources (Mekonnen and Hoekstra 2016). According to the reports of the United Nations International Children's Fund (UNICEF) and WHO in 2015, around two-thirds of the world's population has acquired developed sanitation, and developing countries are still facing a lack of sanitation or sewer connections (Kjellén et al. 2012). In some of the developing countries, there is no central facility to provide the connections to a water supply; generally the people depend on the self-provided services or nongovernment organization (NGO) support (http://unesdoc.unesco.org/images/0024/002471/247153e.pdf).

Wastewater is a major source of water pollution in the developing countries and is generally classified into two main categories (Bora and Dutta 2014): (a) residential wastewater (municipal wastewater) and (b) nonresidential wastewater (urban or industrial wastewater). According to UN-Water Report in 2015, wastewater is roughly composed of 99% water and 1% suspended colloidal and dissolved solids (http://unesdoc.unesco.org/images/0024/002471/247153e.pdf). Residential wastewater originates from the public residence and contains 99.90% water and 10% solids, organic and inorganic content, nutrients, microorganism, and biodegradable organic materials (Bora and Dutta 2014). The organic and inorganic contents (metals) in wastewaters are considered to be highly toxic for humans as well as the environment (Goutam et al. 2018; Bharagava et al. 2017c; Saxena and Bharagava 2015, 2017; Gautam et al. 2017; Saxena et al. 2016). However, the organic content of wastewater can be measured in terms of biochemical oxygen demand (BOD) and chemical oxygen demand (COD), the main water quality parameters, and represent the degradability of wastewater (Henze and Comeau 2008).The amount of residential wastewater is influenced by the lifestyle and living standard of people (Chandrappa and Das 2012). Further, nonresidential or urban wastewater mainly originates from industries, agricultural fields, and commercial actions, and its composition depends on their sources.

To provide a safe and clean water supply to all, the sustainable management of water and wastewaters is an urgent need of the present time. Currently, many conventional and nonconventional wastewater treatment technologies are available to remove pollutants from various types of wastewaters, but these technologies are sometimes not effective and require high fund investment, specific operative conditions, and maintenance cost and high energy necessities (Amin et al. 2014; Baruah et al. 2012; Bora and Dutta 2014). On contrary, nanotechnology offers a great potential for the development of next-generation water and wastewater treatment technologies and could replace the conventional wastewater treatment technologies (Goutam et al. 2018). Nanoparticles (NPs) are increasingly applied for the wastewater treatment and purification due to their high reactivity and degree of functionalization, large surface area, size-dependent properties, etc. (Gawande et al.

2011). Recently, nanomaterials offered many advantages in wastewater treatment owing to the unique characteristics which elicited from nanoscale sizes, such as improved catalysis and adsorption properties as well as high reactivity (Gawande et al. 2011; Guo et al. 2014).

Various nanoparticles have been successfully reported for water and wastewater treatment such as zerovalent metal nanoparticles (silver nanoparticles, iron nanoparticles, zinc nanoparticles), metal oxide nanoparticles (titanium dioxide nanoparticles, iron oxide nanoparticles, zinc oxide nanoparticles), carbon nanotubes, and nanocomposites (Mueller and Nowack 2009). NPs, especially, the chemically synthesized titanium dioxide (TiO_2), have been extensively applied in the photocatalytic treatment of industrial wastewaters (Filipponi and Sutherland 2010). Nano-sized TiO_2-based photocatalytic treatment is a highly effective method for the degradation and detoxification of recalcitrant organic and inorganic pollutants from industrial wastewaters (Ray and Ray 2009; Lu et al. 2015). The chemical synthesis of NPs under specific conditions requires highly expensive and toxic and hazardous chemicals, and their release in the environment creates a serious ecotoxicological concern (Bhattacharya et al. 2013). Nonetheless, the green synthesis is a simple, eco-friendly, and less toxic way of synthesizing NPs from biodegradable materials such as plant extracts, microbes, and enzymes (Bhattacharya et al. 2013; Tiwari et al. 2008). However, the synthesis of NPs using plant extracts is the most useful strategy as it lessens the chance of associated contamination while reducing the reaction time and maintaining the cell structure (Tiwari et al. 2008). Moreover, the application of green NPs for industrial wastewater treatment could be an excellent strategy to cope with environmental pollution.

To date, only very few studies reported the application of green synthesized NPs in wastewater treatment (Bhattacharya et al. 2013; Savage and Diallo 2005). Therefore, this chapter is mainly focused on the green synthesis of nanoparticles using plant extract and their applications in dye removal and water and wastewater treatment. In this chapter, we discussed the green synthesis approach for NPs and their applications in water and wastewater treatment and dye degradation from wastewaters. Further, challenges and issues related to the use of green synthesized NPs in water and wastewater treatment are also discussed.

16.2 Nanotechnology: An Overview

The science and technology associated with various nanostructures has become a wide area of research for the sustainable development of human society, and the technology is growing up incredibly day by day to have a great impact in the future for commercialization. The term "nanotechnology" was first introduced by Taniguchi in 1974. According to him, nanotechnology is a technology which comprises the processes of separation, consolidation, and deformation of materials by single atom or molecule (Iqbal et al. 2012). In other words, nanotechnology is also used for the study of nanomaterials that exhibit amazing properties, functionalities, and phenomena due to the influence of small size (nanoscale) (Khan et al. 2017). It is the ability to understand, fabricate, and manipulate the materials at the nanoscale (Roco 2004). The word "nano" comes from the Greek word *Nanos* which refers to small animals or plants. The one-nanometer scale is equal to 10^{-9}m. The aim of nanotechnology is to measure and manipulate the matters at nanoscale; the scale of its dimension lies in the range of 1–100 nm (Mansoori and Soelaiman 2005). At nano-level, material shows unique chemical, physical, and biological properties due to higher surface to volume ratio where the surface area of the nanoparticle is inversely proportional to particle size, which probably occurs due to the increasing order of surface atoms with respect to the decreasing order of particle size (El Saliby et al. 2008). Thus, as a result of the extremely large surface area along with high surface energy, the nanoparticle reveals distinctive optical, electrical, and magnetic properties than the bulk one (Ichinose et al. 1992).

Nanoscience and nanotechnology have various promising applications due to the affectedly changed and excellent properties of materials at the nanoscale (https://royalsociety.org/~/media/Royal_Society_Content/policy/publications/2004/9693.pdf). It covers a wide spectrum of various technologies, which are based on various types of physical, chemical, and biological processes relied on nano-level, not only to create nanomaterials but also to operate them or to influence them or to use them according to their projected motives (https://www.eolss.net/sample-chapters/C05/E6-152-01.pdf). As the size of materials rises from bulk to nanostructure, the properties and functionalities of the materials may be different. Conclusively, if two parameters of the materials, structural arrangement of atoms or molecules and the length scale of the materials, are molded appropriately, then variation in the properties of materials can easily be achieved (Heath 1995). Nanotechnology provided some of the promising techniques for wastewater treatment such as (a) photocatalysis, (b) nanofiltration, and (c) nanosorbents (Bora and Dutta 2014).

16.2.1 Photocatalysis

Photocatalysis is a process defined as "change in the rate of a chemical reaction or its initiation under the action of ultraviolet, visible, or infrared radiation in the presence of a photocatalyst, a substance that absorbs the light and involved in the chemical transformation of the reaction partners" (McNaught and Wilkinson 1997). It is a surface phenomenon which commonly includes the following five mechanistic steps: (i) diffusion of reactants to the surface of the catalyst, (ii) adsorption of the reactants on the surface of the catalyst, (iii) reaction at the surface of the catalyst, (iv) desorption of the products from the surface of the catalyst, and (v) diffusion of the products from the surface of the catalyst (Pirkanniemi and Sillanpaa 2002).

Photocatalysis is currently being applied in wastewater treatment as two different types of applications: one is solar photocatalysis, and another is photocatalytic application system equipped with artificial UV-visible light (https://www.sswm.info/sites/default/files/reference_attachments/MULLER%20et%20al%202010%20

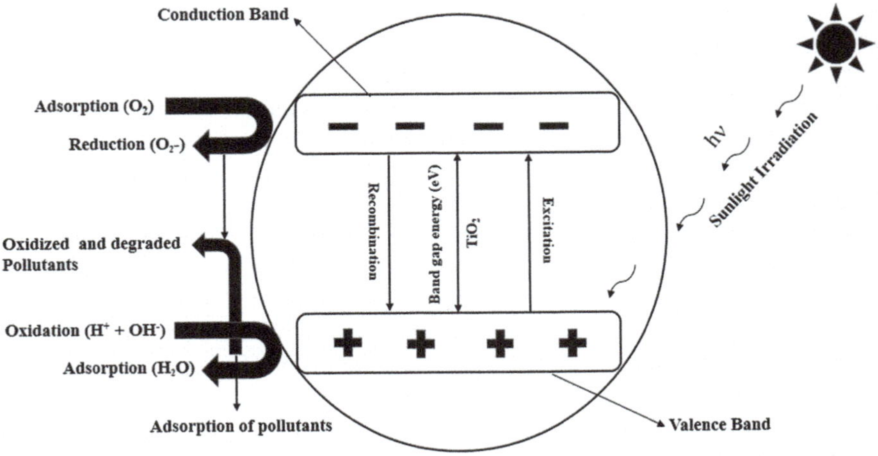

Fig. 16.1 Mechanism of TiO_2-based photocatalysis. (Adapted from Goutam et al. 2018)

Photocatalysis%20for%20Water%20Treatment.pdf). Nanostructured semiconductor materials are the most suitable photocatalyst because they have the most of the photogenerated electrons and holes, which are available at the surface. Photocatalysis plays an important role in the removal of organic and inorganic contaminants such as heavy metal ions and microbes.

A recent study by Goutam et al. (2018) presents an excellent example of the photocatalytic treatment/degradation of tannery wastewater using green synthesized TiO_2 nanoparticles, although some other previously reported literatures by Devatha et al. (2016), Wang et al. (2014), Ghaly et al. (2011), and Bordes et al. (2015) have also shown the significance of photocatalysis in wastewater treatment. Nanoparticle (NP)-based photocatalytic treatment is a highly effective method for the degradation and detoxification of recalcitrant organic and inorganic pollutants from wastewaters (Mahlambi et al. 2015). A general mechanism of TiO_2-NP-based photocatalysis is shown in Fig. 16.1.

The photocatalytic reduction of inorganic contaminants (Cr^{+6} to Cr^{+3}) from wastewater along with photooxidation of organic pollutants can be easily understood by the following reaction mechanisms (Goutam et al. 2018; Malakootian and Mansuri 2015):

$$TiO_2\,(\text{catalyst}) + h\nu \rightarrow h^+ + e^- \tag{16.1}$$

$$Cr_2O_7^{2-} + 14H^+ + 6e^- \rightarrow +2Cr^{3+} + 7H_2O \tag{16.2}$$

$$2H_2O + 4h^+ \rightarrow O_2 + 4H^+ \tag{16.3}$$

$$H_2O + h^+ \rightarrow OH + H^+ \tag{16.4}$$

$$OH + organics \rightarrow CO_2 + H_2O \tag{16.5}$$

$$H^+ + organics \rightarrow CO_2 + H_2O \tag{16.6}$$

16.2.2 Nanofiltration

The second popular technique for wastewater treatment is nanofiltration. It is a type of membrane process and has emerged as a promising area for extended membrane applications. Membrane filtration is a pressure-driven process in which membrane acts as selective barriers to restrict the passage of pollutants such as organics, nutrients, turbidity, microorganisms, inorganic metal ions, and other oxygen-depleting pollutants and permit the pass-through of comparatively clear water (Drioli et al. 2015; Mulder 1997). With technological advances and the ever-increasing stringency of water quality criteria, membrane processes are becoming a more attractive solution to the challenge of water quality and water reuse (Shannon et al. 2008).

Numerous studies have previously reported on the application of microfiltration and ultrafiltration for wastewater treatment and reuse (Seo et al. 1996, 1997; Snoeyink et al. 2000; Shon et al. 2013; Visvanathan et al. 2000; Ben Aim and Semmens 2001; Matsui et al. 2001aa, b). Generally, membrane process has been classified into four broad categories depending on their pore sizes: microfiltration (MF), ultrafiltration (UF), nanofiltration (NF), and reverse osmosis (RO) membranes. As per previously reported in literature, NF membranes are considered to be the best alternative to water purification. NF membrane is generally polyamide membranes and has lower salt rejection facilitating a diverse cut-off based on the molecular weight of a species with more than 92% rejection (Shon et al. 2013).The NF membrane is the relatively newly introduced technology in wastewater treatment system. Due to the less size of NF membrane (~1 to 5 nm), very small uncharged solutes from wastewater are becoming highly rejected (Kim et al. 2001; Shon et al. 2013). Over the last few years, NF technology is developing as a feasible approach for the removal of pollutants from pulp-bleaching effluents and textile effluents, separation of pharmaceuticals from fermentation broths, demineralization in the dairy industry, and metallic pollutants and viruses from wastewaters (Bowen et al. 2002). A general mechanism of nanofiltration for the removal of different pollutants from wastewater is shown in Fig. 16.2.

16.2.3 Nanosorbents

Nanoparticles (NPs) as nanosorbents have been widely used for the removal of microbes, organic dyes, and heavy metal pollutants from water and wastewaters. Nanoparticles have specific properties like extremely small size, high surface area

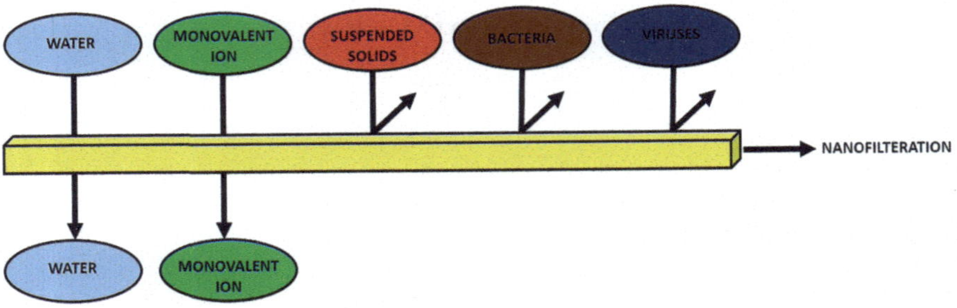

Fig. 16.2 Schematic mechanism of nanofiltration of contaminants from wastewaters

to volume ratio, specificity for pollutants, biocompatibility, and adaptability which attributed to the relevant technology for the enhanced and efficient removal of pollutants from wastewaters. Metal oxide at nanoscale shows the higher level of adsorption as compared to bulk size because of formation of metal-ligand precipitation (Stietiya and Wang 2014). As reported previously, pH value of the wastewater affects the efficiency of metal pollutant removal by adsorption mechanism. The removal efficiency of metallic pollutants (heavy metal ions) from the wastewater increased with a pH value of wastewater. Deprotonation at nanosorbent surface is enhanced with an increased pH value of wastewater and forces of attraction between positively charged metallic ions, and the negative sites on adsorbents also increased due to increased negatively charged sites. Nassar (2012) reported a brief review on iron oxide nano-adsorbent for the removal of various pollutants from wastewater (Nassar 2012). Zhang et al. (2014) reported the synthesis of Fe-La composite oxide for the removal of As (III) from wastewater (Zhang et al. 2014).

According to Khajeh et al. (2013), nanomaterials can be classified in different groups on the basis of their role or surface property in adsorption applications such as (a) carbon nanomaterials, (b) silicon nanomaterials, (c) nanomaterials, and (d) nanoparticles as adsorbents. Although, TiO_2 NPs and other metal oxide nanoparticles such as ZrO_2, Al_2O_3, and SiO_2, ZnO have been reported with very high adsorption capacity because of their specific surface chemistry (He et al. 2008; Lian et al. 2005; Hadjiivanov et al. 1991; Bolis et al. 1991; Cui et al. 2006). A general mechanism of nano-adsorbent for the removal of different pollutants from wastewater is shown in Fig. 16.3.

16.3 Techniques for Nanoparticle Synthesis

Recently, the development of several steadfast methods to synthesize nanoparticles by controlling their sizes, shapes, and chemical composition becomes a marvelous area of research interest in the field of nanotechnology. Nanoparticles in association with organic molecules facilitate collective nature and cause the formation of

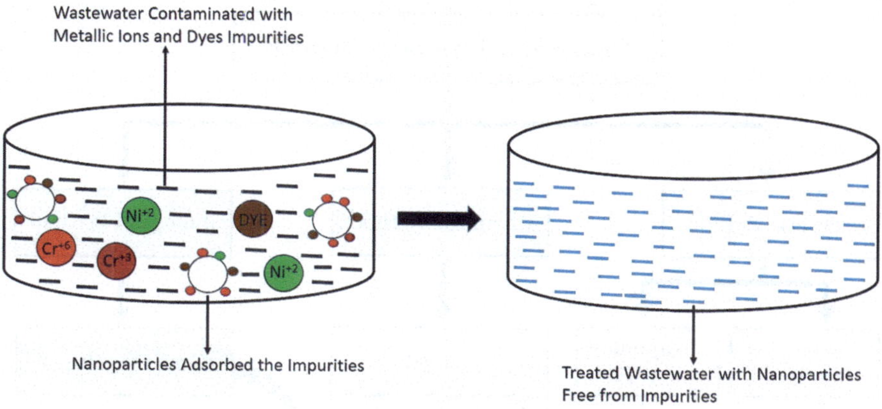

Fig. 16.3 Schematic mechanism of nanosorbents for pollutant removal from wastewaters

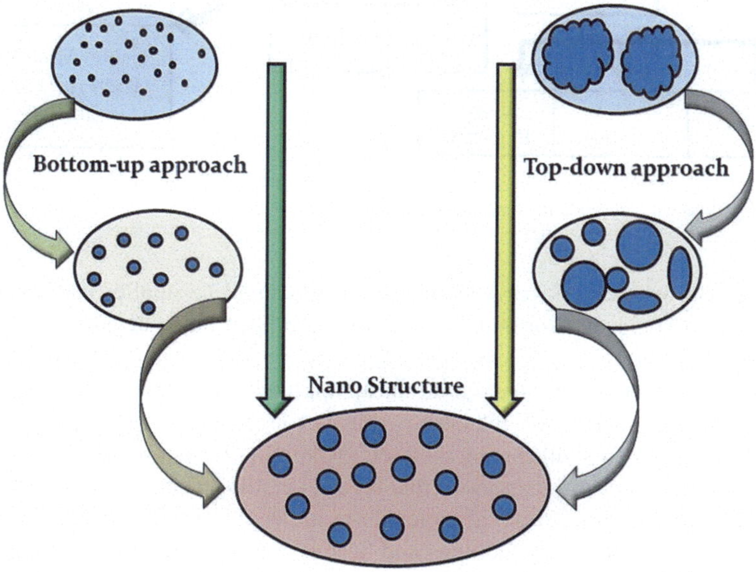

Fig. 16.4 Basic approaches for nanoparticle fabrication

different dimensional (1-D, 2-D, 3-D) mesostructures, which also gave a great interest in research. Nanoparticle synthesis methods generally use precursors in the form of solid, liquid, or gas phase, following chemical and physical deposition processes to construct various nanostructured materials.

The nanoparticle synthesis methods are primarily categorized into two approaches: top-down and bottom-up approaches (Hu and Shaw 1999) as shown in Fig. 16.4. Top-down approach includes the physical processes such as milling,

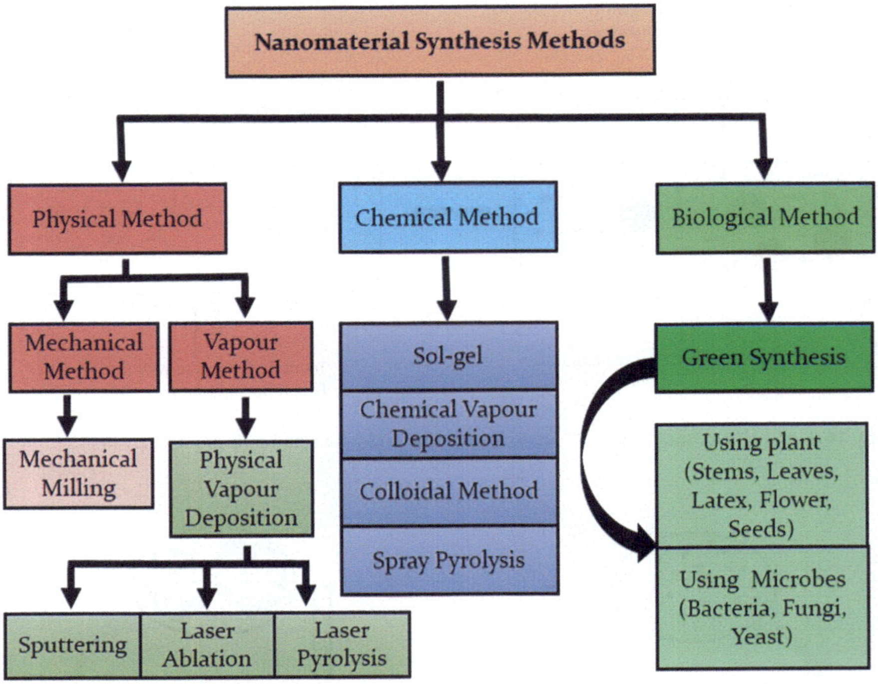

Fig. 16.5 Synthesis methods for nanoparticles

crushing, etc. where bulk particles break up into small to smaller particles; as a result, nanoparticles are formed. Though this approach has a negative impact on the surface structure due to crystallographic injury during the formation of the nanostructure, on the other hand, bottom-up approach includes chemical and biological processes such as sol-gel, laser pyrolysis, plasma spraying process, aerosol-based process, green synthesis method, etc. where the formation of nanoparticles starts from the basic atomic or molecular levels, i.e., the nanostructures are formed by the addition of atom to atom, molecule to molecule, or cluster to cluster. The bottom-up approach is very fabulous and best as compared to top-down approach to produce nanostructured materials with uniform structures and distributions (Saravanan et al. 2008). Various techniques that are used to synthesize nanomaterials are shown in Fig. 16.5.

16.3.1 Mechanical Milling Method

Mechanical milling method mainly associated with the production of powders where fine particles are created from the surfaces of small particles reduced from bulk materials during the rubbing with one another by the introduction of an agitator medium or high-speed air jets to produce enormous pressure and friction together (Ichinose et al. 1992). It is very high energy consumption route where structural defects, chemical disorders, and elastic strain are introduced to parent bulk material during the whole process. As a result of which, nanostructures are produced (Saravanan et al. 2008). Ball milling is an appropriate example of mechanical milling.

16.3.2 Physical Vapor Deposition

Physical vapor deposition is a technique to produce nanostructures at gas phase using a focused beam of electrons that heat the parent bulk material. This method does not depend on any catalytic interaction during the synthesis process, and there occurs no chemical reaction from start to end of the process during the formation of final production from parent entity, namely, sputtering, laser ablation, laser pyrolysis, etc.

16.3.3 Sputtering

Sputtering is one of the most influential physical vapor deposition techniques and useful for nanostructured film production where surface atoms of evaporated material become separated to generate that occurrence (https://sci-hub.tw/10.1533/9781845696689.58).

16.3.4 Laser Ablation

Laser ablation is one type of heating technique where materials like metal and compounds are melted and evaporated by using a laser beam to generate nanostructures under high vacuum system (Ichinose et al. 1992).

16.3.5 Laser Pyrolysis

Laser pyrolysis is a laser-supported deposition technique where the vaporized reactant material is decomposed by using a potent laser beam in some inert gas environment. In this method, the reactant vapor atoms interact with inert gas atoms under collision, then deposited on a substrate to grow nanostructured film (Kulkarni 2015).

16.3.6 Sol-Gel

Sol-gel is a chemical process to synthesize nanoparticles by gelation followed by precipitation and calcination, respectively. Generally, the gel is configured from stabilized sols where sols are often present in the form of colloidal aggregates of small metal oxy-hydroxy particles in an aqueous solvent. However, it is possible to break up this aggregation in the presence of some capping agents. Hydrolysis and condensation in a controlled way play a pivotal role during gelation of the system. The relative rates of hydrolysis and condensation depend on various physical and chemical parameters such as temperature, pH, the concentration of metal ions precursor solution, etc. The change in pH value affects the surface properties of the synthesized material remarkably, which occur due to the change in porosity of gel. As usual for the preparation of nanostructures, metal alkoxides are taken as precursors in some organic solvent. Sometimes in place of alkoxide, chelate also can use to be a precursor which has the ability to stabilize the metal cations. Chelating facilitates mainly for the preparation of multicomponent gel, for example, $TiO_2 - SiO_2$ mixed gel, etc. (Kung and Ko 1996).

16.3.7 Chemical Vapor Deposition

Chemical vapor deposition technique is used to generate a highly pure nanostructural thin film having high performance. In this method, the precursor is deposited on the surface of a substrate by heating followed by evaporation to generate vapor, and the deposition occurs through a chemical reaction under vacuum in such a way to create a difference in chemical properties between precursor material and product. After exposing the surface of a substrate to the vaporized precursor, primarily a template is formed on the surface; thereafter, nanostructures are formed on that template (Filipponi and Sutherland 2010). The formation of the nanostructure is always influenced by the reaction temperature, the reaction rate, and also the precursor concentration during the deposition occurrence (Kim et al. 2004). Overall, this method facilitates for uniform coating of nanostructures on the substrate surface, though there are some restriction due to the higher temperature requirement in the process.

16.3.8 Colloidal Method

Colloidal method is a chemical precipitation method where different ionic solutions are mixed up to get precipitated by the adjustment of the reagent's concentration and capping agents and also by using the temperature and pressure in a controlled way. This method requires a stabilizer to prevent agglomeration that arises mainly due to the presence of van der Waals forces between colloidal nanoparticles which can be stabilized electrically or sterically by Coulombic repulsions and adsorption of surfactants onto the nanoparticles surfaces, respectively (Fendler 2001). This method is very popular to produce metal and metal oxide nanoparticles and useful in various organic and pharmaceutical fields.

16.3.9 Spray Pyrolysis

Spray pyrolysis is a chemical deposition technique which is extensively used for the preparation of nanomaterial thin film. The formation of thin film depends on numerous parameters such as the spray rate, the temperature of the substrate, concentration of chemical solution, etc. The deposition efficiency can be improved by controlling the sizes of droplets and their distribution on the substrate during spraying. It is a very easy approach to prepare nanostructures on the substrate as a film by taking facilities for doping of several elements with any proportions in spraying solution. There are no limitations on quality, dimension, and surface properties of the substrate, and thus, the method has certain positive impacts compared with the other vapor deposition techniques to have opportunities for industrialization (Patil 1999).

16.4 Green Synthesis Approach for Nanoparticles Fabrication

Conventional methods are bound with various limitations such as being expensive, the generation of hazardous toxic chemicals, the requirement of high temperature and pressure, etc. (Sharma et al. 2015). Due to these drawbacks of conventional methods, researchers are currently focusing on the biological systems and giving preference toward green synthesis.

Green synthesis is an environment-friendly, cost-effective process to synthesize nanostructural material to have adjustable structures, morphologies, and particle size distributions (Khan et al. 2017). Recent progress displays an important role in biological synthesis for the production of nanomaterials. Green synthesis methods employ organisms like plant, bacteria, fungi, etc. for the synthesis of nanomaterials, and it has become a very quickly rising area of research due to their less toxic or nontoxic nature, environment-friendly behavior, and low cost of preparation (Das et al. 2016; Goutam et al. 2017, 2018). For plant-mediated synthesis method, the plants are used to synthesize nanoparticles that have excessive potential in heavy metal accumulation mechanism and detoxification (Iravani 2011; Goutam et al. 2018). Plant extract can be produced from leaves, stems, flower, and seeds of various plants. The extract contains biomolecules such as protein, amino acid, enzymes, vitamins, terpenoids, flavonoids, alkaloids, phenolic acids, etc., which act as capping and reducing agents that can reduce metal ions during the bioreduction process to produce nanoparticles or nanostructures with different shapes and sizes (Goutam et al. 2018; Iravani 2011; Shiv Shankar et al. 2003, 2004). For example, *Pelargonium graveolens* extract reduces gold ions into nanoparticles of size 20–40 nm; *Trigonella foenum-graecum* extract can reduce gold ions into nanoparticles of size 15–25 nm (Shiv Shankar et al. 2003; Aromal and Philip 2012). *Alternanthera dentata* leaf extract reduces silver ions into nanoparticles of size 50–100 nm (Kumar et al. 2014), while leaf extracts of *Ziziphora tenuior*, *Ficus carica*, *Cymbopogan citratus*, and *Premna herbacea* reduce the silver nanoparticles to the sizes 8–40 nm, 13 nm, 32 nm, and 10–30 nm, respectively (Sadeghi and Gholamhoseinpoor 2015; Ulug et al. 2015; Geetha et al. 2014; Masurkar et al. 2011; Kumar et al. 2013).

Terpenoid and phenolic are the organic polymers which exhibit strong antioxidant activities and act as stabilizing agents for synthesizing nanomaterials (Shiv Shankar et al. 2003). Phenolate ions were capable of transferring electrons to the metal ions during nanoparticle formation (Lukman et al. 2010). For example, eugenol is the leading phenolic of *clove* extract which plays the foremost role in bioreduction of $AgNO_3$and $HAuCl_4$to form nanoparticles (Singh et al. 2010). Flavonoids such as anthocyanin, flavonol, flavone, flavanone, isoflavonoid, chalcone, etc. are also composed of polyphenolic compounds that are capable of reducing the metal ions during the formation of nanoparticles. For example, *Ocimum basilicum* plant extract that contains flavonoids like luteolin, apigenin, etc. plays a significant role in the formation of silver nanoparticles by reducing silver ions (Ahmad et al. 2010). Amino acids are also found to be able to bind metal ions and reduce them into nanoparticles. It is observed that amino acids such as cysteine, arginine, methionine, lysine help to bind silver ions for the formation of silver nanoparticles (Gruen 1975).

Nanoparticle formation during bioreduction by plant extracts depends on various chemical and physical parameters such as pH, temperature, etc. (Iravani 2011). It is found that at low pH value, the rate of nucleation of metal ions becomes very low such that agglomeration occurs in metal nanoparticles, for which there may be a chance of formation of large-sized nanoparticles due to low pH value, while the higher pH value can help to the formation of small-sized nanoparticles (Thanh et al. 2014). The shape of synthesized nanoparticles also depends on pH. At higher pH, spherical and decahedral types of nanoparticles are generated. It is observed that the case of synthesizing the silver nanoparticles using the plant extract of *Curcuma longa*, alkaline pH associates with large numbers of functional groups in comparison with acidic pH, which facilitates binding huge numbers of silver ions to form greater numbers of silver nanoparticles having a very small diameter (Sathishkumar et al. 2010).

The preeminent activities of biomolecules that are present in plant extract are influenced by the pH of the surrounding medium which highly affects the reaction of biomolecules with the metal ions during nanoparticles production. As per the previous report on the formation of silver nanoparticles using *Medicago sativa* extract, at pH value 11, a high monodispersity is found during the reaction that causes the formation of nanoparticles with average size 11.5 nm, but at pH 2 no such reaction happens (Lukman et al. 2010).The temperature also affects the nucleation rate during nanoparticles synthesis process using plant extract. The sizes, shapes, and production rate of nanoparticles change with the variation of reaction temperature during synthesis (Sathishkumar et al. 2010). Incubation temperature also affects the metal ion reduction process and indicates a variation of colors due to surface plasmon resonance. For example, during synthesis, with a variation of temperature, gold nanoparticles show different color effects such as yellow-brown, purplish-pink, and pinkish-brown colors at the temperatures 60 °C, 80 °C, and 100 °C, respectively (Bankar et al. 2010). According to a previous report, it is necessary to maintain the temperature greater or equal to 30 °C to synthesize silver nanoparticles using *M. Sativa* plant seed extract (Lukman et al. 2010). The structural form of synthesized silver nanoparticles using *Cassia fistula* plant extract varies with temperature. At room temperature due to the accumulation of silver nanoparticles in a linear manner which causes recrystallization, as a result, formation of silver nanowires occurs. But, when calcined at 400 °C, then raising the temperature, the interaction between biomolecules and surfaces of silver nanoparticles also changes, which hampers the coalescence of nanoparticles; as a result, some irregular nanorods and spherical silver nanoparticles are perceived (Lin et al. 2010). Nucleation depends on reaction temperature such that at high temperature, there action rate becomes high and maximum gold ions are used to form nuclei which prevent the secondary reduction process on the surfaces of that nuclei. As a result, spherical nanoparticles are made. Generally, the secondary nucleation occurs at low temperature (Das et al. 2011). A schematic representation of green synthesis mechanism of nanoparticles using leaf extract of the plant is shown in Fig. 16.6.

16.5 Application of Green Synthesized Nanoparticles in Wastewater Treatment

Green nanotechnology has diverse applications in different fields (Fig. 16.7). However, green synthesized metal or metal oxide nanoparticles have been extensively used in the treatment of different type of wastewaters (Table 16.1). In a study, silver nanoparticles have been synthesized using *P. thonningii* leaf extract and used for the effective removal of heavy metals (magnesium, copper, lead, iron) from simulated wastewater (Shittu and Ihebunna 2017). Green synthesis of silver nanocomposite using leaf extract of *Ocimum tenuiflorum* (Black Tulsi) has been reported by Banerjee et al. (2014) for the treatment of textile dye. Treated wastewater can be efficiently reused for industrial and domestic purposes.

Ehrampoush et al. (2015) reported the efficient cadmium removal from wastewater by green synthesized iron oxide nanoparticles using peel extract of tangerine. In this study, peel extract of tangerine was used as a stabilizing agent, which attributed to the inexpensive and eco-friendly synthesis of iron oxide nanoparticles as it also plays an important role to control the size and morphology of nanoparticles

Fig. 16.6 Workflow for green synthesis of nanoparticles using plant leaf extract (Source: authors own)

during synthesis. Ehrampoush et al. (2015) showed that the concentration of peel extract of tangerine impact the size of nanoparticles. The synthesized iron oxide nanoparticles were applied as a good adsorbent for the removal of cadmium from wastewater. The average size of synthesized iron oxide nanoparticles observed decreased when the concentration of peel extract increases. In the study of Ehrampoush et al. (2015), a maximum of 90% removal of cadmium ions has been reported (Ehrampoush et al. 2015).

In the study of Rosales et al. (2017), green synthesis of zerovalent iron nanoparticles using two different extracts, green tea (*Camellia sinensis*) and rooibos (*Aspalathus linearis*), has been performed. In this study, it was found that the reactivity of synthesized nanoparticles of zerovalent iron is higher when using rooibos (*Aspalathus linearis*) extract, although the antioxidant content was highest in green tea (*Camellia sinensis*). Further, green synthesized zerovalent iron nanoparticles were applied in the degradation of textile dye and demonstrated a better performance in the treatment of the wastewater.

In the study of Goutam et al. (2018), anatase phase of green titanium dioxide (TiO_2) NPs using *Jatropha curcas* L. leaf extract was successfully synthesized to evaluate their performance for the photocatalytic treatment of TWW after the secondary treatment process. The synthesized anatase phase of the spherical TiO_2

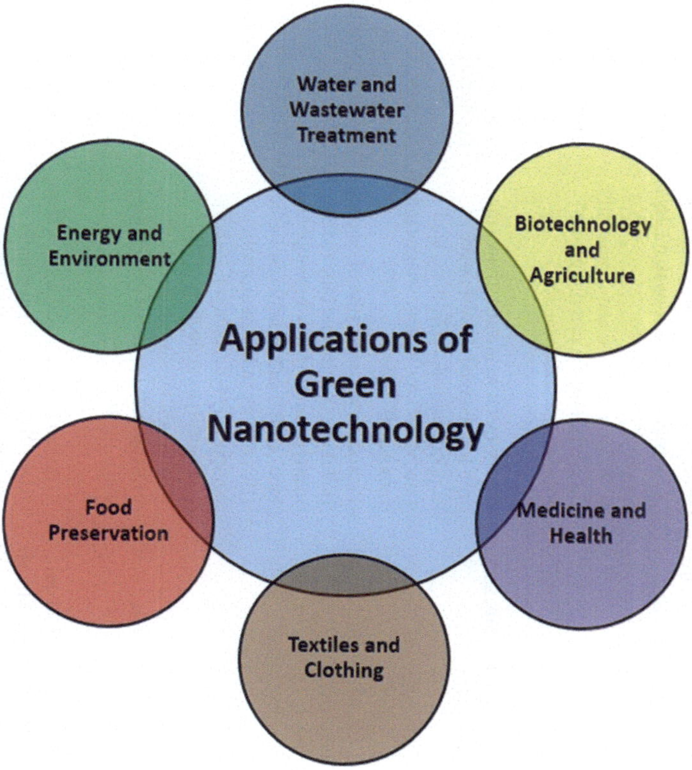

Fig. 16.7 Applications of green nanotechnology

NPs unfolds the presence of phytochemicals in leaf extract, which might involve in the capping/stabilization of NPs. Moreover, the green synthesized TiO_2 NPs were applied for the first time to testify its potential for the simultaneous removal of chemical oxygen demand (COD) and chromium (Cr) from secondary treated TWW. In this study, treatment of tannery wastewater using green synthesized TiO_2 NPs in a self-designed and fabricated parabolic trough reactor (PTR) resulted in the 82.26% removal of COD and 76.48% removal of Cr from TWW (Goutam et al. 2018).

Chaudhary et al. (2017) reported a rapid method of green chemistry approach for the synthesis of gold nanoparticles (AuNPs) using *Lagerstroemia speciosa* leaf extract. The green approach for the synthesis of AuNPs showed strong photocatalytic activity in the reduction of dyes, viz., methylene blue, methyl orange, bromophenol blue, and bromocresol green, and 4-nitrophenol under visible light in the presence of $NaBH_4$ (Choudhary et al. 2017). This study proposed the nontoxic and cost-effective approach for the synthesis of AuNPs using plant extract.

Table 16.1 Application of green synthesized nanoparticles in dye removal and water and wastewater treatment

Sr. no.	Nanoparticles	Biological materials	Biological agents	Characteristics	Application in dye removal and water and wastewater treatment	References
1.	TiO_2 NPs	Plant leaf extract	*Jatropha curcas* L.	Size, 10–20 nm; shape, spherical; phase, anatase	Photocatalytic treatment of tannery wastewater (COD removal, 82.26%; Cr removal, 76.48%)	Goutam et al. (2018)
	Fe NPs	Plant leaf extracts	Green tea (*Camellia sinensis*) and pomegranate (*Punica granatum*)	–	95% color removal and almost 80% dissolved organic carbon removal from textile wastewater	Ozkan et al. (2017)
2.	Ag NPs	Plant leaf extract	*Piliostigma thonningii*	Size, 50–114 nm; shape, spherical	Heavy metal removal activity (max. iron ion removal, 96.9%; max. copper removal, 89%; max. lead removal, 97.89%; max. magnesium removal, 93.6%)	Shittu and Ihebunna (2017)
3.	Au NPs	Plant leaf extract	*Lagerstroemia speciosa*	Size, 41–91 nm; shape, hexagonal	Photocatalytic reduction of organic pollutants (methylene blue (MB), methyl orange (MO), bromophenol blue (BPB), and bromocresol green (BCG) dyes and nitro aromatic compound (4-nitrophenol (4-NP); photocatalytic reduction of dyes with a reduction efficiency of ≥90%	Choudhary et al. (2017)
4.	Ag NPs	Plant leaf extract	*Ficus benjamina*	Size, 60–105 nm	Maximum Cd removal from aqueous solution, 85%	Khairia M. Al-Qahtani (2017)
5.	Fe NPs	Plant leaf extracts	*Azadirachta indica* (AI), *Magnolia champaca* (MC), *Mangifera indica* (MI), and *Murraya koenigii* (MK)	Size (AI, 96–110 nm); (MC, 99–129 nm); (MI and MK, 100–150 nm); shape, spherical	Treatment of domestic wastewater; AI-Fe NPs showed maximum removal efficiency (phosphate removal, 98.1%; ammonium nitrogen removal, 84.3%; and COD removal, 82.4%)	Devatha et al. (2016)

6.	Silver nanocomposite hydrogel (SNC)	Plant leaf extract	*Mukia maderaspatana*	Size, 11–20 nm	Removal of dye (methylene blue)	Devi et al. (2016)
7.	Fe_3O_4	Seaweeds (algae)	*Padina pavonica (PP)* and *Sargassum acinarium (SA)*	Size, PP (10–19.5 nm) and SA (21.6–27.4 nm); shape, spherical	Bioremoval of Pb using PP, 91%; Pb removal using SA, 78%	El-Kassas et al. (2016)
8.	Iron NPs	Tea extract	*Tieguanyin*	Size, 6.58 ± 0.76 nm; shape, spherical	Degradation of dye (bromothymol blue removal, more than 90%)	Xin et al. (2016)
9.	Ag NPs	Plant leaf extract	*Zanthoxylum armatum*	Size, 15–50 nm; shape, spherical	Degradation of dyes (degradation rate constant value of safranin O, 1.02×10^{-3} min^{-1}; degradation rate constant value of methyl red, 1.03×10^{-3} min^{-1}; degradation rate constant value of methyl orange, 1.86×10^{-3} min^{-1}; 10. Degradation rate constant value of methylene blue, 1.44×10^{-3} min^{-1}	Kumari and Singh (2016)
10.	ZnO, CuO, Co_3O_4, NiO, and Cr_2O_3	–	Sunlight irradiation	Size, ZnO, <35 nm; CuO, 7–50 nm; Co_3O_4, 45–90 nm; NiO, 2–25 nm; Cr_2O_3, ~ 17 nm; shape, ZnO (nanotubes); CuO (nanorods); Co_3O_4 (triangles and hexagons); NiO (needle-shaped); and Cr_2O_3 (nanobeads)	Treatment of simulated water containing hazardous dyes: removal dyes mixture (alizarin red S (ARS) + methylene blue (MB) (removal using Cr_2O_3, 88.24%; removal using ZnO, 87.96%; removal using CuO, 86.86%; removal using NiO, 85.89%; removal using Co_3O_4, 80.35%)	Shanker et al. (2016)

(continued)

Table 16.1 (continued)

Sr. no.	Nanoparticles	Biological materials	Biological agents	Characteristics	Application in dye removal and water and wastewater treatment	References
11.	Fe NPs	Plant extracts and juices	Extracts of *Camellia sinensis* (green tea, GT), *Syzygium aromaticum* (clove, CL), *Mentha spicata* (spearmint, SM), *Punica granatum* juice (pomegranate, PG), and red wine (RW)	Size, 50–60 nm; shape, spherical	Reduction of hexavalent Cr	Mystrioti et al. (2016)
12.	FeO NPs	Plant leaf extract	*Amaranthus spinosus*	Maximum particle size, 91 nm; shape, spherical	Decolorization of dyes (methyl orange, 75%, and methylene blue, 69%)	Muthukumar and Manickam (2015)
13.	Iron oxide NPs	Peel extract	*Tangerine*	Size, ~ 50 nm; shape, spherical	Treatment of contaminated solution (Cd removal, 90%)	Ehrampoush et al. (2015)
14.	ZnO NPs	Lemon juice	Lemon fruits	Size, ~ 21.5 nm; shape, spherical	Photocatalytic degradation of dyes (methyl orange, methyl red, and methylene blue)	Davar et al. (2015)
15.	Potassium Zinc Hexacyanoferrate nanocubes	Natural surfactant	*Sapindus mukorossi*	Size, 33–192 nm; shape, cubic	Photocatalytic degradation of organic dyes (malachite green (MG), 94.15%, and eriochrome black T (EBT), 76.13%)	Jassal et al. (2015)
16.	Cu NPs	Peel extract	*Citrus grandis*	Size, 22–27 nm; shape, spherical	Degradation of methyl red, 96%	Sinha and Ahmaruzzaman (2015)
17.	Ag NPs	Petal extract	*Rosa* 'Andeli'	Size, 4–29 nm; shape, spherical	Degradation of commercial dye Putnam sky blue 39 with an efficiency of 95%	Suarez-Cerda et al. (2015)

18.	Fe NPs	Plant leaf extracts	*Eucalyptus* sp.	Size, 20–80 nm; shape, spheroidal	Treatment of eutrophic wastewater (N removal, 71.7%; P removal, 30.4%; and COD removal 84.5%)	Wang et al. (2014)
19.	Reduced graphene oxide (RGO)/Fe_3O_4 nanocomposites	Plant leaf extract	*Solanum trilobatum*	Size, 18 nm; shape, spherical	Degradation of methylene blue, 95.9%	Vinothkannan et al. (2015)
20.	Ag NP soil nanocomposite	Plant leaf extract	*Ocimum tenuiflorum*	Size, 20–40 nm; average size, 32.58 nm	Treatment of textile dye (turquoise blue dye removal, 96.8%)	Banerjee et al. (2014)
21.	ZnO NPs	Plant leaf extract	*Plectranthus amboinicus*	Average size, 88 nm; shape, rod shape	Degradation of methyl red (MR)	Fu and Fu (2015)
22.	α-Fe_2O_3 NPs	Plant leaf extract	*Curcuma* and *Tea*	Crystallite size, using curcuma (4 nm); using tea (5 nm); shape, spherical	Degradation of methylene orange	Alagiri and Hamid (2014)
23.	Fe NPs	Tea extract	Green tea (GT), oolong tea (OT), and black tea (BT)	Size, GT-Fe NPs (20–40 nm); shape, irregular spherical	Treatment of wastewaters by removal of monochlorobenzene (MCB) (MCB removal using GT, 69%; MCB removal using OT, 53%; and MCB removal using BT, 39%)	Kuang et al. (2013)
24.	Au and Ag NPs	Plant stem extract	*Breynia rhamnoides*	Au, ~25 nm; Ag, ~64 nm; shape, spherical	Catalytic conversion of 4-nitrophenol (4-NP) to 4-aminophenol (4-AP)	Gangula et al. (2011)
25.	Nano zerovalent iron (nZVI)	Tea extract	*Camellia sinensis*	Size, 5–15 nm; shape, spherical	Degradation of bromothymol blue	Hoag et al. (2009)

Devatha et al. (2016) reported the green synthesis of iron nanoparticles using plants. The motive of this study was to prepare leaf extracts, precursor, and synthesis of iron nanoparticles and to assess its efficacy in the treatment of domestic wastewater. The synthesis of iron nanoparticles was performed using leaf extracts of *Mangifera indica*, *Murraya koenigii*, *Azadirachta indica*, and *Magnolia champaca* to check its potential for the treatment of domestic wastewater. A significant removal in total phosphates, ammonia nitrogen, and chemical oxygen demand from domestic wastewater was achieved upon treatment with the synthesized nanoparticles. Among the different plant-mediated synthesized iron nanoparticles, *Azadirachta indica* showed the satisfactory results in the removal of 98.08% of phosphate, 84.32% of ammonia nitrogen, and 82.35% of chemical oxygen demand (Devatha et al. 2016).

Al-Qahtani (2017) reported the synthesis of zerovalent silver nanoparticles by using an environmentally eco-friendly method without using hazardous compounds with the help of *Ficus benjamina* leaf extract. The synthesized silver nanoparticles were satisfactorily capable to remove the Cd^{2+} from contaminated solution, and various parameters like adsorbent dose, heavy metal concentration, pH, agitation speed, and contact time were studied (Al-Qahtani 2017).

Davar et al. (2015) synthesized ZnO NPs as a photocatalyst using lemon juice and zinc acetate as precursors and also investigated the effect of sucrose addition on the initial precursors. In this study, it has been showed that the photocatalytic activity of synthesized material was tested for the degradation of methyl orange, methyl red, and methylene blue solutions and resulted in the photocatalytic degradation of organic dyes. Moreover, synthesized ZnO NPs were used in decolorization processes and the treatment of textile dyes (Davar et al. 2015).

According to the study of El-Kassas et al. (2016), Fe_3O_4-NPs were successfully synthesized using two seaweeds *Padina pavonica* L. Thivy and *Sargassum acinarium* L. Setchell 1933 water extracts. Further, the algal extract was used as a reducing agent for $FeCl_3$ causing the photosynthesis of Fe_3O4-NPs. The biosynthesized Fe_3O_4-NPs were entrapped in calcium alginates beads and used in Pb adsorption experiments. Green synthesized Fe_3O_4-NPs alginate beads via *P. pavonica* L. Thivy had a high capacity for 91% removal of Pb while that of *S. acinarium* L. Setchell 1933 had 78% removal capacity after 75 min (El-Kassas Hala et al. 2016).

Mystrioti et al. (2016) successfully synthesized nano-iron suspensions for application in Cr(VI) reduction. In this study, five plant extracts and juices, i.e., extracts of *Camellia sinensis* (green tea, GT), *Syzygium aromaticum* (clove, CL), *Mentha spicata* (spearmint, SM), *Punica granatum* juice (pomegranate, PG), and red wine (RW), have been taken to study their effectiveness in the synthesis of nano-iron suspensions. In the extract of plant and juice, polyphenols contained in extracts act as reducing agents for iron ions in aqueous solutions and stabilize the nanoparticles produced from additional agglomeration and oxidation. In this study, the concentration of non-iron in suspensions was estimated and observed. The maximum concentration of nano-iron in suspensions was estimated using RW and PG at a mixing ratio of iron solution to extract equal to 2 was obtained

22 mM and lower concentrations, were obtained using GT and CL extracts up to 18 mM. Thus, PG juice and RW showed the effective nature for the synthesis of nano-iron. PG juice and RW with GT extracts effectively reduce the Cr(VI) with removal capacity as high as 500 mg Cr(VI) per g of iron in nanoparticles (Mystrioti et al. 2016).

Wang et al. (2014) reported the first time synthesis of spheroidal iron nanoparticles through a one-step room-temperature biosynthetic route using eucalyptus leaf (EL) extracts. In this study with the help of X-ray diffraction (XRD) and Fourier transform infrared (FTIR) spectrometer, it has been showed that some polyphenols are bound to the surfaces of EL-Fe NPs as a capping/stabilizing agent. Reactivity of EL-Fe NPs was evaluated for the treatment of swine wastewater, and results indicated that 71.7% of total N, 30.4% of total P, and 84.5% of COD were removed, respectively (Wang et al. 2014).

Kuang et al. (2013) reported the successful synthesis of iron nanoparticles (Fe NPs) using tea extracts as a catalyst for the Fenton-like oxidation of monochlorobenzene (MCB), where 69%, 53%, and 39% of MCB were, respectively, degraded by Fe NPs synthesized using green tea extracts along with tea extracts and black tea extracts. Further, it has been observed that Fe NPs synthesized using green tea extracts (GT-Fe NPs) demonstrated the best degradation since green tea contains a high concentration of caffeine/polyphenols used as both reducing and capping agents in the synthesis of Fe NPs. In this study, results showed that the oxidation of MCB and the removal of chemical oxygen demand (COD) using GT-Fe NPs were 81% and 31%, respectively, at optimal conditions, where dosages were 0.6 g/L GT-Fe NPs, 0.045 mol/L H_2O_2, and preliminary pH of 3.0. Matched to homogeneous Fenton oxidation of MCB, GT-Fe NPs as a heterogeneous catalyst indicated that Fe^{2+} and Fe^{3+} filtered from GT-Fe NPs nanoparticles and consequently reduced the formation of iron sludge. Conclusively, GT-Fe NPs have shown satisfactory removal of MCB and possible Fenton-like oxidative mechanism of MCB from wastewaters (Kuang et al. 2013).

16.6 Stability and Reusability of Green Synthesized Nanoparticles

From the literature survey, it has been clearly observed that the green synthesized NPs have greater potential in terms of efficient, safe, nontoxic, clean, and environment-friendly synthesis for higher pollutant removal performance. The application perspective of nanoparticles in pollutant removal/water and wastewater treatment cannot be ignored in terms of regeneration and reusability as these factors are important in the cost-benefit analysis of the nano-based water and wastewater treatment technology and its sustainability (Venkateswarlu et al. 2016). It has been reported in many studies that nanoparticles are applied as adsorbent and produce zero effluent (Wang et al. 2013; Dinesh and Pittman Jr 2006; Srivastava et al. 2017).

Moreover, it is worth to consider that NPs have the ability to produce zero effluent owing to the presence of organic functional groups on the surface of adsorbent which could eventually degrade after a certain period of time and same characteristic causes to minimize the reusability of NPs (Lunge et al. 2014).

The reusability of NPs, repeated recycling, and adsorbent stability are essential key factors to economic concerns and commercial success. Srivastava et al. (2017) reported that the spherical magnetic NPs with an average diameter of 8.76 nm synthesized using *Lagerstroemia speciosa* bark (LB) extract by co-precipitation method showed significant Cr(VI) removal from aqueous solution. In this study, adsorption kinetics and equilibrium have been well described by Langmuir isotherm and pseudo-second-order model separately. After the maximum adsorption of Cr(VI), magnetic NPs were collected effortlessly from the aqueous solution by a magnet, and the adsorbed Cr(VI) by synthesized magnetic NPs was found to be more than 93.72% after 11 successive adsorption-desorption cycles (Srivastava et al. 2017). Martinez et al. (2016) synthesized the iron oxide NPs using *Eucalyptus globulus* plant extract at different ratios of iron and plant extract. In this study, synthesized iron oxide NPs were found effective in arsenic adsorption. However, the desorption studies of As(V) clearly showed that synthesized iron oxide NPs can be easily regenerated using moderate concentration basic solutions (Martínez-Cabanas et al. 2016). Further, Lingamdinne et al. (2017) synthesized magnetic inverse spinel iron oxide nanoparticles (MISFNPs) using seed extract of *Cnidium monnieri* L. Cuss (CLC). The synthesized MISFNPs was further used for the removal of Pb(II) and Cr(III) from aqueous solutions through batch studies. From this study, it has been concluded that green synthesized NPs synthesized by a green route are capable of recycling and removal of heavy metals without loss of its stability, and the synthesized MISFNPs were easily recoverable and reusable for the removal of heavy metals up to at least five times without a loss of stability (Lingamdinne et al. 2017). Further research is required in this direction to use green synthesized nanoparticles effectively in water and wastewater treatment.

16.7 Challenges and Future Prospects

Green synthesized NPs are nontoxic in nature and eco-friendly and can be synthesized by simple green chemistry approach using plant extract that contains essential metabolites, which act as reducing/capping agents during NP synthesis (Ali et al. 2017). Green synthesized NPs have been successfully applied in the pollutant removal from water and wastewater during treatment and, thus, can produce zerovalent effluent discharge. However, some challenges and future prospects should be considered (Ali et al. 2017):

(a) The toxicity of long-term exposure of the green synthesized NPs to human beings should be assessed before application from desorbed pollutants.

(b) Despite the simple synthesis method, the use of solution extract volume, temperature, solvent type, pH, strength of precursor, and functional groups from plant metabolites should be optimized to avoid any change in the magnetic behavior and saturation magnetization value of the green synthesized NPs.
(c) Efforts are required to devise the NP morphology and saturation magnetization value by optimizing the synthesis method to maintain the stability of NPs for efficient pollutant removal from water and wastewaters and NP magnetic separability.
(d) There is a need to synthesize new NPs with a wide range of organic functional groups by manipulating the plant metabolites and synthesis method for the selective as well as multi-pollutant removal from water and wastewaters.
(e) Cost-benefit analysis should be performed for commercial purposes as there is no cost data available to date.
(f) Further, biocompatibility test should also be performed for biomedical applications.

16.8 Conclusions

Green nanotechnology is rapidly contributing to the removal of emerging pollutants, organic/inorganic contaminants such as dyes and heavy metal ions from water and wastewaters. Nanoparticles can greatly influence the domain of wastewater treatment in the coming future; however, focusing on the improvement in the existing treatment methods by increasing the efficiency of the processes and enhancing the reusability of nanomaterials can save the cost of operation of the plant or processes. Current wastewater treatment technologies can remove organic and inorganic pollutants from water and wastewater, but these methods are energy intensive and uneconomical because of their inability to completely purify water/wastewater as well as to reuse the retentates. However, the green synthesis of NPs is a very promising approach of green nanotechnology, and green synthesized NPs can easily be separated and reused because of their specific nature and high stability. Moreover, the use of green synthesized NPs in water and wastewater treatment is not only an environment-friendly option but also a promising technology for low-economy countries that can also fulfill the concept of zero effluent discharge after wastewater treatment using low cost and energy. Thus, green synthesized NPs could be an essential component of water and wastewater treatment systems in the future. Further, future research is warranted in this direction to commercialize the use of green synthesized NPs in pollutant removal from wastewater/wastewater treatment.

Acknowledgment University Grant Commission (UGC) Fellowship from UGC, Government of India (GOI), New Delhi, India, awarded to Mr. Surya Pratap Goutam and Mr. Diptarka Roy for doctoral research work is duly acknowledged.

References

Ahmad N, Sharma S, Alam MK, Singh VN, Shamsi SF, Mehta BR, Fatma A (2010) Rapid synthesis of silver nanoparticles using dried medicinal plant of basil. Colloids Surf B: Biointerfaces 81:81–86

Ali I, Peng C, Naz I, Khan ZM, Sultan M, Islam T, Abbasi IA (2017) Phytogenic magnetic nanoparticles for waste water treatment: a review. RSC Adv 7:40158–40178

Alagiri M, Hamid SBA (2014) Green synthesis of a-Fe2O3 nanoparticles for photocatalytic application. J Mater Sci Mater Electron 25(8):3572–3577

Al-Qahtani KM (2017) Cadmium removal from aqueous solution by green synthesis zero valent silver nanoparticles with *Benjamina* leaves extract. Egypt J Aquat Res 43:269–274

Amin MT, Alazba AA, Manzoor U (2014) A review of removal of pollutants from water/wastewater using different types of nanomaterials. Adv Mater Sci Eng 2014:1–24

Aromal SA, Philip D (2012) Green synthesis of gold nanoparticles using Trigonella foenum-graecum and its size dependent catalytic activity. Spectrochim Acta A Mol Biol Spectrosc 97:1–5

Banerjee P, Sau S, Das P, Mukhopadhyay A (2014) Green Synthesis of Silver - Nanocomposite for Treatment of Textile Dye. Nanosci Technol 1(2):1–6

Bankar A, Joshi B, Ravi Kumar A, Zinjarde S (2010) Banana peel extract mediated synthesis of gold nanoparticles. Colloids Surf B: Biointerfaces 80:45–50

Baruah S, Pal SK, Dutta J (2012) Nanostructured zinc oxide for water treatment. Nanosci Nanotechnol Asia 2:90

Ben Aim R, Semmens MJ (2001) Membrane bioreactors for wastewater treatment and reuse: a success story, Proceedings and key-note speech of the IWA-Asian Environmental Technology Conference. IWA, Singapore

Bharagava RN, Chowdhary P, Saxena G (2017a) Bioremediation: an ecosustainable green technology: its applications and limitations. In: Bharagava RN (ed) Environmental pollutants and their bioremediation approaches, 1st edn. CRC Press, Taylor & Francis Group, Boca Raton, pp 1–22. https://doi.org/10.1201/9781315173351-2

Bharagava RN, Saxena G, Chowdhary P (2017b) Constructed wetlands: an emerging phytotechnology for degradation and detoxification of industrial wastewaters. In: Bharagava RN (ed) Environmental pollutants and their bioremediation approaches, 1st edn. CRC Press, Taylor & Francis Group, Boca Raton, pp 397–426. https://doi.org/10.1201/9781315173351-15

Bharagava RN, Saxena G, Mulla SI, Patel DK (2017c) Characterization and identification of recalcitrant organic pollutants (ROPs) in tannery wastewater and its phytotoxicity evaluation for environmental safety. Arch Environ Contam Toxicol. https://doi.org/10.1007/s00244-017-0490-x

Bhattacharya S, Saha I, Mukhopadhyay A, Chattopadhyay D, Ghosh UC, Chatterjee D (2013) Role of nanotechnology in water treatment and purification: potential applications and implications. Int J Chem Sci Technol 3(3):59–64

Bolis V, Fubini B, Giamello E (1991) Effect of form on the surface chemistry of finely divided solids. Mater Chem Phys 29:153–164

Bora T, Dutta J (2014) Applications of nanotechnology in wastewater treatment-a review. J Nanosci Nanotechnol 14:613–626

Bordes MC, Vicent M, Moreno R, Garcia-Montano J, Serra A, Sanchez E (2015) Application of plasma-sprayed TiO2 coatings for industrial (tannery) wastewater treatment. Ceram Int 41:14468–14474

Bowen WR, J S, Welfoot JS (2002) Modelling the performance of membrane nanofiltration – critical assessment and model development. Chem Eng Sci 57:1121–1137

Chandrappa R, Das DB (2012) Solid waste management. Environ Sci Eng. https://doi.org/10.1007/978-3-642-28681-0_2

Choudhary BC, Paul D, Gupta T, Tetgure SR, Garole VJ, Borse AU, Garole DJ (2017) Photocatalytic reduction of organic pollutant under visible light by green route synthesized gold nanoparticles. J Environ Sci 55:236–246

Cui Y, Chang X, Zhai Y, Zhu X, Zheng H, Lian N (2006) ICP-AES determination of trace elements after preconcentrated with *p* dimethylaminobenzaldehyde-modified nanometer $SiO_2$from sample solution. Microchem J 83:35–41

Das AJ, Kumar R, Goutam SP (2016) Sunlight Irradiation Induced Synthesis of Silver Nanoparticles using Glycolipid Biosurfactant and Exploring the Antibacterial Activity. J Bioeng Biomed Sci 06(05)

Das RK, Gogoi N, Bora U (2011) Green synthesis of gold nanoparticles using Nyctanthes arbortristis flower extract. Bioprocess Biosyst Eng 34:615–619

Davar F, Majedi A, Mirzaei A (2015) Green synthesis of ZnO nanoparticles and its application in the degradation of some dyes. J Am Ceram Soc 98:1739–1746

Devatha CP, Thalla AK, Katte SY (2016) Green synthesis of iron nanoparticles using different leaf extracts for treatment of domestic waste water. J Clean Prod 139:1425–1435

Devi GK, Kumar PS, Kumar KS (2016) Green synthesis of novel silver nanocomposite hydrogel based on sodium alginate as an efficient biosorbent for the dye wastewater treatment: prediction of isotherm and kinetic parameters. Desalin Water Treat 2016:1–14

Dinesh M, Pittman CU Jr (2006) Activated carbons and low cost adsorbents for remediation of tri- and hexavalent chromium from water. J Hazard Mater B137:762–811

Drioli E, Ali A, Macedonio F (2015) Membrane distillation: recent developments and perspectives. Desalination 356:56–84

Ehrampoush MH, Miria M, Salmani MH, Mahvi AH (2015) Cadmium removal from aqueous solution by green synthesis iron oxide nanoparticles with tangerine peel extract. J Environ Health Sci Eng 13(84)

El Saliby IJ, Shon H K, Kandasamy J, Vigneswaran S (2008) Nanotechnology for wastewater treatment: in brief, water and wastewater treatment technologies, EOLSS. https://www.eolss.net/sample-chapters/C07/E6–144-23.pdf

El-Kassas Hala Y, Aly-Eldeen Mohamed A, Gharib Samiha M (2016) Green synthesis of iron oxide (Fe3O4) nanoparticles using two selected brown seaweeds: characterization and application for lead bioremediation. Acta Oceanol Sin 35:89–98

Fendler JH (2001) Colloid chemical approach to nanotechnology. Korean J Chem Eng 18(1):1–13

Filipponi L, Sutherland D (2010) Introduction to nanoscience and nanotechnologies. http://nanoyou.eu/attachments/188_Module-1-chapter-1.pdf

Fu L, Fu Z (2015) Plectranthusamboinicus leaf extract–assisted biosynthesis of ZnO nanoparticles and their photocatalytic activity. Ceram Int 41:2492–2496

Gangula A, Podila R, Ramakrishna M, Karanam L, Janardhana C, Rao AM (2011) Catalytic Reduction of 4-Nitrophenol using Biogenic Gold and Silver Nanoparticles Derived from Breynia rhamnoides. Langmuir 27(24):15268–15274

Gautam S, Kaithwas G, Bharagava RN, Saxena G (2017) Pollutants in tannery wastewater, pharmacological effects and bioremediation approaches for human health protection and environmental safety. In: Bharagava RN (ed) Environmental pollutants and their bioremediation approaches, 1st edn. CRC Press, Taylor & Francis Group, Boca Raton, pp 369–396. https://doi.org/10.1201/9781315173351-14

Goutam SP, Yadav AK, Das AJ (2017) Coriander Extract Mediated Green Synthesis of Zinc Oxide Nanoparticles and Their Structural, Optical and Antibacterial Properties. J Nanosci Technol 3:249–252

Gawande MB, Branco PS, Parghi K, Shrikhande JJ, Pandey RK, Ghumman CAA, Bundaleski N, Teodoro O, Jayaram RV (2011) Synthesis and characterization of versatile MgO-ZrO_2 mixed metal oxide nanoparticles and their applications. Cat Sci Technol 1:1653–1664

Geetha N, Geetha TS, Manonmani P, Thiyagarajan M (2014) Green synthesis of silver nanoparticles using Cymbopogan Citratus(Dc) Stapf. Extract and its antibacterial activity. Aus J Basic Appl Sci 8(3):324–331

Ghaly MY, Jamil TS, El-Seesy IE, Souaya ER, Nasr RA (2011) Treatment of highly polluted paper mill wastewater by solar photocatalytic oxidation with synthesized nano TiO2. Chem Eng J 168:446–454

Goutam SP, Saxena G, Singh V, Yadav AK, Bharagava RN, Thapa KB (2018) Green synthesis of TiO_2 nanoparticles using leaf extract of *Jatropha curcas* L. for photocatalytic degradation of tannery wastewater. Chem Eng J 336:386–396. https://doi.org/10.1016/j.cej.2017.12.029

Gruen LC (1975) Interaction of amino acids with silver (I) ions. Biochim Biophys Acta 386:270–274

Guo D, Xie G, Luo J (2014) Mechanical properties of nanoparticles: basics and applications. J Phys D Appl Phys 47:1–25

Hadjiivanov K, Klissurski K, Kantcheva M, Davydov A (1991) State and localization of cobalt, nickel and copper ions adsorbed on titania (anatase). J Chem Soc Faraday Trans 87:907–911

He Q, Chang X, Huang X, Hu Z (2008) Determination of trace elements in food samples by ICP-AES after preconcentration with *p*-toluenesulfonylamide immobilized on silica gel and nanometer SiO_2. Microchim Acta 160:147–152

Heath JR (1995) The chemistry of size and order on the nanometer scale. Science 270:1315–1316. https://doi.org/10.1126/science.270.5240.1315

Henze M, Comeau Y (2008) Wastewater characterization. In: Henze M, van Loosdrecht MCM, Ekama GA, Brdjanovic D (eds). ISBN: 9781843391883Biological wastewater treatment: principles modelling and design. IWA Publishing, London

Hoag GE, Collins JB, Holcomb JL, Hoag JR, Nadagoudab MN, Varma RS (2009) Degradation of bromothymol blue by 'greener' nano-scale zero-valent iron synthesized using tea polyphenols. J Mater Chem 19:8671–8677

Hu EL, Shaw DT (1999) Synthesis and assembly. Nanostructure Sci Tech, Chapter-2, pp 15–33

Ichinose N, Ozaki Y, Kashū S (1992) Superfine particle technology. Springer, New York ISBN:13:978-1-4471-1810-7. https://doi.org/10.1007/978-1-4471-1808-4

Iqbal P, Preece JA, Mendes PM (2012) Nanotechnology: the "top-down" and "bottom-up" approaches. (Book chapter). Wiley, UK. https://doi.org/10.1002/9780470661345.smc195

Iravani S (2011) Green synthesis of metal nanoparticles using plants. Green Chem 13:2638

Jassal V, Shanker U, Kaitha BS, Shankarb S (2015) Green synthesis of potassium zinc hexacyanoferrate nanocubes and their potential application in photocatalytic degradation of organic dyes. RSC Adv 5(33):26141–26149

Khajeh M, Laurent S, Dastafkan K (2013) Nanoadsorbents: classification, preparation, and applications (with emphasis on aqueous media). https://doi.org/10.1021/cr400086v

Khan I, Saeed K, Khan I (2017) Nanoparticles: properties, applications and toxicities. Arab J Chem. https://doi.org/10.1016/j.arabjc.2017.05.011

Kim HS, Katayama H, Takizawa S, Ohgaki S (2001) Removal of coliphage Qb and organic matter from synthetic secondary effluent by powdered activated carbon-microfiltration (PACMF) process. Proceedings of IWA Specialized Conference on Membrane Technology, Israel, 211–219

Kim CS, Okuyama K, Nakaso K, Shimada M (2004) Direct measurement of nucleation and growth modes in titania nanoparticles generation by CVD method. J Chem Eng Japan 37:1379–1389

Kjellén M, Pensulo C, Nordqvist P, Fogde M (2012) Global review of sanitation system trends and interactions with menstrual management practices report for the menstrual management and sanitation systems project. Stockholm Environment Institute Kräftriket 2B 106 91, Stockholm, Web: www.sei-international.org

Kulkarni SK (2015) Nanotechnology: principles and practices. https://doi.org/10.1007/978-3-319-09171-6__3

Kumar S, Daimary RM, Swargiary M, Brahma A, Kumar S, Singh M (2013) Biosynthesis of silver nanoparticles using Premna herbacea leaf extract and evaluation of its antimicrobial activity against bacteria causing dysentery. Int J Pharm Bio Sci 4(4):378–384

Kumar DA, Palanichamy V, Roopan SM (2014) Green synthesis of silver nanoparticles using Alternanthera dentata leaf extract at room temperature and their antimicrobial activity. Spectrochim Acta A Mol Biol Spectrosc 127:168–171

Kumari J, Singh A (2016) Green synthesis of nanostructured silver particles and their catalytic application in dye degradation. J Genetic Eng Biotechnol 14:311–317

Kung HH, Ko EI (1996) Preparation of oxide catalyst supports- a review of recent advances. Chem Eng J 64:203–214

Lian N, Chang X, Zheng H, Wang S, Cui Y, Zhai Y (2005) Application of Dithizone-modified TiO_2 nanoparticles in the preconcentration of trace chromium and lead from sample solution and determination by inductively coupled plasma atomic emission spectrometry. Microchim Acta 151:81–88

Lin L, Wang W, Huang J, Li Q, Sun D, Yang X, Wang H, He N, Wang Y (2010) Nature factory of silver nanowires: plant-mediated synthesis using broth of Cassia fistula leaf. Chem Eng J 162:852–858

Lingamdinne LP, Young Chang Y, Yang JK, Singh J, Ha Choi E, Shiratani M, Koduru JR, Attri P (2017) Biogenic reductive preparation of magnetic inverse spinel iron oxide nanoparticles for the adsorption removal of heavy metals. Chem Eng J 307:74–84

Lu H, Xue Z, Saikaly P, Nunes SP, Bluver TR, Liu WT (2015) Membrane biofouling in a wastewater nitrification reactor: microbial succession from autotrophic colonization to heterotrophic domination. Water Res. https://doi.org/10.1016/j.watres.2015.10.013

Lukman AI, Gong B, Marjo CE, Roessner U, Harris AT (2010) Facile synthesis, stabilization, and anti-bacterial performance of discrete Ag nanoparticles using Medicago sativa seed exudates. J Colloid Interface Sci 353:433–444

Lunge S, Singh S, Sinha A (2014) Magnetic iron oxide (Fe3O4) nanoparticles from tea waste for arsenic removal. J Magn Magn Mater 356:21–31

Mahlambi MM, Ngila CJ, Mamba BB (2015) Recent developments in environmental photocatalytic degradation of organic pollutants: the case of titanium dioxide nanoparticles-a review. J Nanomater 2015:1–29

Malakootian M, Mansuri F (2015) Hexavalent chromium removal by titanium dioxide photocatalytic reduction and the effect of phenol and humic acid on its removal efficiency. Int J Environ Health Eng 4:1–8

Mansoori GA, Soelaiman TAF (2005) Nanotechnology – an introduction for the standards community. J ASTM Int 2(6): Paper ID JAI13110

Martínez-Cabanas M, Lopez-García M, Barriada R, Herrero JL, Sastre de Vicente ME (2016) Green synthesis of iron oxide nanoparticles. Development of magnetic hybridmaterials for efficient As (V) removal. Chem Eng J 301:83–91

Masurkar SA, Chaudhari PR, Shidore VB, Kamble SP (2011) Rapid biosynthesis of silver nanoparticles using Cymbopogan Citratus (Lemongrass) and its antimicrobial activity. Nano-Micro 3(3):189–194

Matsui Y, Colas F, Yuasa A (2001a) Removal of a synthetic organic chemical by PAC-UF systems – II: model application. Water Res 35:464–470

Matsui Y, Yuasa A, Ariga K (2001b) Removal of a synthetic organic chemical by PAC-UF systems – I: theory and modeling. Water Res 35:455–463

McNaught AD, Wilkinson A (1997) IUPAC gold book. Blackwell Scientific Publications, Oxford

Mekonnen MM, Hoekstra AY (2016) Four billion people facing severe water scarcity. Sci Adv 2:e1500323–e1500323. https://doi.org/10.1126/sciadv.1500323

Mueller NC, Nowack B (2009) Nanotechnology developments for the environment sector. Report of the Observatory NANO

Mulder M (1997) Basic principle of membrane technology, 2nd edn. Kluwer academic, The Netherlands

Muthukumar H, Manickam M (2015) Amaranthus spinosus leaf extract mediated FeO nanoparticles:Physicochemical traits, photocatalytic and antioxidant activity. ACS Sustain Chem Eng 3(12):3149–3156

Mystrioti C, Xanthopoulou TD, Papassiopi N, Xenidis A (2016) Comparative evaluation of five plant extracts and juices for nanoiron synthesis and application for hexavalent chromium reduction. Sci Total Environ 539:105–113

Nassar NN (2012) Iron oxide nanoadsorbents for removal of various pollutants from wastewater: an overview. Appl Adsorbents Water Pollut Control 2012:81–118

Ozkan ZY, Cakirgoz M, Kaymak ES, Erdim E (2017) Rapid decolorization of textile wastewater by green synthesized iron nanoparticles. Water Sci Technol 77(2):511–517

Patil PS (1999) Versatility of chemical spray pyrolysis technique. Mater Chem Phys 59:185–198

Pirkanniemi K, Sillanpaa M (2002) Heterogeneous water phase catalysis as an environmental application: a review. Chemosphere 48:1047–1060
Qadir M, Sharma BR, Bruggeman A, Choukr-Allah R, Karajeh F (2007) Non-conventional water resources and opportunities for water augmentation to achieve food security in water scarce countries. Agric Water Manag 87(2–2):2
Qu X, Alvarez PJJ, Li Q (2013) Applications of nanotechnology in water and wastewater treatment. Water Res 47:3931–3946
Ray SA, Ray MK (2009) Bioremediation of heavy metal toxicity-with special reference to chromium. Al Ameen J Med Sci 2(2):57–63
Roco MC (2004) Nanoscale science and engineering: unifying and transforming tools. AICHE J 50:890–897
Rosales E, Meijide J, Pazos M, Sanromán MA (2017) Challenges and recent advances in biochar as low-cost biosorbent: From batch assays to continuous-flow systems. Bioresour Technol 246:176–192
Sadeghi B, Gholamhoseinpoor F (2015) A study on the stability and green synthesis of silver nanoparticles using Ziziphora tenuior (Zt) extract at room temperature. Spectrochim Acta A Mol Biomol Spectrosc 134:310–315
Saravanan P, Gopalan R, Chandrasekaran V (2008) Synthesis and characterisation of nanomaterials. Def Sci J 58:504–516
Sathishkumar M, Sneha K, Yun YS (2010) Immobilization of silver nanoparticles synthesized using Curcuma longa tuber powder and extract on cotton cloth for bactericidal activity. Bioresour Technol 101(20):7958–7965
Savage N, Diallo MS (2005) Nanomaterials and water purification: opportunities and challenges. J Nanopart Res 7:331–342
Saxena G, Bharagava RN (2015) Persistent organic pollutants and bacterial communities present during the treatment of tannery wastewater. In: Chandra R (ed) Environmental waste management, 1st edn. CRC Press, Taylor & Francis Group, Boca Raton, pp 217–247. https://doi.org/10.1201/b19243-10
Saxena G, Bharagava RN (2017) Organic and inorganic pollutants in industrial wastes, their ecotoxicological effects, health hazards and bioremediation approaches. In: Bharagava RN (ed) Environmental pollutants and their bioremediation approaches, 1st edn. CRC Press, Taylor & Francis Group, Boca Raton, pp 23–56. https://doi.org/10.1201/9781315173351-3
Saxena G, Chandra R, Bharagava RN (2016) Environmental pollution, toxicity profile and treatment approaches for tannery wastewater and its chemical pollutants. Rev Environ Contam Toxicol 240:31–69. https://doi.org/10.1007/398_2015_5009
Seo GT, Suzuki Y, Ohgaki S (1996) Biological powdered activated carbon (BPAC) microfiltration for wastewater reclamation and reuse. Desalination 106:39–45
Seo GT, Ohgaki S, Suzuki Y (1997) Sorption characteristics of biological powdered activated carbon in BPAC-MF (biological activated carbon-microfiltration) system for refractory organic removal. Water Sci Technol 35:163–170
Shanker U, Jassal V, Manviri R (2016) Catalytic removal of organic colorants from water using some transition 2 metal oxide nanoparticles synthesized under sunlight RSC Adv, vol 6, pp 94989–94999
Shannon A, Bohn PW, Elimelech M, Georgiadis JG, Marinas BJ, Mayes AM (2008) Science and technology for water purification in the coming decades Mark. Nature 452:301–310
Sharma D, Kanchi S, Bisetty K (2015) Biogenic synthesis of nanoparticles: a review. Arab J Chem. https://doi.org/10.1016/j.arabjc.2015.11.002
Shittu KO, Ihebunna O (2017) Purification of simulated waste water using green synthesized silver nanoparticles of Piliostigma thonningii aqueous leave extract. Adv Nat Sci Nanosci Nanotechnol 8:1–9
Shiv Shankar S, Ahmad A, Pasricha R, Sastry MJ (2003) Bioreduction of chloroaurate ions by geranium leaves and its endophytic fungus yields gold nanoparticles of different shapes. Mater Chem 13:1822–1846

Shiv Shankar S, Rai A, Ahmad A, Sastry M (2004) Rapid synthesis of Au, Ag, and bimetallic Au core–Ag shell nanoparticles using Neem (Azadirachta indica) leaf broth. J Colloid Interface Sci 275:496–502

Shon HK, Phuntsho S, Chaudhary DS, Vigneswaran S, Cho J (2013) Nanofiltration for water and wastewater treatment – a mini review. Drink Water Eng Sci 6:47–53

Sinha T, Ahmaruzzaman M (2015) Biogenic synthesis of Cu nanoparticles and its degradation behavior for methyl red. Mater Lett 159:168–171

Singh A, Talat M, Singh D, Srivastava ON (2010) Biosynthesis of gold and silver nanoparticles by natural precursor clove and their functionalization with amine group. J Nanopart Res 12:1667–1675

Snoeyink VL, Campos C, Marinas BJ (2000) Design and performance of powdered activated carbon/ultrafiltration systems. Water Sci Technol 42:1–10

Srivastava S, Agrawal SB, Mondal MK (2017) Synthesis, characterization and application of Lagerstroemia speciosa embedded magnetic nanoparticle for Cr(VI) adsorption from aqueous solution. J Environ Sci 55:283–293

Stietiya MH, Wang JJ (2014) Zinc and cadmium adsorption to aluminum oxide nanoparticles affected by naturally occurring ligands. J Environ Qual 43:498–506

Suarez-Cerda J, Alonso-Nunez G, Espinoza-Gomez H, Flores-Lopez LZ (2015) Synthesis, kinetics and photocatalytic study of "ultra-small" Ag-NPs obtained by a green chemistry method using an extract of Rosa 'Andeli' double delight petals. J Colloid Interface Sci 458:169–177

Thanh NTK, Maclean N, Mahiddine S (2014) Mechanisms of nucleation and growth of nanoparticles in solution. Chem Rev 114:7610–7630

The United Nations world water development report (2017) Wastewater: the untapped resource; 2017. http://unesdoc.unesco.org/images/0024/002471/247153e.pdf

Tiwari DK, Behari J, Sen P (2008) Application of nanoparticles in waste water treatment. World Appl Sci J 3(3):417–433

Ulug B, HalukTurkdemir M, Cicek A, Mete A (2015) Role of irradiation in the green synthesis of silver nanoparticles mediated by fig (Ficus carica) leaf extract. Spectrochim A: Mol Biomol Spectrosc 135:153–161

Venkateswarlu S, Lee D, Yoon M (2016) Bioinspired 2D-carbon flakes and Fe_3O_4 nanoparticles composite for arsenite removal. ACS Appl Mater Interfaces 8:23876–23885

Vinothkannan M, Karthikeyan C, Kumar GG, Kim AR, Yoo DJ (2015) One-pot green synthesis of reduced graphene oxide (RGO)/Fe3O4 nanocomposites and its catalytic activity toward methylene blue dye degradation. Spectrochim Acta A Mol Biomol Spectrosc 136:256–264

Visvanathan C, Ben Aim R, Parameshwaran K (2000) Membrane separation bioreactors for wastewater treatment. Critical reviews. Environ Sci Technol 30:1–48

Wang S, Sun H, Ang HM, Tade MO (2013) Adsorptive remediation of environmental pollutants using novel graphene-based nanomaterials. Chem Eng J 226:336–347

Wang T, Jin X, Chen Z, Megharaj M, Naidu R (2014) Green synthesis of Fe nanoparticles using eucalyptus leaf extracts for treatment of eutrophic wastewater. Sci Total Environ 466–467:210–213

World Health Organization (WHO) and United Nations Children's Fund (UNICEF) (2012) Progress on drinking water and sanitation: 2012 update. United Nations Children's Fund/World Health Organization, New York/Geneva http://whqlibdoc.who.int/publications/2012/9789280646320_eng_full_text.pdf

Xin H, Yang X, Liu X, Tang X, Weng L, Han Y (2016) Biosynthesis of Iron Nanoparticles Using Tie Guanyin Tea Extract for Degradation of Bromothymol Blue. J Nanotechnol 2016:1–8

Kuang Y, Wang Q, Chen Z, Megharaj M, Naidu R (2013) Heterogeneous Fenton-like oxidation of monochlorobenzene using green synthesis of iron nanoparticles. J Colloid Interface Sci 410:67–73

Zhang W, Fu J, Zhang G, Zhang X (2014) Enhanced arsenate removal by novel Fe–La composite (hydr) oxides synthesized via coprecipitation. Chem Eng J 251:69–79

Chapter 17
Environmental Hazards and Toxicity Profile of Organic and Inorganic Pollutants of Tannery Wastewater and Bioremediation Approaches

Gaurav Saxena, Diane Purchase, and Ram Naresh Bharagava

Contents

Abstract Undoubtedly, leather industries are the key contributors in the economy of many developing countries. However, unfortunately these are also one of the major polluters worldwide, generating large volumes of high-strength wastewater having high pH, dark brown color, unpleasant odor, biological oxygen demand, chemical oxygen demand, total dissolved solids, and a blend of highly toxic envi-

---

G. Saxena · R. N. Bharagava (✉)
Laboratory of Bioremediation and Metagenomics Research (LBMR), Department of Microbiology (DM), Babasaheb Bhimrao Ambedkar University (A Central University), Lucknow, Uttar Pradesh, India
e-mail: ramnaresh_dem@bbau.ac.in

D. Purchase
Department of Natural Sciences, Faculty of Science and Technology, Middlesex University, London, UK
e-mail: Purchase@mdx.ac.uk

G. Saxena, R. N. Bharagava (eds.), *Bioremediation of Industrial Waste for Environmental Safety*, https://doi.org/10.1007/978-981-13-1891-7_17

ronmental pollutants. It reduces sunlight penetration in aquatic resources which in turn decreases both photosynthetic activity and dissolved oxygen concentration affecting aquatic life; however, on land, it causes reduction in soil alkalinity and inhibition of seed germination. Moreover, it may also create serious toxicity in living beings upon exposure. Therefore, it becomes necessary to adequately treat/detoxify the tannery wastewater to protect the environment and living beings. Therefore, this chapter provides an overview on the environmental impacts, toxicity profile of tannery wastewater, and existing and emerging bioremediation strategies for environmental safety.

Keywords Tannery wastewater · Pollutants · Pollution · Toxicity · Bioremediation · Environmental safety

17.1 Introduction

Leather industries (LIs) are one of the major contributors in the economy of many developing nations due to ever increasing demand of leather and leather products. Around 2108.94 M mt^2 of leather is annually produced worldwide, and the world trade for the leather sector is estimated as US$100 billion per year (FAO 2008; UNIDO 2010). The major countries like Germany, the United States, and other European countries are the major importers of leather or leather products, whereas the countries like India, China, Pakistan, Egypt, Brazil, Thailand, and Indonesia are the major exporters of leather and leather products (Saxena et al. 2016).

Regrettably, LIs are also the major pollution causing industries in the world. LIs are mainly concerned with the production finished leather or leather products. During leather production process, a variety of highly toxic chemicals are beings used in tanning process and often discharged along with wastewater into the environment where they create serious ecotoxicological effects in the environment and the toxicity in living beings upon exposure (Gautam et al. 2017; 2018; Bharagava et al. 2018a; Saxena et al. 2016; Saxena and Bharagava 2015; UNIDO 2000). Tannery wastewater (TWW) is mainly characterized by high biochemical oxygen demand (COD), total dissolved solids (TDS), chromium (Cr), phenolics, and a variety of toxic chemicals (Bharagava et al. 2018a; Saxena et al. 2016), and therefore, its adequate treatment is necessary to protect the environment and public health.

Physicochemical treatment approaches are currently being applied to treat the TWW, but these are environmentally destructive and costly and may cause secondary pollution and, hence, are not economically feasible. However, biological approaches such as bioremediation can be a suitable approach to effectively treat and detoxify the highly toxic and hazardous TWW for environmental protection and sustainable development of the society. Therefore, this chapter provides an overview of the environmental impacts, toxicity of organic and inorganic pollutants of TWW, and its bioremediation approaches with their merits, demerits, and challenges.

17.2 Overview of Tannery Industry

Leather industry is a wide and common industry in the world and is a major key player in the world leather trade (Thanikaivelan et al. 2005; Durai and Rajasimman 2011; Lofrano et al. 2013). In India, the LIs play an important role in the national economy as they significantly earn the foreign exchange through leather export, generate employment opportunity for economically weaker sections with 30% of women predominance, and contribute 15% of the total production capacity of the world (Alam et al. 2009). It is the third largest sector known for its production capacity and fifth largest exporter of leather goods and accessories in the world (ILTIP 2010). There are about 3000 listed LIs, which are located in the following states of India: Tamil Nadu, West Bengal, Uttar Pradesh, Maharashtra, Punjab, Karnataka, Andhra Pradesh, Haryana, and Bihar (CPCB 2009). About 75% of LIs are in the cottage and small-scale sector whereas about 20% in medium and only about 5% in the large sector (TGMT 2010).

LIs are specialized in the processing of hide (skins of large animals such as cows, buffaloes, and horses) and skins (skins of small animals such as sheep, goats, and calves) for leather production. The hide/skins are available from animals that died naturally or are the by-products of meat and meat products industry. The prime objective of tanning process is to convert the hide/skins (a highly putrescible material) into stable and imputrescible products termed as leather that is used for various purposes. Tanning processes are classified into vegetable and chrome-tanning process depending on the type of tanning reagent (chromium or tannins) used (Aravindhan et al. 2004; ILTIP 2010). The production of leather from raw hide/skins is accomplished into four major groups of sub-processes: beamhouse operation, tanyard processes, retanning, and finishing (Thanikaivelan et al. 2005; ILTIP 2010; Durai and Rajasimman 2011). However, the tanning process involves different steps and chemicals for different end products, and the kind and amount of waste generated may vary in a wide range of quantity and nature (Saxena et al. 2016; Dixit et al. 2015; Lofrano et al. 2013). Further, the detail information on tanning process can be found in the review article entitled "Recent Trends in Leather Making: Processes, Problems, and Pathways" (Thanikaivelan et al. 2005).

17.3 Nature and Characteristics of Tannery Wastewater

A large amount of water is used in the leather industry as it acts as a transport medium for the chemicals used in the processing of raw leather. An average of 30–35 m^3 of wastewater is generated per ton of raw hide/skins processed (Lofrano et al. 2013). Characteristically, tannery wastewater (TWW) is a basic, dark brown-colored liquid having a high content of organic and inorganic pollutants, which may vary according to the type of chemicals and their amount used (Kongjao et al. 2008).

However, its physicochemical characteristics may vary according to, from industry to industry depending upon the size of leather industry, raw materials used, chemicals used for a specific process, amount of water consumed, type of final product, and the production processes applied in the LIs (Rameshraja and Suresh 2011; Lofrano et al. 2013).

TWW is mainly characterized by high chemical oxygen demand (COD), biochemical oxygen demand (BOD), total dissolved solids (TDS), phenolics, and chromium (III) with high pH, strong odor, and dark brown color (Aravindhan et al. 2004; Rameshraja and Suresh 2011; Durai and Rajasimman 2011; Lofrano et al. 2013; Tripathi et al. 2011). The physicochemical properties of TWW are presented in Table 17.1. TWW, if discharged into the receiving environment without adequate treatment, may cause serious soil and water pollution, and hence, its proper treatment is required for environmental sustainability.

17.4 Environmental Hazards and Toxicity Profile of Organic and Inorganic Pollutants of Tannery Wastewater

TWW is of serious environmental concern to pollution control authorities worldwide due to its highly toxic and hazardous nature. The highly toxic nature of TWW is mainly due to the presence of highly toxic organic and inorganic pollutants, which causes serious environmental pollution and toxicity in living beings. Table 17.2 represents the toxicity profile of various organic and inorganic pollutants used in leather processing.

TWW causes serious soil and water pollution. It (Saxena et al. 2016; Bhattacharya et al. 2016) (a) causes a huge foaming problem on surface waters; (b) inhibits the nitrification process; (c) blocks sunlight penetration due to its dark brown color and, thus, reduces the photosynthetic activity and oxygenation of receiving water bodies and, hence, becomes detrimental to aquatic life; (d) causes depletion in dissolved oxygen that encourages the anaerobic condition, which leads to the putrefying odor of receiving water bodies; (e) causes eutrophication of water bodies, thus adversely affecting the ecological functioning of aquatic resources; (f) causes salinization of soil (acidification, reduce soil fertility) and water due to its highly saline nature (salt concentration); (g) causes groundwater pollution due to leaching of highly toxic chromium (Cr) to the deeper layer of soil; (h) causes the deficiency of some micronutrients in soil such as zinc (Zn), copper (Cu), iron (Fe), etc.; (i) alters the structure of soil microbial communities and reduces their growth and finally retards the bioremediation process due to its high chromium content; and (j) support the growth of pathogenic microbes as it is highly rich in organic susbstances that may act as nutrient source for them (Saxena et al. 2015, 2016).

TWW also causes serious toxicity in living beings. It has been reported to:

(a) Cause genotoxicity and mutagenicity in fish, *Oreochromis niloticus* (Matsumoto et al. 2006)
(b) Disturb protein metabolism in fresh water teleost, *Cirrhinus mrigala* (Ham.) (Afaq and Rana 2009)

Table 17.1 Physicochemical characteristics of tannery wastewater

References	pH	Cond.	COD	BOD_5	TDS	TSS	TS	SS	VSS	Alk.	Chloride	Sulfate	P	Fe	Mg	TKN	NH_4-N	Sulfide	Phenolics	Cr
Bharagava et al. (2018a)	8.2		1428	436	4064	2216	6280	–				6.75	118.66						8.68	6.88
Mannucci et al. (2014)	7.3		2530	–	–					–		1810	14.4			–	126	–	–	–
Anjali and Sabumon (2014)	7.8–5	–	3000–6000	1200–2700	10,000–21,000	–		2000–3000			6000–9500	1000–3000		–		250–400	100–300	25–220	–	60–75
Faouzi et al. (2013)	10.5	–	8944	2000	–			4870		–	1244	–	62			–		–	6045	–
Bhatnagar et al. (2013)	8.4	16,141	926	294	5557	295	5786	–		874	763		–	3.19	241		–		8.66	5.56
Mondal et al. (2012)	7.2	–	2800	1122	1290	–					–			32.9				–		78
Chandra et al. (2011)	8.5–		2933	12,466	–		60,180			–		–	427			578			1.48	38
Mandal et al. (2010)	7.9–9.2	20,042	2533	977	21,620	1244	–	–	–		6528		62	2.56	–		118	860		258
Tripathi et al. (2011)	7.3		490	104	2366	1102	3468			340								9.43	11.93	8.89
Apaydin et al. (2009)	7.4	–	3700	1470	–			2690	1260	–	–	–	–	–	–	–	180	440	–	–
Munz et al. (2009)	7.2	–	2102		–	576	–	–	–	–	3260	–	–	–	–		118	–		–
Verma et al. (2008)	8.85	11,000	458	267	2219	258	2477			729	354	2400			234	229			10.5	19.57
Kongjao et al. (2008)	7.08–8.7	–	4100–6700	630–975	13.300–19,700			600–955		–	–	–	–		–	144–170	–	–	–	11.5–14.3

(continued)

Table 17.1 (continued)

References	pH	Cond.	COD	BOD_5	TDS	TSS	TS	SS	VSS	Alk.	Chloride	Sulfate	P	Fe	Mg	TKN	NH_4-N	Sulfide	Phenolics	Cr
Munz et al. (2008)	7.2		1904			576			459		3260					146	82	27	125	
Kurt et al. (2007)	7.2	19,950	2810	910	–	1520	–	–		–	6400		–	0.62			130	89		62
Lofrano et al. (2006)	6.6	8600	6855	2700		2865				1010	2835	745					70.5			140
Ganesh et al. (2006)	7.08	–	–	–	–	–	10,265	2820	1505	–	–	–				225	128	–	–	90–100
Lefebvre et al. (2006)	7.7	–	2200		36,800			5003	1300	–	–	–	–	–			–	–		–
Szpyrkowicz et al. (2005)	7.7	–	2426	–	–	–	–	–	–	–	–	–				370	335	286		29.3
Thanigavel (2004)	8.2-8.5	–	5665	–	14,750	–	19,775	5025	–	–	–	–	–	–		–	–	–		–
Leta et al. (2004)	10.72	–	11,153	2906	6810	–	–	–	–	–	–	–				–	162	507		32.87
Dantas et al. (2003)	8	–	1803	106	9435	526					2251			–			70			
Song et al. (2000)	7.5–9	–	5000–10,000	1500–2000	–	–	–	–		–	–	–	–	–			–	–	–	100

Note:

*Except pH, all the parameters are expressed in (mgL^{-1}), but conductivity is expressed in (μmho/cm)

**Cond.* conductivity, *COD* chemical oxygen demand, *BOD* biochemical oxygen demand, *TDS* total dissolved solids, *TSS* total suspended solids, *TS* total solids, *SS* suspended solids, *VSS* volatile suspended solids, *Alk.* alkalinity

Table 17.2 Toxicity profile of organic and inorganic pollutants used in leather processing

Name of chemicals	Applications	LD_{50} in rats (oral mg/kg)	Target organs
Pentachlorophenol (PCP) (a carcinogen)	Applied as a biocide in preservative for raw hides/skins	2000	Eyes, nose, skin, respiratory tract, blood, kidney, liver, immune system, and reproductive system
Dibutyl phthalate (DBP) (a endocrine disrupting chemical)	Applied as a plasticizer in artificial leather manufacturing	7499	Eyes, lungs, gastrointestinal (GI) tract, and testes
Benzyl butyl phthalate (BBP) (a endocrine disrupting chemical)	Applied in preparation of microporous artificial leather coating/water vapor-permeable sheet materials	2330	Eyes, lungs, liver, and reproductive system
Bis(2-ethylhexyl) phthalate (DEHP) (a endocrine disrupting chemical)	Applied as a plasticizer in artificial leather manufacturing	30.000	Liver and testes
Short-chain, chlorinated paraffins	Additive for leather treatment (gives smoothness), leather clothing and belts, and as oiling agent	3090	Liver, kidney, and thyroid
Anthracene (a carcinogen)	Additive during tanning	16,000	Kidneys and liver
Nonylphenol (a endocrine disrupting chemical and xenoestrogen)	Applied during finishing	1475	Blood, lungs, eyes, skin, central nervous system (CNS), kidneys, and testes
N-methyl pyrrolidone	Applied as a coalescence, plasticizers, and wetting agents	3914	Eyes, kidneys, lymphatic system, liver, lung, and testes
Methylisothiazolinone (a carcinogen)	Applied as biocide	1800	Skin and eyes
Organotin compounds (dibutyltin) (a carcinogen)	Applied as a catalyst	175	GI tract and liver
Azo dyes (Orange II) (a carcinogen)	Applied as a dyeing agent	3418	Blood, liver, and testes
Chloromethyl benzene or benzyl chloride	Applied to inhibit the growth of molds and mildews on hides/skins, facilitates leather softening and wetting, and to color finished leather in combination with dyes	–	Moderately aquatic toxicant and identified as a Group 2A carcinogen
4-Chloro-3-methyl phenol or *p*-chloro-*m*-cresol (PCMC)	Applied as biocide in raw hide/skins preservation	–	Potential endocrine disruptor by the US Environmental Protection Agency

(continued)

Table 17.2 (continued)

Name of chemicals	Applications	LD_{50} in rats (oral mg/kg)	Target organs
Hexachlorobenzene (a carcinogen)	Applied for raw hide/skins preservation	10,000	Reproductive system
Chromium (a carcinogen)	Applied as a tanning agent	3250	Kidneys, CNS, and hematopoietic system
Formaldehyde (a carcinogen)	Applied in finishing of leather	100	Eyes and lungs
Arsenic (a carcinogen)	Applied in finishing of leather	763	Liver, kidneys, skin, lungs, and lymphatic system
Sodium dichromate	Applied in preparation of chrome-tanning salts	–	Blood, kidneys, heart, lungs, and eyes
Cobalt dichloride	Applied in dyeing and finishing	80	Skin, lungs, liver, kidney, and heart
Cadmium sulfate (pigment)	Applied as fastening agents and used in marking and surfacing of material	280	Lungs, liver, tissues, and reproductive system
Lead chromate (pigment)	Applied as fastening agents and used in marking and surfacing of material	1000	Lungs, liver, tissues, and reproductive system

Adapted from Saxena et al. (2016), Dixit et al. (2015), and Kumar et al. (2008)

(c) Cause hematotoxicity in the common fish *Tilapia mossambica* and fresh water fish, *Labeo rohita* (Hamilton) (Sounderraj et al. 2012; Praveena et al. 2013)
(d) Interfere metabolic processes by altering the activity of oxidative enzymes in different organs of guppy fish, *Poecilia reticulata* (Aich et al. 2011, 2015)
(e) Disrupt the several physiological and cytological processes in plants (Saxena et al. 2016)
(f) Disturb the delicate hormonal balance (endocrine disruption in rats) and compromise the reproductive fitness of living beings (Kumar et al. 2008)
(g) Cause embryonic toxicity, coagulation of fertilized eggs, detachment of tail bud from the yolk sac, yolk sack edema, malformation of the tail, scoliosis, and deformation of swim bladder in the embryos of zebra fish, *Danio rerio* (Rocha and De Oliveira 2017)
(h) Cause reduction in the diversity of macroinvertebrates (Wosnie and Wondie 2014)
(i) Cause acute embryotoxicity and developmental defects in the sea urchins (*Paracentrotus lividus* and *Sphaerechinus granularis*) and serious toxicity in *Daphnia magna* (Oral et al. 2007)
(j) Cause detrimental changes in the biochemical parameters, damage to gonad and mantle tissues, and also genotoxic effects in the snail, *Pila globosa* (Bhattacharya et al. 2016)

Such toxic effects caused by TWW make it a serious pollutant, and hence, its adequate treatment to its final disposal into the environment is required to combat the environmental threats and protect the public health.

17.5 Bioremediation Approaches for Tannery Wastewater Containing Organic and Inorganic Pollutants

Bioremediation approaches are viewed as the eco-friendly alternatives to adequately treat and detoxify the highly toxic TWW as these are self-driven and sustainable and do not create secondary pollution as compared to physicochemical approaches currently applied. An overview of various bioremediation approaches used to treat TWW is provided below:

17.5.1 Microbial Bioremediation

Microbial bioremediation is the use of various microorganisms such as bacteria, fungi, and algae to degrade and detoxify TWW for environmental safety (Bharagava et al. 2018b; Saxena and Bharagava 2017; Bharagava et al. 2017a; Saxena and Bharagava 2016). Various research reports are available on the bioremediation of TWW using different microbes in the public domain (Kim et al. 2014; Noorjahan 2014; Yusuf et al. 2013; El-Bestawy et al. 2013; Sivaprakasam et al. 2008). Table 17.3 also list various microbes reported in the bioremediation of TWW.

Kim et al. (2014) reported the 98.3% of COD and 88.5% of Cr removal from TWW. Noorjahan (2014) reported 90% of COD, 90% of BOD, and 63.8% of Cr removal from TWW using *E. coli* whereas 95.4% of COD and BOD and 73.5% of Cr removal from TWW using *Bacillus* sp. Yusuf et al. (2013) reported 87.6% of COD from TWW using *B. subtilis* whereas 85.2% of COD from TWW using *P. fragi*. El-Bestawy et al. (2013) reported 79.16 of COD, 94.14 of BOD, and 93.66 of Cr from TWW using an optimized bacterial consortium containing *Providencia vermicola* W9B-11, *Escherichia coli* O7:K1 CE10, *Bacillus* sp. 58, *Bacillus amyloliquefaciens* T004, *Pseudomonas stutzeri* M15-10-3, and *Bacillus* sp. PL47. Sivaprakasam et al. (2008) also reported 80% of COD removal from TWW using a bacterial consortium (*P. aeruginosa*, *B. flexus*, *E. homiense*, and *S. aureus*).

However, the highly saline nature of TWW causes difficulties in its bioremediation due to (Saxena et al. 2016) (a) limited adaptation of conventional microbial cultures to higher salt concentration (>3–5% w/v), (b) changes in the ionic strength (salt concentration from 0.5% to 2% w/v) that also may cause cell disruption even with the acclimatized cultures, (c) salt adaptation of cultures that is easily lost when subjected to salt-free medium, and (d) the presence of poorly biodegradable compounds such as tannins and other toxic metals like Cr in wastewater that may inhibit

Table 17.3 Microorganisms used for the bioremediation of tannery wastewater

References	Microorganisms	COD removal (%)	BOD removal (%)	Cr removal (%)
Kim et al. (2014)	*Brachymonas denitrificans*	98.3	–	88.5
Noorjahan (2014)	*E. coli*	90	90	63.8
	Bacillus sp.	95.4	95.4	73.5
Elmagd and Mahmoud (2014)	Mixed culture	98.3	98.4	98.3
Sharma and Malviya (2014)	*Fusarium chlamydosporium* SPFS2-g	71.80	–	–
Yusuf et al. (2013)	*B. subtilis*	87.6	–	–
	P. fragi	85.2		
El-Bestawy et al. (2013)	*Providencia vermicola* W9B-11, *Escherichia coli* O7:K1 CE10, *Bacillus sp.* 58, *Bacillus amyloliquefaciens* T004, *Pseudomonas stutzeri* M15-10-3, *Bacillus sp.* PL47	79.16	94.14	93.66
Mandal et al. (2010)	*Thiobacillus ferrooxidans*	69	72	5
Nanda et al. (2010)	*Nostoc* sp.	37.8	48.6	–
Ramteke et al. (2010)	*E. coli*	98.46	90	–
	Vibrio sp.	87.5		
	Pseudomonas sp.	96.15		
Sivaprakasam et al. (2008)	*P. aeruginosa*, *B. flexus*, *E. homiense*, *S. aureus*	80	–	–
Vankar and Bajpai (2008)	*Trichoderma* sp.	–	–	97.93
Onyancha et al. (2008)	*S. condensate*	–	–	>75
	R. hieroglyphicum			
Srivastava et al. (2007)	*Acinetobacter* sp.	–	–	90
Rajasimman et al. (2007)	Mixed culture	46–85	65–93	–
Wang et al. (2007)	*A. thiooxidans*	–	–	99.7
Srivastava and Thakur (2006)	*Aspergillus* sp.	–	–	–
	Hirsutella sp.			70
Lefebvre et al. (2005)	Halophiles	95	–	–
Thanigavel (2004)	Mixed culture	89.5	–	–
Shakoori et al. (2000)	Bacterial strain	–	–	87

the biological treatment processes; hence, to overcome such limitations, proper enrichment of a particular microbial species is required for the effective bioremediation of TWW.

17.5.2 Phytoremediation

Phytoremediation is the use of green plants or trees and their associated microbiota to treat and detoxify the wastewater or contaminated soils/sites for environmental sustainability (Chandra et al. 2015). There are only few research reports are available on the phytoremediation of TWW in public domain (Gupta et al. 2018; Kassaye et al. 2017; Gregorio et al. 2015).

Gupta et al. (2018) studied the microbe-assisted phytoremediation of tannery effluent-contaminated agricultural soils. They isolated a Cr^{6+}-resistant plant growth-promoting *Pseudomonas* sp. (strain CPSB21) from the tannery effluent-contaminated agricultural soils and evaluated the various plant growth-promoting activities, oxidative stress tolerance, and Cr^{6+} bioremediation. Further, they applied the isolated strain for microbe-assisted phytoremediation and reported that the inoculation of strain CPSB21 alleviated the Cr^{6+} toxicity and enhanced the plant growth parameters and nutrient uptake in sunflower plant during pot experiment. Kassaye et al. (2017) reported the phytoremediation potential of swamp smartweed (*Polygonum coccineum*), para grass (*Brachiaria mutica*), and papyrus (*Cyprus papyrus*) for Cr-containing TWW. They reported that all the three plants exhibited a significant transfer of Cr from wastewater to roots and shoots, but removal efficiency of Cr for swamp smartweed was relatively low as compared to para grass and papyrus and further suggested the use of para grass and papyrus for effective phytoremediation of TWW. Gregorio et al. (2015) reported the bacterial-assisted phytoremediation of organic pollutants in TWW received from a conventional tannery wastewater treatment plant. They bioaugmented a plant growth-promoting rhizobacteria (PGPR) belonging to the *Stenotrophomonas* species in the rhizosphere of *P. australis* reported 88%, 84%, and 71% degradation of 4-*n*-nonylphenol, mono-ethoxylated nonylphenol, and di-ethoxylated nonylphenol as compared to control (simple phyto-based approach).

17.5.3 Electrobioremediation

Electrobioremediation is an emerging hybrid remediation technology coupling bioremediation to electrokinetics. It involves both the advance oxidation process and biological treatment process and is the best solution to achieve the regulatory standard for TWW discharge. Kanagasabi et al. (2012) reported the electrobioremediation of TWW that involved an electrooxidation reactor connected to a column packed with immobilized bacterial biomass. They reported that the electrooxidation technique was found to obtain maximum COD reduction up to 73.1% at 1.5 A dm^{-2} using raw TWW and *Bacillus* strain B proved to be superior in terms of COD reduction up to 91.5% with diluted TWW. However, the intimate coupling of electro- and biooxidation recorded good degradation for both raw and diluted TWW samples and achieved 66.2% and 76.6% COD degradation, respectively.

17.5.4 Anammox

Anammox is an emerging technology used for the anaerobic removal of ammonia from industrial wastewaters due to its low-cost and less-energy consuming nature. It involves the anoxic oxidation of ammonia with nitrite as a preferred electron acceptor and consumes 50% less oxygen and 100% less organic carbon and saves 90% of operational costs in sludge disposal as compared to the conventional nitrification/denitrification processes (Anjali and Sabumon 2014). Therefore, industries, producing wastewaters having a high concentration of ammonia, are showing increased interest in the anammox process. However, the long start-up time and inhibitive nature in the presence of organic carbon and NH_4-N limit its field applications (Anjali and Sabumon 2014). Therefore, it is imperative to develop the mixed consortium capable of anammox in the presence of organic compounds. Further, the development of mixed microbial consortium consisting of ammonia-oxidizing bacteria, anammox bacteria, and denitrifying bacteria is also expected to treat the wastewaters containing both ammonia and organic carbon.

17.5.5 Constructed Wetlands

Constructed wetlands (CWs) are the man-engineered, eco-friendly systems designed to provide the eco-technological solution for the degradation and detoxification of pollutants from highly polluted industrial and municipal wastewaters (Bharagava et al. 2017b). These systems are vegetated with wetland and terrestrial plants to treat and manage wastewater in an effective way. However, the proper functioning of a wetland system depends on the complex relationship between the plants, microorganisms, soil, wastewater characteristics, and operational parameters (Aguilar et al. 2008).

Various studies are available in public domain on the biological treatment of TWW using CWs (Mant et al. 2004; Calheiros et al. 2007, 2012). The phytoremediation potential of *Pennisetum purpureum*, *Brachiaria decumbens*, and *Phragmites australis* in CWs has been reported for the phytoremediation (phytoextraction) of Cr from TWW (Mant et al. 2004). Calheiros et al. (2007) have also reported the phytoremediation potentials of *Canna indica*, *Typha latifolia*, *P. australis*,

Stenotaphrum secundatum, and *Iris pseudacorus* in CWs for the treatment of TWW under two different hydraulic loading rates at 3 and 6 cm/day, and it was found that only *P. australis* and *T. latifolia* were able to establish successfully. Further, they also evaluated *Arundo donax* and *Sarcocornia fruticosa* in two series of horizontal subsurface flow CWs that used to treat TWW received from a conventional biological treatment plant and reported the removal of COD (51% and 80%) and BOD_5 (53% and 90%) for COD inlet: 68–425 mg L^{-1} and for BOD_5 inlet: 16–220 mg L^{-1} (Calheiros et al. 2012).

17.6 Challenges in Bioremediation of Tannery Wastewater

Tannery wastewater is a highly toxic soup of various organic and inorganic pollutants, and thus, its adequate treatment is required for environmental sustainability. However, several limitations make the treatment of TWW more challenging, which include (Saxena et al. 2016) (a) the presence of non-biodegradable metal line Cr that inhibits the growth of remediating microbes, (b) the presence of poorly biodegradable tannins or syntans that makes a high load of organic pollutant, (c) the presence of high salt load that is too difficult to be acclimatized by microbes, and (d) the presence of sulfide that has high antibacterial activity and, thus, cause system failure. However, search for new microbes with high bioremediation potential and application of combined treatment technologies like electrobioremediation may provide a better treatment solution for TWW. Further research efforts are required to expand the efficacy of the existing bioremediation methods for TWW treatment and management.

17.7 Conclusion

This chapter was mainly focused on the various bioremediation technologies for the treatment and management of hazardous TWW. It was evidenced from the literature that bioremediation methods can be the suitable alternative to the physicochemical methods, which are costly and environmentally destructive and create secondary pollution. However, bioremediation is a time-consuming process, and its proper monitoring is required for the better treatment and management of TWW.

Acknowledgment The financial support as "Major Research Projects" (Grant No.: EEQ/2017/000407) from "Science and Engineering Research Board" (SERB), Department of Science & Technology (DST), Government of India (GOI), New Delhi, India, and University Grant Commission (UGC) Fellowship received by Mr. Gaurav Saxena for doctoral studies is duly acknowledged.

References

Afaq S, Rana KS (2009) Impact of leather dyes on total protein of fresh water teleost, Cirrhinus mrigala (Ham.). Asian J Exp Sci 23(1):299–302

Aguilar JRP, Cabriales JJP, Vega MM (2008) Identification and characterization of sulfuroxidizing bacteria in an artificial wetland that treats wastewater from a tannery. Int J Phytoremediation 10(5):359–370

Aich A, Chattopadhyay B, Datta S, Mukhopadhyay SK (2011) Impact of composite tannery effluent on the amino-transferase activities in a fish biosystem, using Guppy fish (Poecilia reticulata) as an experimental model. Toxicol Environ Chem 93(1):85–91

Aich A, Goswami AR, Roy US, Mukhopadhyay SK (2015) Ecotoxicological assessment of tannery effluent using guppy fish (Poecilia reticulata) as an experimental model: a biomarker study. J Toxicol Environ Health A 78(4):278–286

Alam MZ, Ahmad S, Malik A (2009) Genotoxic and mutagenic potential of agricultural soil irrigated with tannery effluents at Jajmau (Kanpur), India. Arch Environ Contam Toxicol 57(3):463–476

Anjali G, Sabumon PC (2014) Unprecedented development of anammox in presence of organic carbon using seed biomass from a tannery Common Effluent Treatment Plant (CETP). Bioresour Technol 153:30–38

Apaydin O, Kurt U, Gonullu MT (2009) An investigation on tannery wastewater by electrocoagulation. Glob Nest J 11(4):546–555

Aravindhan R, Madhan B, Rao R, Nair B, Ramasami T (2004) Bioaccumulation of chromium from tannery wastewater an approach for chrome recovery and reuse. Environ Sci Technol 38(1):300–306

Bharagava RN, Saxena G, Chowdhary P (2017a) Constructed wetlands: an emerging phytotechnology for degradation and detoxification of industrial wastewaters. In: Bharagava RN (ed) Environmental pollutants and their bioremediation approaches, 1st edn. CRC Press, Taylor & Francis Group, Boca Raton, pp 397–426. https://doi.org/10.1201/9781315173351-15

Bharagava RN, Chowdhary P, Saxena G (2017b) Bioremediation: an ecosustainable green technology: its applications and limitations. In: Bharagava RN (ed) Environmental pollutants and their bioremediation approaches, 1st edn. CRC Press, Taylor & Francis Group, Boca Raton, pp 1–22. https://doi.org/10.1201/9781315173351-2

Bharagava RN, Saxena G, Mulla SI, Patel DK (2018a) Characterization and identification of recalcitrant organic pollutants (ROPs) in tannery wastewater and its phytotoxicity evaluation for environmental safety. Arch Environ Contam Toxicol 75(2):259–272. https://doi.org/10.1007/s00244-017-0490-x

Bharagava RN, Purchase D, Saxena G, Mulla SI (2018b) Applications of metagenomics in microbial bioremediation of pollutants: from genomics to environmental cleanup. In: Das S, Dash H (eds) Microbial diversity in the genomic era, 1st edn. Academic Press, Elsevier. https://doi.org/10.1016/B978-0-12-814849-5.00026-5

Bhatnagar MK, Singh R, Gupta S, Bhatnagar P (2013) Study of tannery effluents and its effects on sediments of river Ganga in special reference to heavy metals at Jajmau, Kanpur, India. J Environ Res Dev 8(1):56–59

Bhattacharya P, Swarnakar S, Mukhopadhyay A, Ghosh S (2016) Exposure of composite tannery effluent on snail, *Pila globosa*: a comparative assessment of toxic impacts of the untreated and membrane treated effluents. Ecotoxicol Environ Saf 126:45–55

Calheiros CSC, Rangel AOSS, Castro PML (2007) Constructed wetland systems vegetated with different plants applied to the treatment of tannery wastewater. Water Res 41(8):1790–1798

Calheiros CSC, Quiterio PVB, Silva G, Crispim LFC, Brix H, Moura SC, Castro PML (2012) Use of constructed wetland systems with Arundo and Sarcocornia for polishing high salinity tannery wastewater. J Environ Manag 95(1):66–71

Chandra R, Bharagava RN, Kapley A, Purohit HJ (2011) Bacterial diversity, organic pollutants and their metabolites in two aeration lagoons of common effluent treatment plant

(CETP) during the degradation and detoxification of tannery wastewater. Bioresour Technol 102(3):2333–2341

Chandra R, Saxena G, Kumar V (2015) Phytoremediation of environmental pollutants: an eco-sustainable green technology to environmental management. In: Chandra R (ed) Advances in biodegradation and bioremediation of industrial waste, 1st edn. CRC Press, Taylor & Francis Group, Boca Raton, pp 1–30. https://doi.org/10.1201/b18218-2

CPCB (2009) Recovery of better quality reusable salt from soak liquor of tanneries in solar evaporation pans. Central Pollution Control Board (CPCB) Ministry of Environment & Forests Control of Urban Pollution Series. Accessed online at http://cpcb.nic.in/displaypdf.php?id=UmVjb3Zlcl9zYWx0LnBkZg==

Dantas TLP, Jose HJ, Moreira RFPM (2003) Fenton and photo-fenton oxidation of tannery wastewater. Acta Sci Technol 25(1):91–95

Dixit S, Yadav A, Dwivedi PD, Das M (2015) Toxic hazards of leather industry and technologies to combat threat: a review. J Clean Prod 87:39–49

Durai G, Rajasimmam M (2011) Biological treatment of tannery wastewater: a review. J Environ Sci Technol 4:1–17

El-Bestawy E, Al-Fassi F, Amer R, Aburokba R (2013) Biological treatment of leather-tanning industrial wastewater using free living bacteria. Adv Life Sci Technol 12:46–65

Elmagd AM, Mahmoud MS (2014) Tannery wastewater treatment using activated sludge process system (lab scale modeling). Int J Eng Tech Res 2(5):21–28

FAO (2008) Management of waste from animal product processing. Food and Agricultural Organisation of United Nations, Rome

Faouzi M, Merzouki M, Benlemlih M (2013) Contribution to optimize the biological treatment of synthetic tannery effluent by the sequencing batch reactor. J Mater Environ Sci 4(4):532–541

Ganesh R, Balaji G, Ramanujam RA (2006) Biodegradation of tannery wastewater using sequencing batch reactor-respirometric assessment. Bioresour Technol 97(15):1815–1821

Gautam S, Kaithwas G, Bharagava RN, Saxena G (2017) Pollutants in tannery wastewater, pharmacological effects and bioremediation approaches for human health protection and environmental safety. In: Bharagava RN (ed) Environmental pollutants and their bioremediation approaches, 1st edn. CRC Press, Taylor & Francis Group, Boca Raton, pp 369–396. https://doi.org/10.1201/9781315173351-14

Gautam SP, Saxena G, Singh V, Yadav AK, Bharagava RN (2018) Green synthesis of TiO2 nanoparticles using leaf extract of Jatropha curcas L. for photocatalytic degradation of tannery wastewater. Chem Eng J 336:386–396. https://doi.org/10.1016/j.cej.2017.12.029

Gregorio SD, Giorgetti L, Castiglione MR, Mariotti L, Lorenzi R (2015) Phytoremediation for improving the quality of effluents from a conventional tannery wastewater treatment plant. Int J Environ Sci Technol 12(4):1387–1400

Gupta R, Rani R, Chandra A, Kumar V (2018) Potential applications of Pseudomonas sp. (strain CPSB21) to ameliorate Cr6+ stress and phytoremediation of tannery effluent contaminated agricultural soils. Sci Rep 8:4860. https://doi.org/10.1038/s41598-018-23322-5

ILTIP (2010) Indian leather and tanning industry profile: Italian Trade Commission, pp 1–43

Kanagasabi S, Kang YL, Manickam M, Ibrahim S, Pichiah S (2012) Intimate coupling of electro and biooxidation of tannery wastewater. Desalin Water Treat 51:34–36

Kassaye G, Gabbiye N, Alemu A (2017) Phytoremediation of chromium from tannery wastewater using local plant species. Water Pract Technol 12(4):894–901

Kim IS, Ekpeghere KI, Ha SY, Kim BS, Song B, Kim JT, Kim HG, Koh SC (2014) Full scale biological treatment of tannery wastewater using the novel microbial consortium BM-S-1. J Environ Sci Health A Tox Hazard Subst Environ Eng 49(3):355–364

Kongjao S, Damronglerd S, Hunsom M (2008) Simultaneous removal of organic and inorganic pollutants in tannery wastewater using electrocoagulation technique. Korean J Chem Eng 25(4):703–709

Kumar V, Majumdar C, Roy P (2008) Effects of endocrine disrupting chemicals from leather industry effluents on male reproductive system. J Steroid Biochem Mol Biol 111(3–5):208–216

Kurt U, Apaydin O, Gonullu MT (2007) Reduction of COD in wastewater from an organized tannery industrial region by electro-fenton process. J Hazard Mater 143(1–2):33–40

Lefebvre ON, Vasudevan N, Torrijos M, Thanasekaran K, Moletta R (2005) Halophilic biological treatment of tannery soaks liquor in a sequencing batch reactor. Water Res 39(8):1471–1480

Lefebvre ON, Vasudevan N, Torrijosa M, Thanasekaran K, Moletta R (2006) Anaerobic digestion of tannery soak liquor with an aerobic post-treatment. Water Res 40(7):1492–1500

Leta S, Assefa F, Gumaelius L, Dalhammar G (2004) Biological nitrogen and organic matter removal from tannery wastewater in pilot plant operations in Ethiopia. Appl Microbiol Biotechnol 66(3):333–339

Lofrano G, Belgiorno V, Gallo M, Raimo A, Meric S (2006) Toxicity reduction in leather tanning wastewater by improved coagulation flocculation process. Glob Nest J 8(2):151–158

Lofrano G, Meriç S, Zengin GE, Orhon D (2013) Chemical and biological treatment technologies for leather tannery chemicals and wastewaters: a review. Sci Total Environ 461-462:265–281

Mandal T, Dasgupta D, Mandal S, Datta S (2010) Treatment of leather industry by aerobic biological fenton oxidation process. J Hazard Mater 180(1–3):204–211

Mannucci A, Munz G, Mori G, Lubello C (2014) Factors affecting biological sulfate reduction in tannery wastewater treatment. Environ Eng Manag J 13(4):1005–1012

Mant C, Costa S, Williams J, Tambourgi E (2004) Phytoremediation of chromium by model constructed wetland. Bioresour Technol 97(15):1767–1772

Matsumoto ST, Mnlovani SM, Malaguttii MIA, Dias AL, Fonseca IC, Morales MAM (2006) Genotoxicity and mutagenicity of water contaminated with tannery effluent, as evaluated by the micronucleus test and comet assay using the fish Oreochromis niloticus and chromosome aberrations in onion root tips. Genet Mol Biol 29(1):148–158

Mondal A, Banerjee PK, Bhattacharjee C, Saha PD (2012) Treatment of chromium present in tannery wastewater using chemical & biological techniques. Elixir Pollut 49:9832–9835

Munz G, Gori R, Cammilli L, Lubello C (2008) Characterization of tannery wastewater and biomass in a membrane bioreactor using respirometric analysis. Bioresour Technol 99(18):8612–8618

Munz G, De Angelis D, Gori R, Mori G, Casarci M, Lubello C (2009) The role of tannins in conventional angogated membrane treatment of tannery wastewater. J Hazard Mater 164(2–3):733–739

Nanda S, Sarangi PK, Abraham J (2010) Cyanobacterial remediation of industrial effluents Tannery effluents. New York Sci J 3(12):32–36

Noorjahan CM (2014) Physicochemical characteristics, identification of bacteria and biodegradation of industrial effluent. J Bioremed Biodegr 5:229

Onyancha D, Mavura W, Ngila J, Ongoma P, Chacha J (2008) Studies of chromium removal from tannery wastewaters by algae biosorbents, Spirogyra condensate and Rhizocolonium hieroglyphicum. J Hazard Mater 158(2–3):605–614

Oral R, Meric S, De Nicola E, Petruzzelli D, Rocca CD, Pagano G (2007) Multi-species toxicity evaluation of a chromium-based leather tannery wastewater. Desalination 211:48–57

Praveena M, Sandeep V, Kavitha N, Jayantha Rao K (2013) Impact of tannery effluent, chromium on hematological parameters in a fresh water fish, Labeo Rohita (Hamilton). Res J Animal Veterinary Fishery Sci 1(6):1–5

Rajasimman M, Jayakumar M, Ravindranath E, Chitra K (2007) Treatment of solid and liquid wastes from tanneries in an UASB reactor. Proceedings of 60th Annual Session of Indian Institute of Chemical Engineers, CHEMCON-2007, Kolkata, India

Rameshraja D, Suresh S (2011) Treatment of tannery wastewater by various oxidation and combined processes. Int J Environ Res 5(2):349–360

Ramteke PW, Awasthi S, Srinath T, Joseph B (2010) Efficiency assessment of common effluent treatment plant (CETP) treating tannery effluents. Environ Monit Assess 169(1–4):125–131

Rocha OP, De Oliveira DP (2017) Investigation of a Brazilian tannery effluent by means of zebra fish (*Danio rerio*) embryo acute toxicity (FET) test. J Toxicol Environ Health Part A Curr Issues 80:1078–1085

Saxena G, Bharagava RN (2015) Persistent organic pollutants and bacterial communities present during the treatment of tannery wastewater. In: Chandra R (ed) Environmental waste management, 1st edn. CRC Press, Taylor & Francis Group, Boca Raton, pp 217–247. https://doi.org/10.1201/b19243-10

Saxena G, Bharagava RN (2016) Ram Chandra: advances in biodegradation and bioremediation of industrial waste. Clean Techn Environ Policy 18:979–980. https://doi.org/10.1007/s10098-0151084-9

Saxena G, Bharagava RN (2017) Organic and inorganic pollutants in industrial wastes, their ecotoxicological effects, health hazards and bioremediation approaches. In: Bharagava RN (ed) Environmental pollutants and their bioremediation approaches, 1st edn. CRC Press, Taylor & Francis Group, Boca Raton, pp 23–56. https://doi.org/10.1201/9781315173351-3

Saxena G, Bharagava RN, Kaithwas G, Raj A (2015) Microbial indicators, pathogens and methods for their monitoring in water environment. J Water Health 13(2):319–339

Saxena G, Chandra R, Bharagava RN (2016) Environmental pollution, toxicity profile and treatment approaches for tannery wastewater and its chemical pollutants. Rev Environ Contam Toxicol 240:31–69. https://doi.org/10.1007/398_2015_5009

Shakoori AR, Makhdoom M, Haq RU (2000) Hexavalent chromium reduction by a dichromate resistant gram-positive bacterium isolated from effluents of tanneries. Appl Microbiol Biotechnol 53(3):348–351

Sharma S, Malaviya P (2014) Bioremediation of tannery wastewater by chromium resistant fungal isolate Fusarium chlamydosporium SPFS2-g. Current World Envi 9:721–727

Sivaprakasam S, Mahadevan S, Sekar S, Rajakumar S (2008) Biological treatment of tannery wastewater by using salt-tolerant bacterial strains. Microb Cell Fact 7:15

Song Z, Williams CJ, Edyvean GJ (2000) Sedimentation of tannery wastewater. Water Res 34(7):2171–2176

Sounderraj SFL, Lesley N, Senthilkumar P (2012) Studies on the effect of tannery effluent and chromium accumulation in common crop Tilapia mossambica. Int J Pharm Biol Arch 3(4):978–985

Srivastava S, Thakur IS (2006) Isolation and process parameter optimization of Aspergillus sp. for removal of chromium from tannery effluent. Bioresour Technol 97:1167–1173

Srivastava S, Ahmad AH, Thakur IS (2007) Removal of chromium and pentachlorophenol from tannery wastewaters. Bioresour Technol 98(5):1128–1132

Szpyrkowicz L, Kaul SN, Neti Rao N, Satyanarayan S (2005) Influence of anode material on electrochemical oxidation for the treatment of tannery wastewater. Water Res 39(8):1601–1613

Thanigavel M (2004) Biodegradation of tannery effluent in fluidized bed bioreactor with low density biomass support. M. Tech. thesis. Annamalai University, Tamil Nadu, India

Technical Guidance Manual for Tanneries (TGMT) (2010) Ministry of environment and forests, government of India. Accessed online at http://environmentclearance.nic.in/writereaddata/Form-1A/HomeLinks/TGM_Tannery_010910_NK.pdf

Thanikaivelan P, Rao JR, Nair BU, Ramasami T (2005) Recent trends in leather making: processes, problems, and pathways. Crit Rev Environ Sci Technol 35(1):37–79

Tripathi M, Vikram S, Jain RK, Garg SK (2011) Isolation and growth characteristics of chromium (VI) and pentachlorophenol tolerant bacterial isolate from treated tannery effluent for its possible use in simultaneous bioremediation. Indian J Microbiol 51(1):61–69

UNIDO (2000) United Nations Industrial Development Organization (UNIDO): pollutants in tannery effluent, definitions and environmental impact, limits for discharge into water bodies and sewers

UNIDO (2010) United Nations Industrial Development Organization (UNIDO): future trends in the world leather and leather products industry and trade, Vienna

Vankar PS, Bajpai D (2008) Phytoremediation of chrome-VI of tannery effluent by Trichoderma species. Desalination 222(1–3):255–262

Verma T, Ramteke PW, Garg SK (2008) Quality assessment of treated tannery wastewater with special emphasis on pathogenic E. coli detection through serotyping. Environ Monit Assess 145(1–3):243–249

Wang YS, Pan ZY, Lang JM, Xu JM, Zheng YG (2007) Bioleaching of chromium from tannery sludge by indigenous, Acidithiobacillus thiooxidans. J Hazard Mater 147(1–2):319–334

Wosnie A, Wondie A (2014) Assessment of downstream impact of Bahir Dar tannery effluent on the head of Blue Nile River using macroinvertebrates as bioindicators. Int J Biodivers Conserv 6:342–350

Yusuf RO, Noor ZZ, Abu Hassan MA, Agarry SE, Solomon BO (2013) A comparison of the efficacy of two strains of Bacillus subtilis and Pseudomonas fragi in the treatment of tannery wastewater. Desalin Water Treat 51(16–18):3189–3195

Chapter 18
Bioremediation: An Eco-friendly Cleanup Strategy for Polyaromatic Hydrocarbons from Petroleum Industry Waste

M. S. Dhanya and Arun Kalia

Contents

M. S. Dhanya (✉) · A. Kalia
Department of Environmental Sciences and Technology, School of Environment and Earth Sciences, Central University of Punjab, Bathinda, Punjab, India

G. Saxena, R. N. Bharagava (eds.), *Bioremediation of Industrial Waste for Environmental Safety*, https://doi.org/10.1007/978-981-13-1891-7_18

Abstract The polycyclic aromatic hydrocarbons (PAHs) are a major pollutant from petroleum industry and oil refineries. This group of organic xenobiotics is produced either by pyrolytic or petrogenic sources. The physicochemical properties of PAHs such as hydrophobicity and electrochemical stability increase its persistence in the environment and add to the carcinogenicity and other health impacts. The major degradation processes of PAH in the environment are adsorption, volatilization, photolysis, chemical oxidation, bioaccumulation, and microbial degradation. The microbial communities like bacteria, fungi, and algae play a significantly important role in the biological removal of PAHs. The natural attenuation, bioaugmentation, biostimulation, phytoremediation, rhizoremediation, and composting are major bioremediation approaches for PAHs from contaminated environment. The enzymes linked with the PAH degradation are mainly oxygenases, manganese peroxidases, lipases, and laccases. The surfactant production in the microbes increases the bioavailability of PAH and enhanced the removal process. The PAH degradation depends on the various environmental conditions such as temperature, pH, aeration, moisture content, nutrient availability, absence of toxic compounds, and the type and number of degrading microbial population. The bacterial and fungal degradation pathways produce intermediate metabolites and mineralization to carbon dioxide. The molecular techniques like gene engineering and protein engineering improve the removal of PAHs. Thus the biodegradation of PAHs is linked to the carbon cycle and remediation of these persistent organo-molecules from the environment.

Keywords Petroleum industry · Polycyclic aromatic hydrocarbons · Biodegradation · Enzymes · Biosurfactant · Metabolites

18.1 Introduction

The environmental contamination by polycyclic aromatic hydrocarbons (PAHs) and its health effects are of great concern as these organic contaminants are resistant to biodegradation, persist in the environment for long periods, have bioaccumulation potential and carcinogenic and toxic effects (Jacob et al. 1986; Dipple and Bigger 1991). The semi-volatility of PAHs results in formation of diones, nitro- and dinitro-PAHs, and sulfonic acids, respectively by reaction with ozone, nitrogen oxides, and sulfur dioxide. The 16 PAH compounds were recommended as priority pollutants by the US Environmental Protection Agency (US EPA) and as persistent organic pollutants (POPs) by the United Nations Environment Programme (UNEP) due to its persistence and toxicological nature (Saxena and Bharagava 2015; Keith and

Telliard 1979). The human health is under threat of PAHs due to its carcinogenic, mutagenic, and teratogenic nature as reported by the Environmental Protection Agency (USEPA 2011).

Petroleum industry contributes to the energy demands of the world but is one of the major polluting industries. As the petroleum consists of saturates, aromatics, resins, and asphaltenes, there is a chance of its release to the environment from the petroleum industries. The serious environmental problems of emission of polycyclic hydrocarbons are associated with oil spills during exploration, transportation, and refining (Zhang et al. 2011). So these petrochemical plants and oil refineries produce a large amount of hazardous waste. The estimates show petrochemical plants and oil refineries produce a great quantity of solid oily waste of around 10,000 m^3 per day which are grouped under hazardous waste due to its flammability, corrosivity, toxicity, and reactivity (Gafarov et al. 2006). The polycyclic aromatic hydrocarbons are one of the major pollutants in that hazardous waste group along with alkanes, cycloalkanes, BTEX (benzene, toluene, ethylbenzene, and xylene), and phenol. The crude oils on an average contain 0.2–7% PAHs, with configurations ranging from two to six rings.

The cleanup of soil can be accomplished using different remediation technologies, but bioremediation plays one of the most efficient means to restore original ecosystem conditions (Haritash and Kaushik 2009). The PAH-degrading microorganisms include algae, bacteria, and fungi. The use of microorganisms for bioremediation of PAH-contaminated environments seems to be an attractive technology for restoration of polluted sites.

18.2 IUPAC Nomenclature of Polycyclic Aromatic Hydrocarbons (PAHs)

Polycyclic aromatic hydrocarbons (PAHs) are a group of compounds containing carbon and hydrogen, composed of two or more fused aromatic rings in linear, angular, and cluster arrangements. The nomenclature of PAHs is based on the rules and recommendations of the International Union of Applied and Pure Chemistry (IUPAC) from 1957. The step-by-step procedure for naming a PAH are:

Step 1: Find within the molecule the polycyclic substructure which corresponds to the highest priority (usually largest) parent compound.

Step 2: Determine the number and type of ring systems fused to this parent structure.

Step 3: Preference is hereby given to the most simple priority substructures by choosing the maximum number of first-order attachments and the maximum number of identical attached components.

The chemical structure of 16 priority PAHs is as follows (Fig. 18.1). Table 18.1 represents the chemical formula and CAS registry no. of 16 priority PAHs.

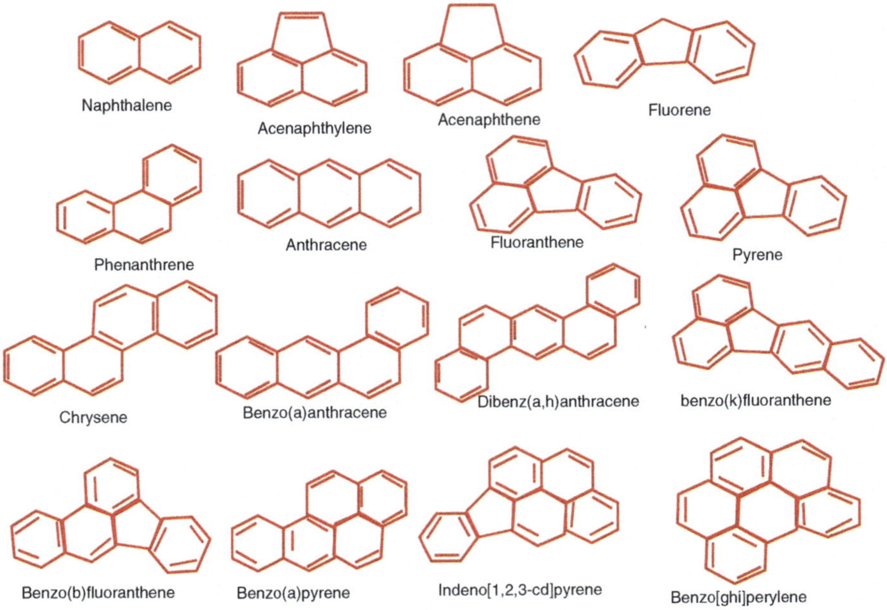

Fig. 18.1 Two-dimensional structure of 16 priority PAHs

Table 18.1 IUPAC name, chemical formula and CAS registry no. of 16 priority PAHs

Preferred IUPAC name	Other common name	Molecular weight (g)	Formula	CAS registry no.
Naphthalene	Naphthalene	128	$C_{10}H_8$	91-20-3
Acenaphthylene	Acenaphthylene	152	$C_{12}H_8$	208-96-8
Acenaphthene	Acenaphthene	154	$C_{12}H_{10}$	83-32-9
9H-fluorene	9H-fluorene	166	$C_1 3H_{10}$	86-73-7
Phenanthrene	Phenanthrene	178	$C_{14}H_{10}$	85-01-8
Anthracene	Anthracene	178	$C_{14}H_{10}$	120-12-7
Fluoranthene	Fluoranthene	202	$C_{16}H_{10}$	206-44-0
Pyrene	Pyrene	202	$C_{16}H_{10}$	129-00-0
Tetraphene	Benz(a)anthracene	228	C_{18} H_{12}	56-55-3
Chrysene	Chrysene	228	C18 H12	218-01-9
Benzo(pqr)tetraphene	Benz(a)pyrene	252	C_{20} H_{12}	50-32-8
Benzo(e) acephenanthrylene	Benzo(b)fluoranthene	252	C_{20} H_{12}	205-99-2
Benzo(k)fluoranthene	Benzo(k)fluoranthene	252	C_{20} H_{12}	207-08-9
Indeno[1,2,3-cd]pyrene	Indeno[1,2,3-cd] pyrene	276	C_{20} H_{12}	193-39-5
Benzo[ghi]perylene	Benzo[ghi]perylene	276	C_{22} H_{12}	192-24-2
Benzo(k)tetraphene	Dibenz(ah)anthracene	276	C_{22} H_{14}	215-58-7

18.3 Characteristics of Priority PAHs

Polycyclic aromatic hydrocarbons (PAHs) consist of two or more benzene rings fused in a linear, angular, or cluster arrangement. These compounds have two or more fused carbon rings that have hydrogen or an alkyl (C_nH_{2n+1}) group attached to each carbon. Compounds range from naphthalene ($C_{10}H_8$, two rings) to coronene ($C_{24}H_{12}$, seven rings). The molecular size (i.e., their number of aromatic rings) and molecule topology (i.e., their pattern of ring linkage) of PAHs are directly related to chemical properties of PAHs. The larger size and higher angularity may lead to persistence from increase of hydrophobicity and electrochemical stability (Kanaly and Harayama 2000). These compounds have low solubility in water, high melting and boiling points, and low vapor pressure. With an increase in molecular weight, their solubility in water decreases, melting and boiling point increase, and vapor pressure decreases (Jones 1991; Wild et al. 1994). The physical and chemical properties of PAH are given in Table 18.2. Based on molecular weight, PAHs are divided into two categories: low molecular weight (LMW) and high molecular weight (HMW) PAHs. The low molecular weight PAHs are with two or three rings and are relatively volatile, soluble, and more degradable in nature. The high molecular weight (HMW) PAHs have four or more rings and hydrophobicity; these PAHs are very toxic to whole cells (Yucheng et al. 2008).

18.4 Sources of Petrogenic PAHs (Pollutants)

PAHs enter into the environment components, such as air, water, and soils, from natural and anthropogenic sources. The sources of PAHs are divided into categories: pyrolytic and petrogenic. The pyrolytic sources of PAHs include (1) incomplete fuel combustion, wood and coal burning, car emissions, tobacco-related activities, and meat grilling (Freeman and Cattell 1990; Jacob et al. 1986; Piskonen and Itävaara 2004; Farhadian et al. 2010) and (2) coal tar, creosote, roofing tar, and parking lot seal coats (ATSDR 2010). However, the sources of petrogenic PAHs are mainly derived from crude oil, unburned fuel, and refinery products (Elias et al. 2007).

The petrochemical industry is one of the major contributors of PAHs into the environment (Tsibart and Gennadiev 2013). The petroleum exploration and refining operations release PAHs (Fetzer 2000). The petroleum spills, discharges, seepages, industrial and municipal wastewater, urban and suburban surface runoff, and atmospheric deposition add PAH to the water environment. The accidents during oil exploration and transportation also add PAH to the marine environment (Souza et al. 2014). The PAHs enter to soil by petroleum spills and discharges, forest and grass fires, volcanoes, industrial and municipal solid waste, and atmospheric deposition. The cooking and tobacco smoking (Mumford et al. 1989) also release PAH to the environment. The industries that manufacture aluminum, creosote, cement, and asphalt also contribute PAH into the environment (Tsibart and Gennadiev 2012).

Table 18.2 Physical and chemical properties of 16 priority PAHs

PAHs	Molecular weight	Formula	Melting point (°C)	Boiling point (°C)	Vapor pressure (Pa at 25 °C)	Log Kow	Henry constant (atm m^3 mol^{-1})	Resonance energy (eV)	Ionization potential (eV)
Naphthalene	128	$C_{10}H_8$	80.2	218	10.9–11.9	3.37	4.5×10^{-3}	1.325	–
Acenaphthylene	152	$C_{12}H_8$	95	265	9×10^{-1}	4	–	1.325	–
Acenaphthene	154	$C_{12}H_{10}$	96.2	278	30×10^{-1}	3.92	2.4×10^{-4}	–	–
Fluorene	166	C_13H_{10}	115–116	295	9×10^{-2}	4.18	7.4×10^{-5}	–	–
Phenanthrene	178	$C_{14}H_{10}$	100.5	339	2×10^{-2}	4.57	2.7×10^{-4}	1.924	8.03
Anthracene	178	$C_{14}H_{10}$	216.4	340	1.0×10^{-3}	4.54	–	–	7.43
Fluoranthene	202	$C_{16}H_{10}$	108.8	375	1.2×10^{-3}	5.22	1.9×10^{-3}	2.184	
Pyrene	202	$C_{16}H_{10}$	150	360	6.0×10^{-4}	5.18	1.3×10^{-5}	2.099	7.53
Benz(a)anthracene	228	$C_{18}\ H_{12}$	162	435	2.8×10^{-5}	5.91	1.2×10^{-5}	2.313	<7.35
Chrysene	228	C18 H12	254	448	5.7×10^{-7}	1.65	6.7×10^{-7}	2.477	–
Benz(a)pyrene	252	$C_{20}\ H_{12}$	179	495	7×10^{-7}	6.04	2×10^{-7}	2.579	<7.45
Benzo(b)fluoranthene	252	$C_{20}\ H_{12}$	168.3	481	1.07×10^{-5}	5.80	–	–	7.7
Benzo(k)fluoranthene	252	$C_{20}\ H_{12}$	215.7	481	5.2×10^{-8}	6.0	–	–	7.48
Indeno[1,2,3-cd] pyrene	276	$C_{20}\ H_{12}$	150–153	536	N/A	6.58	–	–	–
Benzo[ghi]perylene	276	$C_{22}\ H_{12}$	277	525	6×10^{-8}	6.50	2×10^{-7}	3.098	7.31
Dibenzo(a,h)anthracene	276	$C_{22}\ H_{14}$	262–265	524	3.7×10^{-10}	6.75	2×10^{-9}	2.958	–

Adapted from IARC (1983), Mackay et al. (1992), WGPAH (2001), ISPAC (2003), Nollet (2006); Ionization potential from Pysh and Yang (1963) and Simonsick and Hites (1986)

18.5 Transport and Fate of PAHs in Terrestrial and Aquatic Ecosystem

The movement of PAHs in the environment depends on properties such water solubility and volatility (Agency for Toxic Substances and Disease Registry). The PAH compounds once entered into the environment result in adsorption, volatilization, photolysis, chemical degradation, microbial degradation, and mineralization (Fig. 18.2). The PAHs in soils may get adsorbed on mineral surfaces of clay particles and organic matter (i.e., humic and fulvic acids) (Bogan and Sullivan 2003). The PAHs trapped in soil micropores increase persistence and limit the biodegradability, leachability, and volatility of PAHs (Bamforth and Singleton 2005; Wick et al. 2011). The degradation rate of PAHs depends on soil structure, particle size, and organic matter content in the soil (Manilal and Alexander 1991; Sims et al. 1990; Steiber et al. 1990). The researchers estimated the total PAH concentrations between 1 μg kg^{-1} and 300 g kg^{-1} in soil (Bamforth and Singleton 2005; Kanaly and Harayama 2000).

The PAHs can undergo volatilization of either the parent compound or their breakdown products from the contaminated environment to the atmosphere (Bossert and Bartha 1984; Park et al. 1990; Civilini 1994; Cousins and Jones 1998). The polycyclic aromatic hydrocarbons released into the atmosphere travel over distances by air currents and finally reached earth as wet or dry particulate deposition. The PAHs exposed to sunlight and in air, soil, or water are oxidized to polar oxidized

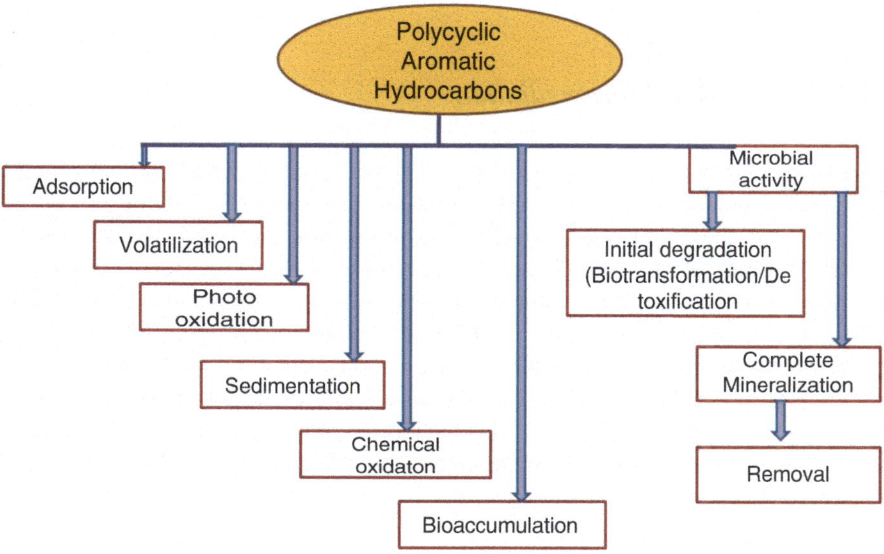

Fig. 18.2 Environmental fate of polycyclic aromatic hydrocarbons. (Adapted from Cerniglia 1992)

compounds by the process of photooxidation. The oxidation of PAHs also occurs by chemical oxidation process. The terrestrial vegetation is capable of adsorbing PAH from air or absorption by plant tissues. The bioaccumulation of PAHs is inversely linked to the ability of organisms to metabolize it. The biodegradation or biotransformation by microbes converts PAHs into less complex metabolites. The mineralization process results in conversion of PAHs into inorganic minerals, H_2O, CO_2 under aerobic conditions, or CH_4 under anaerobic conditions.

The presence of PAH residues was reported in surface water, groundwater, road runoff (Boxall and Maltby 1997; Pitt et al. 1995), vegetation (Wagrowski and Hites 1997), and foods (Edwards 1983). The breakdown of PAHs in soil and water generally takes days, weeks to months depending on the type of PAH compounds, reaction with sunlight, chemicals, and bioavailability for microbial degradation.

18.6 Bioremediation Techniques of PAHs from Petroleum Industry

Bioremediation plays one of the most efficient means to restore original ecosystem conditions (Haritash and Kaushik 2009). Biodegradation by microorganisms is usually the dominant and most important way to utilize living organisms such as bacteria and plants or their enzymes to degrade and detoxify the organic and inorganic pollutants (Bharagava et al. 2017a, b; Saxena and Bharagava 2017; Chandra et al. 2015; Nnamchi et al. 2006). Bisht et al. (2015) reviewed the different approaches of bioremediation of PAHs.

18.6.1 Intrinsic Bioremediation

This is also known as natural attenuation or bioattenuation where the indigenous microbial populations degrade PAHs based on their natural metabolic processes (Kuiper et al. 2004; Widada et al. 2001). The intrinsic bioremediation processes include a variety of physical, chemical, and biological processes that act to reduce the mass, toxicity, mobility, volume, or concentration of contaminants (USEPA 1999). The aerobic and anaerobic biodegradation, dilution, sorption, volatilization decay, chemical or biological stabilization, and transformation of contaminants are involved in this technique of bioremediation (Gentry et al. 2004). The time consumption is a challenging factor of this technology. But if no other techniques are applicable, this can be used for on-site remediation (Kuiper et al. 2004).

18.6.2 Bioaugmentation Through PAH-Degrading Microbes

In the bioaugmentation process, microbial strains having PAH-degrading capability are inoculated to the contaminated environment. This method introduces either wild-type or genetically modified microbes into the environment to increase degradation and transformation (Kuiper et al. 2004). The advances in the bioaugmentation technology include cell encapsulation, gene bioaugmentation, rhizosphere bioaugmentation, and phyto-augmentation (Gentry et al. 2004). The mineralization of phenanthrene in contaminated soil by inoculation of phenanthrene-utilizing bacteria was reported by Madsen and Kristensen (1997). The reduction in population of introduced microorganisms over the time of inoculation on to the site is the limiting factor of this method (Gentry et al. 2004). It may be resulted from abiotic factors like extreme temperature, pH, nutrient availability, the presence of toxic substances, and biotic stresses like antagonistic interactions from other microbes, protozoans, fungi, etc. (Rahman et al. 2002; Kim et al. 2004). Bogan et al. (2003) and Kim et al. (2004) reported *Mycobacterium* sp. is capable of degrading a great variety of low and high molecular weight PAHs.

18.6.3 Biostimulation of PAH-Degrading Microbes

The biostimulation is the addition of supplementing carbon sources or other nutrients to stimulate the activity of indigenous or inoculated PAH-degrading strains. The increase in oil biodegradation under aerobic conditions with nutrient amendments was reported by Jobson et al. (1974), Venosa et al. (1996), Lee et al. (1997), Brook et al. (2001), and Vinas et al. (2005).

18.6.4 Phytoremediation

It is the process in which the plants are used for removal of xenobiotics from the environment (Salt et al. 1998; Singh and Jain 2003). The phytoextraction of faba bean or alfalfa with humic acid addition to soil is capable of remediating PAHs in crude oil-contaminated soil (Halim et al. 2003). The interaction between humic acid and PAH concentrated on one or two of the three phases was studied by Liang et al. (2008). Pilon-Smits (2005) reported PAH remediation by the plants. The limitation of this method is the stress faced by the plants due to non-availability of nutrients or climatic factors (Harvey et al. 2002).

18.6.5 Rhizoremediation

Rhizoremediation is type of phytoremediation with both plant roots and their associated rhizospheric microbes. This technique harbors large population of bacteria on plant roots and transports the root-colonizing, remediating microorganism to pollutants 10–15 m deep in the soil (Kingsley et al. 1994). This remediation works on the mechanism of the synergistic effect of plant roots and rhizospheric microbial communities like the secretion of organic acids followed by reduction in soil pH and production of siderophores, phytochelains, amino acids, and 1-aminocyclopropane-1-carboxylate (ACC) deaminase by plant growth-promoting rhizobacteria (PGPR), supply bacteria with cofactors required for the activation of bacterial enzymes involved in the pollutant degradation pathway (Anderson et al. 1993; Lugtenberg et al. 1997).

Bisht et al. (2010, 2014) studied rhizospheric and endophytic bacteria for rhizoremediation of PAH using *Populus* sp. The proper agronomic techniques improved the removal efficiency of rhizoremediation (Zhuang et al. 2007). Aprill and Sims (1990) concluded the effectiveness of prairie grasses in the removal of four PAHs in a sandy loam soil. Qiu et al. (1994) also reported grass-enhanced bioremediation for clay soils contaminated with polynuclear aromatic hydrocarbons.

18.6.6 Composting

The bioremediation of PAH-contaminated soil using composting technology was studied by Crawford et al. (1993), Joyce et al. (1998), Kirchmann and Ewnetu (1998), Jørgensen et al. (2000), Namkoong et al. (2002), and Antizar-Ladislao et al. (2006). The removal of anthracene in soil by compost addition was reported by Kastner et al. (1995) and Kastner et al. (1999). Qiu and McFarland (1991) studied composting of PAH-contaminated soil using *Phanerochaete chrysosporium. Phanerochaete chrysosporium* was used for removal of benzo(a)pyrene in soil composting systems by McFarland and Qiu (1995). Wellmann et al. (2001) concluded from his study that manure addition improved degradation of PAHs in soil by 81% in 41 days. Wong et al. (2002) and Atagana (2004) studied removal by co-composting of PAH-contaminated soil with pig manure and poultry manure.

18.7 Role of Microbial Diversity in Degradation of PAHs

Many microbial strains have been isolated and characterized for biodegradation of PAHs including bacteria, fungi, and microalgae. Biodegradation of polycyclic aromatic hydrocarbons was carried out either by bacteria (Arulazhagan and Vasudevan 2011; Mao et al. 2012; Sun et al. 2014; Cébron et al. 2015; Darmawan et al. 2015;

Ferreira et al. 2015; Okai et al. 2015), fungi (Wang et al. 2014; Li et al. 2005; Chan et al. 2006; Elisabet Aranda 2009; Hadibarata et al. 2009; Hadibarata and Kristanti 2014; Cébron et al. 2015; Mineki et al. 2015; Young et al. 2015), or algae (Chan et al. 2006; Diaz et al. 2014; Luo et al. 2014).

18.7.1 Biodegradation of PAHs by Bacteria

Table 18.3 lists the bacteria involved in biodegradation of major PAH (Cerniglia 1992, 1997; Kanaly and Harayama 2000). The biodegradation efficiency of some bacteria is given in Table 18.4. *Pseudomonas cepacia* was found to simultaneously

Table 18.3 Bacteria involved in biodegradation of PAHs

PAH	Bacterial species
Naphthalene	*Acinetobacter calcoaceticus*, *Alcaligenes denitrificans*, *Brevundimonas vesicularis*, *Burkholderia cepacia*, *Comamonas testosteroni*, *Corynebacterium renale*, *Cycloclasticus* sp., *Neptunomonas naphthovorans*, *Moraxella* sp., *Mycobacterium* sp., *Pseudomonas* sp., *P. fluorescens*, *P. marginalis*, *P. putida*, *P. saccharophila*, *P. stutzeri*, *Sphingomonas paucimobilis*, *Streptomyces* sp., *Rhodococcus* sp.
Acenaphthene	*Alcaligenes eutrophus*, *Alcaligenes paradoxus*, *Beijerinckia* sp., *Burkholderia cepacia*, *Cycloclasticus* sp., *Neptunomonas naphthovorans*, *Pseudomonas* sp., *P. fluorescens*, *P. putida*
Anthracene	*Acinetobacter calcoaceticus*, *Arthrobacter* sp., *Beijerinckia* sp., *Burkholderia cepacia*, *Comamonas testosteroni*, *Cycloclasticus pugetii*, *Cycloclasticus* sp., *Flavobacterium* sp., *Gordonia* sp., *Mycobacterium* sp., *P. fluorescens*, *P. marginalis*, *P. putida*, *Rhodococcus* sp., *Sphingomonas* sp., *S. paucimobilis*, *S. yanoikuyae*
Pyrene	*Acinetobacter calcoaceticus*, *Alcaligenes denitrificans*, *Burkholderia cepacia*, *Flavobacterium* sp., *Gordonia* sp., *Mycobacterium* sp., *P. putida*, *P. saccharophila*, *Rhodococcus* sp., *Sphingomonas* sp., *Stenotrophomonas maltophilia*
Benzo[a]anthracene	*Agrobacterium* sp., *Alcaligenes denitrificans*, *Bacillus* sp., *Beijerinckia* sp., *Burkholderia cepacia*, *Burkholderia* sp., *Pseudomonas* sp., *P. putida*, *P. saccharophila*, *Sphingomonas* sp., sp. *paucimobilis*, sp. *yanoikuyae*, *Stenotrophomonas maltophilia*
Chrysene	*Acinetobacter calcoaceticus*, *Agrobacterium* sp., *Bacillus* sp., *Burkholderia* sp., *Pseudomonas* sp., *P. marginalis*, *P. saccharophila*, *Rhodococcus* sp., *Sphingomonas* sp., *S. paucimobilis*, *Stenotrophomonas maltophilia*
Benzo[b]fluoranthene	*Alcaligenes denitrificans*, sp. *paucimobilis*
Benzo[a]pyrene	*Beijerinckia* sp., *Mycobacterium* sp., *Pseudomonas NCIB 9816*, *Sphingomonas paucimobilis*
Dibenzo[a,h]anthracene	*S. paucimobilis*, *Stenotrophomonas maltophilia*

Adapted from Cerniglia and Gibson (1979), Cerniglia (1992, 1997), Kanaly and Harayama (2000), and Antizar-Ladislao et al. (2004)

Table 18.4 Bacterial efficiency in biodegradation of PAHs

PAHs	Microorganism	Biodegradation (%)	References
Naphthalene	*Bacillus* sp.	86	Kafilzadeh et al. (2011)
	Pseudomonas sp.	80	
	Corynebacterium sp.	77	
	Staphylococcus sp.	69	
	Micrococcus sp.	58	
	Bacillus sp.	33	Kafilzadeh et al. (2012)
	Pseudomonas sp.	42	Ahirwar and Dehariya (2013)
	Corynebacterium sp. and *Pseudomonas* sp.	100	Kafilzadeh and Pour (2012)
Phenanthrene	*Pseudomonas* sp. ARP26	93	Coral and Karagoz (2005)
	Pseudomonas sp. ARP28	98	
	Sphingomonas sp. GY2B	99.8	Tao et al. (2007)
	Mycobacterium sp. BB1	82	Boldrin et al. (1993)
	Pseudomonas stutzeri ZP2	96	Zhao et al. (2009)
	Corynebacterium sp.	91.5	Kafilzadeh and Pour (2012)
	Pseudomonas sp.	83.2	
Pyrene	*Pseudomonas putida PL2*	50	Ping et al. (2011)
	Pseudomonas putida MUB1	88.2	Miya and Firestone (2000)
	Bacillus sp. *PK-12*	64	Khanna et al. (2012)
	Bacillus sp. *PK-13*	55	
	Bacillus sp. *PK-14*	53	
	Mycobacterium sp.	48	Heitkamp et al. (1988a, b)
		60	
	Mycobacterium sp.	89.1	Kafilzadeh and Pour (2012)
Fluoranthene	*Mycobacterium* sp. *KR20*	60	Rehmann et al. (2001)
	Pseudomonas alcaligenes PA-10	—	Gordon and Dobson (2001)
	Acinetobacter sp. *PSM11*	61	Kumar et al. (2011)
	Bacterial strains PSM10	48	
	Bacterial strains PSM6	42	
	Bacterial strains PSM7	41	

degrade a mixture of three-, four-, five-, and seven-benzene-ring PAHs when inoculated at high cell densities, and it was observed that the degradation of five- and seven-benzene-ring PAHs improved when low molecular weight PAHs were present (Juhasz and Naidu 2000; Silva et al. 2014).

18.7.2 Biodegradation of PAHs by Fungi

Ligninolytic fungi too have the property of PAH degradation. *Phanerochaete chrysosporium*, *Bjerkandera adusta*, and *Pleurotus ostreatus* are the common PAH-degrading fungi. The fungi produce hydroxylated intermediates that help in mineralization of PAH-polluted sites in comparison to bacteria (Cerniglia 1997). The metabolites produced by fungi are with higher water solubility and enhanced chemical reactivity. The fungi belonging to genera *Agrocybe*, *Bjerkandera*, *Coriolopsis*, *Crinipellis*, *Flammulina*, *Kuehneromyces*, *Laetiporus*, *Marasmiellus*, *Nematoloma*, *Phanerochaete*, *Pleurotus*, *Ramaria*, *Rhizoctonia*, *Rhodotorula*, *Trametes*, *Trichosporon*, *Aspergillus*, *Candida*, *Chrysosporium*, *Fusarium*, *Neurospora*, *Penicillium*, *Saccharomyces*, *Trichoderma*, *Cunninghamella*, *Mortierella*, *Mucor*, and *Syncephalastrum* are related with PAH bioremediation (Alexopoulos et al. 1996; James and O'Donnell 2004).

Phanerochaete chrysosporium has metabolized a wide variety of PAHs under ligninolytic and nonligninolytic conditions and converted them to CO_2 (Bumpus et al. 1985; Bumpus 1989; Haemmerli et al. 1986; Sanglard et al. 1986). The detailed list of fungal species involved in degradation of PAHs is presented in Table 18.5. Table 18.6 shows the efficiency of PAH biodegradation by fungi.

18.7.3 Biodegradation of PAHs by Microalgae

Different researchers (Cerniglia et al. 1979, 1980a, b; Narro et al. 1992a, b; Safonova et al. 2005; Chan et al. 2006; El-Sheekh et al. 2012) reported the degradation ability of microalgae for PAH degradation. The free cells of *Selenastrum capricornutum* were also found to have a higher ability to degrade PAHs (Lei et al. 2000, 2007). The PAH removal by microalgal species is mainly due to passive physicochemical biosorption. Tam et al. (2010) reported *Selenastrum capricornutum* immobilized in alginate beads able to remove 1 mg l-1 phenanthrene, 0.25 mg l^{-1} fluoranthene and 0.1 mg l^{-1} pyrene. They found the immobilization enhanced the biodegradation potential of microalgae.

18.8 Biosurfactants in Degradation of PAHs

The surfactants help to overcome the non-availability of PAH due to their insolubility in water. The solubilization and degradation improved with the surfactant addition. The biosurfactants are preferable to chemical surfactants because of its biodegradability, compatibility with the environment, low toxicity, high selectivity,

Table 18.5 Fungal species involved in biodegradation of PAHs

PAHs	Fungal species
Naphthalene	*Absidia glauca*, *Aspergillus niger*, *Basidiobolus ranarum*, *Candida utilis*, *Choanephora campincta*, *Circinella* sp., *Claviceps paspali*, *Cokeromyces poitrassi*, *Conidiobolus gonimodes*, *C. bainieri*, *C. elegans*, *C. japonica*, *Emericellopsis* sp., *Epicoccum nigrum*, *Gilbertella persicaria*, *Gliocladium* sp., *Helicostylum piriforme*, *Hyphochytrium catenoides*, *Linderina pennispora*, *Mucor hiemalis*, *Neurospora crassa*, *Panaeolus cambodginensis*, *Panaeolus subbalteatus*, *Penicillium chrysogenum*, *Pestalotia* sp., *Phlyctochytrium reinboldtae*, *Phycomyces blakesleeanus*, *Phytophthora cinnamomi*, *Psilocybe cubensis*, *Psilocybe strictipes*, *Psilocybe stuntzii*, *Psilocybe subaeruginascens*, *Rhizophlyctis harderi*, *Rhizophlyctis rosea*, *Rhizopus oryzae*, *Rhizopus stolonifer*, *S. cerevisiae*, *Saprolegnia parasitica*, *Smittium culicis*, *Smittium culisetae*, *Smittium simulii*, *Sordaria micola*, *Syncephalastrum racemosum*, *Thamnidium anomalum*, *Zygorhynchus moelleri*
Acenaphthene	*Cunninghamella elegans*, *T. versicolor*
Fluorene	*Cunninghamella elegans*, *Phanerochaete chrysosporium*, *Pleurotus ostreatus*
Anthracene	*Bjerkandera* sp., *Bjerkandera adusta*, *Ceriporiopsis subvermispora*, *Cladosporium herbarum*, *Coriolopsis polyzona*, *Curvularia lunata*, *Curvularia tuberculata*, *Cryphonectria parasitica*, *Cylindrocladium simplex*, *C. elegans*, *Daedalea quercina*, *Drechslera spicifera*, *Flammulina velutipes*, *Fusarium moniliforme*, *Laetiporus sulphureus*, *Marasmiellus* sp., *Monosporium olivaceum*, *Oxysporus* sp., *Penicillium* sp., *Pleurotus ostreatus*, *P. chrysosporium*, *P. laevis*, *Ramaria* sp., *Rhizopus arrhizus*, *R. solani*, *Trametes versicolor*, *Verticillium lecanii*
Pyrene	*Agrocybe aegerita*, *Bjerkandera adusta*, *C. elegans*, *Dichomitus squalens*, *Flammulina velutipes*, *Kuehneromyces mutabilis*, *Laetiporus sulphureus*, *Phanerochaete chrysosporium*, *Penicillium* sp., *P. janthinellum*, *P. glabrum*, *Pleurotus ostreatus*, *Pleurotus* sp., *Syncephalastrum racemosum*, *Trametes versicolor*
Benzo[a] anthracene	*C. elegans*, *P. laevis*, *P. janthinellum*, *Trametes versicolor*
Chrysene	*Penicillium* sp., *P. janthinellum*, *Syncephalastrum racemosum*
Benzo[a] pyrene	*Chrysosporium pannorum*, *Cunninghamella elegans*, *Phanerochaete chrysosporium*, *Stropharia coronilla*
Dibenzo[a,h] Anthracene	*Trametes versicolor*, *P. janthinellum*

Adapted from Cerniglia and Gibson (1979), Pothuluri et al. (1990, 1993, 1995), Cerniglia (1992, 1997). Kanaly and Harayama (2000), and Antizar-Ladislao et al. (2004)

and their activity even in extreme temperature, pH, and salinity conditions (Banat et al. 2010). The increase in PAH degradation by addition of surfactants produced by bacteria and fungi was reported by many researchers (Wong et al. 2004; Jacques et al. 2005; Batista et al. 2010; Gusmão et al. 2010; Luna et al. 2011; Sobrinho et al. 2013).

The mechanisms of biodegradation by biosurfactants is either due to increase of availability of PAH to the microbial activity by reduction of surface tension of the medium around the bacterium and interfacial tension reduction between the bacterial

Table 18.6 Efficiency of some fungi in biodegradation of PAHs

PAH	Fungal species	Biodegradation (%)	References
Pyrene	*T. versicolor*	34.1	Sack et al. (1997)
Anthracene	*Bjerkandera* sp. Bos 55	99.2	Field et al. (1992)
Benzo(a)pyrene	*Bjerkandera* sp. Bos 55	83.1	
Anthracene	*Cryphonectria parasitica*	96	Krivobok et al. (1998)
	Ceriporiopsis subvermispora	88	
	Oxysporus sp.	94	
	Cladosporium herbarum	85	
	Rhizopus arrhizus	95	
	P. chrysosporium	40	Jové et al. (2016)
	Irpex lacteus	38	
	P. ostreatus	30	
Fluoranthene	*Ceriporiopsis subvermispora*	8 7	Salicis et al. (1999)
	Ceriporiopsis subvermispora	71	
	Oxysporus sp.	79	
	Aspergillus terreus	89	
	Cicinobolus cesatii	86	
	Cryptococcus albidus	93	
	Cunninghamella elegans	98	

cell wall and hydrocarbon molecule or promoting modifications in the membrane, facilitating hydrocarbon adherence (hydrophobicity increase) and reducing the lipopolysaccharide index of the cell wall without damaging the membrane (Pirôllo et al. 2008; Aparna et al. 2011). The monomers of the biosurfactants form a micelle with hydrophobic part at the center and hydrophilic part on the surface, thereby increasing exposure of hydrocarbon to microbes (Soberón-Chávez and Maier 2010; Tondo et al. 2010). The membrane modifications like change in composition of proteins or increase in cell wall hydrophobicity improve microbial cell accessibility and hydrocarbon dispersion by encapsulation in micelles, spherical or irregular vesicles, and lamellar structures (Aparna et al. 2012).

Grotenhuis et al. (1999) reported increase in PAH degradation by *Bjerkandera* by surfactant addition. Sobisch et al. (2000) reported the surfactant T1 and T5 increased the solubility and degradation of PAHs. Wong et al. (2004) detailed improvement in phenanthrene biodegradation by surfactants produced by *Pseudomonas aeruginosa.* Jacques et al. (2005) investigated enhancement of degradation of anthracene by surfactant from *P. citronellolis.* The nine bacterial groups belonging to genera *Acinetobacter*, *Arcobacter*, *Bacillus*, *Halanaerobium*, *Leuconostoc*, *Marinobacter*, *Streptomyces*, *Propionibacterium*, and *Streptococcus* are found in oil samples from oil reservoirs (Sette et al. 2007).

18.9 Enzymatic Biodegradation of PAHs

Enzymes involved in the degradation of PAHs are oxygenase, dehydrogenase, and ligninolytic enzymes. The ligninolytic system contains the principal enzyme groups, i.e., lignin peroxidase (LiP), Mn-dependent peroxidase (MnP), phenol oxidase (laccase, tyrosinase), and H_2O_2 producing enzymes (Novotný et al. 2004). The other fungal enzymes with PAH-degrading ability are cytochrome P450 monooxygenase, epoxide hydrolases, lipases, proteases, and dioxygenases (Bezalel et al. 1997; Balaji et al. 2014).

Penicillium chrysogenum produce both lipases and laccases for degrading PAHs. *P. chrysogenum*, *Aspergillus fumigates*, *Mucor racemose*, and *Rhizopus stolonifer* produce laccases during PAH degradation (Balaji et al. 2014). *Lasiodiplodia theobromae* is also capable of producing lipase (Thiyagarajan et al. 2008; Venkatesagowda et al. 2012; Balaji et al. 2014). The type of fungal strain directly linked to the enzyme efficiency of PAH degradation. The white rot fungi with high ligninolytic potential such as *P. chrysosporium*, *P. ostreatus*, and *Bjerkandera adusta* are reported to have high potential to metabolize PAHs (Haritash and Kaushik 2009).

Fungal ligninolytic enzymes involved in PAH degradation are lignin peroxidase, laccase, and manganese peroxidase which are extracellular and catalyze radical formation by oxidation to destabilize bonds in a molecule (Ball et al. 1995; Hofrichhter et al. 1999). The optimum temperature for laccase activity is 45 °C, and for Mn-dependent peroxidase activity is 75–80 °C (Farnet et al. 2000). *Pleurotus* laccase produces hydroxyl radicals in PAH degradation pathway, while Mn-dependent peroxidase of *Nematoloma frowardii* degrades a broad spectrum of PAHs directly to carbon dioxide and polar fission products (Hofrichter et al. 1998; Guillen et al. 2000).

The naphthalene was degraded more efficiently by Toluene *ortho*-monooxygenase (TOM) by hydroxylation (Canada et al. 2002). The initial oxidation of polycyclic aromatic hydrocarbon occurs by dioxygenase (Fig. 18.3) enzyme. Naphthalene is initially oxygenated by naphthalene dioxygenase to produce1,2-dihydroxynaphthalene. It is further converted to salicylate. It is further metabolized through catechol or gentisic acid. Phenanthrene is oxidized by phenanthrene dioxygenase to produce 3,4 dihydroxyphenanthrene. This follows two metabolic pathways. One is the formation of salicylate, and the other one forms protocatechuic acid. Anthracene by the dioxygenase produces 1,2 dihydroxyanthracene which also metabolized to salicylate and protocatechuic acid. The pyrene is dioxygenated and forms phenanthroic acid and further follows the similar metabolism.

The extracellular peroxidases of fungi are also responsible for the initial oxidation of PAHs (Acevedo et al. 2011; Betts 2012; Li et al. 2014; Zhang et al. 2015). Fungal lignin peroxidases oxidize a number of PAHs directly, while fungal manganese peroxidases co-oxidize them indirectly through enzyme-mediated lignin peroxidation. Vyas et al. (1994) have tested several white rot fungi and suggested that all of them oxidize anthracene to anthraquinone.

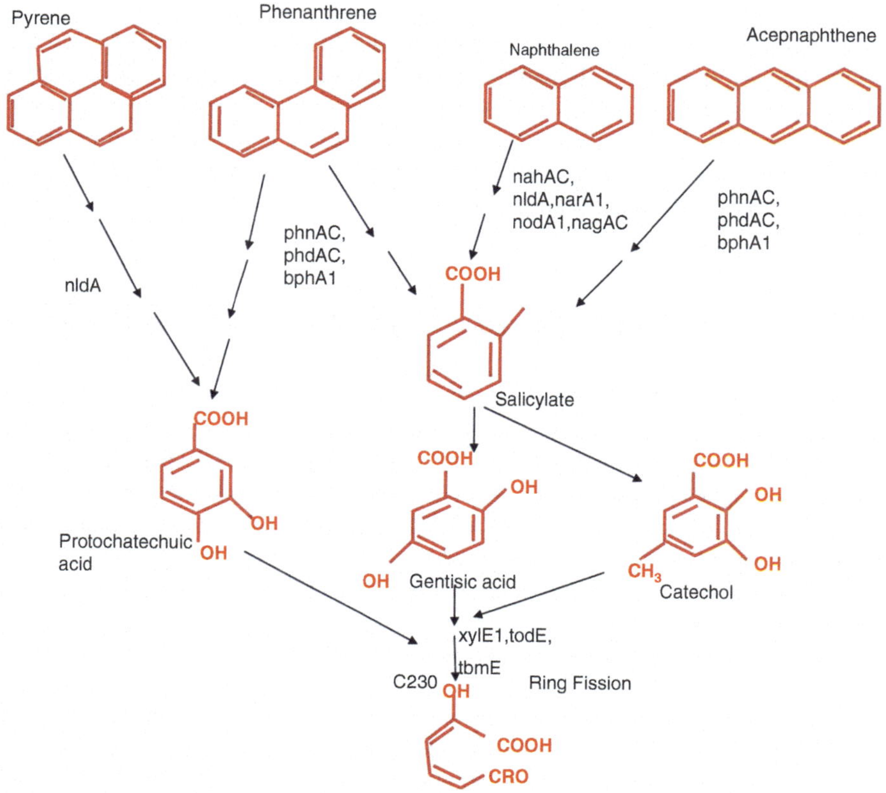

Fig. 18.3 Schematic presentation of genes involved in the microbial degradation of PAHs. (Adapted from Williams and Sayers 1994 and Habe and Omori 2003)

18.10 Molecular Techniques in Bioremediation of PAHs

18.10.1 Role of Genetically Engineered Microorganisms in PAHs Biodegradation

The genetic modification of the bacteria is done for developing metabolic pathways for PAH degradation, and Sayler et al. (1999), Sayler and Ripp (2000) reported field application of genetically engineered microorganisms for polycyclic aromatic hydrocarbon bioremediation process monitoring and control. Generally genes from native bacterial strains are used for engineering potential microbes. *Pseudomonas fluorescens* (BC-45) isolated from PAH-contaminated soil was genetically engineered with lux gene from *Vibrio fischeri*. This genetically engineered bacterium is capable of emitting light in the presence of PAHs (Gupta 2008). Several *Pseudomonas* species are engineered for oil spill degradation. Ananda Chakrabarty developed the first genetically engineered oil-eating bacteria belonging to *Pseudomonas* sp. to remove oil spills. Schell (1983) cloned *E. coli* with naphthalene-degrading genes from plasmid NAH7. Ripps et al. (2000) reported GEM *Pseudomonas fluorescens* HK44 for polyaromatic hydrocarbon (PAH) bioremediation process. Khan et al. (2001) cloned the genes encoding polycyclic aromatic ring dioxygenase from *Mycobacterium* sp. strain PYR-1. Isiodu et al. (2016) studied the role of plasmid-borne genes in the biodegradation of polycyclic aromatic hydrocarbons (PAHs) by a consortium of aerobic heterotrophic bacteria.

18.10.2 Role of Protein Engineering in Degradation of PAHs

The low water solubility makes high molecular weight PAHs less bioavailable, and the lack of naturally occurring microbial enzyme systems resists the biodegradation in the environment (Heitkamp et al. 1988a, b; Cerniglia 1992; Schneider et al. 1996). The engineering of proteins with its selectivity and specificity helps in biocatalysis of bioremediation. Hydroxylation and epoxidation reactions are catalyzed by monooxygenases. The protein engineering helps the oxygenases to overcome its limitations such as complexity, poor stability, low expression, or low activity on unnatural substrates.

The initial oxidation of polycyclic aromatic hydrocarbons by P450 monooxygenases and non-heme iron dioxygenases acts as the biodegradation rate-limiting step in microbes (Herbes and Schwall 1978; Heitkamp et al. 1988a, b; Cerniglia 1992). The role of P450 enzyme for oxidation of pyrene to 1,6- and 1,8-pyrenequinone was reported in the detoxification pathway of pyrene in the fungus *Cunninghamella elegans* (Cerniglia et al. 1986). The hydrophobic substitutions at Y96 mutant broaden the $P450_{cam}$ substrate range by increasing active site hydrophobicity and should promote the binding of organic molecules and also reduce uncoupling by disfavoring the entry of water molecules into the active site during the catalytic cycle (Stevenson et al. 1996; England et al. 1998). The increase in PAH oxidation by BM-3 mutant was reported with application of site-directed mutagenesis at residues in the active site and at the entrance of the substrate access channel (Carmichael and Wong 2001). Harford-Cross et al. (2000) investigated the activity of cytochrome $P450_{cam}$ (CYP101) from *Pseudomonas putida* for the oxidation of the phenanthrene, fluoranthene, pyrene, and benzo[a]pyrene with mutations of the active site residues F87 and Y96. The NADH oxidation rates of the mutants with fluoranthene and phenanthrene were 144 min^{-1} and 374 min^{-1}, respectively. Li et al. (2001) engineered cytochrome P450 BM-3 from *Bacillus megaterium* for oxidation of polycyclic aromatic hydrocarbons. The improved activities and coupling efficiencies of this mutant that were reported for naphthalene, fluorene, acenaphthene,

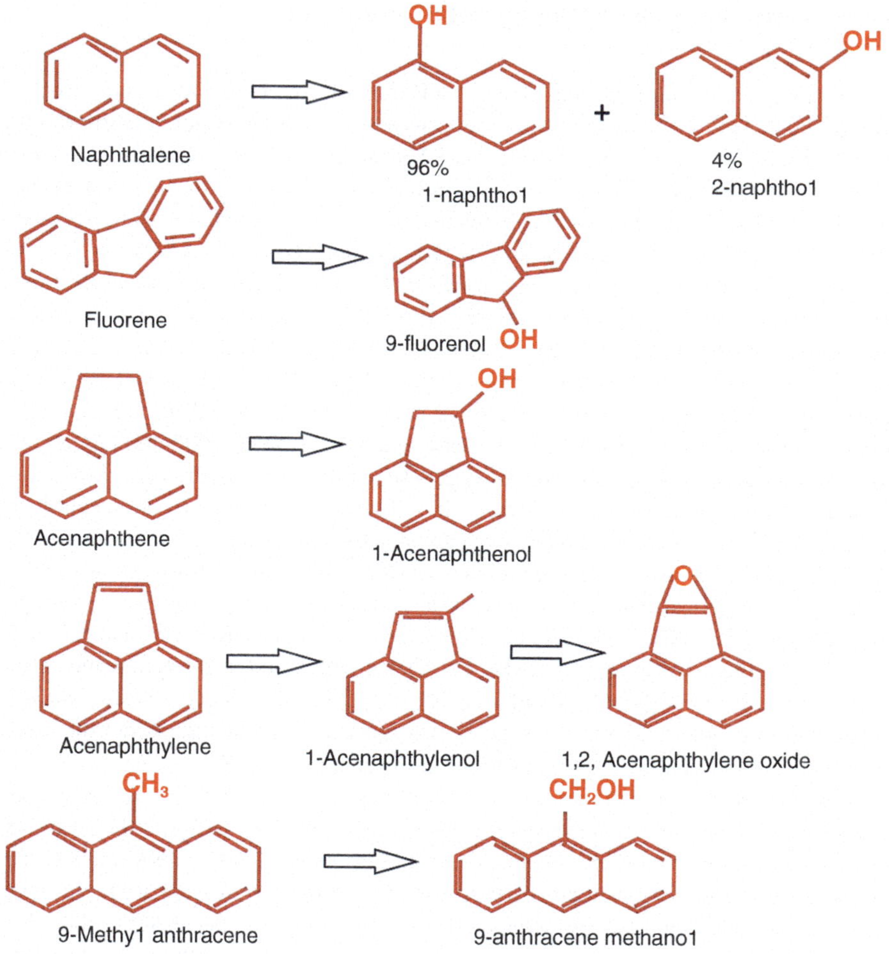

Fig. 18.4 Proposed reaction catalyzed by P450 BM-3 mutants. (Adapted from Li et al. 2001)

acenaphthylene, and 9-methylanthracene were 160, 53, 109, 287, and 22/min, respectively, and 11, 26, 5.4, 15, and 3.2%, respectively.

Wong et al. (2004) reported the oxidation of pyrene, phenanthrene, and fluoranthene by mutants of P450 BM-3. The regioselectivity and enantioselectivity of naphthalene dioxygenase were influenced by multiple mutations (Yu et al. 2001). The proposed reaction catalyzed by P450 BM-3 mutants as per Li et al. (2001) is given in Fig. 18.4.

18.11 Mechanism of Biodegradation of PAHs

The first step in the aerobic catabolism of a PAH molecule by bacteria is the oxidation of the PAH to a dihydrodiol by a multicomponent enzyme system incorporating both atoms of molecular oxygen into the PAH nucleus (Gibson et al. 1975). After oxidation, the dehydrated intermediates followed either an *ortho*- or a *meta*-cleavage type of pathway, leading to intermediates that are further converted to intermediates (Fig. 18.5) of the tricarboxylic acid cycle (Kanaly and Harayama 2000).

The ligninolytic fungi can co-metabolize PAHs to form *trans*-dihydrodiols, phenols, quinines, and dihydrodiol epoxides (Fig. 18.6). The enzymes such as lignin peroxidases, manganese peroxidases, and laccases help in the degradation of PAHs (Harayama 1997). The schematic presentation of initial microbial attack and ring cleavage and genes involved in the respective reactions for PAHs are given in Fig. 18.3 (Williams and Sayers 1994; Habe and Omori 2003). Naphthalene is initially oxygenated by a naphthalene dioxygenase to yield 1,2-dihydroxynaphthalene, which is further degraded to salicylate. It is further metabolized via catechol or gentisic acid. Phenanthrene is mostly oxygenated by a phenanthrene dioxygenase to yield 3,4-dihydroxyphenanthrene,which is further metabolized through two pathways: following naphthalene biodegradation pathway with the formation of salicylate or leading to protocatechuic acid. Anthracene is dioxygenated at the 1,2 position to yield 1,2-dihydroxyanthracene,which is later metabolized to salicylate and catechol. Pyrene is mostly dioxygenated at the 1,2 position. A successive derivative, 4-phenanthroic acid, undergoes a second dioxygenase reaction and further metabolized via catabolic pathways similar to those of phenanthrene.

The anaerobic degradation process of naphthalene and phenanthrene (Young and Phelps 2005; Chakraborty and Caotes 2004) is given in Fig. 18.7, and the addition of carbon dioxide produces benzoic acid and subsequent transformation to benzoyl-CoA. For naphthalene, carbon dioxide addition yields 2-naphthoic acid which is further degraded by sequential reduction steps to decalin-2-carboxylic acid. The phenanthrene is also directly carboxylated to phenanthrene carboxylic acid following mineralization.

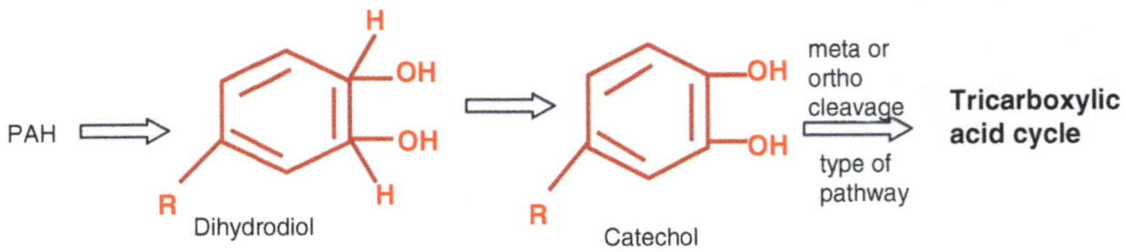

Fig. 18.5 Pathways of PAH degradation by bacteria. (Adapted from Bamforth and Singleton 2005 and Cerniglia 1992)

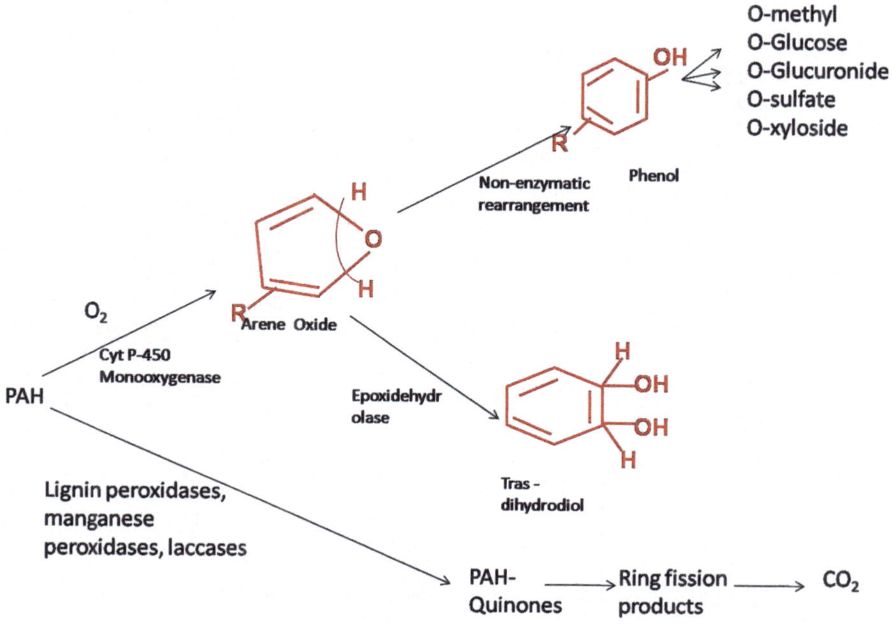

Fig. 18.6 Pathways for the fungal metabolism of PAHs. (After Cerniglia 1997)

18.12 Metabolites of Degradation of PAHs

The metabolites of acenaphthene degradation by *Cunninghamella elegans* are 1-acenaphthenone, 1,2-acenaphthenedione, cis-1,2-dihydroxyacenaphthene, trans-1,2-dihydroxyacenaphthene, 1,5-dihydraxyacenaphthene, and 6-hydroxyacenaphthenon (Pothuluri et al. 1992a, b). Anthracene was metabolized to anthracene trans-1,2-dihydrodiol 1-anthrol, 9,10-anthraquinone, phthalate, glucuronide, sulfate, and xyloside conjugates of hydroxylated intermediates by microbial strains such as *Bjerkandera* sp., *Cunninghamella elegans*, *Nematoloma frowardii*, *Phanerochaete chrysosporium*, *Phanerochaete laevis*, *Pleurotus ostreatus*, *Pleurotus sajor-caju*, *Ramaria* sp., *Rhizoctonia solani*, and *Trametes versicolor* (Bezalel et al. 1996a, b, c; Bogan and Lamar 1996; Cerniglia and Yang 1984; Hammel et al. 1992; Johannes and Majcherczyk 2000; Kotterman et al. 1998; Sac and Gunther 1993). The metabolites of phenanthrene produced by *Aspergillus niger*, *Cunninghamella elegans*, *Nematoloma frowardii*, *Phanerochaete chrysosporium*, *Phanerochaete laevis*, *Pleurotus ostreatus*, *Syncephalastrum racemosum*, and *Trametes versicolor* are phenanthrene trans-1,2-dihydrodiol; phenanthrene trans-3,4-dihydrodiol; phenanthrene trans-9,10-dihydrodiol; glucoside conjugate of 1-phenanthrol; 1-, 2-, 3-, 4-, and 9- phenanthrol; 1-methoxyphenanthrene; phenanthrene-9,10-quinone; and 2,2-diphenic acid (Bezalel et al. 1996a, b, c; Bogan and Lamar 1996; Bumpus 1989; Cerniglia 1997; Hammel et al. 1992; Kotterman et al. 1998; Sac and Gunther 1993).

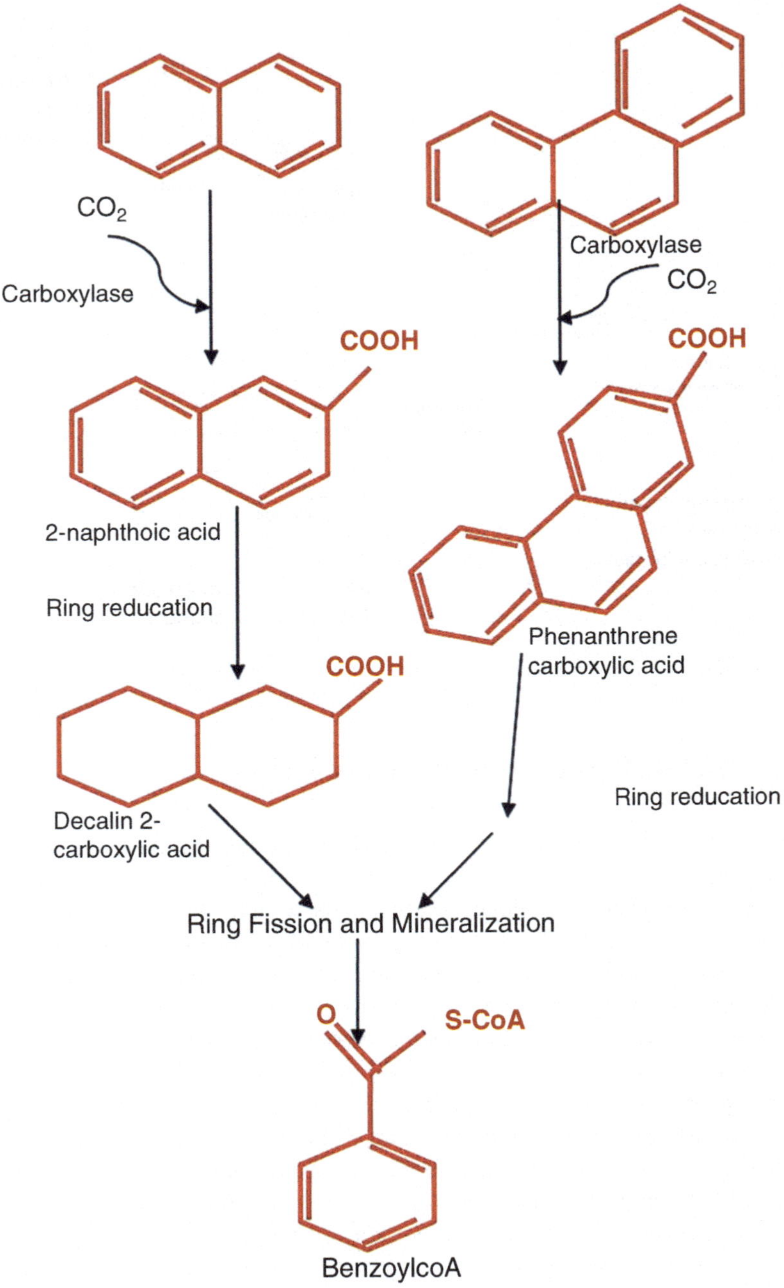

Fig. 18.7 Schematic representation for the anaerobic degradation of naphthalene and phenanthrene. (Adapted from Young and Phelps 2005 and Chakraborty and Caotes 2004)

Cunninghamella elegans, *Laetiporus sulphureus*, *Phanerochaete chrysosporium*, *Pleurotus ostreatus*, and *Trametes versicolor* produce 9-fluorenone, 9-fluorenol, and 1-hydroxy-9 fluorenone as the metabolites of fluorine (Bezalel et al. 1996a, b, c; Bogan et al. 1996a, b; Bogan and Lamar 1996; Sac and Gunther 1993). Fluoranthene metabolites produced by *Cunninghamella elegans*, *Nematoloma frowardii*, *Laetiporus sulphureus*, *Penicillium* sp., and *Pleurotus ostreatus* were fluoranthene trans-2,3- dihydrodiol, 8- and 9-hydroxyfluoranthene trans-2,3-dihydrodiols, and glucoside conjugates of hydroxylated intermediates (Pothuluri et al. 1990; Sac and Gunther 1993). *Aspergillus niger*, *Agrocybe aegerita*, *Candida parapsilosis*, *Crinipellis maxima*, *Crinipellis perniciosa*, *Crinipellis stipitaria*, and *Crinipellis zonata* produced 1,6-pyrenequinone, 1,8-pyrenequinone, glucoside conjugates, and 1-pyrenol as metabolites of pyrene (Bezalel et al. 1996a, b, c; Hammel et al. 1986; Manilla-Perez et al. 2011)

18.13 Factors Affecting Biodegradation

The environmental factors such as soil pH, nutrient availability, and the bioavailability of the contaminant can influence the process of bioremediation by inhibiting growth of the pollutant-degrading microorganisms. The microbial population and type of the microorganisms, degree of acclimation, accessibility of nutrients, chemical structure of the compound, cellular transport properties, and chemical partitioning in growth medium affect the biodegradation ability of the microbes. The biological degradation in contaminated soil and water ecosystem is controlled by factors like carbon source, nutrients, temperature, pH, redox potential, and oxygen content that are needed for microbial growth (Adams et al. 2015; Ghosal et al. 2016).

18.13.1 Temperature

The solubility of PAHs increases with an increase in temperature (Lim et al. 1999) that increases the bioavailability of the PAH molecules. In addition, oxygen solubility decreases with increasing temperature, which will reduce the metabolic activity of aerobic microorganisms.

The enzymatic activity is directly related with the temperature. With increasing temperatures, the solubility of PAHs increases, which increases their bioavailability. In addition, increasing temperatures also decrease the solubility of oxygen, thereby reducing the metabolic activity of aerobic microorganisms (Bamforth and Singleton 2005). The rate of hydrocarbon degradation at different temperatures was studied by Beaudin et al. (1999) and Chung et al. (2000). The temperature as an indicator of microbial activity in PAH degradation was studied by Amir et al. (2005) in composting of lagooning sewage sludge. The high PAH degradation was observed at the stabilization phase of composting following presence of thermophilic communities.

Biodegradation of PAHs can occur over a wide temperature range, but mesophilic temperatures had higher efficiency of transformations than at very low or high temperatures. The microorganisms adapted to metabolize PAHs at extreme temperatures, for example, naphthalene and phenanthrene degradation, were also reported from crude oil in seawater at temperatures as low as 0 °C (Simon et al. 1993). In comparison, the laccase and manganese peroxidase enzymes of ligninolytic fungi were reported to have a temperature optimum of ~50 °C and >75 °C, respectively, in spent mushroom compost during the degradation of PAHs (Lau et al. 2003) with over 90% degradation of the contaminating PAHs occurring at the set temperature.

18.13.2 pH

The pH of the environment plays a crucial role in bioremediation process. The retired gaswork sites result in leaching of demolition wastes that increases the pH that in turn makes the environment unfavorable for the PAH-degrading microbes. Similarly oxidation and leaching of coal spoil with sulfide oxidation result in acidic conditions. The pH of the area controls microbial biotransforming ability of PAHs (Alexander 1995). The phenanthrene removal was 80% at neutral pH, and only 40% removal was reported at pH 5.5 after16 days of incubation of *Burkholderia cocovenenans* (Wong et al. 2002).

The degradation of phenanthrene and anthracene by *Sphingomonas paucimobilis* was inhibited at pH 5.2 (Kastner et al. 1998). Kim et al. (2005) studied effects of pH on the degradation of phenanthrene and pyrene by *Mycobacterium vanbaalenii* PYR1.The degradation of naphthalene by acidophilic native strains at coal spoil with pH 2 was 50% and 10–20% for phenanthrene and anthracene.

18.13.3 Oxygen

The metabolism of PAH-degrading microbes can occur in aerobic and anaerobic conditions. The initial oxidation of aromatic ring requires oxygen under aerobic degradation (Gibson et al. 1968). Genthner et al. (1997) found the PAH degradation was limited under denitrifying, sulfate-reducing, and methanogenic conditions. McNally et al. (1999) reported that anaerobic degradation under denitrifying conditions was comparable to aerobic degradation. The aeration and moisture content are directly linked to PAH bioremediation (Vinas et al. 2005). It enhanced heterotrophic and PAH-degrading microbial populations resulting in the highest rate of PAH biodegradation.

18.13.4 PAH Concentration

The rate of change in contaminant concentration is proportional to the contaminant concentration in the soil, and the time prediction tool in degradation depends on the microorganism, the contaminant (Cutright 1995). *Trichoderma asperellum* degraded 74% of phenanthrene, 63% of pyrene, and 81% of benzo[a]pyrene after 14 days of incubation at concentration of 1000 mg/kg (Zafra et al. 2015a).

Lee et al. (2014) reported *Peniophora incarnata* KUC8836 was able to degrade up to 95.3% of phenanthrene and 97.9% of pyrene after 2 weeks of incubation. The concentration of contaminant had a selective pressure on hydrocarbon-degrading organisms, and higher PAHs are growth-limiting for microorganisms which developed a response against PAHs regarding cell membrane structure, mycelia pigmentation, and sporulation alterations (Zafra et al. (2015b).

18.13.5 Nutrients

Mineki et al. (2015) investigated the degradation of PAHs with *Trichoderma/Hypocrea* genus using pyrene as sole source of carbon, and pyrene-degrading activity was enhanced to 24–25% in 14 days by adding 0.02% yeast extract, 0.1% sucrose, or 0.1% lactose. Carbon and nitrogen are essential for enzyme activity and thereby PAH degradation (Hofrichter et al. 1998).

The microbial population, number and type of the microorganisms, degree of acclimation, accessibility of nutrients, chemical structure of the compound, cellular transport properties, and chemical partitioning in growth medium also affect the biodegradation of PAHs.

18.14 Commercial Applications of Bioremediation in Petroleum Industry Waste Management

In the bioremediation of contaminated sites, in situ, on-site, and bioreactor techniques are commonly practiced techniques for PAH removal. Luna et al. (2013) reported *Candida sphaerica* UCP0995 for application in the petroleum industry.

Straube et al. (2003) reported the use of *Pseudomonas aeruginosa* strain 64 for a pilot-scale land farming treatment of PAH-contaminated soil from a wood treatment facility. Along with bioaugmentation, they have also undertaken biostimulation of the soil with water, ground rice hulls as a bulking agent, and palletized dried blood as a nitrogen source. The total PAH reduction was ~86% along with substantial reduction in high molecular weight PAHs in 1 year.

Lundstedt et al. (2003) found higher reduction in low molecular weight PAHs during treatment of an aged gaswork soil contaminated with PAHs. Guerin (2000) found significant PAH removal from the soil at a tar-contaminated site is remediated by composting along with conventional land treatment process.

Bewley and Webb (2001) reported bioaugmentation and biostimulation processes for in situ bioremediation of an aquifer contaminated with PAHs with nutrients (a commercial mixture of urea and diammonium phosphate), a commercially available phenol-degrading mixed bacterial inoculum (PHENOBAC, Microbac Ltd., Durham), and sodium nitrate (an oxygen source). The PAH reduction obtained was 0.9 $\mu g\ L^{-1}$ from 11 $\mu g\ L^{-1}$ after 2.5 years of treatment. Sasek et al. (2003) also reported the remediation of a manufactured gas plant soil contaminated with PAHs via composting in a thermally insulated chamber using mushroom compost containing wheat straw, chicken manure, and gypsum. The PAH removal obtained after 100 days was ~37 to 80% of individual PAHs; degradation was observed after 100 days of composting. Zein et al. (2006) reported more than 99% PAH removal after ex situ aerobic biotreatment of groundwater contaminated with PAHs, gasoline hydrocarbons, and methyl tert-butyl ether.

18.15 Future Perspectives

The newly developed strategies for remediation of PAH by microbes include increase in PAH bioavailability by development of efficient biosurfactant production and the use of bio-emulsifiers and bioluminescence-based biosensors to detect and monitor the PAH in the environment, and improvement in genetic engineering to overcome the limitations of conventional remediation approaches. The development on alternate mechanisms, modifying structures and functions by engineering genes, enzymes, or microorganisms, may help in efficient degradation pathway for PAHs. The bioremediation of PAH was advanced by biomolecular engineering with rational design and directed evolution approaches. More research work should be focused on improvising the bioremediation techniques for PAH removal in an efficient and eco-friendly manner.

18.16 Conclusion

The bioremediation approach for cleaning up of the environment contaminated with the hydrophobic polycyclic aromatic hydrocarbons in petroleum wastes generated from the petroleum industry by the PAH-degrading bacteria has exert less pressure on the environment.

- Sixteen priority polycyclic aromatic hydrocarbons are identified by US EPA based on its health impacts.
- The low solubility in water reduces its bioavailability and increases persistence.

- Many bacterial, fungal and algal species are identified as potential PAH degraders.
- The low molecular weight PAH compounds are more accessible and degradable by the microbial community.
- The conventional bioremediation approaches have its limitation in PAH degradation mechanisms.
- The production of enzymes and biosurfactants by genetic and protein engineering helps to improve biodegradability of high molecular weight PAHs.

Acknowledgment The authors acknowledge the Central University of Punjab, Bathinda, for providing the facilities for the preparation of this manuscript and research work.

References

Acevedo F, Pizzul L, Castillo MD, Cuevas R, Diez MC (2011) Degradation of polycyclic aromatic hydrocarbons by the Chilean white-rot fungus *Anthracophyllum discolor*. J Hazard Mater 185:212–219

Adams GO, Fufeyin PT, Okoro SE, Ehinomen I (2015) Bioremediation, biostimulation and bioaugmention: a review. Int J Environ Biorem Biodegrad 3:28–39

Ahirwar S, Dehariya K (2013) Isolation and characterization of hydrocarbon degrading microorganisms from petroleum oil contaminated soil sites. Bull Environ Sci Res 2(4):5–10

Alexander M (1995) How toxic are toxic chemicals in soil? Environ Sci Technol 29:2713–2717

Alexopoulos CJ, Mims CW, Blackwell M (1996) Introductory mycology. Wiley, New York

Amir S, Hafidi M, Merlina G, Hamdi H, Revel JC (2005) Fate of polycyclic aromatic hydrocarbons during composting of lagooning sewage sludge. Chemosphere 58:449–458

Anderson TA, Guthrie EA, Walton BT (1993) Bioremediation in the rhizosphere. Environ Sci Technol 27:2630–2636

Antizar-Ladislao B, Lopez-Real J, Beck A (2004) Bioremediation of polycyclic aromatic hydrocarbon (PAH)-contaminated waste using composting approaches. Crit Rev Environ Sci Technol 34:249–289

Antizar-Ladislao B, Lopez-Real J, Beck A (2006) Degradation of polycyclic aromatic hydrocarbons (PAHs) in an aged coal tar contaminated soil under in-vessel composting conditions. Environ Pollut 141(3):459–468

Aparna A, Srinikethan G, Hedge S (2011) Effect of addition of biosurfactant produced by Pseudomonas ssp. on biodegradation of crude oil. In: International proceedings of chemical, biological & environmental engineering. Proceedings of the 2nd International Proceedings of Chemical, Singapore, Singapore, 26–28 February 2011; Volume 6, p 71–75

Aparna A, Srinikethan G, Smitha H (2012) Production and characterization of biosurfactant produced by a novel *Pseudomonas* sp. 2B. Colloids Surf B Biointerfaces 95:23–29

Aprill W, Sims RC (1990) Evaluation of the use of prairie grasses for stimulating polycyclic aromatic hydrocarbon treatment in soil. Chemotherapy 20:253–265

Arulazhagan P, Vasudevan N (2011) Role of nutrients in the utilization of polycyclic aromatic hydrocarbons by halotolerant bacterial strain. J Environ Sci (China) 23:282–287

Atagana HI (2004) Co-composting of PAH-contaminated soil with poultry manure. Lett Appl Microbiol 39:163–168

ATSDR (2010) In US Agency for Toxic Substances and diseases Registry (ed) Toxicological profile for plutonium, 320 pp. US Department of Health and Human Services, Public Health Service, Atlanta

Ball AS, Jackson AM (1995) The recovery of lignocellulose-degrading enzymes from spent mushroom compost. Bioresour Technol 54:311–314

Balaji V, Arulazhagan P, Ebenezer P (2014) Enzymatic bioremediation of polyaromatic hydrocarbons by fungal consortia enriched from petroleum contaminated soil and oilseeds. J Environ Biol 35:521–529

Bamforth SM, Singleton I (2005) Bioremediation of polycyclic aromatic hydrocarbons: current knowledge and future directions. J Chem Technol Biotechnol 80:723–736

Banat IM, Franzetti A, Gandolfi I, Bestetti G, Martinotti MG, Fracchia L, Smyth TJ, Marchant R (2010) Microbial biosurfactants production, applications. Appl Microbiol Biotechnol 87:427–444

Batista RM, Rufino RD, Luna JM, Souza JEG, Sarubbo LA (2010) Effect of medium components on the production of a biosurfactant from *Candida tropicalis* applied to the removal of hydrophobic contaminants in soil. Water Environ Res 82:418–425

Beaudin N, Caron RF, Legros R, Ramsay J, Ramsay B (1999) Identification of the key factors affecting composting of a weathered hydrocarbon contaminated soil. Biodegradation 10:L 127–L 133

Betts WB (2012) Biodegradation: natural and synthetic materials. Springer, New York

Bewley RJF, Webb G (2001) In situ bioremediation of groundwater contaminated with phenols, BTEX and PAHs using nitrate as electron acceptor. Land Contam Reclam 9:335–347

Bezalel L, Hadar Y, Cerniglia CE (1996a) Mineralization of polycyclic aromatic hydrocarbons by the white rot fungus *Pleurotus ostreatus*. Appl Environ Microbiol 62:292–295

Bezalel L, Hadar Y, Fu PP, Freeman JP, Cerniglia CE (1996b) Initial oxidation products in the metabolism of pyrene, anthracene, fluorene, and dibenzothiophene by the white rot fungus *Pleurotus ostreatus*. Appl Environ Microbiol 62:2554–2559

Bezalel L, Hadar Y, Fu PP, Freeman JP, Cerniglia CE (1996c) Metabolism of phenanthrene by the white rot fungus *Pleurotus ostreatus*. Appl Environ Microbiol 62:2547–2553

Bezalel L, Hadar Y, Cerniglia CE (1997) Enzymatic mechanisms involved in Phenanthrene degradation by the white rot fungus *Pleurotus ostreatus*. Appl Environ Microbiol 63:2495–2501

Bharagava RN, Chowdhary P, Saxena G (2017a) Bioremediation: an ecosustainable green technology: its applications and limitations. In: Bharagava RN (ed) Environmental pollutants and their bioremediation approaches, 1st edn. CRC Press/Taylor & Francis Group, Boca Raton, pp 1–22. https://doi.org/10.1201/9781315173351-2

Bharagava RN, Saxena G, Chowdhary P (2017b) Constructed wetlands: an emerging phytotechnology for degradation and detoxification of industrial wastewaters. In: Bharagava RN (ed) Environmental pollutants and their bioremediation approaches, 1st edn. CRC Press/Taylor & Francis Group, Boca Raton, pp 397–426. https://doi.org/10.1201/9781315173351-15

Bisht S, Pandey P, Sood A, Sharma S, Bisht NS (2010) Biodegradation of naphthalene and anthracene by chemotactically active rhizobacteria of *Populus deltoides*. Braz J Microbiol 41(4):922–930

Bisht S, Pandey P, Kaur G, Aggarwal H, Sood A, Sharma S, Kumar V, Bisht NS (2014) Utilization of endophytic strain *Bacillus* sp. SBER3 for biodegradation of polyaromatic hydrocarbons (PAH) in soil model system. Eur J Soil Biol 60:67–76

Bisht S, Pandey P, Bhargava B, Sharma S, Kumar V, Sharma SD (2015) Bioremediation of polyaromatic hydrocarbons (PAHs) using rhizosphere technology. Braz J Microbiol 46(1):7–21

Bogan BW, Lamar RT (1996) Polycyclic aromatic hydrocarbon-degrading capabilities of hanerochaete laevis HHB-1625 and its extracellular ligninolytic enzymes. Appl Environ Microbiol 62:1597–1603

Bogan BW, Sullivan WR (2003) Physicochemical soil parameters affecting sequestration and mycobacterial biodegradation of polycyclic aromatic hydrocarbons in soil. Chemosphere 52:1717–1726

Bogan BW, Schoenike B, Lamar RT, Cullen D (1996a) Manganese peroxidase mRNA and enzyme activity levels during bioremediation of polycyclic aromatic hydrocarbon-contaminated soil with *Phanerochaete chrysosporium*. Appl Environ Microbiol 62:2381–2386

Bogan BW, Lamar RT, Hammel KE (1996b) Fluorene oxidation in vivo by *Phanerochaete chrysosporium* and in vitro during manganese peroxidase-dependent lipid peroxidation. Appl Environ Microbiol 62:1788–1792

Bogan BW, Lahner LM, Sullivan WR, Paterek JR (2003) Degradation of straight-chain aliphatic and high-molecular-weight polycyclic aromatic hydrocarbons by a strain of *Mycobacterium austroafricanum*. J Appl Microbiol 94:230–239

Boldrin B, Tiehm A, Fritzsche C (1993) Degradation of phenanthrene, fluorene, fluoranthene, and pyrene by a *Mycobacterium sp*. Appl Environ Microbiol 59:1927–1930

Bossert I, Bartha R (1984) The fate of petroleum in soil ecosystems. In: Atlas RM (ed) Petroleum microbiology. Macmillan, New York, pp 435–473

Boxall ABA, Maltby L (1997) The effects of motorway runoff on freshwater ecosystems. (3). Toxicant confirmation. Arch Environ Contam Toxicol 33:9–16

Brook TR, Stiver WH, Zytner RG (2001) Biodegradation of diesel fuel in soil under various nitrogen addition regimes. Soil Sediment Contam: Int J 10:539–553

Bumpus JA (1989) Biodegradation of polycyclic aromatic hydrocarbons by *Phanerochaete chrysosporium*. Appl Environ Microbiol 55:154–158

Bumpus JA, Tien M, Wright D, Aust SD (1985) Oxidation of persistent environmental pollutants by a white-rot fungus. Science 228:1434–1436

Canada KA, Iwashita S, Shim H, Wood TK (2002) Directed evolution of toluene ortho-monooxygenase for enhanced 1-naphtholsynthesis and chlorinated ethene degradation. J Bacteriol 184:344–349

Carmichael AB, Wong LL (2001) Protein engineering of *Bacillus megaterium* CYP102—the oxidation of polycyclic aromatic hydrocarbons. Eur J Biochem 268:3117–3125

Cébron A, Beguiristain T, Bongoua-Devisme J, Denonfoux J, Faure P, Lorgeoux C, Ouvrard S, Parisot N, Peyret P, Leyval C (2015) Impact of clay mineral, wood sawdust or root organic matter on the bacterial and fungal community structures in two aged PAH-contaminated soils. Environ Sci Pollut Res 22(18):13724–13738

Cerniglia CE (1992) Biodegradation of polycyclic aromatic hydrocarbons. In: Microorganisms to combat pollution, Springer, Dordrecht, pp 227–244

Cerniglia CE (1997) Fungal metabolism of polycyclic aromatic hydrocarbons: past, present and future applications in bioremediation. J Microbiol Biotechnol 19:324–333

Cerniglia CE, Gibson DT (1979) Oxidation of benzo [a] pyrene by the filamentous fungus *Cunninghamella elegans*. J Biol Chem 254:12174–12180

Cerniglia CE, Yang SK (1984) Stereoselective metabolism of anthracene and phenanthrene by the fungus *Cunninghamella elegans*. Appl Environ Microbiol 47:119–124

Cerniglia CE, Gibson DT, Van Baalen C (1979) Algal oxidation of aromatic hydrocarbons: formation of 1-naphthol from naphthalene by *Agmenellum quadruplicatum*, strain PR-6. Biochem Biophys Res Commun 88:50–58

Cerniglia CE, Mahaffey W, Gibson DT (1980a) Fungal oxidation of benzo [a] pyrene: formation of (−)-trans-7, 8-dihydroxy-7,8-dihydrobenzo [a] pyrene by *Cunninghamella elegans*. Biochem Biophys Res Commun 94:226–232

Cerniglia CE, Dodge RH, Gibson DT (1980b) Studies on the fungal oxidation of polycyclic aromatic hydrocarbons. Bot Mar 23:121–124

Cerniglia CE, Kelly DW, Freeman JP, Miller DW (1986) Microbial metabolism of pyrene. Chem Biol Interact 57:203–216

Chakraborty R, Coates JD (2004) Anaerobic degradation of monoaromatic hydrocarbons. Appl Microbiol Biotechnol 64:437–446

Chan SMN, Luan T, Wong MH, Tam NFY (2006) Removal and biodegradation of polycyclic aromatic hydrocarbons by *Selenastrum capricornutum*. Environ Toxicol Chem 25:1772–1779

Chandra R, Saxena G, Kumar V (2015) Phytoremediation of environmental pollutants: an eco-sustainable green technology to environmental management. In: Chandra R (ed) Advances in biodegradation and bioremediation of industrial waste, 1st edn. CRC Press/Taylor & Francis Group, Boca Raton, pp 1–30. https://doi.org/10.1201/b18218-2

Chung KH, Lee JH, Ro KS (2000) Composting of kerosene-contaminated soil: fate of kerosene. J Environ Sci Health A 35:1183–1194

Civilini M (1994) Fate of creosote compounds during composting. Microbiol Eur 2:16–24

Coral G, Karagoz S (2005) Isolation and characterization of phenanthrene-degrading bacteria from a petroleum refinery soil. Ann Microbiol 55:255–259

Cousins IT, Jones KC (1998) Air–soil exchange of semi-volatile organic compounds (SOCs) in the UK. Environ Pollut 102:105–118

Crawford SL, Johnson GE, Goetz FE (1993) The potential for bioremediation of soils containing PAHs by composting. Compost Sci Util 1:41–47

Cutright TJ (1995) Polycyclic aromatic hydrocarbon biodegradation and kinetics using *Cunninghamella echinulata* var. *elegans*. Int Biodeterior Biodegrad 35:397–408

Darmawan R, Nakata H, Ohta H, Niidome T, Takikawa K, Morimura S (2015) Isolation and evaluation of PAH degrading bacteria. J Bioremed Biodegr 6:3. https://doi.org/10.4172/2155-6199.1000283

Diaz M, Mora V, Pedrozo F, Nichela D, Baffico G (2014) Evaluation of native acidophilic algae species as potential indicators of polycyclic aromatic hydrocarbon (PAH) soil contamination. J Appl Phycol 27:321–325

Dipple A, Bigger CAH (1991) Mechanism of action of food-associated polycyclic aromatic hydrocarbon carcinogens. Mutat Res 259:263–276

Edwards NT (1983) Polycyclic aromatic hydrocarbons (PAHs) in the terrestrial environment: a review. J Environ Qual 12:427–441

Elias M, Wood A, Hashim Z, Siong W, Hamzah M, Rahman S, Salim N, Talib A (2007) Polycyclic aromatic hydrocarbons (PAH) contamination in the sediments of east coast peninsular Malaysia. Malays J Anal Sci 11:70–75

Elisabet Aranda RU (2009) Conversion of polycyclic aromatic hydrocarbons, methyl naphthalenes and dibenzofuran by two fungal peroxygenases. Biodegradation 21:267–281

El-Sheekh MM, Ghareib MM, Abou-El-Souod GW (2012) Biodegradation of phenolic and polycyclic aromatic compounds by some algae and cyanobacteria. J Bioremed Biodegr 3:133

England PA, Harford-Cross CF, Stevenson JA, Rouch DA, Wong LL (1998) The oxidation of naphthalene and pyrene by cytochrome P450(cam). FEBS Lett 424:271–274

Farhadian A, Jinap S, Abasand F, Sakar Z (2010) Determination of polycyclic aromatic hydrocarbons in grilled meat. Food Control 21:606–610

Farnet AM, Criquet S, Tagger S, Gil G, Petit JL (2000) Purification, partial characterization, and reactivity with aromatic compounds of two laccases from *Marasmius quercophilus* strain 17. Can J Microbiol 46:189–194

Ferreira L, Rosales E, Sanromán MA, Pazos M (2015) Preliminary testing and design of permeable bioreactive barrier for phenanthrene degradation by *Pseudomonas stutzeri* CECT 930 immobilized in hydrogel matrices. J Chem Technol Biotechnol 90:500–506

Fetzer JC (2000) The Chemistry and Analysis of the Large Polycyclic Aromatic Hydrocarbon, 2nd edn. Wiley Publishers, New York, pp 27–29

Field JA, De Jong E, Costa GF, De Bont JA (1992) Biodegradation of polycyclic aromatic hydrocarbons by new isolates of white rot fungi. Appl Environ Microbiol 58(7):2219–2226

Freeman DJ, Cattell CR (1990) Wood burning as a source of atmospheric polycyclic aromatic hydrocarbons. Environ Sci Technol 24:1581–1585

Gafarov AB, Panov AV, Filonov AE, Boronin AM (2006) Change in the composition of a bacterial association degrading aromatic compounds during oil sludge detoxification in a continuous-flow microbial reactor. Appl Biochem Microbiol 42:160–165

Genthner BRS, Townsend GT, Lantz SE, Mueller JG (1997) Persistence of polycyclic aromatic hydrocarbon components of creosote under anaerobic enrichment conditions. Arch Environ Contam Toxicol 32:99–105

Gentry TJ, Rensing C, Pepper IL (2004) New approaches for bioaugmentation as a remediation technology. Crit Rev Environ Sci Technol 34:447–494

Ghosal D, Ghosh S, Dutta TK, Ahn Y (2016) Current state of knowledge in microbial degradation of polycyclic aromatic hydrocarbons (PAHs):a review. Front Microbiol 7:1369. https://doi.org/10.3389/fmicb.2016.01369

Gibson DT, Koch JR, Kallio RE (1968) Oxidative degradation of aromatic hydrocarbons by microorganisms. I. Enzymatic formation of catechol from benzene. Biochemistry 7:2653–2661

Gibson DT, Mahadevan V, Jerina RM, Yagi H, Yeh HJC (1975) Oxidation of the carcinogens benzo[a]pyrene and benzo[a]anthracene to dihydrodiols by a bacterium. Science 189:295–297

Gordon L, Dobson AD (2001) Fluoranthene degradation in *Pseudomonas alcaligenes* PA-10. Biodegradation 12(6):393–400

Grotenhuis T, Field J, Wasseveld R, Rulkens W (1999) Biodegradation of polyaromatic hydrocarbons (PAH) in polluted soil by the white-rot fungus *Bjerkandera*. J Chem Technol Biotechnol 71:359–360

Guerin TF (2000) Long-term performance of a land treatment facility for the bioremediation of non-volatile oily wastes. Resour Conserv Recycl 28:105–120

Guillén F, Gómez-Toribio V, Martínez MJ, Martínez AT (2000) Production of hydroxyl radical by the synergistic action of fungal laccase and aryl alcohol oxidase. Arch Biochem Biophys 382:142–147

Gupta PK (2008) Molecular biology and genetic engineering. Rastogi Publications, New Delhi, p 614

Gusmão CAB, Rufino RD, Sarubbo LA (2010) Laboratory production and characterization of a new biosurfactant from *Candida glabrata* UCP1002 cultivated in vegetable fat waste applied to the removal of hydrophobic contaminant. World J Microbiol Biotechnol 26:1683–1692

Habe H, Omori T (2003) Genetics of polycyclic aromatic hydrocarbon metabolism in diverse aerobic bacteria. Biosci Biotech Bioch 67:225–243

Hadibarata T, Kristanti RA (2014) Potential of a white-rot fungus *Pleurotus eryngii* F032 for degradation and transformation of fluorene. Fungal Biol 118:222–227

Hadibarata T, Tachibana S, Itoh K (2009) Biodegradation of chrysene, an aromatic hydrocarbon by *Polyporus* sp. S133 in liquid medium. J Hazard Mater 164:911–917

Haemmerli SD, Leisola MS, Sanglard D, Fiechter A (1986) Oxidation of benzo (a) pyrene by extracellular ligninases of *Phanerochaete chrysosporium*. Veratryl alcohol and stability of ligninase. J Biol Chem 261:6900–6903

Halim M, Conte P, Piccolo A (2003) Potential availability of heavy metals to phyto extraction from contaminated soils induced by exogenous humic substances. Chemosphere 52:265–275

Hammel KE, Kalyanaraman B, Kirk TK (1986) Oxidation of polycyclic aromatic hydrocarbons and dibenzo[a]dioxins by Phunerochucre chrysosporium Iigninase. J Biol Chem 26(1):16948–16952

Hammel KE, Gai WZ, Green B, Moen MA (1992) Oxidative degradation of phenanthrene by the ligninolytic fungus *Phanerochaete chrysosporium*. Appl Environ Microbiol 58:1832–1838

Harayama S (1997) Polycyclic aromatic hydrocarbon bioremediation design. Curr Opin Biotechnol 8:268–274

Harford-Cross CF, Carmichael AB, Allan FK, England PA, Rouch DA, Wong LL (2000) Protein engineering of cytochrome P450cam (CYP101) for the oxidation of polycyclic aromatic hydrocarbons. Protein Eng 13:121–128

Haritash AK, Kaushik CP (2009) Biodegradation aspects of polycyclic aromatic hydrocarbons (PAHs): a review. J Hazard Mater 169:1–15

Harvey PJ, Campanella BF, Castro PM, Harms H, Lichtfouse E, Schäffner AR, Smrcek S, Werck-Reichhart D (2002) Phytoremediation of polyaromatic hydrocarbons, anilines and phenols. Environ Sci Pollut Res Int 9:29–47

Heitkamp MA, Franklin W, Cerniglia CE (1988a) Microbial metabolism of polycyclic aromatic hydrocarbons: isolation and characterization of a pyrene degrading bacterium. Appl Environ Microbiol 54:2549–2555

Heitkamp MA, Freeman JP, Miller DW, Cerniglia CE (1988b) Pyrene degradation by a *Mycobacterium sp.*: identification of ring oxidation and ring fission products. Appl Environ Microbiol:2556–2565

Herbes SE, Schwall LR (1978) Microbial transformation of polycyclic aromatic hydrocarbons in pristine and petroleum-contaminated sediments. Appl Environ Microbiol 35:306–316

Hofrichter M, Scheibner K, Schneega I, Fritsche W (1998) Enzymatic combustion of aromatic and aliphatic compounds by manganese peroxidase from *Nematoloma frowardii*. Appl Environ Microbiol 64:399–404

Hofrichter M, Scheibner K, Bublitz F, Schneegaß I, Ziegenhagen D, Martens R, Fritsche W (1999) Depolymerization of straw lignin by manganese peroxidase from Nematoloma frowardii is accompanied by release of carbon dioxide. Holzforschung 53(2):161–166

International Agency of Research on Cancer (IARC) (1983) Polynuclear aromatic compounds, IARC monographs on the evaluation of carcinogenic risks to humans. IARC, Lyon

International Society for Polycyclic Aromatic Compounds (2003) PAH structure/properties. www.ispac.org

Isiodu GG, Stanley OH, Victor E, Okerentugba PO (2016) Role of plasmid-borne genes in the biodegradation of polycyclic aromatic hydrocarbons (PAHs) by consortium of aerobic heterotrophic bacteria. J Pet Environ Biotechnol 7:264. https://doi.org/10.4172/2157-7463.1000264

Jacob J, Karcher W, Beiliardo JJ, Wagstaffe PJ (1986) Polycyclic aromatic hydrocarbons of environmental and occupational importance. Fresenius Z Anal Chem 323:1–10

Jacques RJS, Santos EC, Bento FM, Peralba MCR, Selbach PA, Sa ELS, Camargo FAO (2005) Anthracene biodegradation by *Pseudomonas sp.* isolated from a petrochemical sludge landfarming site. Int Biodeterior Biodegradation 56:143–150

James TY, O'Donnell K (2004) Zygomycota, microscopic 'pin' or 'sugar' molds. Available at: http://tolweb.org/Zygomycota/20518/2004.12.21 (Version21 December 2004), part of The Tree of Life Web Project, http://tolweb.org

Jobson AM, Cook FD, Westlake DWS (1974) Effect of amendments on the microbial utilization of oil applied to soil. Appl Microbiol Biotechnol 27:166–171

Johannes C, Majcherczyk A (2000) Natural mediators in the oxidation of polycyclic aromatic hydrocarbons by laccase mediator systems. Appl Environ Microbiol 66:524–528

Jones KC (1991) Contaminant trends in soils and crops. Environ Pollut 69:311–325

Jørgensen KS, Puustinen J, Suortti AM (2000) Bioremediation of petroleum hydrocarbon-contaminated soil by composting in biopiles. Environ Pollut 107:245–254

Jové P, Olivella MÀ, Camarero S, Caixach J, Planas C, Cano L, De Las Heras FX (2016) Fungal biodegradation of anthracene-polluted cork: a comparative study. J Environ Sci Health A 51:70–77

Joyce JF, Sato C, Cardenas R, Surampalli RY (1998) Composting of polycyclic aromatic hydrocarbons in simulated municipal solid waste. Water Environ Res 70:356–361

Juhasz AL, Naidu R (2000) Bioremediation of high molecular weight polycyclic aromatic hydrocarbons: a review of the microbial degradation of benzo[*a*]pyrene. Int Biodeterior Biodegrad 45:57–88

Kafilzadeh F, Pour FP (2012) Degradation of naphthalene, phenanthrene and pyrene by *Pseudomonas* sp. and *Corynebacterium* sp. in the landfills. Int J Bios 2(9):77–84

Kafilzadeh F, Sahragard P, Jamali H, Tahery Y (2011) Isolation and identification of hydrocarbons degrading bacteria in soil around Shiraz Refinery. Afr J Microbiol Res 4(19):3084–3089

Kafilzadeh F, Hoshyaripour F, Tahery Y, Azad HN (2012) Bioremediation of pyrene by isolated bacterial strains from the soil of the landfills in Shiraz, Iran. Ann Biol Res 3(1):486–494

Kanaly RA, Harayama S (2000) Biodegradation of high-molecular-weight polycyclic aromatic hydrocarbons by bacteria. J Bacteriol 182(8):2059–2067

Kastner M, Lotter S, Heerenklage J, Breuer-Jammali M, Stegmann R, Mahro B (1995) Fate of 14C-labeled anthracene and hexadecane in compost-manure soil. Appl Microbiol Biotechnol 43:1128–1135

Kastner M, Breuer-Jammali M, Mahro B (1998) Impact of inoculation protocols, salinity, and pH on the degradation of polycyclic aromatic hydrocarbons (PAHs) and survival of PAH-degrading bacteria introduced into soil. Appl Environ Microbiol 64:359–362

Kastner M, Streibich S, Beyrer M, Richnow HH, Fritsche W (1999) Formation of bound residues during microbial degradation of [14C]anthracene in soil. Appl Environ Microbiol 65(5):1834–1842

Keith LH, Telliard WA (1979) ES & T special report—priority pollutants: I-a perspective view. Environ Sci Technol 13:416–423

Khan AA, Wang R-F, Cao W-W, Doerge DR, Wennerstrom D, Cerniglia CE (2001) Molecular cloning, nucleotide sequence, and expression of genes encoding polycyclic aromatic ring dioxygenase from *Mycobacterium* sp. strain PYR-1. Appl Environ Microbiol 67:3577–3585

Khanna P, Goyal D, Khanna S (2012) Characterization of pyrene utilizing bacillus spp. from crude oil contaminated soil. Braz J Microbiol 43(2):606–617

Kim SJ, Jones RC, Cha CJ, Kweon O, Edmondson RD, Cerniglia CE (2004) Identification of proteins induced by polycyclic aromatic hydrocarbon in *Mycobacterium vanbaalenii* PYR-1 using two-dimensional polyacrylamide gel electrophoresis and de novo sequencing methods. Proteomics 4(12):3899–3908

Kim YH, Freeman JP, Moody JD, Engesser KH, Cerniglia CE (2005) Effects of pH on the degradation of phenanthrene and pyrene by *Mycobacterium van*baalenii PYR1. Appl Microbiol Biotechnol 67:275–285

Kingsley MT, Fredrickson JK, Metting FB, Seidler RJ (1994) Environmental restoration using plant-microbe bioaugmentation. In: Hinchee RE, Leeson A, Semprini L, Ong SK (eds) Bioremediation of chlorinated and polyaromatic hydrocarbon compounds. Lewis Publishers, Boca Raton, pp 287–292

Kirchmann H, Ewnetu W (1998) Biodegradation of petroleum-based oil wastes through composting. Biodegradation 9:151–156

Kotterman MJ, Vis EH, Field JA (1998) Successive mineralization and detoxification of benzo[a] pyrene by the white rot fungus *Bjerkandera* sp. strain BOS55 and indigenous microflora. Appl Environ Microbiol 64:2853–2858

Krivobok S, Miriouchkine E, Seigle-Murandi F, Benoit-Guyod JL (1998) Biodegradation of anthracene by soil fungi. Chemosphere 37(3):523–530

Kuiper EL, Lagendijk GV, Lugtenberg B (2004) Rhizoremediation: a beneficial plant microbe interaction. Mol Plant Microbiol Int 17:6–15

Kumar S, Upadhyay SK, Kumari B, Tiwari S, Singh SN, Singh PK (2011) In vitro degradation of fluoranthene by bacteria isolated from petroleum sludge. Bioresour Technol 102:3709–3715

Lau KL, Tsang YY, Chiu SW (2003) Use of spent mushroom compost to bioremediate PAH-contaminated samples. Chemosphere 52:1539–1546

Lee K, Tremblay GH, Gauthier J, Cobanli SE, Griffin M (1997) Bioaugmentation and biostimulation: a paradox between laboratory and field results. Proceedings of 1997 international oil spill conference, American Petroleum Institute, Washington, DC, pp 697–705

Lee H, Jang Y, Choi YS, Kim MJ, Lee J, Lee H, Hong JH, Lee YM, Kim GH, Kim JJ (2014) Biotechnological procedures to select white rot fungi for the degradation of PAHs. J Microbiol Methods 97:56–62

Lei AP, Wong YS, Tam NFY (2000) Removal of pyrene by different microalgal species. Water Sci Technol 46:195–201

Lei AP, Hu ZL, Wong YS, Tam NFY (2007) Removal of fluoranthene and pyrene by different microalgal species. Bioresour Technol 98:273–280

Li QS, Ogawa J, Schmid RD, Shimizu S (2001) Engineering cytochrome P450BM-3 for oxidation of polycyclic aromatic hydrocarbons. Appl Environ Microbiol 67:5735–5739

Li P, Li H, Stagnitti F, Wang X, Zhang H, Gong Z, Liu W, Xiong X, Li L, Austin C, Barry DA (2005) Biodegradation of pyrene and phenanthrene in soil using immobilized fungi *Fusarium* sp. Bull Environ Contam Toxicol 75:443–450

Li X, Wang Y, Wu S, Qiu L, Gu L, Li J, Zhang B, Zhong W (2014) Pecularities of metabolism of anthracene and pyrene by laccase producing fungus *Pycnoporus sanguineus* H1. Biotechnol Appl Biochem 61:549–554

Liang Y, Sorensen DL, McLean JE, Sims RC (2008) Pyrene fate affected by humic acid amendment in soil slurry systems. J Biol Eng 2:11

Lim LH, Harrison RM, Harrad S (1999) The contribution of traffic to atmospheric concentrations of polycyclic aromatic hydrocarbons. Environ Sci Technol 33:3538–3542

Lugtenberg B, van der Bij A, Bloemberg G, Woeng TCA, Dekkers L, Kravchenko L, Mulders I, Phoelich C, Simons M, Tikhonovich I, de Weger L, Wijffelman C (1997) Towards the molecular basis of plant root colonization by Pseudomonas bacteria. Pseudomonas'97, VI international congress on pseudomonas: molecular biology and biotechnology, Madrid, 1997

Luna JM, Rufino RD, Sarubbo LA, Rodrigues LRM, Teixeira JAC, Campos-Takaki GM (2011) Evaluation antimicrobial and antiadhesive properties of the biosurfactant lunasan produced by *Candida sphaerica* UCP 0995. Curr Microbiol 62:1527–1534

Luna JM, Rufino RD, Sarubbo LA, Campos-Takaki GM (2013) Characterisation, surface properties and biological activity of a biosurfactant produced from industrial waste by *Candida sphaerica* UCP0995 for application in the petroleum industry. Colloids Surf B Biointerfaces 102:202–209

Lundstedt S, Haglund P, Oberg L (2003) Degradation and formation of polycyclic aromatic compounds during bioslurry treatment of an aged gasworks soil. Environ Toxicol Chem 22:1413–1420

Luo L, Wang P, Lin L, Luan T, Ke L, Tam NFY (2014) Removal and transformation of high molecular weight polycyclic aromatic hydrocarbons in water by live and dead microalgae. Process Biochem 49:1723–1732

Mackay D, Shiu WY, Ma KC (1992) Illustrated handbook of physical-chemical properties and environmental fate for organic chemicals. Vol. II. Polynuclear aromatic hydrocarbons, polychlorinated dioxins and dibenzofurans. Lewis Publishers, Boca Raton

Madsen T, Kristensen P (1997) Effects of bacterial inoculation and nonionic surfactants on degradation of polycyclic aromatic hydrocarbons in soil. Environ Toxicol Chem 16:631–637

Manilal VB, Alexander M (1991) Factors affecting the microbial degradation of phenanthrene in soil. Appl Microbiol Biotechnol 35:401–405

Manilla-Pérez E, Lange AB, Luftmann H, Robenek H, Steinbüchel A (2011) Neutral lipid production in *Alcanivorax borkumensis* SK2 and other marine hydrocarbonoclastic bacteria. Eur J Lipid Sci Technol 113:8–17

Mao J, Luo Y, Teng Y, Li Z (2012) Bioremediation of polycyclic aromatic hydrocarbon-contaminated soil by a bacterial consortium and associated microbial community changes. Int Biodeterior Biodegrad 70:141–147

McFarland MJ, Qiu XJ (1995) Removal of benzo (a) pyrene in soil composting systems amended with the white rot fungus *Phanerochaete chrysosporium*. J Hazard Mater 42:61–70

McNally DL, Mihelcic JR, Lueking DR (1999) Biodegradation of mixtures of polycyclic aromatic hydrocarbons under aerobic and nitrate-reducing conditions. Chemosphere 38:1313–1321

Mineki S, Suzuki K, Iwata K, Nakajima D, Goto S (2015) Degradation of polyaromatic hydrocarbons by fungi isolated from soil in Japan. Polycycl Aromat Compd 35:120–128

Miya RK, Firestone MK (2000) Phenanthrene biodegradation in soil by slender oar root exudates and root debris. J Environ Qual 30:1911–1918

Mumford JL, Chapman RS, Narris DB, He XZ, Cao SR, Xian YL, Li XM (1989) Indoor air exposure to coal and wood combustion emissions associated with a high lung cancer in Xuan Wei. China Environ Int 15:315–320

Namkoong W, Hwang EY, Park JS, Choi JY (2002) Bioremediation of diesel contaminated soil with composting. Environ Pollut 119:23–31

Narro ML, Cerniglia CE, Van Baalen C, Gibson DT (1992a) Evidence for an NIH shift in oxidation of naphthalene by the marine cyanobacterium *Oscillatoria* sp. strain. JCM Appl Environ Microbiol 58:1360–1363

Narro ML, Cerniglia CE, Van Baalen C, Gibson DT (1992b) Metabolism of phenanthrene by the marine cyanobacterium *Agmenellum quadruplicatum* PR-6. Appl Environ Microbiol 58:1351–1359

Nnamchi C, Obeta J, Ezeogu L (2006) Isolation and characterization of some polycyclic aromatic hydrocarbon degrading bacteria from Nsukka soils in Nigeria. Int J Environ Sci Technol 3:181–190

Nollet LML (2006) Chromatographic analysis of the environment, Series: Chromatographic science series, vol 93, 3rd edn. CRC Press, Boca Raton, p 1320

Novotný Č, Svobodová K, Erbanová P, Cajthaml T, Kasinath A, Lang E, Sasek V (2004) Ligninolytic fungi in bioremediation: extracellular enzyme production and degradation rate. Soil Biol Biochem 36:1545–1551

Okai M, Kihara I, Yokoyama Y, Ishida M, Urano N (2015) Isolation and characterization of benzo [a] pyrene-degrading bacteria from the Tokyo Bay area and Tama River in Japan. FEMS Microbiol Lett 362:fnv143

Park KS, Sims RC, Dupont RR, Doucette WJ, Matthews JE (1990) Fate of PAH compounds in two soil types. Influence of volatilization, abiotic loss and biological activity. Environ Toxicol Chem 9:187–195

Pilon-Smits E (2005) Phytoremediation. Ann Rev Plant Biol 56:15–39

Ping L, Zhang C, Zhu Y, Wu M, Hu X, Li Z, Zhao H (2011) Biodegrading of pyrene by a newly isolated pseudomonas putida PL2. Biotechnol Bioprocess E 16:1000. https://doi.org/10.1007/s12257-010-0435-y

Pirôllo MP, Mariano AP, Lovaglio RB, Costa SG, Walter V, Hausmann R, Contiero J (2008) Biosurfactant synthesis by *Pseudomonas aeruginosa* LBI isolated from a hydrocarbon-contaminated site. J Appl Microbiol 105:1484–1490

Piskonen R, Itävaara M (2004) Evaluation of chemical pretreatment of contaminated soil for improved PAH bioremediation. Appl Microbiol Biotechnol 65:627–634

Pitt R, Field R, Lalor M, Brown M (1995) Urban storm water toxic pollutants: assessment, sources and treatability. Water Environ Res 67:260–275

Pothuluri JV, Freeman JP, Evans FE, Cerniglia CE (1990) Fungal transformation of fluoranthene. Appl Environ Microbiol 56:2974–2983

Pothuluri JV, Freeman JP, Evans FE, Cerniglia CE (1992a) Fungal metabolism of acenaphthene by *Cunninghamella elegans*. Appl Environ Microbiol 58:3654–3659

Pothuluri JV, Heflich RH, Fu PP, Cerniglia CE (1992b) Fungal metabolism and detoxification of fluoranthene. Appl Environ Microbiol 58:937–941

Pothuluri JV, Freeman JP, Evans FE, Cerniglia CE (1993) Biotransformation of fluorene by the fungus *Cunninghamella elegans*. Appl Environ Microbiol 59:1977–1980

Pothuluri JV, Selby A, Evans FE, Freeman JP, Cerniglia CE (1995) Transformation of chrysene and other polycyclic aromatic hydrocarbon mixtures by the fungus *Cunninghamella elegans*. Can J Bot 73:1025–1033

Pothuluri JV, Evans FE, Heinze TM, Cerniglia CE (1996) Formation of sulfate and glucoside conjugates of benzo[e]pyrene by *Cunninghamella elegans*. Appl Microbiol Biotechnol 45:677–683

Pysh ES, Yang NC (1963) Polarographic oxidation potentials of aromatic compounds. J Am Chem Soc 85:2124–2130

Qiu XJ, McFarland MJ (1991) Bound residue formation in PAH contaminated soil composting using *Phanerochaete chrysosporium*. Hazard Waste Hazard Mater 8:115–126

Qiu X, Shah SI, Kendall EW, Sorensen DL, Sim RC, Engelke MC (1994) Grass enhanced bioremediation for clay soils contaminated with polynuclear aromatic hydrocarbons. In: Anderson TA, Coates JR (eds) Bioremediation through rhizosphere technology. American Chemical Society, Washington, DC, pp 142–157

Rahman KS, Rahman TP, Banat IM (2002) Occurrence of crude oil degrading bacteria in gasoline and diesel station soils. J Basic Microbiol 42:284–291

Rehmann K, Hertkorn N, Kettrup AA (2001) Fluoranthene metabolism in *Mycobacterium* sp. strain KR20 identity of pathway intermediates during degradation and growth. Microbiology 147(10):2783–2794

Ripps S, Nivens DE, Ahn Y, Werner C, Jarrel J, Easter JP, Cox CD, Burlage RS, Sayler GS (2000) Controlled field release of a bioluminescent genetically engineered microorganism for bioremediation process monitoring and control. Environ Sci Technol 34:846–853

Sack U, Günther T (1993) Metabolism of PAH by fungi and correlation with extracellular enzymatic activities. J Basic Microbiol 33:269–277
Sack U, Heinze TM, Deck J, Cerniglia CE, Martens R, Zadrazil F, Fritsche W (1997) Comparison of phenanthrene and pyrene degradation by different wood-decaying fungi. Appl Environ Microbiol 63(10):3919–3925
Safonova E, Kvitko K, Kuschk P, Möder M, Reisser W (2005) Biodegradation of phenanthrene by the green alga *Scenedesmus obliquus* ES-55. Eng Life Sci 5:234–239
Salicis F, Krivobok S, Jack M, Benoit-Guyod JL (1999) Biodegradation of fluoranthene by soil fungi. Chemosphere 38(13):3031–3039
Salt DE, Smith RD, Raskin I (1998) Phytoremediation. Ann Rev Plant Physiol Plant Mol Biol 49:643–668
Sanglard D, Leisola MS, Fiechter A (1986) Role of extracellular ligninases in biodegradation of benzo (a) pyrene by *Phanerochaete chrysosporium.* Enzym Microb Technol 8:209–212
Sasek V, Bhatt M, Cajthaml T, Malachova K, Lednicka D (2003) Compost- mediated removal of polycyclic aromatic hydrocarbons from contaminated soil. Arch Environ Contam Toxicol 44:336–342
Saxena G, Bharagava RN (2015) Persistent organic pollutants and bacterial communities present during the treatment of tannery wastewater. In: Chandra R (ed) Environmental waste management, 1st edn. CRC Press/Taylor & Francis Group, Boca Raton, pp 217–247. https://doi.org/10.1201/b19243-10
Saxena G, Bharagava RN (2017) Organic and inorganic pollutants in industrial wastes, their ecotoxicological effects, health hazards and bioremediation approaches. In: Bharagava RN (ed) Environmental pollutants and their bioremediation approaches, 1st edn. CRC Press/Taylor & Francis Group, Boca Raton, pp 23–56. https://doi.org/10.1201/9781315173351-3
Sayler GS, Ripp S (2000) Field applications of genetically engineered microorganisms for bioremediation process. Curr Opin Biotechnol 11:286–289
Sayler GS, Cox CD, Burlager RS, Nivens DE, Werner C, Ahn Y, Matrubutham U (1999) Field application of a genetically engineered microorganism for polycyclic aromatic hydrocarbon bioremediation process monitoring and control. In: Fass R, Flashner Y, Reuveny S (eds) Novel approaches for bioremediation of organic pollution. Kluwer Academic Plenum Publishers, New York, pp 241–254
Schell MA (1983) Cloning and expression in E. coli of naphthalene degrading genes from plasmid NAH7. J Bacteriol 153:822–829
Schneider J, Grosser R, Jayasimhulu K, Xue W, Warshawsky D (1996) Degradation of pyrene, benz[a]anthracene and benzo[a]pyrene by *Mycobacterium* sp. strain RJGII-135, isolated from a former coal gas station site. Appl Environ Microbiol 62:13–19
Sette LD, Simioni KCM, Vasconcellos SP, Dussan LJ, Neto EVS, Oliveira VM (2007) Analysis of composition of bacterial communities in oil reservoirs from a southern offshore Brazilian basin. Antonie Leeuwenhoek 91:253–266
Silva EJ, Rocha e Silva NMP, Rufino RD, Luna JM, Silva RO, Sarubbo LA (2014) Characterization of a biosurfactant produced by *Pseudomonas cepacia* CCT6659 in the presence of industrial wastes and its application in the biodegradation of hydrophobic compounds in soil. Colloids Surf B Biointerfaces 117:36–41
Simon MJ, Osslund TD, Saunders R, Ensley BD, Suggs S, Harcourt A, Suen WC, Cruden DL, Gibson DT, Zylstra GJ (1993) Sequences of genes encoding naphthalene dioxygenase in pseudomonas putida strains G7 and NCIB 9816-4. Gene 127(1):31–37
Simonsick WJ, Hites RA (1986) Characterization of high molecular weight polycyclic aromatic hydrocarbons by charge exchange chemical ionization mass spectrometry. Anal Chem 58:2114–2121
Sims JL, Sims RC, Matthews JE (1990) Approach to bioremediation of contaminated soil. Hazard Waste Hazard Mater 7:117–149
Singh OV, Jain RK (2003) Phytoremediation of toxic aromatic pollutants from soil. Appl Microbiol Biotechnol 63:128–135

Soberón-Chávez G, Maier RM (2010) Biosurfactants: a general overview. In: Soberón-Chávez G (ed) Biosurfactants: from genes to applications. Springer, Münster, pp 1–11
Sobisch T, He ßH, Niebelschutz H, Schmidt U (2000) Effect of additives on biodegradation of PAH in soils. Colloids Surf A Physicochem Eng Asp 162(2000):1–14
Sobrinho HB, Luna, JM, Rufino RD, Porto ALF, Sarubbo LA (2013) Biosurfactants: classification, properties and environmental applications. In: Recent developments in biotechnology, 1st edn, vol 11. Studium Press LLC, Houston, pp 1–29
Souza EC, Vessoni-Penna TC, Souza Oliveira RP (2014) Biosurfactant-enhanced hydrocarbon bioremediation: an overview. Int Biodeterior Biodegrad 89:88–94
Steiber M, Bockle K, Werner P, Frimmel FH (eds) (1990) Biodegradation of polycyclic hydrocarbons (PAH) in the subsurface. Kluwer Academic Publisher, Dordrecht
Stevenson JA, Westlake ACG, Whittock C, Wong LL (1996) The catalytic oxidation of linear and branched alkanes by cytochrome P450cam. J Am Chem Soc 118:12846–12847
Straube WL, Nestler CC, Hansen LD, Ringleberg D, Pritchard PJ, Jonesmeehan J (2003) Remediation of polyaromatic hydrocarbons (PAHs) through land farming with biostimulation and bioaugmentation. Acta Biotechnol 2(3):179–196
Sun GD, Xu Y, Liu Y, Liu ZP (2014) Microbial community dynamics of soil mesocosms using *Orychophragmus violaceus* combined with *Rhodococcus ruber* Em1 for bioremediation of highly PAH-contaminated soil. Appl Microbiol Biotechnol 98(24):10243–10253
Tam NFY, Chan MN, Wong YS (2010) Removal and biodegradation of polycyclic aromatic hydrocarbons by immobilized microalgal beads. Trans Ecol Environ 140:391–402
Tao XQ, Lu GN, Dang Z, Xang C, Yi X (2007) A phenanthrene-degrading strain Sphingomonas sp. GY2B isolated from contaminated soils. Process Biochem 42:401–408
Thiyagarajan A, Saravanakumar K, Kaviyarasan V (2008) Optimization of extracellular peroxidase production from *Coprinus* sp. Indian J Sci Technol 1:1–5
Tondo DW, Leopolino EC, Souza BS, Micke GA, Costa ACO, Fielder HD, Bunton CA, Nome F (2010) Synthesis of a new zwitterionic surfactant containing an imidazolium ring: evaluating the chameleon-like behavior of zwitterionic micelles. Langmuir 26(20): 15754–15760
Tsibart AS, Gennadiev AN (2012) Polycyclic Aromatic Hydrocarbons in Soils: Sources, Behavior, and Indication Significance (A Review). Eura Soil Sci 46(7):728–741
Tsibart AS, Gennadiev AN (2013) Polycyclic aromatic hydrocarbons in soils: sources, behavior, and indication significance (a review). Eurasian Soil Sci 46:728–741
United States Environmental Protection Agency: Integrated Risk Information System USEPA (2011) Environmental Protection Agency. Available from: http://www.epa.gov/IRIS
USEPA (1999) Use of monitored natural attenuation at superfund, RCRA corrective action, and underground storage tank sites. Fed Regist 64:25039–25040
Venkatesagowda B, Ponugupaty E, Barbosa AM, Dekker RFH (2012) Diversity of plant oil seed-associated fungi isolated from seven oil-bearing seeds and their potential for the production of lipolytic enzymes. World J Microbiol Biotechnol 28:71–80
Venosa AD, Suidan MT, Wrenn BA, Strohmeier KL, Haines JR, Eberhart BL, King DW, Holder E (1996) Bioremediation of experimental oil spillon the shoreline of Delaware Bay. Environ Sci Technol 30:1764–1775
Vinas M, Sabate J, Espuny MJ, Solanas AM (2005) Bacterial community dynamics and polycyclic aromatic hydrocarbon degradation during bioremediation of heavily creosote-contaminated soil. Appl Environ Microbiol 71:7008–7018
Vyas BRM, Bakowski S, Sasek V, Matucha M (1994) Degradation of anthracene by selected white rot fungi. FEMS Microbiol Ecol 14:65–70
Wagrowski DM, Hites RA (1997) Polycyclic aromatic hydrocarbon accumulation in urban, suburban, and rural vegetation. Environ Sci Technol 31:279–282
Wang C, Sun H, Liu H, Wang B (2014) Biodegradation of pyrene by *Phanerochaete chrysosporium* and enzyme activities in soils: effect of SOM, sterilization and aging. J Environ Sci 26:1135–1144

Wellman DE, Ulery AL, Barcellona MP, DuerrAuster S (2001) Animal waste-enhanced degradation of hydrocarbon-contaminated soil. Soil Sediment Contam 10(6):511–523

WGPAH (2001) Working group on polycyclic aromatic hydrocarbons (WGPAH) Ambient air pollution by polycyclic aromatic hydrocarbons: position paper Annexes

Wick A, Haus N, Sukkariyah B, Haering K, Daniels W (2011) Remediation of PAH-contaminated soils and sediments: a literature review. Virginia Polytechnic Institute and State University Department of Crop and Soil Environmental Sciences Blacksburg, VA 24061 Online source: http://landrehab.org/

Widada J, Nojiri H, Omori T (2001) Recent developments in molecular techniques for identification and monitoring of xenobiotic-degrading bacteria and their catabolic genes in bioremediation. Appl Microbiol Biotechnol 60:45–59

Wild SR, Hard SJ, Jones KC (1994) The influence of sewage sludge applications to agricultural land on human exposure to polychlorinated dibenzo-p-dioxins (PCDDs) and furans (PCDFs). Environ Pollut 83:357–369

Williams PA, Sayers JR (1994) The evolution of pathways for aromatic hydrocarbon oxidation in *Pseudomonas*. Biodegradation 5(3–4):195–217

Wong JWC, Wan CK, Fang M (2002) Pig manure as a co-composting material for biodegradation of PAH-contaminated soil. Environ Technol 23:15–26

Wong JWC, Fang M, Zhao Z, Xing B (2004) Effect of surfactants on solubilization and degradation of phenanthrene under thermophilic conditions. J Environ Qual 33:2015–2025

Young LY, Phelps CD (2005) Metabolic biomarkers for monitoring in situ anaerobic hydrocarbon degradation. Environ Health Perspect 113(1):62–67

Young D, Rice J, Martin R, Lindquist E, Lipzen A, Grigoriev I, Hibbett D (2015) Degradation of bunker C fuel oil by white-rot fungi in sawdust cultures suggests potential applications in bioremediation. PLoS One 10:e0130381

Yu CL, Parales RE, Gibson DT (2001) Multiple mutations at the active site of naphthalene dioxygenase affect regioselectivity and enantioselectivity. J Ind Microbiol Biotechnol 27:94–103

Yucheng W, Ying T, Zhengao L, Xuewei L, Luo Y (2008) Potential role of polycyclic aromatic hydrocarbons PAHs oxidation by fungal laccase in the remediation of an aged contaminated soil. Soil Biol Biochem 40(789):796

Zafra G, Absalón AE, Cortés-Espinosa DV (2015a) Morphological changes and growth of filamentous fungi in the presence of high concentrations of PAHs. Braz J Microbiol 46:937–941

Zafra G, Moreno-Montaño A, Absalón ÁE, Cortés-Espinos DV (2015b) Degradation of polycyclic aromatic hydrocarbons in soil by a tolerant strain of *Trichoderma asperellum*. Environ Sci Pollut Res 22:1034–1042

Zein MM, Pinto PX, Garcia-Blanco S, Suidan MT, Venosa AD (2006) Treatment of groundwater contaminated with PAHs, gasoline hydrocarbons, and methyl tert-butyl ether in a laboratory biomass-retaining bioreactor. Biodegrad 17:57–69

Zhang Z, Hou Z, Yang C, Ma C, Tao F, Xu P (2011) Degradation of n-alkanes and polycyclic aromatic hydrocarbons in petroleum by a newly isolated *Pseudomonas aeruginosa* DQ8. Bioresour Technol 102:4111–4116

Zhang S, Ning Y, Zhang X, Zhao Y, Yang X, Wu K, Yang S, La G, Sun X, Li X (2015) Contrasting characteristics of anthracene and pyrene degradation by wood rot fungus *Pycnoporus sanguineus* H1. Int Biodeterior Biodegrad 105:228–232

Zhao HP, Wu QS, Wang L, Zhao XT, Gao HW (2009) Degradation of phenanthrene by bacterial strain isolated from soil in oil refinery fields in shanghai China. J Hazard Mater 164:863–869

Zhuang X, Chen J, Shim H, Bai Z (2007) New advances in plant growth-promoting rhizobacteria for bioremediation. Environ Int 33(3):406–413

The manufacturer's authorised representative in the EU is Springer Nature Customer Service Centre GmbH, Europaplatz 3, 69115 Heidelberg, Germany. If you have any concerns regarding our products, please contact ProductSafety@springernature.com

Printed and bound by CPI Group (UK) Ltd, Croydon, CR0 4YY
30/09/2026
02225446-0016